Technikzukünfte, Wissenschaft und Gesellschaft / Futures of Technology, Science and Society

Reihe herausgegeben von

Armin Grunwald, ITAS, Karlsruhe Institute of Technology, Karlsruhe, Germany

Reinhard Heil, ITAS, Karlsruhe Institute of Technology, Karlsruhe, Germany

Christopher Coenen, ITAS, Karlsruhe Institute of Technology, Karlsruhe, Germany

Martin Sand, TU Delft, Delft, The Netherlands

Diese interdisziplinäre Buchreihe ist Technikzukünften in ihren wissenschaftlichen und gesellschaftlichen Kontexten gewidmet. Der Plural „Zukünfte“ ist dabei Programm. Denn erstens wird ein breites Spektrum wissenschaftlich-technischer Entwicklungen beleuchtet, und zweitens sind Debatten zu Technowissenschaften wie u. a. den Bio-, Informations-, Nano- und Neurotechnologien oder der Robotik durch eine Vielzahl von Perspektiven und Interessen bestimmt. Diese Zukünfte beeinflussen einerseits den Verlauf des Fortschritts, seine Ergebnisse und Folgen, z. B. durch Ausgestaltung der wissenschaftlichen Agenda. Andererseits sind wissenschaftlich-technische Neuerungen Anlass, neue Zukünfte mit anderen gesellschaftlichen Implikationen auszudenken. Diese Wechselseitigkeit reflektierend, befasst sich die Reihe vorrangig mit der sozialen und kulturellen Prägung von Naturwissenschaft und Technik, der verantwortlichen Gestaltung ihrer Ergebnisse in der Gesellschaft sowie mit den Auswirkungen auf unsere Bilder vom Menschen.

This interdisciplinary series of books is devoted to technology futures in their scientific and societal contexts. The use of the plural “futures” is by no means accidental: firstly, light is to be shed on a broad spectrum of developments in science and technology; secondly, debates on technoscientific fields such as biotechnology, information technology, nanotechnology, neurotechnology and robotics are influenced by a multitude of viewpoints and interests. On the one hand, these futures have an impact on the way advances are made, as well as on their results and consequences, for example by shaping the scientific agenda. On the other hand, scientific and technological innovations offer an opportunity to conceive of new futures with different implications for society. Reflecting this reciprocity, the series concentrates primarily on the way in which science and technology are influenced social and culturally, on how their results can be shaped in a responsible manner in society, and on the way they affect our images of humankind.

Christian Hugo Hoffmann
Editor

Artificial Intelligence, Entrepreneurship and Risk

Reflections and Positions at the Crossroads between Philosophy and Management

Springer VS

Editor
Christian Hugo Hoffmann
Hoffmann Economics
Freienbach, Switzerland

ISSN 2524-3764 ISSN 2524-3772 (electronic)
Technikzukünfte, Wissenschaft und Gesellschaft / Futures of Technology, Science and Society
ISBN 978-3-658-45543-9 ISBN 978-3-658-45544-6 (eBook)
https://doi.org/10.1007/978-3-658-45544-6

Planung/Lektorat: Frank Schindler
This Springer VS imprint is published by the registered company Springer Fachmedien Wiesbaden GmbH, part of Springer Nature.
The registered company address is: Abraham-Lincoln-Str. 46, 65189 Wiesbaden, Germany

Introduction to the Volume

It takes all the running you can do, to keep in the same place.

If you want to get somewhere else, you must run at least twice as fast as that!

By the outstanding logician Lewis Carroll, 27 December 1871, dedicated to the wonder of life Alice, born on 22 December 2023

In the rapidly evolving landscape of the twenty-first century with technological advancements and entrepreneurial endeavors, the intertwined forces of Artificial Intelligence (AI), Entrepreneurship, and Risk Management have sparked a captivating dialogue that transcends conventional boundaries. This first opening sentence was proposed by ChatGPT on 29-05-2023. In this light, we may ask ourselves: How can we (as opinion formers and society as a whole) determine, assess, and mitigate the societal risks posed by AI products such as ChatGPT or GPT-4? These creations of OpenAI, a company just over seven years old at the time of writing, and the leading entrepreneurs behind it challenge our understanding of interaction and cognition. In this era of AI, are we—subjects equipped with a mind, empathy and the capability to understand—interacting with another mind or are we talking to a bot which "simply", but at the same time somewhat "intelligently"[1] exploits correlations between words? How can we know that e.g.,

[1] For what the I in AI stands for, the reader is referred to Hoffmann (2022).

this text has not been entirely written by ChatGPT? What are the consequences in the face of these pitfalls? What can and should we do about it?

The latter is surely a question that can and will translate into market opportunities for astute entrepreneurs. As the three domains of AI, Entrepreneurship and Risk Management intersect, they give rise to a myriad of such profound questions, confronting us with the daunting difficulty to explore their philosophical underpinnings while also embracing the practical implications for management and innovation. Voilà, another sentence by ChatGPT which (not who) boldly "speaks" of "**us** […] to explore"—yet, does the "us" include ChatGPT or not? Does it have an idea of what philosophy, philosophical thinking or underpinnings are?

This book, *Artificial Intelligence, Entrepreneurship and Risk: Reflections and Positions at the Crossroads between Philosophy and Management*, embarks on a thought-provoking journey that seeks to illuminate the intricate connections between those dynamic realms. By merging some of the many applications and disciplines located between philosophy and management, we aim to provide a comprehensive study of the philosophical assumptions that underlie the entrepreneurial utilization of AI and its inherent risks, among others. It is also time to delve into the philosophical foundations of AI, examine the fundamental beliefs surrounding AI's nature, and its effects for the human condition. Drawing on the works of eminent philosophers, economists and business leaders alike, we engage in inspirational discussions on ethics, philosophy of technology, and the potential societal ramifications of advancing AI technologies. By grounding our exploration in philosophical reflections, we set the stage for a comprehensive understanding of AI's role in entrepreneurship and the inherent risks it entails. Continuing our journey, we pivot to the dynamic realm of entrepreneurship, where visionaries and risk-takers breathe life into transformative ideas. We highlight the entrepreneurial mindset, revealing the inherent drive, creativity, and determination required to navigate the crowded and uncertain terrain of innovation. Through insightful case studies and personal accounts, we shed light on the symbiotic relationship between entrepreneurship and AI, showcasing how these concepts are connected to drive disruptive change and shape the business world of tomorrow.

As we unfold the dynamic interplay between AI and entrepreneurship, we cannot overlook the critical role of risk. Many of the authors in this volume uncover the multifaceted nature of risk-taking in the context of AI and entrepreneurship, fusing the perspectives of philosophy and management. Moreover, we learn about the cognitive biases that influence decision-making under uncertainty, while also examining the ethical dilemmas and responsibility that arise when AI-powered ventures push boundaries and call societal norms into question. By embracing risk as an integral part of innovation, we navigate the intricate landscape where philosophy and management converge. Throughout the book, we present a diverse range of perspectives, expert opinions, and intriguing reflections, bridging the realms of philosophy and management. We invite readers to embark on a stimulating intellectual journey, one that pushes the boundaries of conventional thinking and encourages critical examination of the ethical, societal, and strategic impact of AI, entrepreneurship, and risk thinking.

Artificial Intelligence, Entrepreneurship and Risk serves as a guide for philosophers, business persons, academics, and anyone captivated by the manifold ramifications of AI in entrepreneurship. It invites readers to engage in a multidimensional investigation of these concepts, encouraging inter- and multidisciplinary thinking and fostering a deeper understanding of the multi-level interconnections between technology, human agency, and the risks inherent in shaping our future. Let us unravel the complex insights stemming from this study and embrace the transformative potential that emerges when philosophy and management intersect in the face of technological progress.

This book is embedded in the renowned book series *Futures of Technology, Science and Society* which "is devoted to technology futures in their scientific and societal contexts. The use of the plural 'futures' is by no means accidental: firstly, light is to be shed on a broad spectrum of developments in science and technology; secondly, debates on technoscientific fields such as biotechnology, information technology, nanotechnology, neurotechnology and robotics are influenced by a multitude of viewpoints and interests. On the one hand, these futures have an impact on the way advances are made, as well as on their results and consequences, for example by shaping the scientific agenda. On the other hand, scientific and technological innovations offer an opportunity to conceive of new futures with different implications for society. Reflecting this reciprocity, the series concentrates primarily on the way in which science and technology are influenced social and culturally, on how their results can be shaped in a responsible manner in society, and on the way they affect our images of humankind."[2]

The collection of articles at hand not only follows the mission statement of the interdisciplinary book series of *Futures of Technology* and integrates two or more perspectives in terms of academic fields (in our case: philosophy, management science, computer science, finance, etc.), but also makes room for the crucial point of view of practitioners and brings them into dialogue. Thereby, it draws

---

[2] Source: https://www.springer.com/series/13596 (29-05-23).

on a wealth of first-hand experience of entrepreneurship and AI to (from the editor's point of view) advocate an appreciation of entrepreneurship and more enthusiasm towards innovation creation, which startups stand for more than any other form of business. And this rather more personal purpose set by the editor, which is key for individuals as well as for the economy as a whole, is served from an unusual perspective in this volume, combining different points of views and connecting different fields of endeavor. The following contributions are written by world experts from both academia and the economic practice. The authors from both camps have in common that they are highly appreciated by the editor of the book at hand on both a personal and an intellectual level—the *who's who* of Christian Hugo Hoffmann's network.

Artificial Intelligence, Entrepreneurship and Risk is divided into four main parts. The first main part tackles general conceptual and methodological issues concerning the field of tension between philosophy and management mentioned in the subtitle of this book, while the three subsequent, more application-oriented main parts are devoted to topics that are of particular relevance and topicality within this field. The second and third main sections contain contributions to the philosophy of technology, on the one hand, and contributions to the ethics of technology and business practice, on the other, which deal thematically with the so-called new technologies and, in particular, with AI concepts as well as applications. Finally, the fourth and last main section deals with the figure of the entrepreneur, which is of interest for philosophical anthropology, and the relevant facets of action theory in this context, i.e., in particular aspects of creativity, decision-making competence and risk management.

Conceptual and Methodological Contributions at the Crossroads between Philosophy and Management

1. "Philosophy" and "management": two family-like concepts?

In Part I.1, Rolf Pfister from the Munich Center for Mathematical Philosophy (LMU) as well as the Lab42 AI Research Institute of the Mindfire Foundation in Davos, Switzerland, (which is also supported by the editor) opens the door by going beyond theory and offering hands-on strategies for accommodating philosophical methods in management. He grounds his work in the philosophy of science which is remarkable because most predecessors who thought about philosophy and management only bring the two together by reducing the former to applied ethics. His chapter is followed by my personal friend's quest for so-called reflexive rhetoric in tech. Céline Pieters, who has worked in academia and sold her gaming company, prompts the question in her chapter: How can rhetorical strategies are mobilized with the aim to engage tech entrepreneurs' practice of moral imagination? The next investigation is undertaken by my former colleague at "Philosophie + Management" in Lucerne, Dr. Hartmut Westermann, wondering how to conceptually clarify the two terms "philosophy" and "management" by means of philosophical reflections on language.

2. Ancilla oeconomiae? On the usefulness (and disadvantage?) of philosophy for managers

In Part I.2, my colleague Dr. Cristian Erk from the University of St. Gallen, where I completed my first PhD, is building on that prior work, looking behind the façade of normative management. Inspired by systems thinking and the *St. Galler Management-Modell* (Ulrich & Probst, 1990), he makes a passionate plea for managers to become "philosopher managers" by developing and cultivating a habit of philosophical curiosity about the true causes of things. In other words, philosophy is a *conditio sine qua non* for leading businesses—I could not have thought of a better punch line to open this section of the book. After, my friend and business partner Deepak Bansal, who is the Founder and CEO of MQ Learning, a Swiss-based EduQua-certified academy dedicated to teaching humanistic skills in the digital age, elaborates on the role of ethical decision-making intelligence to guide responsible technology implementation.

3. Field reports: Which philosophy for which management?

In Part I.3, we turn to practitioners and, first, come across Dr. Michael Heumann from Arete Ethik Invest in Zurich, a specialist for ethical and sustainable investments that was advised by this editor. In his text, Michael is particularly addressing methodological challenges of computationalist thought and of treating specific quantification demands in maintaining an integrative ethical approach in the daily business of investments. Next is Jitka Perina, who is a chemist by training, worked in the pharmaceutical industry, and later completed a Master of Advanced Studies in Philosophy and Management. In her contribution to this volume, she implements her learnings as consultant and coach. The last article of this Part I.3 is made by my brother in spirit, Dr. Benjamin Lange, Visiting Researcher at Google, Chief Operating Officer at Babl AI & Director of Compass Ethics for the GSA region.

Contributions of Academic Philosophy and Ethics to the so-called new Technologies and AI Concepts

Part II is inaugurated by a tricephalic academic excellence. The first is Prof. Armin Grunwald from the Karlsruhe Institute of Technology, a famous philosopher of technology, investigating the role of Technology Assessment (TA) as contribution to societal AI management. The second is Prof. Andreas Suchanek, holding the Dr. Werner Jackstädt Chair of Economic and Business Ethics at the Leipzig Graduate School of Management. He is making the case for looking more closely at opportunities and risks in using AI assistants for decision-making. And the third is Suchanek's former student, Prof. Nick Lin-Hi who received the Max Weber Prize for Business Ethics, and was one of the "4 times 40 under 40—Germany's young elite" in the science category in 2010 and 2011 for Capital magazine. His chapter is entitled "Robots, Artificial Intelligence, and Ingroup-Outgroup Phenomena: Between Acceptance and Discrimination". He is followed by my colleague Dr. Lea Chilian from the Institute of Social Ethics at the University of Zurich where we, together, offered a seminar on Ethics of Digital Transformation to students of theology, philosophy and economics. She kicks-off her article with the observation that people love technology and derives lessons for human-human relationships from how humans interact with machines. After, my sparring partner Dr. Jörg Noller from the LMU in Munich is analyzing the human-machine interaction concerning AI by introducing the notion of transsubjectivity. I welcome his contribution very much because, among other things, it represents a critical reply to the chapter by Jitka Perina who made a case for the ideological account, according to which AI will merge with humans. The final words in this Part II are granted to my colleague in the Ethics Group at the Digital Society Initiative in Zurich, Prof. Walther Zimmerli. Walther adds to Jörg's invaluable critical perspective by paying attention to AI hypes and myths. He reflects upon the question of whether there is a more fundamental feature at all underlying the recurring boom of AI.

Hands-on Contributions to the so-called new Technologies and AI Concepts in terms of Business Opportunities

Part III again makes room for practitioners where representatives of the startup world bring in their point of view: On the one hand, my colleague at Technopark Zurich, Dr. Matthias Hölling, equips us with unique findings from our joint work on how to set up an internationally renowned cluster of excellence for AI and data science startups. On the other hand, Dr. Christian Kusmitsch from Mindset Technologies in Austria and his large consortium of co-authors even handed in a double chapter. After, we change gears, and hear from Prof. Yash Raj Shrestha, a trained computer and data scientist and the Academic Director of the Strategy and Artificial Intelligence Laboratory (SAIL) at ETH Zurich (where the editor completed his postdoc) why recent advancements of AI raise questions on managerial accountability and create transparency issues. Among the startup voices, we also find Armin Ribis and his coauthor from Artificially SA, a Lugano-based AI company which creates value through innovative, customized, and maintainable AI solutions that promise to increase real business performances in a demonstrable and sustainable way. The last contribution to this Part III is made by David Rohrmann, Managing Partner at the 1492 The Collective Intelligence consulting boutique and think tank in Berlin.

Action Theoretical Contributions to Entrepreneurship, Creativity, Decision Competence and Risk Management

In my own chapter on the subject matter, I develop a twofold thesis, namely that, on the one hand, there are tipping points in our founder cultures, according to which both the way firms are established and the specific founder personalities can change fundamentally. On the other hand, I argue that we face two global tipping points in our startup cultures. This paper is complemented and put to a test by the two following chapters: While my close friend Sebastian Klein (University of Bonn) puts my proposal in the legal context, my other close friend Dr. Sebastian Hersberger in Basel, who is involved in several startups as well, puts emphasis on the power of execution along the stages of de-risking a venture and building a business. Dr. Dr. Erich Prem, a leading research and technology strategy consultant in Vienna with a focus on entrepreneurial ethics & AI is scrutinizing AI risks and entrepreneurial ethics in the perspective of digital humanism. Another important voice at the Swiss forefront of digitization, entrepreneurship and ecosystems is expressed by the text of my highly esteemed friend Dr. Frederik Flöther, Lead Quantum & Deputy CEO at QuantumBasel, a leading innovation hub for quantum computing and AI. He and his partner Dr. Maria Fay from SAP shed light on the role of ethics and sustainability in business innovation. Both Part IV and the whole book are closed by my good friend and startup colleague of the Baoba AG, Dr. Serge Pelap, who was inspired in his text by the great Yuval Noah Harari and perhaps also by the successful British series *Black Mirror*. His plea for the power of social connections and collective intelligence in the age of AI, narrating the story behind Baoba, is strong and will have a lasting impact. Merci bien mon ami!

Before going in medias res, a few thank yous are in order. Much and special gratitude for this book to emerge is owed to Professor Dr. Armin Grunwald who is a coryphaeus in the field of philosophy of technology, who supervised my work (Hoffmann, 2022) at the Karlsruhe Institute of Technology (KIT), who is heading both the ITAS (the *Institut für Technikfolgenabschätzung und Systemanalyse*) at KIT, the "leading institute for technology assessment in Germany and worldwide", and the Office of Technology Assessment at the German Bundestag (TAB). I am deeply grateful for his invaluable advice and encouragement.

As announced, the subsequent chapters combine a variety of highly relevant and popular topics such as startups, entrepreneurship, risk management, or the societal impact of AI in an original way to form a clearly delineated object of inquiry. In the systems movement, going back to thought leaders like Wiener (1961) or Ashby (1956), this performance can be circumscribed as "synthesis". Another key concept (apart from "synthesis") that systems thinking has coined is the one of a *feedback loop* which occurs when outputs of a system are routed back as inputs as part of a chain of cause-and-effect that forms a circuit (Hoffmann, 2017: Fig. 1; Schwaninger, 2005). By embracing this feedback view of the world, I also wish to thank my contact person at the publishing house Frank Schindler for his help and the review of my manuscript. In this spirit, I dedicate the final word of thank you to the reader, in general, and cordially invite every single one of you to close the feedback loop. Please reach out to me and/or the other contributors to this book with questions, comments, and stimulating thoughts. Feedback is highly appreciated.

References

Ashby, W.R. 1956. An Introduction to Cybernetics. London: John Wiley & Sons.

Hoffmann, C.H. 2022. The Quest for a Universal Theory of Intelligence. The Mind, the Machine, and Singularity Hypotheses. Berlin: De Gruyter.

Hoffmann, C.H. 2017. Assessing Risk Assessment: Towards Alternative Risk Models for Complex Financial Systems. Berlin: Springer.

Schwaninger, M. 2005. Systemorientiertes Design—ganzheitliche Perspektive in Innovationsprozessen. In. B. Schäppi, M.M. Andreasen, M. Kirchgeorg, & F.-J. Radermacher (Eds.). Handbuch Produktentwicklung. München: Hanser: 29–56.

Ulrich, H., & Probst, G.J.B. 1990. Anleitung zum ganzheitlichen Denken und Handeln. Ein Brevier für Führungskräfte. Bern/Stuttgart: Haupt.

Wiener, N. 1961. Cybernetics, or Control and Communication in the Animal and the Machine. Cambridge, MA: MIT Press.

Contents

Conceptual and Methodological Contributions at the Crossroads between Philosophy and Management—Field Reports: Which Philosophy for Which Management?

Contributions of Academic Philosophy and Ethics to the so-called new Technologies and AI Concepts

Editor and Contributors

About the Editor

Dr. Christian Hugo Hoffmann is the Managing Director & Founder of Hoffmann Economics, an investment and investment advisory boutique focusing on Venture Capital. He holds a PhD in Management from the University of St. Gallen (Switzerland), used to work as Assistant Professor of Finance at the University of Liechtenstein, and is currently pursuing his second PhD in the Ethics of Artificial Intelligence. Apart from being passionate about Artificial Intelligence (AI) in academia, Christian is a tech entrepreneur by heart with three software start-ups in Germany, Switzerland, and Malawi under his belt. Moreover, Christian served as Deputy Director of and Head of AI at the Swiss Fintech Innovation Lab in Zurich, as Director of Startup Grind Geneva, and continues to fulfill his role as cluster-preneur, start-up investor, coach and mentor in various programs, and with involvements in several tech start-ups. More information at: https://www.christian-hugo-hoffmann.com/.

Contributors

Deepak Bansal MQ Learning, Zurich, Switzerland

Ralf Buechsenschuss University of Mannheim, Mannheim, Germany

Celine Bökemeyer Universität Vechta, Vechta, Deutschland

Johanna Böttcher Universität Vechta, Vechta, Deutschland

Lea Chilian Universität Zürich, Zurich, Switzerland

Ulla Coester HHL Leipzig Graduate School of Management, Leipzig, Deutschland

Marlene Dettmer Universität Vechta, Vechta, Deutschland

Christian Erk Universität St. Gallen, St. Gallen, Switzerland

Maria Fay University of Münster, Münster, Germany

Frederik F. Flöther QuantumBasel, Arlesheim, Switzerland

Armin Grunwald Karlsruhe Institute of Technology, Karlsruhe, Germany

Michael Hengl 1492 webrain GmbH, Altaussee, Austria

Sebastian David Hersberger sh Ventures GmbH, Allschwil, Switzerland

Michael Heumann Arete Ethik Invest AG, Zurich, Switzerland

Christian Hugo Hoffmann Hoffmann Economics, Freienbach, Switzerland

Matthias Hölling Stiftung Technopark Zürich, Zurich, Switzerland

Sebastian Klein Universität Bonn, Bonn, Deutschland

Benjamin Lange Ludwig-Maximilians-Universität München, Munich, Germany;
Munich Center for Machine Learning, München, Germany. e-mail: benjamin.lange@lmu.de

Nick Lin-Hi Universität Vechta, Vechta, Deutschland

Alexander Neuhaus Frequentis AG, Wien, Austria

Jörg Noller Ludwig-Maximilians-Universität München, Munich, Germany

Serge Pelap Baoba, Zurich, Switzerland

Jitka Perina Master Training Group, Zug, Schweiz

Rolf Pfister Munich Center for Mathematical Philosophy (MCMP), Ludwig-Maximilians-Universität München, Munich, Germany;
Lab42 AI Research Institute, Davos, Switzerland

Céline Pieters Group of Philosophy of Media and Technology, Universität Wien, Wien, Austria

Erich Prem Institute of Philosophy, Philosophy of Media and Technology, University of Vienna, Vienna, Austria;
eutema GmbH, Vienna, Austria. e-mail: prem@eutema.com

Armin Ribis makeAIwork, digitell GmbH, Zürich, Schweiz

David Rohrmann Spaze GmbH, Berlin, Germany

Yash Raj Shrestha University of Lausanne, Lausanne, Switzerland

Fabian Simmank Mindset Technologies, Hergiswil, Switzerland

Andreas Suchanek HHL Leipzig Graduate School of Management, Leipzig, Deutschland

Simone Carlo Surace Artificialy SA, Lugano, Schweiz;
Flock Labs AG, Bern, Schweiz

Patrick Nicolas Tinguely ETH Zurich, Zurich, Switzerland

Helmut Unger UNGER Performance Advisory, Pressbaum, Austria

Hartmut Westermann Philosophie & Management, Luzern, Schweiz

Walther Ch. Zimmerli Institute for Advanced Study, TU Munich, Garching/Munich, Germany

Part I

Conceptual and Methodological Contributions at the Crossroads between Philosophy and Management—"Philosophy" and "Management": Two Family-like Concepts?

Beyond Theory: A Philosophical Framework for Decision-Making in Management

Rolf Pfister

1 Introduction

Philosophy offers an extensive collection of knowledge and methods that allow to successfully analyse complex issues and gain insights. This knowledge can be of great help in management to improve understanding and decision-making. However, there is a large gap between the theoretical methods developed in philosophy and other sciences and their practical application in management. This chapter aims to bridge this gap by providing a philosophical framework for decision-making in management. The goal is to provide a better understanding of how optimal solutions are found and how appropriate methods can be identified and applied to support decision-making. The aim is not to discuss specific philosophical methods such as game theory or utilitarianism and their application, but to show how a philosophical framework facilitates critical analyses of relevant issues, identification of important aspects, and efficient and practically realisable reasoning processes.

The chapter is structured as follows: Sect. 2 analyses the roles of philosophy and management and their relationship. Section 3 shows that goals are a fundamental component of decision-making processes in management and which properties they must possess for optimal decision outcomes. Section 4 discusses the relevance and required properties of means, i.e. actions that allow the desired

R. Pfister (✉)
Munich Center for Mathematical Philosophy (MCMP), Ludwig-Maximilians-Universität München, Munich, Germany
e-mail: rolf.pfister@posteo.de

Lab42 AI Research Institute, Davos, Switzerland

C. H. Hoffmann (ed.), *Artificial Intelligence, Entrepreneurship and Risk*, Technikzukünfte, Wissenschaft und Gesellschaft / Futures of Technology, Science and Society, https://doi.org/10.1007/978-3-658-45544-6_1

goals to be reached. Section 5 examines the importance and crucial characteristics of models, which allows complex issues to be simplified to focus on relevant aspects and to make predictions. Section 6 evaluates the use of scenarios, which make it possible to address uncertain events and compare various developments in terms of the efficiency of means and the fulfilment of goals. Section 7 analyses the selection and application of philosophical and economic methods, such as game theory, in management and how they contribute to successful decision-making. Section 8 provides a summary and concluding remarks on the benefits of applying philosophical methods in management.

2 The Relation Between Philosophy and Management

While there is no universally accepted definition of philosophy (cf. Overgaard et al., 2013, chap. 2), Iniguez (2022, p. 1) argues that "[t]he purpose of philosophy is to provide insights that help us to better understand the world, answers to the question of who we are, or reasoning to justify how we should act." With regard to the definition of management, Kaehler et al. (2019, p. 20), based on a literature review, conclude that:

> Management is a steering influence on market, production and/or resource operations in an organization and its units that may address both people and non-people issues and is exerted by multiple organizational actors through either anticipatory norm-setting (= constitutive or strategic management) or situational intervention (= operational management) with the aim of achieving the unit's objectives. To manage a unit is synonymous with "directing" or "leading" it.

The two statements highlight the close relationship between philosophy and management: Management is exerting influence on people and circumstances to achieve certain goals. Philosophy provides the means to do so by offering methods that enable a thorough analysis of the issue and identification of the best possible means to achieve the desired goals.

To be successful in this regard, it is necessary to overcome the gap between the theoretical methods of philosophy and the practical aspects of management. In management, concrete issues are addressed, such as how to react to a new, better product from a competitor or how to deal with the outbreak of war in a country relevant to the company. It is often not clear which philosophical methods should be applied in such cases—especially as decision-making in management often affects many areas at once, economic ones as well as political, social and ethical ones.

This problem can be addressed by using a philosophical framework in management decision-making. The philosophical framework represents an objective and result-oriented approach for optimal understanding. It enables issues to be perceived in such a way that they are suitable for critical analysis, that their core elements can be identified, and that they are framed in such a way that appropriate philosophical methods can be identified and applied. In the following sections, individual components of the philosophical framework are presented, which together form a comprehensive picture and allow an optimal understanding of complex issues in management.

3 Goals

The definition of management above highlights that management is about achieving specific goals. This aspect is of fundamental importance and determines whether the practice of management is successful. All management decisions serve the purpose of achieving the set goals and all activities that are directed by management serve as the means to achieve these goals. For successful management, it is therefore necessary to properly define the goals to be achieved. This requires considering several aspects which are analysed in the following.

First, it is important to ensure that the defined goals represent the desired end state. Otherwise, a goal may be chosen that is only a means to another goal. For example, a commonly set goal of companies is to sell more products—however, their actual intrinsic motivation is not to sell more products but to make more profit. Focusing on means instead of the actual goal causes other means to be disregarded, even if they are better suited to achieving the goal. For example, reducing production costs or setting up a different sales approach could lead to a greater increase in profit than selling more products. In extreme cases, the means can have the opposite effect to the actual goal—for example, to sell more products, the price of the products could be lowered, which could reduce profits instead of increasing them. The distinction between intermediate goals and final goals is often made in management using the two terms 'objectives' and 'goals'.

Goals are end states that are to be achieved and that constitute a value in themselves. Objectives, by contrast, are concrete and measurable sub-goals that must be met to ultimately achieve a goal.[1]

Second, it may be that not just one goal, but several goals need to be fulfilled concurrently (cf. Goodwin and Wright, 2014, chaps. 2–4). For example, a company can have the goals of making a profit, achieving a high level of employee satisfaction, and having a positive impact on society. In this case, the relationship between the goals must be considered. Goals can be synergetic, neutral or contradictory to one another. A goal is synergetic to another if it partially contributes to its fulfilment; for example, a high level of employee satisfaction represents a socially positive effect. A goal is neutral to another if it has no influence on it. A goal is contradictory to another if it impedes its fulfilment; for example the sale of cigarettes makes it more difficult to support the health of customers. It must be noted that the relationship does not have to be bidirectional. For example, high employee satisfaction has a positive effect on society, but a positive effect on society does not necessarily increase employee satisfaction.[2]

In this context, the significance of the goals in relation to each other must also be taken into account. Often, not all goals can be completely fulfilled due to limited resources or due to other constraints. In this case, the goals must be weighted in relation to each other, and it must be decided which goals should be fulfilled to what degree (cf. Jafarzadeh et al., 2022). For goals to be compared with each other, goals must be converted into a joint metric. Such a metric can be, for example, a financial value that is attributed to the fulfilment of a goal. In the example above, a high level of employee satisfaction could be valued at two million euros for the company and a positive social impact at one million euros. If the company has limited resources and these only allow it to fulfil one of the two goals, it will opt for employee satisfaction accordingly.[3]

---

[1] The terminology used here therefore differs from the terminology used in Kaehler et al. (2019, p. 20)'s definition of management above. There, the aim of management is described as the achievement of 'objectives' – but these are to be understood here in the sense of 'goals'. In the practical execution of management, however, it may well be that the manager of a department is given a goal for his department that is the realisation of a means, of an objective, for the company – for example, a company wants to increase its profits and the manager of the sales department is tasked with implementing the company's objective of developing a new sales strategy.

[2] For a practical example of how the relations of goals can be analysed, see Christensen (2022).

[3] For a study of a practical example on how goals are weighed against each other, see e.g. Christensen et al. (2018).

Third, goals must fulfil various qualitative conditions to be useful. A framework that is frequently used in practice is that of SMART goals: Goals must be specific, measurable, attainable, relevant and time-bound (Cothran and Wysocki, 2005). If one or more of these points are not met, this makes it difficult to implement the goals or measure their success. The criterion of specificity should be emphasised in particular. If a goal is not clearly defined, the possible means cannot be assessed in terms of their suitability because it is not clear what is to be achieved.

Concise examples of the relevance of this criterion are provided particularly in the political sphere, in which the undefined nature of goals is utilised. For example, almost all political parties are in favour of justice. However this goal is without meaning, as there are three different types of justice: Equality-based justice means giving everyone the same—as is the case in public schools, where all children receive the same education. Needs-based justice means giving everyone what they need—as is the case with many health insurance schemes. Performance-based justice means giving everyone what they have achieved—as is often the case with salaries. This case of justice as an unspecified goal shows how important it is to clearly define a goal to be able to fulfil it.

A clear definition of a goal is related to its measurability, but is not identical to it. This is because measurable variables that indicate the fulfilment of a goal are often not the same as the goal itself, but merely have a stronger or weaker relationship with the goal. For instance, while the metric monetary value of the revenue surplus is equivalent to the goal of profit, the metric sales only correlates with the goal of profit to a limited extent. In such cases it is possible that a company achieves the metric but not the goal. In extreme cases, it is even possible that the fulfilment of the metrics is at the expense of the goal—for example, if a company offers services at a loss to generate more sales and profits are further reduced in the process. This phenomenon can often be observed in everyday business practice, regardless of whether it is assumed that a reduction in CO2 emissions automatically leads to greater environmental protection, that applicants with better grades perform better, or that fewer customer complaints indicate higher product quality.

Fourth, goals must be differentiated from externally defined framework conditions. Such framework conditions are, for example, laws that stipulate certain employee regulations or product specifications. They also include other externally imposed processes and conditions that must be adhered to, such as social order or customary means of payment. Although most of the conditions are in general not particularly relevant, they can be of great importance in individual cases. For instance, the maximum speed allowed on roads is irrelevant for most management decisions—unless, it is a car manufacturer's decision regarding the development of a new sports car. Framework conditions differ from goals in that their fulfilment is not intrinsically motivated but rather externally imposed and compliance with them is, at least in general, necessary.

Overall, the above considerations emphasise that it is necessary to think comprehensively about the goals to be achieved and to define them accurately. Only if the goals are well-defined and fulfil all requirements can it be ensured that the decisions to achieve them can be made on a sound basis and are effective. Otherwise, there is a risk that management decisions will not serve the actual goals to be achieved and that the wrong goals or no goals at all will be achieved.

4 Means

Means are actions that can be carried out to fulfil specific goals (cf. Converse, 2023). A large part of management is concerned with determining suitable means and ensuring their successful implementation. In general, many different means can be used to achieve a specific goal. For example, the goal of increasing a company's profit can be achieved by reducing production costs or operating costs, or by increasing revenue. These means can be further subdivided into various, more precisely defined means—revenue can be increased, for example, by raising prices, through a more efficient sales model, or by tapping into new markets.

The various means differ according to several factors. Among others, different means fulfil goals to different degrees (cf. Kruglanski et al., 2015). For example, a diesel-powered delivery vehicle may be cheaper to deliver goods to the customer, but an electric delivery vehicle could be perceived more favourably from an environmental perspective. Means can also be compared on the basis of the costs needed, the time needed to realise them and the time needed to achieve the goals, as well as other factors that may be relevant to individual companies. For this purpose, it may be necessary to determine the relevant criteria and to evaluate the various means on the basis of these criteria (cf. Goodwin and Wright, 2014).

Often not just one means is implemented, but several means at the same time. Equivalent to goals, means can also be synergetic, neutral or contradictory to other means. Accordingly, it is necessary to examine the relationship between means. This allows to determine the set of means that offers the best overall fulfilment of goals within the limits of the maximum possible costs (cf. Goodwin and Wright, 2014, chap. 14). For example, if a company wants to convert its sales fleet to electric vehicles to fulfil the goal of environmental friendliness, the construction of a solar power system is a synergetic means, as its electricity can be used directly for vehicles. The sale of new products with the aim of increasing profits, on the other hand, would be neutral to this. At the same time, the company must consider whether it can implement all three means at once or whether, for example, it will have to forego the construction of the solar power system because it does not have sufficient funds.

The analysis shows that means must be carefully evaluated—both in terms of their own properties and how they relate to each other. Only a well-considered and well-chosen set of means allows the given goals to be achieved as effectively and efficiently as possible. If the best available means are not selected, the goals will be achieved only at excessive cost or to a lesser extent.

5 Models

As shown in Sect. 2, the main task of management is to make decisions to achieve certain goals. This involves recognising the various possible options for action and selecting the one that is most suitable. In simpler, clearly defined situations, such as a poker game, where one has only a few options available at any given time, one can compare all options with each other and use e.g. statistical methods to determine the most promising action. This method already reaches its limits in a chess game, as this offers more possibilities than can even be calculated with computer systems today—accordingly, it is no longer possible to fully grasp the entire issue and completely think through and evaluate all possible options for action. Instead, one has to simplify the problem and, e.g., use heuristics, i.e. mental shortcuts, to assess certain sets of moves as more or less promising. For example, one can exclude all sets of moves in which one would lose several major chess pieces within a short time.

In daily practice, one is usually confronted with very complex and not clearly defined issues, i.e. issues that cannot be grasped in their full complexity. Specifically, while in games all relevant factors are predefined, e.g. the possible cards and their distribution, this is not the case in practical issues. Equally, while games specify which moves are allowed, in practical issues the possible courses of action are not defined. Since practical problems generally cannot be fully captured, it is necessary to create models of them, which can serve as a basis for decision-making.

Models are simplifications of complex issues in which factors that are not relevant are disregarded (Frigg and Hartmann, 2006). For example, in a marketing campaign, customers are often assessed only on the basis of a few socio-economic factors. All other customer characteristics, such as their name, hair colour or place of birth, are disregarded unless they are of interest for marketing purposes. It is crucial for a successfully applicable model that it is as simple as possible, but that at the same time all those factors that are relevant are taken into consideration. Deciding which factors are relevant and which are not relevant is very complex and many different methods can be used to do this (Frigg, 2022). Examples include more philosophical methods such as statistical analysis and causal models (see e.g. Peterson, 2017), more economic methods such as SWOT analysis[4] (see e.g. Altman, 2017), case studies and expert interviews, and methods from artificial intelligence such as machine learning (see e.g. O'Callaghan, 2023).

By using a model, it is then possible to derive various possible courses of action. A causal model, for instance, makes it possible to identify factors that influence a desired target value (cf. Pearl, 2009). If the model shows for instance a causal relationship between the punctuality of public transport and customer satisfaction, the latter can be increased by improving the former. It should be noted that different models indicate different possible courses of action or only define them to a limited extent. For instance, a causal model shows which factors need to be influenced to have an impact on another factor, but it does not specify how the influence should be exerted. As a concrete example, punctuality can be increased in many ways. Some of the possible courses of action may be specified in the model in the form of further causal factors that influence punctuality, but many other courses of action are also possible (unless the model is complete, which is usually not the case). In addition, not only the causal factors but also the causal relationships themselves can be changed—for example, customers can be influenced to lower their expectations and thus the influence of punctuality on customer satisfaction can be reduced. Nevertheless, identifying relevant factors is already very helpful and numerous methods can be used to determine different options for action. Methods frequently used in this area include consulting experts, case analyses and creative brainstorming.

As described above, models are simplified representations of complex issues in which irrelevant aspects are disregarded. This results in two potential sources of incorrect predictions from the models.

First, the reduction to relevant factors means that factors that are excluded are not considered relevant but may still have a hidden influence on the decision-making process. This also applies to the possible options for action. Here, too, favourable options for action may not be taken into account or excluded too quickly because they are supposedly not relevant. This aspect is evident, for

[4] An introduction to the SWOT analysis (Strengths, Weaknesses, Opportunities, Threats) can be found in Sammut-Bonnici and Galea (2015).

example, when it is assumed in economics that actors act rationally. In principle, this is an ideal of many actors—this text here also ultimately aims to enable the reader to act rationally—but in practice, decision-making is much more complex and is also influenced by other factors. Such other factors can be, for instance, emotions in the case of individuals, and sub-optimal decision-making processes or personal preferences of the decision-makers in the case of companies.

Second, models are merely representations of complex issues, but not the issues themselves. Accordingly, the models can be incorrect and not represent issues as they actually are. This is particularly the case if the underlying mechanisms can only be assumed but are not known with certainty. A concise example of this aspect is provided by the various theories in the field of international relations. Idealism, for example, assumes that states act rationally and therefore strive for peaceful and profitable co-existence to maximise the standard of living of their citizens. Realism, on the other hand, assumes that the central goal of every state is to survive, which it can most certainly achieve by being more powerful than all its possible opponents, where necessary with the help of war. In addition to these two theories, there are many others, all of which assume different underlying mechanisms in the complex field of international relations (Burchill et al., 2022). Although there are good examples for all the theories that can be used to confirm their observations, the underlying mechanisms themselves cannot be observed or can be observed only to a limited extent—accordingly, the theories can be tested only to a limited extent and have to be evaluated, e.g. on the basis of their performance.

In this context, it is relevant that models are not intended to depict truth, but merely aim to support decision-making by enabling predictions to be made (cf. Weirich, 2015). The reason for this is that models only need to fulfil their purpose, which is to make correct predictions—whether they do this through a high correspondence with truth or through alternative, more efficient forms of representation that do not correspond with truth but lead to correct predictions is irrelevant from a results perspective. Models simplify complex issues to the extent that correct predictions are made, but irrelevant aspects are ignored as far as possible and relevant aspects are replaced as far as possible by simpler forms of representation. Accordingly, the quality of models should not be assessed on the basis of their correspondence with truth, but on the basis of how well they are suited to support decision-making processes. Heuristics, which people often use in everyday life, are understood as very simple models (Kahneman et al., 1982). Although these sometimes lead to incorrect results, they allow sufficiently good decisions to be made with few resources. The disadvantages of heuristics are that people have biases and social prejudices that lead to suboptimal decisions.

Nevertheless, using heuristics is the only way to make as many decisions as quickly as everyday life requires.

Models are a central component of philosophy, management and even science in general. Ultimately, the aim of science is to develop models that allow one to understand and manipulate the world as well as possible, i.e., to influence it according to one's own interests. For example, only an understanding of physics or chemistry makes it possible to build aeroplanes that fly across the world or to produce medicines that influence health. This shows clear parallels in the goals of science and management: both are concerned with developing the most useful models possible to fulfil certain goals. The difference lies primarily in the fact that science is concerned with the development of more basic models, whereas management is concerned with using and implementing actions from specific models.

The task of philosophy in this respect is to analyse the creation and use of models on a meta-level. This includes, for example, which ideals models should fulfil, such as simplicity, power or correspondence with truth (cf. Pfister, 2022, Sect. 2.3). Pragmatism, for example, argues that models must above all be useful, but that their consistency with truth is not essential. Realism, on the other hand, argues that scientific models converge to truth. Which ideals are to be fulfilled also depends on the various scientific fields and their application—logic and mathematics require much less correspondence with truth than chemistry and biology, for example. Furthermore, philosophy develops and analyses methods that are suitable for developing and using models. These include methods for obtaining observational data on which the models are based; formal inference methods such as deduction, induction and abduction, which allow new information to be derived from given information and predictions to be made (Pfister, 2022); and methods for comparing and selecting models.

The importance of models is enormous both in science and in management, and the quality of the models is decisive for success (cf. Bailer-Jones, 2009). In science, examples of this include the development of the heliocentric world view, which allowed better predictions to be made about the movement of planets; the development of quantum physics, which made it possible to explain many observations at the particle level that classical physics could not; and Bohr's atomic model, which enabled significant developments in the field of chemistry. The examples show how important the quality of models is in science and that it is crucial for making good predictions—and therefore good decisions.

The importance of the quality of models is also vital in the area of management. Most of the factors that determine the success of a company are predetermined from the outside: Either in the form of external conditions such as laws that apply to all companies or in the form of factual conditions such as available resources that cannot be changed, at least at the moment of decision-making. Models used by companies, on the other hand, can be adapted at will and can represent a competitive advantage if they allow better predictions and decision-making than the models of competitors.

An example at the level of society as a whole of how decisive the quality of models are in the economic sphere is provided by the conflict during the Cold War, in which the economic model of capitalism was able to explain, predict, and therefore utilise economic processes much better than the model of the planned economy, which ultimately contributed to the downfall of the USSR (Loth, 2015; Bollard, 2023). In the corporate world, there are also many examples of bad decisions being made due to poor modelling. These include Kodak's continued focus on analogue rather than digital photography, as well as General Motors' financial problems caused by not recognising changing market demands. Similarly, bad models can often be observed on a small scale in corporate decisions. An indicator of this is often when actual developments do not match expectations. This can happen because an unlikely but considered event occurred, or because the model did not predict the event that occurred.

It is not always simple to determine why expectations have not been met and whether or not this indicates a poor quality of the model, as the coronavirus pandemic shows. Almost all companies were surprised by the outbreak of the coronavirus pandemic, which is understandable given that hardly any company has ever been affected by such a pandemic before. Nevertheless, there have been numerous historical examples, from the Spanish flu to bovine spongiform encephalopathy (BSE) and avian influenza, and there is a consensus among scientists that there is a significant risk of pandemic outbreaks due to globalisation, large-scale livestock farming and population growth. Furthermore, while the probability of a global pandemic is relatively small, the probability that one of the many possible unlikely events will occur—be it an outbreak of war, a natural disaster, a nuclear accident or a pandemic—is not.

Whether the occurrence of an unlikely event indicates that the model is bad or not depends on various factors. For example, the model may be so simplified that it does not take unlikely events into account from the outset. Alternatively, the model may have predicted the event, but the event was not considered further due to its improbability. Or an unlikely event was taken into account in the model and subsequently evaluated, but no means were implemented to address this event due to its improbability. When evaluating the quality of models, it is not automatically the case that the model that takes the unlikely event into account is the best. Instead, the best model is the one that is most suitable for the specific company. For instance, if the company is a global pharmaceutical company, the consideration of possible pandemics is crucial. In contrast, if the company is a local retail business, considering pandemics over the possibility of unlikely events occurring in general is not helpful. This is because too many resources would have to be used for the evaluation of the pandemic and at the same time only a few options for action are available, not least due to the limited predictability of the course of events.

In summary, management often concerns very complex issues that need to be simplified with the help of models to identify relevant aspects and make efficient and accurate predictions. The quality of the models is crucial and determines the success or failure of the decision-making process and, therefore, of the company.

6 Scenarios

Management success depends largely on the ability to predict how a complex issue will develop in the future. Predictions of what the future state will be are made on the basis of the current state and models. Such predictions can be made for different cases—for example, it is possible to compare how the future is likely to be if a certain means is implemented, but also how it will be if an alternative means is implemented. Each of these predictions, for which different assumptions are made, represents a scenario (Cairns and Wright, 2017). Considering different scenarios makes it possible to compare different options for action and to determine how the future develops in the individual cases.

Scenarios are also relevant in that they allow different possible developments of an issue to be taken into consideration. A basic assumption of modelling and all actions—and therefore management—is that the issue to be dealt with follows certain rules, at least in part. Otherwise the future of the issue could not be predicted or influenced.

At the same time, complex issues are contingent, at least in practice. This means that there is a possibility, but not a necessity, that particular events will occur and that an issue will develop in one way or another (Huoranszki, 2022). The contingency can be attributed to two possible reasons. On the one hand, it may be that the issue is intrinsically contingent, i.e. certain aspects of the issue are not subject to rules. This includes, for example, the decay of radioactive atoms or the outbreak of a pandemic. On the other hand, the knowledge of a stakeholder addressing the issue may be limited, and although the issue itself would be predictable, it is not for the stakeholder. This may be due, for example, to the fact that the stakeholder has limited knowledge only or that the model used makes too many simplifications. This includes many statistical models that make it possible, for instance, to predict how much money customers spend on average, but not how much money a particular customer spends. Although it is not entirely clear from a philosophical perspective to what extent contingency is due to intrinsic limitations and to what extent it is due to limitations of the stakeholders involved in the issue, contingency is at least in practical management decision-making the default, as the knowledge of decision-makers is limited.

Scenarios make it possible to address this contingency. Scenarios can be used not only to consider different means in the same development of an issue, but also to consider the same or different means in different developments of an issue. In the area of financial planning, for example, generalised comparisons are made between different scenarios by assuming an optimistic as well as a pessimistic economic development. How many different and which different developments of the issue are to be taken into account is a case-by-case decision that depends both on the decision to be made and on the issue to be investigated. In general, different possible developments of an issue can be considered together as one scenario if they are similar in their development under the given means and if their degrees of goal fulfilment do not differ greatly from one another. However, if there are major deviations in their development or in their degree of goal fulfilment, it can be advantageous to consider the various development options separately.

Equally, it is advantageous for efficient decision-making to consider means in different scenarios if their effect on goal fulfilment strongly depends on how the problem develops. Which means are chosen for implementation at the end of the behavioural process depends on the evaluation method selected. For instance, it can be advantageous to implement the set of means that leads to the highest average goal fulfilment across all scenarios, or to choose the set of means that guarantees a certain minimum fulfilment in each scenario. The latter can be particularly relevant in regard to the financial stability of a company. It may be better to choose a means that leads to little company growth which occurs in every scenario, than to choose a means that leads to high company growth in the most likely scenario but holds the risk of insolvency in some unlikely scenarios.

In summary, scenarios increase the complexity of management decision-making processes, as not only is one possible development of the issue considered but various developments. However, scenarios allow contingent developments of the issue to be taken into account and the extent to which means are suitable for achieving the goals under different scenarios can be evaluated.

7 Methods

Management benefits from a rich set of philosophically founded methods developed in many different scientific disciplines, such as epistemology, economics, mathematics and the social sciences. Frequently used methods are, for example, game theory, cost–benefit analysis, statistical analysis, social network analysis, SWOT analysis, and cognitive behavioural theory (cf. Nermend et al., 2021).

Methods are systematic approaches that are theoretically justified and empirically validated and that allow an issue to be analysed using specific procedures to gain certain insights. Thereby, the gain of insight is the main aspect, and the goal is to answer certain questions that are of relevance to one. The underlying theoretical considerations and the procedures to be used are developed and applied in such a way that they are suitable for understanding the given issue.

As such, the methods strongly parallel the decision-making process in management presented in this chapter. In both cases, the aim is to fulfil certain goals by identifying and applying suitable means. Equally, in both cases, the basic theoretical assumptions form a model that allows an issue to be considered from a certain perspective and reduced to the essential aspects that are important for the relevant insight. Only the type of insight to be gained is different. While decision-making in management has the aim of determining which means need to be implemented to achieve the set goals, methods primarily serve to provide answers to queries that arise in the course of the decision-making process.

Consequently, methods are a means used to support the management decision-making process. Which methods are to be applied in which step depends on the specific question to be answered in the decision-making process. The framework presented in this book provides guidance for the selection of suitable methods, as it allows methods to be categorised according to their area of application and their purpose. For instance, in regard to defining corporate goals, methods that allow goals to be set in relation to one another and compared with one another are particularly relevant. By contrast, in regard to identifying possible means, creative and less formal approaches can be advantageous. Methods that are suitable for quantitative optimisation, on the other hand, can be used in particular to evaluate and compare possible means. Furthermore, methods can be classified according to the topics of interest. If, in addition to financial goals, moral goals are also to be met, ethical approaches can be useful; in regard to influencing human behaviour, psychological and social methods are particularly relevant.

Due to the diversity and complexity of decision-making processes in management, it is not possible to provide precise instructions as to which method should be used when—the nature of the issues to be addressed and the available resources that can be used for the decision-making process are too heterogeneous. In the end, the selection and application of the method always depend on how the decision-making process can be optimally supported. Methods should be used in particular where they offer an additional gain in knowledge and where not using a method would lead to poorer results in implementation. At the same time, their use should be minimised to keep the decision-making process as efficient as possible. A balance must be found between the optimal solution that can be found through the use of methods and the effort required to find this solution.

In addition, different methods require different assumptions and, as they are based on models, can lead to erroneous conclusions; thus their interoperability and reliability should be considered. This finding also shows that a wider range of methods should be used and that it is not advisable to focus on and rely on specific methods (cf. Downes, 2020). Although some methods yield significantly better results than others in certain contexts, it is important to be aware that if the context changes, the method may lead to erroneous results or other methods may provide better results. Ultimately, not least because of their underlying models, methods can be understood as complex heuristics, i.e. as mental shortcuts that allow fast predictions and conclusions at the expense of detail and reliability.

Finally, it should be noted that the framework presented here is also just a method, as the parallels above clearly show. In contrast to other methods, such as statistical analysis or SWOT analysis, the aim of the framework is not to address content-related aspects, but to address methodological ones. The framework should enable suitable methods to be recognised, categorised and evaluated in terms of their usefulness in the decision-making process in management. Like all methods, the framework aims to provide a solution to a question—in this case, how can management decisions be best realised—and should be used precisely when it serves this purpose. If other methods are more suitable than the framework presented here due to certain circumstances, then these should be used—here too, an openness towards other, alternative methods and an orientation towards their usefulness is called for.

8 Conclusion

This chapter presents a philosophical framework for decision-making processes in management. To be able to make good decisions, goals must be defined that have to fulfil certain requirements. Subsequently, means can be identified and compared to evaluate which means are best suited to fulfil the goals as much as possible under the given conditions and resources. The means are evaluated by using models that make it possible to reduce complex issues to their most important aspects and thus to make efficient decisions. The quality of models is of importance because it determines how well complex issues can be understood and evaluated. Scenarios make it possible to address contingency, i.e. the possible occurrence of events, and allow the appropriateness of means and the fulfilment of goals to be evaluated under uncertain developments of the issue. Methods from economics and other disciplines, such as statistical analyses and behavioural theories, provide models that make it possible to answer various detailed questions quickly and easily within the decision-making process; for example, the extent to which a certain means can fulfil a certain goal. In this context, the framework presented here represents a method as well—but not with the aim of answering content-related questions, but rather to provide a practical approach for identifying, evaluating and utilising suitable methods within the decision-making process.

The aim of the chapter is to familiarise managers with the decision-making process from a philosophical perspective and to familiarise them with the approach and the opportunities that arise from it. If the decision-making process can be better organised on the basis of the framework and the function of frequently used methods can be better understood and utilised, this can lead to a higher quality of decisions. It is often difficult to utilise theoretic considerations from philosophy and other scientific disciplines in concrete applications, not least because of the ambiguity and complexity of practical issues. It is hoped that this chapter will make it possible to bridge this gap and make it easier to utilise the theoretical findings. Although dealing with theoretical knowledge and the numerous scientifically based methods can be tedious, their use enables better decisions to be made, so that their application can be profitable for managers—and, in the long term, even enables a simpler, more efficient and more trustworthy approach to address complex issues.

References

Altman, M. 2017. Handbook of behavioural economics and smart decision-making: Rational decision-making within the bounds of reason. Edward Elgar Publishing.

Bailer-Jones, D.M. 2009. Scientific models in philosophy of science. University of Pittsburgh Pre.

Bollard, A. 2023. Economists in the Cold War: How a Handful of Economists Fought the Battle of Ideas. Oxford University Press.

Burchill, S., A. Linklater, J. Donnelly, T. Nardin, M. Paterson, C. Reus-Smit, A. Saramago, T. Haastrup, and A. Sajed. 2022. Theories of international relations. Bloomsbury Publishing.

Cairns, G. and G. Wright. 2017. Scenario thinking: Preparing your organization for the future in an unpredictable world. Springer.

Christensen, J., C.M. Dahlmann, A.H. Mathiasen, D.P. Moynihan, and N.B.G. Petersen. 2018. How do elected officials evaluate performance? Goal preferences, governance preferences, and the process of goal reprioritization. Journal of Public Administration Research and Theory 28(2): 197–211.

Christensen, L.D. 2022. Policy coherence in the nordic bioeconomy? A novel settheoretic approach to studying relations among policy goals. Environmental Policy and Governance 32(5): 390–410.

Converse, B.A. 2023, 07. Problem Solving from a Goal-Systems Perspective, Goal Systems Theory: Psychological Processes and Applications. Oxford University Press. https://doi.org/10.1093/oso/9780197687468.003.0011.

Cothran, H.M. and A.F. Wysocki. 2005. Developing smart goals for your organization: Fe577/fe577, 11/2005. EDIS 2005(14).

Downes, S.M. 2020. Models and modeling in the sciences: A philosophical introduction. Routledge.

Frigg, R. 2022. Models and theories: A philosophical inquiry. Taylor & Francis.

Frigg, R. and S. Hartmann. 2006. Models in science.

Goodwin, P. and G. Wright. 2014. Decision analysis for management judgment. John Wiley & Sons.

Huoranszki, F. 2022. The Metaphysics of Contingency: A Theory of Objects' Abilities and Dispositions. Bloomsbury Publishing.

Iniguez, S. 2022. Part i: Wisdom—why should we practice philosophy?, Philosophy Inc. Applying Wisdom to Everyday Management, 1–27. Springer.

Jafarzadeh, H., J. Heidary-Dahooie, P. Akbari, and A. Qorbani. 2022. A project prioritization approach considering uncertainty, reliability, criteria prioritization, and robustness. Decision Support Systems 156: 113731.

Kaehler, B., J. Grundei, B. Kaehler, and J. Grundei. 2019. The concept of management: In search of a new definition. HR governance: A theoretical introduction: 3–26.

Kahneman, D., P. Slovic, and A. Tversky. 1982. Judgment under uncertainty: Heuristics and biases. Cambridge university press.

Kruglanski, A.W., M. Chernikova, M. Babush, M. Dugas, and B.M. Schumpe. 2015. The architecture of goal systems: Multifinality, equifinality, and counterfinality in means—end relations, Advances in motivation science, Volume 2, 69–98. Elsevier.

Loth, W. 2015. The cold war and the social and economic history of the twentieth century. Asia Review 5(1): 231–242.

Nermend, K., M. Latuszyńska, and E. Thalassinos. 2021. Decision-Making in Management: Methods and Behavioral Tools. Springer Nature.

O'Callaghan, M. 2023. Decision Intelligence: Human–Machine Integration for Decision-Making. CRC Press.

Overgaard, S., P. Gilbert, and S. Burwood. 2013. An introduction to metaphilosophy. Cambridge University Press.

Pearl, J. 2009. Causality. Cambridge university press.

Peterson, M. 2017. An introduction to decision theory. Cambridge University Press.
Pfister, R. 2022. Towards a theory of abduction based on conditionals. Synthese 200(3): 206.
Sammut-Bonnici, T. and D. Galea 2015. SWOT Analysis, pp. 1–8. John Wiley & Sons, Ltd.
Weirich, P. 2015. Models of Decision-Making. Cambridge University Press.

Rolf Pfister is active in the fields of philosophy and artificial intelligence and researches how new insights are gained in science and what role artificial intelligence can have in this. A particular focus is on researching how new theoretical concepts such as gravity are introduced into science and how such methods can be formalised.

In addition to his research activities at the Munich Center for Mathematical Philosophy (MCMP) at LMU Munich, Rolf Pfister oversees as the Research Director at Lab42, a Swiss AI research institute, the development of projects aimed at achieving human-level AI.

In his consultancy role, Rolf Pfister applies his knowledge in AI to help businesses integrate AI technologies into their operations. His consultancy work is grounded in his experience as a founder of several startups and through various consulting and executive positions, as well as his understanding of AI's theoretical and practical applications.

Reflexive Rhetoric in Tech: A Practice of Moral Imagination

Céline Pieters

1 Introduction

While the technology industry makes up a third of the total market and is steadily growing, various concerns about the societal impact of ongoing and emerging technologies are raised accordingly. Broadly speaking, the issues related to robotics, AI or synthetic biology technologies, which are known in Silicon Valley as the "next big thing" to invest in, are formulated via the problematisation of the risks related to the technology itself or through the identification of a series of problematic ethos.

Concerning the risks, ethical concerns are framed in terms of security, safety and societal impacts. This is, for instance, highlighted by Bernadette Bensaude-Vincent about synthetic biology (SB):

> "The ethical debates in the series of SB conferences are essentially about risk issues. From the outset, biosecurity came up as the major concern: as synthetic biology can synthesize pathogen viruses it can be misused to make more effective bioweapons (Kelle 2009). The risk of bioterrorism is enhanced since artificial DNA can be purchased on the Internet. […] Biosafety is also a major concern. Significantly, in Europe, the ethical agenda of synthetic biology has been set up in the program entitled Synbiosafe […]". (Bensaude-Vincent, 2013).

The series of problematic ethos, i.e. the *characters* who could lead to these risks, are readily apparent in pop culture, with, for instance, the ethos of:

C. Pieters (✉)
Group of Philosophy of Media and Technology, Universität Wien, Wien, Austria
e-mail: celine.pieters@univie.ac.at

C. H. Hoffmann (ed.), *Artificial Intelligence, Entrepreneurship and Risk*, Technikzukünfte, Wissenschaft und Gesellschaft / Futures of Technology, Science and Society, https://doi.org/10.1007/978-3-658-45544-6_2

- *The Freaky Genius*, "an exiled scientist such as Dr Frankenstein who, driven by his own curiosity, forgets about responsibility."
- *The Irresponsible Amateur* "[…] who tinkers with science and technology in his garage, without regulatory oversight or ethical standards."
- *The Unethical Businessman*, "a character who has a blind spot for the societal consequences of his business, who lets business drive science, makes big money, and abuses science by playing God." (Anderson, 2016)

Following the identification of these risks and problematic ethos, the standard procedure then involves discussing and implementing potential solutions. Although many are considered and even carried out (such as ethical education, interdisciplinarity or awareness), the typical answer remains the formulation of (more, and clear) regulations as well as ethical frameworks and guidelines. As a matter of fact, once a number of major problems raised by the emerging technologies are identified, the focus spontaneously shifts to formulating prescriptions: we draw up rules, policies and laws, or, when it comes to encouraging ethical behaviours, we draft ethical frameworks, codes of conduct and guidelines (i.e., a series of lists of moral principles or moral theories to draw on when making decisions).

Obviously enough, these normative sources face little dispute regarding their necessity or reasonability; few argue that regulations are a waste of time, and the proliferation of ethical frameworks attests to the credibility and recognition this kind of solution has received. Yet, despite their popularity, the limitations of mitigating risks solely through regulations are also widely acknowledged (see for instance Coeckelbergh, 2006). Concerning rules, regulations and policies, the level of anticipation tends to be low since policies often come when the technology is already available. As for codes of conduct and ethical frameworks, these frequently draw criticism due to uncertainty about how (and to what extent) they actually encourage ethical behaviours and judgements. Considering these commonly known limitations, other paths are thus explored with the ambition to proactively mitigate risks from emerging technologies and cultivate (a sense of) responsibility in the tech industry.

In this paper, I delve into one of these avenues by investigating the practice of 'moral imagination' as a means for moral deliberation. Various sources highlight the importance of our ability to imagine within our moral judgements and deliberations (Johnson, 1993; Fesmire, 2003; Coeckelbergh, 2007; Weil-Dubuc et al., 2022). However, the conditions under which to appeal to one's ability to imagine has no 'one size fits all' recipe, and the questions then arise: how to encourage the stakeholders of tech design, including scientists, engineers, designers, tech entrepreneurs but also society as a whole, to engage in the practice of moral imagination? How to stimulate such a practice so that it actually fosters moral deliberation? Furthermore, how do moral imagination and deliberation contribute to creating a sense of responsibility?

By exploring these issues further, my goal is to highlight the relevance of the practice of moral imagination in tech. Moreover, the contribution of this paper is to introduce rhetoric as a technique for practicing imagination as being constitutive of both collective and internal deliberation.

2 The Need for Moral Imagination

In his work on *Logique juridique* (1979), the philosopher and legal theorist Chaïm Perelman raises a thought-provoking question that is relevant to our problematic:

> "If a sign warns passengers that it is forbidden to enter the station in the company of a dog, must the attendant allow access to a passenger holding a tame bear on a leash? Should the literal meaning or the spirit of the law, i.e., the legislator's intention, prevail?" (Perelman, [1979], 1999)

In essence, Perelman asks whether we should entrust judges with the task of interpreting such signs, or if our focus should be on refining the rule itself. Should we leave the decision to one's judgment, or should we rather work on providing clearer terms, making the rule exhaustive to prevent such situations in the future? Although the spontaneous response is often a call for clarification, Perelman critiques the notion of pursuing absolute clarity in law. According to him, the ambition to make rules exhaustive is a lure:

> "Sometimes, the legislator wants to reduce the judge's discretionary power, and tries to specify the terms of the law by introducing quantitative elements wherever possible, but this is not always feasible." (Perelman, [1979], 1999)

In order to illustrate his point, Perelman gives the example of notions such as equity, public interest, urgency and morality. He insists on the fact that:

> "[...] the 'law' is never exhaustive, nor even entirely prior; even after it has been established, it will always be in the process of being clarified, constituted and modified in its application. In other words, judgment is always constitutive [of the law] as well as it is applicative."

> "[…] Very often, the understanding, and the application of a legal text cannot be conceived without an appreciation, a judgment of importance or value." (Perelman, [1979], 1999)

Incidentally, the very principle of clarity is currently a topic of discussion in the examination of whether and how ethical frameworks effectively foster moral behaviours and judgments in tech. To illustrate: "Ethics are about the organization's identity, not about its rules. The reason why this distinction matter is because it's easy to assume that if you have clear and consistent policies in place then everything will be fine. But this isn't the case—ethical issues come up when we need to make decisions based on our values and beliefs, and can't simply follow a rulebook or checklist of obligations. […] In other words: ethics are about how an organization behaves as a whole—including its values and culture—rather than simply what formal policies it has written down for its employees." (Karwat, 2022).

The general argument is that morality extends beyond following and applying principles in a given situation. As Mark Johnson introduces it: "Morality is not a system of laws but an imaginative exercise based on what we share with others" (Johnson, 1993).

As such, morality often transcends a mere act of choice; it involves the creation of options or courses of action rather than being reduced to a multiple-choice problem. The necessity of imagination becomes evident in the creation of viable choices, and the literature commonly recognises its pivotal role in thinking about the future (see for instance Coeckelbergh, 2006). In addition, the importance of imagination becomes particularly evident when seeking to expand the somewhat 'narrow field of vision' of engineers concerning ethical issues (Davis, 1989), or when addressing the potential risk oversights. On this matter, a practical manifestation is the exercise of the 'Red team *vs* Blue Team' used for instance in tech entrepreneurship or National Defense trainings. The 'game' consists in imagining and simulating a critical situation where the red team is made up of offensive experts who try to attack or misuse a given technology, and the blue team has to defend against and respond to the red team's attack.

Nonetheless, it is crucial here to emphasise a significant nuance in order to avoid oversimplifying the concept of 'moral imagination'. While the preceding examples call upon imagination for its heuristic function—namely, as a tool for identifying possible deficiencies in terms of safety, security or unintended effects that engineers may not have considered—condensing imagination to a heuristic tool poses the risk of losing the 'moral' in the formula 'moral imagination'.

In *John Dewey and Moral Imagination*, Steven Fesmire highlights that imagination allows us to "concretely perceive what is before us in the light of what should be" (Fesmire, 2003). Using imagination as a heuristic tool for identifying ethical issues only fulfils the first part of this quote. The exercise that consists in imagining possible risks doesn't necessarily imply that moral deliberation and judgments are experienced, shaped or re-shaped in the process. Consequently, a critical question arises: which practices of imagination genuinely engage one in reflexivity and moral deliberation (moving beyond the heuristics of risks)?

3 Reflexive Rhetoric in Tech

How to stimulate moral imagination? What activities expand one's perception beyond the immediate environment and foster moral judgment and moral deliberation (Johnson, 1993)?

In this section, I address this problematic in the specific context of tech entrepreneurship. I use a research-action methodology based on the practice of rhetoric, which explores a possible answer to the challenge while aiming to contribute to the knowledge on moral imagination. Before diving into the details of the experiment, let me explain the relevance of considering the discipline of rhetoric for ethical practices.

3.1 Why Rhetoric?

As far as Aristotle was concerned, the discipline of rhetoric was defined as "the art of finding the available means of persuasion in a given case" (Rhet., I.2. 1356a). In that respect, and despite the fact that rhetoric reached our modern society in the caricatured form of manipulation and/or as the art of speaking well and knowing how to compose successful writing and presentations (Danblon, 2013), rhetoric is primarily a technique for the discovery and invention of arguments. It describes the cognitive and the linguistic tasks that consist in passing from concepts and ideas to speaking and acting accordingly in a particular case. In this way, the practice of rhetoric is also an opportunity for moral deliberation, both collective and internal.

According to Perelman, three co-present elements determine a situation as being rhetorical: an orator, the one producing a speech (which can be communicated verbally or in writing); an audience, the one to whom the argumentation is addressed (which can be the orator him/herself within the context of an internal deliberation; *délibération intime* (Perelman 1991[1958]); and a purpose, the adherence to the speaker's thesis (or the increase of intensity in the adherence to the thesis), with the aim to create a disposition for action (Perelman, 1968). Following this, the practice of the technique of rhetoric is not only useful for improving our ability to speak accordingly and produce persuasion. Observing what we do when trying to reach an audience (from what we say to how we say it to what we choose not to say) is also an opportunity for us to look at our own beliefs and representations and to reflect on our decisions and actions. In other words, training our rhetorical abilities is also a means of questioning our values in the sense of what is 'desired' and 'desirable' (Dewey, 1934).

Furthermore, it should not be forgotten that the historical links between rhetorical and ethical education run deep. This association is strongly embedded within the *ethos* of the field, given the classical emphasis on rhetoric as the exercise of 'good citizenship'. As Quintilian stated:

> "the perfect orator […] should be a good man, and consequently we demand of him not merely the possession of exceptional gifts of speech but in all of the excellences of character as well". (Quintilian [trad. Butler] 1922)

In "The Before Exercises; Composition as Training in Virtue", Abigail Johnson highlights that:

> "The aim and objective of classical education [was] to instill wisdom and virtue in students, focusing on the ideal person we are trying to form and the models we can use to accomplish this. This central idea is nowhere clearer than in the *progymnasmata*, the ancient writing method […]". (Johnson A., 2018)

The *progymnasmata* (preliminary rhetorical exercises) are used in the experiments presented in this paper. My initiative is in line with the movement for the re-validation of rhetoric in education that was initiated by Pierre Chiron, Emmanuelle Danblon and Benoit Sans, and carried out by the GRAL group at the *Université Libre de Bruxelles* (ULB) as well as in other research centres in Switzerland and Sweden (Chiron & Sans, 2020; Danblon, 2018). If anything, a relevant question would be more appropriately framed as 'why *not* rhetoric' rather than 'why rhetoric' when contemplating possible tools and techniques for practicing moral imagination and, more broadly, for cultivating a designated 'ethical culture' in tech; a culture where reflecting on the societal impact isn't an extra task to add to technicists' and entrepreneurs' to-do list, but where it is constitutive of their own *ethos*.

3.2 Reflexive Rhetoric: An Experiment

This experiment on the technique of rhetoric consists in placing a series of participants in a situation where they are asked to imagine and produce a speech in a particular case. Based on the ancient model of the *progymnasmata*, the exercises are designed to fit contemporary topics and fields of knowledge. In this case, the first session of the experiment took place in the context of a workshop organised at the University of Vienna in October 2022.[1] The second session was conducted at the InnovationLab of the University of Vienna in September 2023.

Modelled after previous experiments carried out in the context of robotics (Pieters, 2020), the experiment mobilised other rhetorical exercises known for emphasising the construction of one's *ethos*. Following a research-action methodology approach, it was conceived as a means to explore what is commonly 'desired' and found 'desirable' within the ethos of tech entrepreneurs. In other words, 'research' in 'research-action methodology' meant investigating the characters and dispositions that are commonly associated with the figure of a role model in tech businesses. As for 'action', the rhetorical experiments were conceived with the ambition to encourage the participants to become conscious of their own values (i.e., interests, motivations, beliefs (Dewey, 1934)) that are driving them to the tech business world. As such, it was designed with the goal of making the students engage in (internal) deliberations and judgements about their own 'desirable' ethos as tech entrepreneurs; while the rhetorical exercise would constraint them in establishing their credibility as a speaker, they would necessarily reflect on—and deliberate 'with themselves' about—personal as well as shared values.[2]

---

[1] The workshop '*Tech Entrepreneurs in Fiction*' was co-organised with Erich Prem and Mark Coeckelbergh as part of a research project on 'Responsible Tech Entrepreneurs in AI and Synthetic Biology' at the University of Vienna.

[2] This was notably observed in previous experiments such as the one held at the *Cité des Sciences et de l'Industrie* in Paris in 2018 where a series of roboticists and journalists were gathered around the practice of the *suasoria*: an exercise where the orator provides persuasive arguments for or against a proposed course of action. The experiment highlighted the common ideas of roboticists and journalists on the epistemic notions (language, knowledge, reason, etc.) that shape the status of robotics as a discipline (Pieters, 2020). Another experiment was conducted at the *International Conference for Robotics and Automation* (ICRA 2019) where roboticists practiced the exercise of the *ekphrasis* (a vivid description that places an object, person or event "in front of the eyes") (Webb, 2009; Dainville & Donckier, 2021). During the time of the workshop, roboticists were invited to articulate in detail their own representations and sensations towards the robot and their emotions directed to the work done on the machine.

3.2.1 Preliminary Experiment—Ethopoeia

In the list of *progymnasmata*, the *ethopoeia* is an exercise that one spontaneously thinks about when aiming to work on (and observe how) to build an *ethos*. *Ethopoeia* is derived from the Greek *ethos* (character) and *poeia* (representation). It is a rhetorical exercise that consists in imagining the speech that a specific person would deliver in a particular context (Kennedy, 2003; Chiron & Sans, 2020): for instance, what would the Queen of England say to her people as they are preparing themselves to go to war, or what would a father say to his children as they leave home. In other words, it is about imagining a character and embodying that character within a speech.

In the initial experiment, a collective of 13 tech entrepreneurs, investors, researchers and visual artists was convened and divided into groups. The aim of this manoeuvre was to mix diverse profiles for fostering a rich(er) exchange of ideas within groups. Firstly, the participants were provided with a brief introduction emphasising the workshop's objective. They were informed that the workshop was meant to explore how tech entrepreneurs are depicted within common representations and to question how role models' ethos are built (whether in fictional or real contexts). Who are the entrepreneurs driving ethical changes today? Or who could they be? In addition to these questions, the difficulties in imagining and building the ethos of a role model were intended to be part of the discussion.

After the introduction, the groups were invited to proceed with the experiment and received the following instructions:

> *"Imagine that you are the founders of a start-up in the field of technology (imagine the concept of that start-up). Today, you stand before an audience at a Job Day, eager to hire some collaborators. Make a 3-min welcoming speech for these people while explaining why they should come and work for you. You have been told that the audience consists of individuals 'who care about ethics.'"*

In other words, the participants were invited to produce the *ethopoeia* of a role model in tech, i.e. to explore what a 'role model in tech entrepreneurship' would say in a given context. After 30 to 40 min of preparation, each group of participants introduced their imagined start-up, taking into account the fictional yet specific audience. *In fine,* the task for the members of each group came down to agreeing on what would be judged as the most desirable by an audience 'who care about ethics'. They also had to make themselves credible in the sense that their start-up would have to be considered feasible.

On the one hand, the three groups had similar arguments for highlighting their desirability:

- the diversity of backgrounds of the (fictional) team members of the start-up (scientists, artists, philosophers, etc.),
- their focus on environmental issues,
- and their fair profit management (fair share).

On the other hand, they imagined three very different concepts of fictional tech businesses, revealing three ways of approaching the above considerations:

1. *Innovative approach—Imagining a new concept of start-up being 'ethical' by its purpose.* The first group (Group A) imagined the fictional start-up *CleanO*, which was presented as a start-up aiming at depolluting the ocean.
2. *Counter-disruptive approach—Identifying specific ethical problems in a certain field or industry, and proposing alternative ways of managing, producing*, etc. The fictional start-up *Moochenko* imagined by Group B was introduced at the 'Job Day' as a company that would 'do better than its competitors' since it proposed a business plan promoting digital art while valuing local artists.
3. *Consulting approach—Creating a start-up* in *Ethics, i.e., by offering ethical advice to other companies.* The fictional start-up *Third* (Group C) was introduced as a €7 billion funded start-up aiming to 'make a difference' by helping other companies in their 'ethical transformation'.

While the experiment revealed current trends in the understanding of 'ethics' in the context of tech entrepreneurship, this first experiment fell short on the set expectations, especially concerning the proof of *ethos.*

Because of a lack of time and repetition of practical exercises (which was mentioned by the participants in their feedback), the groups barely developed a specific *ethos* (at least consciously). While they focused on finding arguments to convince the audience to come and work for them (highlighting factors like their commitment to sustainability, equal salaries for employees, respect for customer data privacy etc.), they only revealed the typical ethos of a *Problem-solver*, i.e., someone 'who recognizes the problems and challenges that are currently accepted as being societal issues, and who takes them into account in their business plan'. This first attempt thus did not offer much insight into the 'desirable ethos' of tech entrepreneurs.

As for the 'action' part of the methodology, the experiment was valued by the participants for creating an opportunity to think about positive impact in tech.

They noted that the preparation of such a speech compelled them to present their ideas and values to the other members of their group and to reflect on their own visions of what it means to be 'ethical'. Although pleasant, the practical dimension of the rhetorical experiment was not considered especially important by the participants. It was concluded that the experiment was worth trying again, with modifications to the identified failing factors (time, repetition, basic training in rhetoric and eventually the choice of the practiced exercise).

3.2.2 Experiment at the InnovationLab—Praise

Building upon the previous observations, another experiment was thus conducted, this time over 3 weeks at the InnovationLab of the University of Vienna with 3 sessions of 2:30 h each. The students were tech entrepreneurs in the making who were invited to follow a training in 'Storytelling'. The two first sessions aimed to prepare the students for a final task: the rhetorical exercise of praise.

- Session 1: Introduction—*Basic Rhetoric*

The series of workshops was presented to the students as an opportunity to become conscious of their own spontaneous rhetorical strategies when introducing themselves as tech entrepreneurs. They would discover rhetorical tools that would help them to better formulate and think about their own drives and values. Rather than learning 'how to speak well', the primary aim of the course was presented as an opportunity to 'become aware of the various possibilities for building their own *ethos*' as tech entrepreneurs.

The primary focus of this first session was to familiarise participants with the three rhetorical proofs that are constitutive of every discourse, and with their intrinsic connection: the *logos* (arguments and reasonings provided by the speaker); the *ethos* (character, disposition); and the *pathos* (emotions, emotional character aimed at putting the audience into a certain frame of mind). In order to direct their attention towards the technical aspect of the practice, the initial rhetorical exercise intentionally diverged in content from the specific challenges and stakes associated with tech entrepreneurs: the students started with a preliminary exercise that consisted in arguing for or against the possibility of following the fictional character 'Leonie', their group leader, who was urging them to jump in a lake during a hike. The participants were supposed to convince the others why they should jump with her or not, so they based their argumentation on what they knew about the character (from a brief description) and on the *ethos* they imagined for that character.

The second exercise consisted in arguing for or against a series of cases that were not only fictional, but also genuinely absurd (for instance: *'Argue that pickles are frightened'*). The feedback of the participants confirmed that the distance taken with the content of the argumentation enabled them to better grasp rhetoric as a technique (*techne*, τέχνη). Moreover, the quirkiness of the subjects allowed the group to feel at ease expressing themselves in front of each other. Although simple, this aspect is far from trivial in rhetorical practice. After both exercises, the participants were invited to reflect on their own spontaneous rhetorical strategies while analysing their choices at the level of the *logos*, *ethos* and the *pathos*. In this way, they had learned and experienced basic notions of rhetoric.

- Session 2: Practice—*Success stories*

The second session was dedicated to the notion of success, which is strongly tied to the figure of a role model. Again, the objective was to create an opportunity for the students to (better) identify their own motivations, interests and goals and to explore them as a series of tools to be used for *showing*[3] their own character and disposition (*ethos*) within a discourse.

The first hour was devoted to a collective discussion about success where a series of questions were suggested to the students: what is your personal definition of success? Why do you want to become an entrepreneur? How much do you think your definition of success has to do with your culture, personal experience and education? How are money and success related according to you? The discussion took place in an informal meeting, which was appreciated by the participants as they noted that it gave then the chance to 'press pause' in order to think of their personal history (which they normally don't have/take time for). The setting of the discussion allowed for the participants to express and discuss personal stories that they could invoke within their speech in the final task.

Then, the students were invited to practice another preparatory exercise that consisted in presenting their own start-up project to different audiences: to children, to adults, and to investors. They were first asked to structure their speech according to a series of criteria described by Oren Klaff in his book *Pitch* (Klaff, 2011). They were then asked to rework their speech while focusing on their choices at the levels of *logos, ethos,* and *pathos,* and their relation to one another: 'Do I *show* rather than *tell* what I wish to say? Do I use examples, comparisons, etc. to do so? What makes me credible as a speaker? Why would people believe

[3] According to the rhetorical principle of 'show, don't tell' (Webb, 2009; Pernot & Lévy, 1997; Booth, 1983).

what *I* say specifically? Which emotion am I hoping to convey? Do I make others relate to my point of view through shared feelings? etc.

Finally, they presented their speeches in front of each other, and the other students were asked to critique each speech through a rhetorical prism.

- Final task—*Praise*

In the last session, the students started with a preliminary exercise that was meant to let them collect more knowledge on their own drives and values and so acquire possible content for their upcoming speech. After that, they were invited to practice the rhetorical exercise of praise.

The preliminary exercise was one that is common in entrepreneurial trainings (and shared by Marie-Charlotte Pottier, from *Jobyourself*, Brussels). It consisted in presenting the participants with a series of magazines, journals, newspapers, advertising, etc. where they were asked to choose 30 images 'that have an effect on them, either positive or negative'. Leaning on automatic thinking, the only important criterion was that the choice of images attracted their attention and evoked a certain emotion. Then, they were asked to associate each image with a word and to sort those words into 4 to 8 categories to which they then assigned values ('honesty', 'sense of community', 'quality', etc.). They were presented with a list of 300 values from which they could find inspiration in order to identify which ones corresponded to their categories of words. Each participant ended up with 4 to 8 values that they were then asked to illustrate in their praise. The next challenge was to find the accurate *ethos, logos* and *pathos* to embody these values in their praise. They could also lean on the previous discussion on the notion of success, as well as the rhetorical proofs that they had now discovered and practiced.

The final task of the praise exercise was intended to invest the epideictic genre (i.e. the rhetorical genre dedicated to ceremony, commemoration, declamation, demonstration [ἐπιδεικτικός "for show"]. According to Aristotle, the epideictic genre deals with matters of virtue and vice and revolves around praise and blame. Within the *progymnasmata*, praise is known for reaching into the background of the virtuous, seeking to understand their origins and actions and their far-reaching effects on society. As Theon of Alexandria defines it, praise is a discourse that highlights the importance of the virtuous acts and other merits of a particular person and, in this way, unites the community around its interests and values (Theon [trad. Patillon], 1997). As such, and because it relies on the rhetorical principle of amplification while promising to highlight the moral standards of the speakers, the praise exercise appeared to be a good choice for learning more about personal and common representations of a 'desirable' *ethos* in tech entrepreneurship. Yet, the exercise of the *ethopoeia* was not completely discarded as it still inspired the instructions of the exercise:

> *"Imagine a speech that the very first employee of your start-up would give about you as you are retiring. Speak as if you were him or her."*

In terms of the 'research' part of the research-action methodology, the produced speeches were more interesting this time in the sense that they all contained more details. As a matter of fact, the participants didn't only define a role model in tech as 'somebody who takes care of his/her employees', but actually *showed* how a value, for instance 'loyalty', could be embodied in the context of tech entrepreneurship.

Thanks to the data gathered from all the speeches as well as from the comments made by the students after each speech, the experiment facilitates the description of the characters and dispositions of 'desirable' tech entrepreneurs, other than the *Problem-solver* archetype. This time, there were enough elements in the participants' speeches to propose a variation of role models who could contrast with the figure of *The Unethical Businessman*;

- *The Principled Explorer*, who dares to face different challenges and changes, is open to surprises and is able to seize opportunities when they present themselves. However, he/she is also loyal since, even if he/she retires at a very early age, he/she "wouldn't sell the company to a bigger one". He/she also "believes in his/her team" while hoping to promote a feeling of 'joy' within the team.
- *The Growing Anchor*, who "makes the best out of a situation" and stands as a "compass for everybody". He/she is "humble" and always "pushes him/herself", aiming to "never stop growing as a person". He/she also creates the right environment for people to grow "personally and mentally" and allows "time for relaxation", promoting "harmony and comfort" within the team.

Given that the students were instructed to speak from the perspective of their very first employee, the exercise oriented the descriptions towards embodying the qualities of a 'team leader' rather than revealing the entire *ethos* of a 'tech entrepreneur'. A solution for collecting broader descriptions could then be to rephrase the instructions by letting them choose the (fictional) identity of the speaker:

> *"Imagine a speech that your* -employee/associate/customer/parent/partner- *gives about you as you are retiring. Speak as if you were him or her."*

Repeating the experiment would largely enrich the descriptions of the 'desirable' *ethos* mentioned above, as well as enlarging the list of possible figures. However, the experiment shows the potential of using rhetoric as a heuristic method when contemplating ethical concerns in tech (and in other contexts).

Moreover, it is interesting to note that the exercise which consisted in imagining and writing these profiles (based on the students' speeches) was itself heuristic, and this observation leads one to seriously consider the potential role of rhetoricians, philosophers and writers in ethical research. Beyond emphasising the significance of moral imagination, should we also practice it ourselves while leaning on our expertise?

As for knowing if and how moral imagination eventually plays a role in shaping participants' values while putting them in a situation of deliberation (internal and collective), the feedback of the participants reveals a positive answer.

Following the last session of the workshop, the students shared that the practice of rhetorical exercises made them "think about what [they] are doing from a perspective that [they] didn't think existed" and that it helped them "think about what is important to [them]". While they noted that there was "still a long way to go before feeling confident" in their ability to "choose the right words and dispositions" to express themselves as tech entrepreneurs, they also considered the basic training in rhetoric that they received to be reassuring in this matter.

Finally, they highlighted the importance of the practical dimension of rhetoric to becoming aware of and further building their own *ethos*: "The fact that we have to make a speech in front of others changes everything", "[...] I sounded different from what I expected", "[...] When in front of the others, only there, do we become conscious of the emotions that are attached to the arguments we give. Only then can we feel whether or not what we show to others mirrors what we deeply want".

From a technical (i.e. rhetorical) standpoint, the experience had the benefit of highlighting the *ethos* as a rhetorical proof that is constructible (though accounting for our pre-existing *ethos* (Amossy, 2001), and this has ethical as well as psychological implications.

4 Conclusion

The ethical issues arising from technology are usually formulated via the problematisation of the risks related to the technology itself and the identification of a series of problematic *ethos*. While rules, policies, ethical frameworks and guidelines are drafted with the ambition to prevent these risks, it is also acknowledged that normative sources may lead to a mere adherence to codes rather than cultivating an ethical culture in tech. More than that, it is suggested that any law, rule, policy or ethical framework can never be clarified once and for all (see (Perelman, 1991 [1958]) and Dewey on values (Dewey, 1934)), which implies that individual and collective judgments are always part of the constitution and the application of rules and moral principles.

Taking these observations into account, other alternative avenues for navigating potential risks and questionable *ethos* are considered, such as the practice of moral imagination. However, despite the theoretical support for moral imagination in ethical literature, questions persist regarding its practical implementation.

In this paper, I proposed an experimental approach using the technique of rhetoric as a means of practicing moral imagination. The first aim of the experiments was heuristic: it was meant to explore the possible 'desirable' *ethos* of tech entrepreneurs, in contrast with the figures of *The Unethical Businessman* as well as the other "tech monsters" in pop culture, namely *The Freaky Genius* (Frankenstein) and *The Irresponsible Amateur.* The second objective of the experiment was ethical: it aimed to provide participants with an opportunity to become aware of their own spontaneous rhetorical strategies and to deliberate, with others and in the form of 'inner deliberation', about personal and shared values (Dewey, 1934).

While the experiments would benefit from fine-tuning and repetition, these initial attempts have already yielded progress in identifying what is deemed desirable and successful for tech entrepreneurs. Furthermore, considering participants' feedback, it is reasonable to argue that the practice of moral imagination, exercised through the rhetorical technique, offers pedagogical benefits that contribute to addressing challenges related to problematic *ethos* in tech.

However, in demonstrating the potential of employing rhetoric as a heuristic method, one might ask whether rhetoricians, philosophers and writers have a more significant role to fulfil beyond merely emphasising the importance of moral imagination.

References

Aelius Theon, *Progymnasmata*, Trad. Michel Patillon, Paris: Les Belles Lettres, 1997.

Amossy, R. (2001). Ethos at the crossroads of disciplines: Rhetoric, pragmatics, sociology. *Poetics today*, *22*(1), 1–23.

Anderson, K. (2016). *Synthetic Biology in Pop Culture*, Studio 360, New York Public Radio, https://www.wnyc.org/story/synthetic-biology-pop-culture-2/

Aristotle, *Rhetoric* (trad. Roberts W. [1954]). New York: Modern Library.

Bensaude Vincent, B. (2013). Ethical perspectives on synthetic biology. *Biological Theory*, *8*(4), 368–375.

Booth, W. (1983). *The rhetoric of fiction*. University of Chicago Press.

Chiron, P. & Sans, B. (2020). *Practicing the progymnasmata, from ancient times to present*, Ruedulm.

Coeckelbergh M. (2007). *Imagination and Principles: An Essay on the Role of Imagination in Moral Reasoning*. Palgrave Macmillan.

Coeckelbergh, M. (2006). *Regulation or Responsibility? Autonomy, Moral Imagination, and Engineering*. In Science, Technology & Human Values 31(3):1–24

Dainville, J. & Donckier, L. (2021). Les usages de l'ekphrasis. *Exercices de rhétorique* (17).

Danblon E., (2013). *The Reason of Rhetoric, Philosophy and Rhetoric*, Volume 46, Number 4, pp. 493–507.

Danblon, E. (2018). Can moral luck be taught? A rhetorical tool and its political function. *Paradigmi*, *36*(3), 519–530.

Davis, M. (1989). *Explaining wrongdoing*. Journal of Social Philosophy 20, (Spring/Fall): 74–90.

Dewey, J. (1934). *Art as Experience*. Perigee Books.

Fesmire, S. (2003). *John Dewey and Moral Imagination: Pragmatism in Ethics*. Indiana University Press.

Johnson A. (2018). *The Before Exercises; Composition as Training in Virtue*, Memorial Press.

Johnson M. (1993). *Moral imagination: implications of cognitive science for ethics*, University of Chicago Press.

Karwat, D. (2022). Creating a New Moral Imagination for Engineering. *Issues in Science and Technology*.

Kelle A. (2009). Security issues related to synthetic biology. In: *Schmidt M, Kelle A, Ganguli-Mitra A, de Vriend H (eds) Synthetic biology: the technoscience and its societal consequences*. Springer, Dordrecht, pp 101–119.

Kennedy, G.A. (Ed.). (2003). *Progymnasmata: Greek textbooks of prose composition and rhetoric* (10) Brill.

Klaff, O., & Bowlby, S. (2011). *Pitch Anything*. McGraw-Hill Professional Publishing.

Perelman Ch., Olbrechts-Tyteca L. (1991 [1958]). *The New Rhetoric. A Treatise on Argumentation*, University of Notre Dame Press.

Perelman, Ch. (1968). Une théorie philosophique de l'argumentation. *ARSP: Archiv für Rechts-und Sozialphilosophie/Archives for Philosophy of Law and Social Philosophy*, 141–151.

Perelman, C. (1979). Logique juridique: nouvelle rhétorique.

Pernot, L. & Lévy, C. (1997). *Dire l'évidence. Philosophie et rhétorique antiques,* L'Harmattan.
Pieters, C. et al (2020). Roboticists and Reporters: A rhetorical experiment at the Cité des Sciences et de l'Industrie of Paris. *IEEE robotics & automation magazine,* pp. 141–149.
Quintilian, *The Institutio Oratoria* (trad. Butler, H. E. [1922]). Vol. 8, pp. 3–5. Heinemann.
Webb R. (2009). Ekphrasis, Imagination and Persuasion in Ancient Rhetorical Theory and Practice, Ashgate Publishing, Ltd.
Weil-Dubuc, P. et al. (2022). Les pouvoirs de l'imagination. Réflexion sur les rôles de l'imagination en éthique. *Revue française d'éthique appliquée, 12*(1), 17–22.

Semantische Vagheit und pragmatische Präzisierung: sprachphilosophische Reflexionen zur begrifflichen Klärung des Themenfelds „Philosophie und Management"

Hartmut Westermann

Einleitung

„Philosophie und Management? Das scheint mir ein recht umfassendes Thema zu sein! Fast schon so universal wie: Alles und Nichts." Mit diesen Worten reagierte ein befreundeter Sprachphilosoph, als ich ihm – es muss 2002 oder 2003 gewesen sein – von meiner beginnenden Dozententätigkeit an dem just mit „Philosophie und Management" titulierten Nachdiplomstudiengang der Universität Luzern erzählte. Die maliziös intonierte Bemerkung traf nicht nur genau meinen Humor, sie brachte in zugespitzter Form auch ein Problem zum Ausdruck, das mich seinerzeit angesichts der neuen didaktischen Herausforderung auch selbst intensiv beschäftigte: Kann die Kombination zweier vager und semantisch derart unterbestimmter Begriffe wie „Philosophie" und „Management" ein wissenschaftlich traktierbares und im Rahmen eines Studiengangs vermittelbares Themenfeld erschließen, in dem sinnvolle, auf ihren jeweiligen Wahrheitswert hin überprüfbare Aussagen und rationale, hinsichtlich ihrer logischen Folgerichtigkeit analysierbare Argumentationen zumindest prinzipiell möglich sind? Oder inauguriert die Rede von „Philosophie und Management" nicht eher einen trickreich zwischen philosophischer Theorie und ökonomischer Praxis changierenden, als pseudowissenschaftlich zu deklarierenden Diskurs, der sich den etablierten Standards von Logik und Argumentationstheorie in großen Teilen entzieht?

H. Westermann (✉)
Philosophie & Management, Luzern, Schweiz
e-mail: hartmut.westermann@mailbox.org

C. H. Hoffmann (ed.), *Artificial Intelligence, Entrepreneurship and Risk*, Technikzukünfte, Wissenschaft und Gesellschaft / Futures of Technology, Science and Society, https://doi.org/10.1007/978-3-658-45544-6_3

Meine anfängliche Befürchtung, ein unter „Philosophie und Management" firmierender Studiengang könne allzu leicht zu einem expertisebefreiten Schwadronieren und einem – um die treffende Bezeichnung Harry Frankfurts aufzunehmen – *bullshitting* motivieren, das sich am Unterschied zwischen wahren und falschen Aussagen völlig desinteressiert zeigt, hat sich während meiner weiteren Tätigkeit in diesem Bereich als weit überzogen, wenn auch nicht als gänzlich unbegründet erwiesen.[1] Dass die skizzierten, im Grundsatz durchaus bestehenden Gefahren bei entsprechender Sensibilisierung entschärft werden können, ist – grob gesprochen – der Erfüllung bestimmter Kommunikationsbedingungen und dem Einsatz bestimmter pragmatischer Präzisierungsstrategien zu verdanken. Sinnvoll angewendet ermöglichen diese – angesichts einer nicht zu leugnenden, sondern offen anzuerkennenden semantischen Vagheit der Rede von „Philosophie und Management" – eine für den jeweiligen konkreten Verwendungskontext hinreichende Begriffsklärung.

Im ersten Teil meines Beitrags werde ich die semantische Vagheit der Begriffe „Philosophie" und „Management" genauer erläutern, ehe im zweiten Teil die angesprochenen Kommunikationsbedingungen und die pragmatischen Präzisierungsstrategien charakterisiert werden, die aus meiner Sicht als besonders hilfreich gelten dürfen. Der Text versteht sich als eine sprachphilosophische Reflexion der Erfahrungen, die ich im Rahmen einer langjährigen Dozenten-, Betreuungs- und Vortragstätigkeit für den genannten Nachdiplomstudiengang und dessen Alumni-Netzwerk sammeln durfte. Insofern die sprachphilosophischen Ausführungen auf meinen subjektiven Erfahrungswerten beruhen, besitzen sie auch selbst einen nur kontingenten Charakter. Gleichwohl hoffe ich, dass sich die hier vorgestellten Überlegungen für eine fachwissenschaftliche Diskussion als anschlussfähig und vielleicht ja auch für praktische Anwendungszwecke als zuträglich erweisen können.

[1] Vgl. Frankfurt (2005).

1 Die semantische Vagheit der Begriffe „Philosophie“ und „Management“

1.1 Die Präzisierungsbedürftigkeit der Vagheitszuschreibung

Den Begriffen „Philosophie“ und „Management“ scheint, wie man herausfordernd spotten könnte, zunächst nur Eines gemeinsam zu sein: ihre notorische Vagheit. Allerdings zeichnen sich in kritischem Gestus geäußerte Vagheitszuschreibungen, auch wenn sie intuitiv plausibel und nachvollziehbar wirken mögen, häufig selbst durch eine gewisse Vagheit aus, so dass der artikulierte Vorwurf auf sie zurückzufallen droht. Wer beispielsweise an der in universitären und universitätspolitischen Kontexten hyperinflationär gewordenen Verwendung von so unscharfen Begriffen wie „Kultur“, „Kommunikation“ und „Nachhaltigkeit“ Kritik übt, trifft zwar häufig auf breite Zustimmung, ohne damit aber in praktischer Hinsicht auch nur das Geringste zu ändern. Dies liegt vermutlich daran, dass die Kritik keinen konkreten Präzisierungsbedarf anzeigt, den man problemlösungsorientiert angehen könnte, sondern lediglich einem Unbehagen Ausdruck verleiht, das seinerseits im Ungefähren bleibt. Statt die genauer zu bestimmenden Begriffsmerkmale explizit zu benennen, zieht man sich zurück auf eine sprachkritische Version der Bartleby-Sentenz „I would prefer not to“.[2] Eine solche Positionierung führt zu keiner begriffspräzisierenden Sprachhandlung, sondern erschöpft sich in der bloßen Demonstration einer begriffsskeptischen Haltung, die einer Handlungsenthaltung gleichkommt: „Ich würde es vorziehen, den Begriff *x* nicht zu verwenden.“ Um angesichts der angezeigten Problematik nicht selbst allzu vage zu bleiben, gilt es also, die Vagheit der Begriffe „Philosophie“ und „Management“ nicht einfach zu konstatieren, sondern sie genauer zu erläutern, um so den Weg für eine kontextsuffiziente Präzisierung zu eröffnen.

1.2 Äquivozität, Univozität und Analogie

Halten wir zunächst fest, was die angesprochene Vagheit der Begriffe „Philosophie“ und „Management“ nicht ist: Sie ist keine Mehrdeutigkeit im Sinne bloßer Äquivozität. Äquivok sind sprachliche Ausdrücke dann, wenn sie – bei gleichem Laut- und Schriftbild – in verschiedenen Gebrauchszusammenhängen völlig unterschiedliche Bedeutungen besitzen. Im Deutschen sind „Tau“ (mit den

[2] Vgl. Melville (1853).

Bedeutungen „Seil“ und „feuchter Niederschlag“) oder „Bank“ (mit den Bedeutungen „Geldinstitut“ und „Sitzmöbel“) einschlägige Beispiele. Es ist leicht zu sehen, dass die Bedeutungen, die die Begriffe „Philosophie“ und „Management“ in ihren verschiedenen Verwendungssituationen jeweils annehmen können, nicht in einer derart gravierenden Weise voneinander abweichen. Allerdings können beide Begriffe auch keine Univozität für sich beanspruchen. Univok sind sprachliche Ausdrücke nämlich nur dann, wenn sie in dem Sinne eindeutig sind, dass sie – wie im Deutschen beispielsweise der Ausdruck „Primzahl“ – in allen Verwendungszusammenhängen mit strikt derselben Bedeutung gebraucht werden.

Sind die Begriffe „Philosophie“ und „Management“ also weder als äquivoke noch als univoke Ausdrücke zu charakterisieren, so ist es naheliegend, sie gemäß dem klassischen Konzept der Analogizität zu begreifen. Analog ist ein sprachlicher Ausdruck dann, wenn er in verschiedenen Verwendungssituationen in zwar unterschiedlichen, aber gleichwohl ähnlichen und inhaltlich miteinander zusammenhängenden Bedeutungen gebraucht wird. Eben diese Charakterisierung scheint mir, betrachtet man ihre vielfältigen Anwendungsgebiete genauer, sowohl auf den Begriff „Philosophie“ wie auch auf den Begriff „Management“ zuzutreffen. Die sich dann stellende Anschlussfrage richtet sich nach dem, was die verschiedenen analogen Prädikationen von „Philosophie“ und „Management“ jeweils miteinander verbindet: Muss es, da die Begriffsbedeutungen kontextuell ja nur partiell differieren, nicht auch gemeinsame semantische Merkmale geben, die in allen Verwendungssituationen gleich sind? Also – metaphorisch gesprochen – so etwas wie einen univoken Kern, der ein Abdriften der analogen, kontextuell variierenden Begriffsbedeutung in die univoke Mehrdeutigkeit verhindert?

1.3 *Pollachōs legomena* und *pros hen*-Relation

Zur Klärung dieser Frage ist ein Blick auf die Sprachphilosophie des Aristoteles von Nutzen, auf den das Konzept der Analogizität philosophiehistorisch zurückgeht: Begriffe, die auf vielfältige Weise ausgesagt und in zwar unterschiedlicher, aber nicht mehrdeutiger Bedeutung gebraucht werden, nennt Aristoteles *pollachōs legomena*. Anders als bei mehrdeutigen Begriffen beziehen sich im Falle eines *pollachōs legomenon* die unterschiedlichen Bedeutungen stets auf ein- und dasselbe. Durch diese auf ein einziges Relatum ausgerichtete Beziehung, die sogenannte *pros hen*-Relation, wird vermieden, dass sich aus der Unterschiedlichkeit der Bedeutungen eine Mehrdeutigkeit (in der aristotelischen Diktion: eine Homonymie) ergibt. Wird etwa der Begriff des Seienden (*on*) als *pollachōs legomenon* bestimmt, insofern die Begriffsbedeutung je nach kategorialem Aussagekontext (Substanz, Qualität, Quantität etc.) variiert, so folgt hieraus weder eine Mehrdeutigkeit des Begriffs noch ein ontologisches Auseinanderfallen der Wirklichkeit in eine Pluralität von Seinsweisen, da die unterschiedlichen Bedeutungen insofern einander ähnlich und inhaltlich verbunden sind, als sie jeweils in Relation zu demselben Prinzip (in diesem Fall: zur ontologisch priorisierten Substanz) stehen.[3]

Für Aristoteles sind, wie Buch IV seiner „Metaphysik" im Detail zeigt, nicht nur die meisten philosophisch relevanten Begriffe als *pollachōs legomena* zu analysieren, auch alltagssprachliche Begriffe können häufig in diesem Sinne verstanden werden. So wird, wie Aristoteles exemplifiziert, auch der Begriff des Gesunden auf zwar vielfältige Weise, dabei aber stets in einer auf die Gesundheit bezogenen *pros hen*-Relation ausgesagt: Einiges wird „gesund" genannt, weil es die Gesundheit erhält, anderes, weil es Gesundheit bewirkt, wieder anderes, weil es ein Anzeichen für Gesundheit ist, und wiederum anderes, weil es für die Gesundheit aufnahmefähig ist.

Will man die Vagheit der Begriffe „Philosophie" und „Management" genauer verstehen, so scheint es sinnvoll, in Ergänzung des oben skizzierten Konzepts der Analogizität auch den aristotelischen Begriff des *pollachōs legomenon* und der *pros hen*-Relation in klärender Weise zur Anwendung zu bringen. Folgt man nämlich der von Aristoteles eröffneten Methodik, so lässt sich eine Mehrdeutigkeit des Begriffs „Philosophie" wie des Begriffs „Management" genau dann vermeiden, wenn es gelingt, die jeweilige Vielfalt der Aussageweisen relational auf eine einzige, den inhaltlichen Zusammenhang garantierende Brennpunktbedeutung (*focal meaning*) zu beziehen.[4]

1.4 Definierbarkeit, Subsumierbarkeit und Differenzierbarkeit

Die Suche nach einer Brennpunktbedeutung (*focal meaning*) der Begriffe „Philosophie" und „Management" ist nicht zu verwechseln mit dem Versuch einer Definition der genannten Begriffe. Zwar sieht man sich, wenn man vage (oder zumindest vage wirkende) Begriffe gebraucht, nicht selten mit der Aufforderung konfrontiert, vor einer weiteren Verwendung der Begriffe zunächst eine Definition

[3] Vgl. Herzberg (2011).

[4] Zur *focal meaning* vgl. insbes. die Arbeiten von G.E.L. Owen zur *pros hen*-Relation bei Aristoteles.

vorzulegen. Im Falle der Begriffe „Philosophie“ und „Management“ erscheint es in Anbetracht ihres weiten Anwendungsspektrums und ihres analogen Charakters allerdings nicht unproblematisch, dieser Aufforderung *in praxi* nachzukommen. Schließlich droht das folgende Dilemma: Entweder wird die vorgeschlagene Definition so eng gefasst, dass sie relevante Verwendungskontexte samt den zugehörigen Bedeutungen ausblendet und damit selbst stipulativ erscheint. Oder die Definition gerät bei dem Versuch, möglichst alle Aussageweisen zu berücksichtigen, so weit, dass sie auch ihrerseits vage wirkt und keinen nennenswerten Klärungsgewinn erzielt. Statt der Definitionsaufforderung zu folgen und notgedrungen zwischen Skylla und Charybdis zu wählen, empfiehlt es sich daher, mit einer Definitionsaufforderung zweiter Stufe zu kontern und zuallererst eine Definition des Begriffs „Definition“ einzufordern. Ein solcher Zug ist weniger sophistisch als er vielleicht scheinen mag, und er kann, wie gleich zu zeigen ist, hinsichtlich der Vagheit der Begriffe „Philosophie“ und „Management“ durchaus erhellend sein.

Gemäß dem klassischen, auf der aristotelischen Logik und Prädikationstheorie basierenden Konzept operiert eine Begriffsdefinition in zwei Akten: Im ersten Akt wird der Artbegriff (*species*), den es zu definieren gilt (*definiendum*), unter einen Gattungsbegriff nächsthöherer Allgemeinheitsstufe (*genus proximum*) subsumiert. Im zweiten Akt wird der zu definierende Artbegriff mittels eines ihm eigentümlichen (*proprium*) Unterscheidungsmerkmals (*differentia specifica*) von den übrigen Begriffen gleicher Allgemeinheitsstufe abgegrenzt. *Genus proximum* und *differentia specifica* ergeben zusammen die Definition im Sinne des Definierenden (*definiens*). So wird beispielsweise der Begriff „Mensch“ definiert, indem er a) dem Begriff „Lebewesen“ untergeordnet und b) mittels des Differenzmerkmals „vernunftbegabt“ von allen anderen Lebewesen abgegrenzt wird. Entsprechend lautet der gesamte Definitionssatz: „Der Mensch (als *species* zugleich das *definiendum*) ist das vernunftbegabte (*differentia specifica*) Lebewesen (*genus proximum*).“

Es ist offensichtlich, dass es nicht wenige sprachliche Ausdrücke gibt, die sich einer diesem strengen Operationsmodus folgenden Definition entziehen. Nicht definierbar sind etwa Eigennamen wie „Aristoteles“ oder „Athen“ und definite Kennzeichnungen wie „der Begründer des Peripatos“ oder „der bekannteste Lehrer von Alexander dem Großen“, aber auch bestimmte Allgemeinbegriffe wie „Qualität“ und „Quantität“ oder „Ort“ und „Zeit“. Dass diese (dem aristotelischen Kategorienschema zugehörigen) Allgemeinbegriffe nicht definiert werden können, liegt daran, dass es sich bei ihnen (wie bei allen aristotelischen Kategorien und den sogenannten Prädikamenten der mittelalterlichen Logik) um Begriffe höchster Allgemeinheitsstufe handelt. Beinhaltet jede Definition eine

Subsumtion des zu definierenden Begriffs unter einen Begriff höherer Allgemeinheitsstufe, folgt trivialerweise, dass Begriffe höchster Allgemeinheitsstufe zwar für die Definition weniger allgemeiner Begriffe gebraucht werden, ihrerseits aber nicht definiert, da nicht subsumiert werden können.

Unternimmt man nun versuchsweise eine Definition der Begriffe „Philosophie" und „Management", dann stößt man rasch auf eben das Problem, das soeben mit Blick auf die Kategorien aufgezeigt wurde: Es findet sich kein geeigneter Gattungsbegriff, unter den man „Philosophie" oder unter den man „Management" sinnvoll subsumieren könnte. Damit scheint die angezeigte Vagheit der beiden genannten Begriffe eng zusammenzuhängen mit ihrem hohen Allgemeinheitsgrad, der eine plausible Subsumtion und damit auch eine valide Definition nach klassischem Muster verunmöglicht. In heutiger Diktion kann man „Philosophie" und „Management" entsprechend als basale Begriffe bezeichnen, d. i. als Begriffe, die so grundlegend sind, dass sie – sozusagen definitionsgemäß – gar nicht definiert werden können.

Aus dem konstatierten hohen Allgemeinheitsgrad, dem basalen Charakter, der Nichtsubsumierbarkeit und ergo Nichtdefinierbarkeit der beiden Begriffe ist aber – ebenso wenig wie aus ihrer zuvor aufgezeigten Analogizität – eine Opazität abzuleiten, angesichts derer man von einer weiteren Begriffsverwendung besser absehen sollte. Dass eine derartige Konsequenz überzogen wäre, zeigt der Vergleich mit den genannten Begriffen „Qualität", „Quantität", „Zeit" und „Ort", auf die die selbigen Charakteristika zutreffen, ohne dass man deshalb für ihren Verzicht votieren würde. Zudem kann im Falle der Begriffe „Philosophie" und „Management", wie später bei der Darlegung der pragmatischen Präzisierungsstrategien gezeigt wird, gut mit dem zweiten Teilakt des klassischen Definitionsverfahrens, der Differenzierung, operiert werden, indem man mittels des Einsatzes kontextuell sinnvoller Gegenbegriffe einer im jeweiligen Verwendungszusammenhang hinreichenden Bedeutungsklärung zuarbeitet.

1.5 Familienähnlichkeit *(family resemblance)*

Dass sich die Begriffe „Philosophie" und „Management" einer Definition nach klassischem Konzept entziehen, lässt sich nicht nur durch den Hinweis auf ihren hohen Allgemeinheitsgrad erklären. Alternativ kann aufgrund der Analogizität beider Begriffe auch die These vertreten werden, dass ihnen nicht nur die rigide wirkenden Definitionsversuche, sondern sämtliche Explikationsverfahren aristotelischer Provenienz und merkmalssemantischer Methodik schon vom Ansatz her inadäquat sind. Zur Plausibilisierung dieser These ist ein – sich auf die

Extension, d. i. den Begriffsumfang beziehendes – Argument anzuführen. Demnach zeichnet sich das, was unter die genannten Begriffe fällt, nicht – wie eine taxonomische Klassifikation annehmen würde – durch ein den semantischen Merkmalen der Begriffe korrelierendes Set von Eigenschaften aus, die je für sich notwendig und zusammen für eine korrekte Zuordnung des Extensionsobjekts zum Begriff hinreichend wären. Vielmehr handelt es sich – wie der Blick auf ihre vielfältigen Verwendungen und die Differenzen ihrer Referenzobjekte anzeigt – um sogenannte familienähnliche Begriffe.

Der Begriff der Familienähnlichkeit wurde von Ludwig Wittgenstein in seiner Spätphilosophie konzipiert. Als „familienähnlich“ bezeichnet man unscharfe, verschwommene Begriffe, die sich wegen ihrer unklaren Grenzen – nun mit Blick auf ihre Intension, d. i. den Begriffsinhalt formuliert – nicht durch die Angabe von je für sich notwendigen und zusammen hinreichenden semantischen Merkmalen explizieren lassen. Das bekannteste Beispiel für familienähnliche Begriffe, das Wittgenstein selbst ausgeführt hat, ist „Spiel“: Unter diesen Begriff fallen zahlreiche Brettspiele, Kartenspiele, Kampfspiele, Ballspiele, Reigenspiele und andere mehr. Während Vertreter einer klassischen Merkmalssemantik aristotelischer Provenienz präsupponieren, dass es so etwas wie die Begriffszuordnung legitimierende Eigenschaften geben muss, die all diesen Spielen gemeinsam sind, fordert Wittgenstein dazu auf, eine derartige Gemeinsamkeit nicht einfach zu postulieren, sondern sie zu suchen und konkret zu benennen. „Sag nicht:»Es muß ihnen etwas gemeinsam sein, sonst hießen sie nicht ‚Spiele‘« – sondern schau, ob ihnen allen etwas gemeinsam ist.“[5] Dabeistelle man fest, dass es zwar nichts gebe, was allen Spielen gemeinsam ist, wohl aber zeige sich „ein kompliziertes Netz von Ähnlichkeiten, die einander übergreifen und kreuzen. Ähnlichkeiten im Großen und Kleinen.“[6] Zur treffenden Bezeichnung solch inhaltlicher Ähnlichkeiten und – metaphorisch gesprochen – Verwandtschaften nutzt Wittgenstein den von ihm neu in die sprachphilosophische Diskussion eingebrachten Term „Familienähnlichkeit“.

Folgt man Wittgenstein, so gilt für familienähnliche Begriffe, dass es zwar einerseits unmöglich ist, sie nach den Vorgaben einer klassischen Merkmalssemantik aristotelischer Provenienz zu definieren oder zu explizieren, dass aber andererseits eine solche Definition oder Explikation auch gar nicht notwendig ist, um sie in unterschiedlichen Verwendungszusammenhängen auf eine jeweils kommunikativ gelingende und verständliche Weise zu gebrauchen. Dies ist entsprechend auch für die Begriffe „Philosophie“ und „Management“ zu veranschlagen, sofern man

[5] Ludwig Wittgenstein, *Philosophische Untersuchungen* § 66.

[6] Ebd.

diese – wie hier vorgeschlagen – als familienähnliche Begriffe auffasst. Damit scheint es zumindest im Grundsatz möglich, die attestierte semantische Vagheit dieser Begriffe bei Erfüllung bestimmter Kommunikationsbedingungen und unter Verwendung bestimmter pragmatischer Strategien hinreichend zu präzisieren.

1.6 Dichte resp. dicke Begriffe, Deskriptivität vs. Evaluativität und Normativität

Bevor ich zur genaueren Erläuterung der angesprochenen Kommunikationsbedingungen und Präzisierungsstrategien komme, ist noch ein weiteres Charakteristikum der Begriffe „Philosophie" und „Management" zu berücksichtigen, das eng mit ihrer Vagheit zusammenhängt. Dieses liegt darin, dass es sich bei beiden um sogenannte „dichte" resp. „dicke Begriffe" (*thick concepts, thick normative concepts, thick evaluative concepts*) handelt. Der englische Philosoph Bernard Williams, der diese Terminologie konzipiert und in die sprachphilosophische und metaethische Diskussion eingeführt hat, bestimmt die Besonderheit dichter Begriffe darin, dass sie sowohl einen deskriptiven Gehalt besitzen als auch in evaluativer Hinsicht aufgeladen sind.[7] Diese Kombination von beschreibenden und wertenden Aspekten unterscheidet dichte Begriffe einerseits von dünnen deskriptiven Begriffen (*thin descriptive concepts*) wie „Wasser", die nur beschreibende, und andererseits von dünnen evaluativen Begriffen (*thin evaluative concepts*) wie „gut", die nur wertende Aspekte beinhalten.

Betrachtet man die Begriffsgeschichte von „Philosophie", so fällt nicht nur auf, dass die Explikationen dieses Begriffs zumeist von den Philosophen selbst vorgenommen werden, sondern auch, dass dies – wenig überraschend – zuallermeist *pro domo* geschieht. Charakterisiert etwa Platon die Philosophie, indem er ihre Methodik (mit Blick auf die Instrumentarien Dialektik, Dihairetik und Synoptik) analysiert, so spricht er nicht nur in beschreibender Weise über das, was die Philosophie ist, sondern zugleich in wertender Absicht über das, was sie sein soll und was sie von Rhetorik, Politik und vor allem von der Sophistik unterscheidet und – in selbstverständlich positiver Hinsicht – auszeichnet.

Ein schönes Beispiel aus jüngster Zeit für einen Philosophiebegriff, der sowohl beschreibende als auch wertende Aspekte beinhaltet, findet sich in Geert Keils auf die sogenannten analytischen Tugenden rekurrierender Erläuterung des Begriffs der Analytischen Philosophie und in seiner Skizzierung des Professionalisierungsschubs, welchen das Fach Philosophie in den letzten

[7] Vgl. Williams (1985).

Jahren (mit Blick etwa auf Methodenbewusstsein, Arbeitsteiligkeit und Problemkonzentration) erfahren habe: Grenzt Platon die dialektisch und damit methodisch seriös operierenden Philosophen von den eristisch, antilogisch und damit methodisch unseriös verfahrenden Sophisten ab, so unterscheidet Geert Keil zwischen Philosophen, die sich in ihrem Arbeiten an den Standards philosophisch-wissenschaftlicher Solidität ausrichten, und „medial trainierten Philosophiedarstellern", die dies genau nicht tun und daher die in diesem Fall positiv konnotierte Bezeichnung „Philosoph" auch gar nicht verdienen.[8]

Vergleichbare Zuschreibungen von deskriptiven und evaluativ-normativen Aspekten finden sich auch in nicht wenigen Verwendungskontexten des Begriffs „Management". Bei funktionalen Ausdifferenzierungen dieses Begriffs beispielsweise werden die einzelnen Managementfunktionen meist nicht neutral, sondern mit in sich positiv aufgeladenen Begriffen (wie „Planung", „Organisation" und „Kontrolle") bezeichnet, während die naheliegenden Gegen- oder Kontrastbegriffe („Fehlplanung", „Planlosigkeit", „Desorganisation", „geringer Organisationsgrad", „Unkontrolliertheit", „Kontrollverlust" u.ä.) klar negativ konnotiert sind.

Zwar hat die Verwendung von *thick concepts* auch einen nicht zu bestreitenden kommunikativen Wert, so dass die Forderung nach einem sprachanalytischen Reinheitsgebot, das nur entweder deskriptive oder aber evaluativ-normative Begriffe, aber keine Kombination derselben erlaubt, weit überzogen wäre. Doch scheint mir die Feststellung, dass die Begriffe „Philosophie" und „Management" häufig als *thick concepts* gebraucht werden, zweckdienlich, um die besondere Art ihrer Vagheit zu präzisieren, und vor allem auch, um im Einzelfall mittels kritischer Nachfrage klären zu können, wovon gerade die Rede ist: Von dem, was Philosophie ist? Von dem, was Management ist? Oder von dem, was Philosophie und Management jeweils sein sollen?

2 Pragmatische Präzisierungsstrategien

Nachdem im ersten Teil des Beitrags die Rede von der semantischen Vagheit der Begriffe „Philosophie" und „Management" genauer erläutert wurde, geht es im Folgenden um die Frage, wie man in konkreten Verwendungszusammenhängen am besten mit den genannten Begriffen verfährt und wie innerhalb eines

[8] Mythen über die analytische Philosophie: GAP-Präsident Geert Keil im Gespräch. Interview vom 27.03.19 mit Romy Jaster. Vgl. insbes. 5:54 bis 6:59. https://www.youtube.com/watch?v=z0fEjZApHZk Zuletzt abgerufen am 01.02.2024.

mit „Philosophie und Management" titulierten Themengebiets (etwa im Kontext von Studiengängen, Forschungsprojekten, Sammelbänden, Tagungen u. a.)[9] gearbeitet werden kann, ohne dass sich die in der Einleitung angesprochene Gefahr eines die semantische Vagheit seiner Titulierung perpetuierenden und an der Wahrheit von Aussagen und der Folgerichtigkeit von Argumentationen weitgehend desinteressierten Diskurses aktualisiert.

2.1 Kommunikationsbedingungen und Konversationsmaximen

Pragmatische Präzisierungsstrategien können nur dann eine für den jeweiligen Verwendungskontext hinreichende Begriffsklärung erzielen, wenn zugleich bestimmte Kommunikationsbedingungen erfüllt sind. Welche dies im Einzelnen sind, lässt sich am besten anhand der von dem englischen Sprachphilosophen H. P. Grice beschriebenen Konversationsmaximen darstellen.[10]

Grundlage der Konversationsmaximen (bzw. der sogenannten konversationellen Implikaturen) ist das Kooperationsprinzip (*cooperative principle*), dessen allgemeine Regel Grice wie folgt formuliert: „Make your conversational contribution such as is required, at the stage at which it occurs, by the accepted purpose or direction of the talk exchange in which you are engaged." Aus diesem Prinzip ergeben sich vier, in Anlehnung an Kants Urteilstafel benannte Konversationsmaximen: a) Die Maxime der Quantität (*Maxim of quantity/content length and depth*) fordert, einen Kommunikationsbeitrag mit Blick auf den Kommunikationszweck zum einen so informativ wie möglich, zum anderen aber auch nicht informativer als nötig zu gestalten. b) Die Maxime der Qualität (*Maxim of quality/truth*) fordert, in einem Kommunikationsbeitrag nichts zu behaupten, was man selbst für falsch hält oder für dessen Wahrheit man keine zureichenden Rechtfertigungsgründe darlegen kann. c) Die Maxime der Relation (*Maxim of relation/relevance*) fordert, nur solche Beiträge beizusteuern, die mit Blick auf das Ziel, den Zweck und die Ausrichtung der Kommunikation relevant sind. d) Die Maxime der Modalität (*Maxim of manner/clarity*) schließlich fordert die Klarheit des sprachlichen Ausdrucks, wozu etwa die Vermeidung mehrdeutiger

[9] Im Folgenden werde ich hierfür den Ausdruck „Philosophie und Management"-Kontexte verwenden.

[10] Vgl. Grice (1975).

und allzu unscharfer Begriffe, aber auch der Verzicht auf unnötige Ausführlichkeit gehören.[11]

Nach der Auffassung von Grice ist es rational, den genannten Maximen zu folgen, da sie selbst rational motiviert sind. Die Maximen besitzen mithin Geltungskraft für jede Art von Konversation, sofern diese als eine kooperative Interaktion verstanden werden kann. Auch wenn die Maximen also einen recht universalen Charakter aufweisen, scheint mir ihre Beachtung im spezifischen Kontext von „Philosophie und Management" doch besonders geboten. Angesichts der Vagheit und situativen Präzisierungsbedürftigkeit dieses Themenfelds versteht es sich von selbst, der Maxime der Modalität entsprechende Aufmerksamkeit zu schenken: „Avoid obscurity of expression! Avoid ambiguity! Be brief! Be orderly!" In Anbetracht der hier – wie in jeder inter- und transdisziplinären Diskussion – akuten Gefahr eines inhaltlichen Abdriftens empfiehlt es sich aber auch, im Fall der Fälle an die Maxime der Relation zu erinnern: „Be relevant!" Die Sensibilisierung für die Maximen der Quantität und Qualität schließlich scheint mir ein hervorragendes Antidot, um einem so weitverbreiteten wie hochgefährlichen Diskurstoxin, nämlich dem schon in der Einleitung erwähnten *bullshitting,* entgegenzuwirken: „Make your contribution as informative as is required for the current purposes of the exchange! Do not make your contribution more informative than is required!" „Do not say what you believe to be false! Do not say that for which you lack adequate evidence!" Nur bei hinreichender Befolgung der skizzierten Konversationsmaximen können die kommunikativen Bedingungen als erfüllt gelten, die für eine effektive Anwendung der nun genauer zu erläuternden pragmatischen Präzisierungsstrategien notwendig sind.

2.2 Kontextspezifische Begriffsexplikation, Differenzierung und Pluralisierung

Eine Definition der Begriffe „Philosophie" und „Management" ist, wie ich im ersten Teil dieses Beitrags zu zeigen versucht habe, weder möglich noch notwendig. Statt den zwecklosen Versuch einer stipulativ wirkenden Bedeutungsfixierung zu unternehmen, gilt es angesichts der semantischen Vagheit und

[11] In der Sache gehen die von Grice formulierten Kommunikationsmaximen auf das von dem italienischen Renaissance-Humanisten Lorenzo Valla (1406–1457) entwickelte Rhetorikkonzept der *copia* zurück. So ähnelt die Maxime der Quantität Vallas Forderung nach *brevitas*, die Maxime der Qualität der Forderung nach *perspicuitas*, die Maxime der Relation der Forderung nach *utilitas* und die Maxime der Modalität der Forderung nach *ordo*. Vgl. hierzu Westermann (2006, insbes. 44).

des vielfältigen Gebrauchs dieser Begriffe, pragmatische Präzisierungsstrategien einzusetzen, die eine für den jeweiligen Verwendungszusammenhang hinreichende und kommunikativ zufriedenstellende Klärung erlauben. Die wichtigste dieser Präzisierungsstrategien ist die kontextspezifische Explikation, die der Analogizität und der Familienähnlichkeit der genannten Begriffe bestmöglich Rechnung trägt. Im Folgenden werde ich diese Strategie mittels zweier Beispiele veranschaulichen, die aus meiner Sicht exemplarischen Charakter besitzen.

In dem einleitend erwähnten Nachdiplomstudiengang „Philosophie und Management" habe ich die Erfahrung gemacht, dass Personen, die in unterschiedlichen Managementbereichen tätig sind, häufig ein großes Interesse an der philosophischen Durchdringung und Reflexion von Begriffen haben, die für sie in ihrer täglichen beruflichen Praxis implizit oder explizit in besonderer Weise relevant sind, ohne dabei aber zur genuinen Fachterminologie ihrer jeweiligen Expertise zu zählen. Hierzu gehören nicht wenige Begriffe, die auch in der Philosophiegeschichte eine zentrale Rolle spielen und in der heutigen systematischen Philosophie weiterhin intensiv diskutiert werden, etwa die Begriffe „Zeit" und „Macht".

Thematisiert man nun in einem „Philosophie und Management"-Kontext aus philosophischer Perspektive den Begriff „Zeit", so ist zunächst zu präzisieren, was eine philosophische Zeitkonzeption üblicherweise von anderen, nichtphilosophischen Zeittheorien (angefangen bei naturwissenschaftlich-physikalischen Konzeptionen über empirisch-psychologische und neuronalwissenschaftliche Forschungen zur Zeitwahrnehmung bis hin zum sogenannten Zeitmanagement in der Ökonomie) unterscheidet. In vergleichbarer Manier kann auch bei einer Thematisierung des Begriffs „Macht" eine philosophische Theoriebildung durch die Usancen ihrer Verfahrensweise und inhaltlichen Ausrichtung abgegrenzt werden von einer soziologischen, einer politikwissenschaftlichen etc. Zum Einsatz kommt also jeweils die Methode der (gleichsam nach außen wirkenden, externen) Differenzierung, wie sie oben als zweiter Teilakt des klassischen Definitionsverfahrens vorgestellt wurde. Wichtig ist zu sehen, dass hier aber keine Definition des Begriffs „Philosophie" und auch keine Definition der Begriffe „Zeitphilosophie" oder „Machtphilosophie" angestrengt wird, sondern dass nur die typischen (und damit nicht notwendigerweise proprietären) Merkmale einer philosophischen Auseinandersetzung mit einem bestimmten Thema geklärt werden sollen.

Im Anschluss an die skizzierte Differenzierung ist in einem weiteren Schritt deutlich zu machen, dass es nicht nur einen einzigen Begriff der Philosophie, nicht nur eine einzige philosophische Konzeption der Zeit und nicht nur eine einzige philosophische Konzeption der Macht gibt und dass die philosophische Thematisierung dieser Begriffe auch ihrerseits durch eine

Vielzahl unterschiedlicher Herangehensweisen, inhaltlicher Kontextualisierungen und Positionierungen sowie philosophiehistorischer Situierungen geprägt ist. Zur pragmatischen Methode der Begriffsdifferenzierung gesellt sich damit noch eine zweite, nämlich die einer internen Pluralisierung.

Für die konkrete Kommunikation in „Philosophie und Management"-Kontexten bedeutet dies, dass man mit der Behauptung inhaltlicher Fachkonsense großer Reichweite („In der Philosophie versteht man unter Zeit, dass ..." „Der philosophische Machtbegriff zeichnet sich dadurch aus, dass ...") zumindest vorsichtig sein und auf jede Inanspruchnahme vermeintlicher Fachautorität („Als Philosoph kann ich dazu nur dazu sagen, dass ..." „Die Philosophie beantwortet diese Frage mit ...") besser ganz verzichten sollte. Eine gewisse Skepsis ist auch angesagt, wenn vergleichbare Professions-, Expertise- und Autoritätsreklamationen von Personen mit unterschiedlichem beruflichem Managementhintergrund artikuliert werden, da auch hier eine inhaltliche Nivellierung und die Ausblendung interner Dissense nicht auszuschließen sind („Als Banker/Leader/Kreativ-Manager/... kann ich dazu nur sagen, dass ..." „Im Sinne des Personalmanagements und vor allem der Personalführung lautet die Antwort auf diese Frage ...").

Nimmt man die beiden eben skizzierten Methoden – externe Differenzierung und interne Pluralisierung – zusammen, so folgt, dass die kontextbezogenen Begriffsexplikationen stets so spezifisch vorgenommen werden sollten, wie dies im jeweiligen Thematisierungszusammenhang möglich ist. Behandelt man beispielsweise die sogenannte Zeitabhandlung des Aristoteles[12], so ist nicht zu erläutern, was allgemein unter „Philosophie" oder unter „Naturphilosophie", sondern was spezifisch unter „aristotelischer Naturphilosophie" verstanden werden kann. Befasst man sich mit der Machtkonzeption Macchiavellis, so gilt Analoges: Zu erörtern, was allgemein mit „Philosophie" oder „Politischer Philosophie" gemeint ist, ist in diesem Zusammenhang nicht sonderlich hilfreich. Anders verhält es sich mit der spezifischen Darlegung dessen, was Staatsphilosophie unter den besonderen Konditionen des italienischen *rinascimento* bedeutet.

Durch die Anwendung solcher Präzisierungsstrategien gelingt, so jedenfalls meine Erfahrung, in den allermeisten Fällen eine für das Verstehen aller Diskursteilnehmenden hinreichende und für die konkreten kommunikativen Zwecke zufriedenstellende Klärung der relevanten Begriffe. Damit zeigt sich eine Definition der Begriffe, die dem Themenfeld „Philosophie und Management" seinen Namen geben, nicht nur aus theoretischen Überlegungen heraus als etwas zugleich Unmögliches wie Nichtnotwendiges, auch der empirisch erfahrene Blick

[12] Vgl. Aristoteles, *Physik* IV 10–14.

auf die Erfordernisse einer praktischen Arbeit in diesem Feld erweist eine Definition der Begriffe „Philosophie" und „Management" als etwas, das *in concreto* gar nicht gebraucht wird.

2.3 Bedeutungsklärende Begriffsverwendung

Eine weitere pragmatische Strategie besteht darin, nicht *über* die präzisierungsbedürftigen Begriffe zu sprechen, d. i. sie nicht metasprachlich zu thematisieren (*to mention*), sondern sie auf eine bedeutungsklärende Weise zu verwenden, d. i. objektsprachlich zu gebrauchen (*to use*). Eine derartige gleichsam *in praxi* erfolgende Bedeutungspräzisierung ist dann möglich, wenn der Begriffsgebrauch bestimmten Verwendungsregeln nicht nur folgt, sondern deren Geltung zugleich transparent werden lässt.

Die entscheidende sprachphilosophische Intuition, die einer solchen Präzisierungsstrategie zugrunde liegt, geht zurück auf die Überlegungen des späten Wittgenstein zum Verhältnis von Semantik und Pragmatik. Wie Wittgenstein in seinen „Philosophischen Untersuchungen" ausführt, liegt die Bedeutung von Wörtern nicht in den Gegenständen, für welche sie (vermeintlich) stehen, vielmehr wird sie in den meisten, wenn auch nicht in allen Fällen durch den spezifischen Gebrauch dieser Wörter festgelegt: „Die Bedeutung eines Wortes ist sein Gebrauch in der Sprache."[13] Entscheidend ist hierbei, dass die praktische Verwendung eines Wortes nicht chaotisch, sondern geordnet erfolgt. Die Sprachpraxis gehorcht demnach bestimmten Regeln, die zwar mitunter ungenau sind und auch gebrochen werden können, im Falle eines korrekten Wortgebrauchs aber eingehalten werden. Damit verhält sich die Frage nach der Bedeutung eines bestimmten Wortes analog zu der Frage nach der Bedeutung einer bestimmten Schachfigur: Zur Beantwortung sind jeweils diejenigen Regeln zu eruieren, die dem korrekten Gebrauch, das heißt: der richtigen Verwendungspraxis inhärent sind.

In „Philosophie und Management"-Kontexten kann nach meiner Erfahrung die oben charakterisierte Strategie einer kontextspezifischen Begriffsexplikation, wenn sie extensiv und je nach Gusto zu extensiv betrieben wird, auch nachträglich sein. Vor allem bei der Einführung neuer inhaltlicher Themenbereiche mit noch ungewohnter Terminologie droht die Gefahr einer empfundenen Überexplikation der relevanten Begriffe. Dies kann bei manchen zu Ermüdungserscheinungen führen, während sich andere ein rascheres Vorgehen in der Sache wünschen

---

[13] Ludwig Wittgenstein, *Philosophische Untersuchungen* § 43.

und nicht durch langwierige terminologische Vorüberlegungen gebremst werden möchten. In solchen Fällen empfiehlt es sich, auf umfangreiche bedeutungsexplizierende Ausführungen über die noch klärungsbedürftigen Begriffe besser zu verzichten, um ihre Bedeutung, d. i. die spezifischen Regeln ihres Gebrauchs sukzessiv durch eine möglichst regelkonforme Verwendung transparent werden zu lassen. Nicht selten gelingt es auch auf diese Art, eine für den konkreten Anwendungskontext hinreichende und kommunikativ zufriedenstellende Begriffsklärung zu erzielen.

2.4 Graduierung vs. Klassifizierung

Die pragmatische Strategie, die ich in diesem Beitrag abschließend skizzieren möchte, basiert auf der Unterscheidung zwischen einer graduierenden und einer klassifizierenden Begriffsverwendung.

Begriffe werden auf klassifikatorische Weise verwendet, wenn sie einem bestimmten Subjekt entweder zu- oder aber abgesprochen werden, ohne dass es unterschiedliche Grade der Zu- oder Abschreibung geben würde. Als Beispiel kann hier erneut der Begriff „Primzahl" genommen werden: Eine natürliche Zahl ist entweder eine Primzahl oder sie ist keine Primzahl. Tertium non datur. Wenn eine Zahl eine Primzahl ist, dann ist sie nicht „mehr oder weniger" Primzahl als eine andere Primzahl. Sie ist eine Primzahl – oder aber nicht. Demgegenüber werden Begriffe auf graduierende Weise verwendet, wenn sie einem Subjekt nicht schlechthin, sondern in einem bestimmten Maß oder zu einem bestimmten Grad zu- oder abgeschrieben. Als intuitiv naheliegende Beispiele eignen sich alltagssprachliche Ausdrücke wie „klein", „müde" oder „schmutzig". In grammatikalischer Hinsicht zeichnet sich eine graduierende Begriffsverwendung durch die Zulässigkeit komparativischer Formulierungen („kleiner", „müder", „schmutziger") aus, die im Falle einer klassifikatorischen Begriffsverwendung aus leicht verständlichen Gründen ausgeschlossen sind („primzahliger").

In „Philosophie und Management"-Kontexten ist die Differenzierung zwischen einer klassifikatorischen und einer graduierenden Begriffsverwendung von besonderer Relevanz, weil die Nichtbeachtung dieses Unterschieds – vor dem Hintergrund divergierender Usancen beim Begriffsgebrauch – leicht zu gravierenden Missverständnissen führen kann. Wird beispielsweise der Begriff „Selbstbewusstsein" thematisiert, der ja sowohl im Rahmen philosophischer Theoriebildung (insbes. in der Transzendentalphilosophie und dem Deutschen Idealismus) als auch in vielen Managementtheorien (Leadership, „Selbstmanagement" etc.) eine wichtige Rolle spielt, so ist zu beachten, dass dieser Begriff im Management, wie auch in der Alltagssprache, üblicherweise als ein graduierbarer Begriff gebraucht wird („Mitarbeiter wachsen in ihrem Selbstbewusstsein, wenn …"), während in philosophischen Zusammenhängen zuallermeist eine klassifikatorische Verwendung vorgenommen wird („Selbstbewusstsein ist die Wahrheit der Gewissheit seiner selbst.").

Erkennt man derartige Unterschiede in der Begriffsverwendung nicht, droht eine begriffliche Konfusion und wechselseitige Irritation. Eine adäquate Berücksichtigung solcher Differenzen hingegen verspricht einen Klärungsgewinn, der einer entsprechend präzisierten und reflektierten Begriffsverwendung zugutekommt.

Schlusswort

„Die Philosophie", so lautet eine der prägnanten Sentenzen Ludwig Wittgensteins, „ist ein Kampf gegen die Verhexung unseres Verstandes durch die Mittel der Sprache."[14] Wer es – wie die Beiträge des vorliegenden Sammelbandes – unternimmt, die Schnittstellen von Philosophie und Management zu thematisieren, der begibt sich auf ein begrifflich unsicheres und recht heikles Terrain. Sollten die eben vorgestellten sprachphilosophischen Reflexionen dabei helfen, einige der drohenden Gefahren zu vermeiden und der einen oder anderen Verhexung zu entgehen, so haben sie ihren Zweck erfüllt.

Literatur

Frankfurt, Harry G. (2005). *On Bullshit*. Princeton, New Jersey.

Grice, H. P. (1975). Logic and Conversation. In: Peter Cole/Jerry L. Morgan (Hrsg.), *Speech acts* (=*Syntax and semantics* 3), New York u. a., 41–58.

Herzberg, Stephan (2011). Pros hen-Einheit und Analogie. Die Vielheit des Seienden und das ontologisch Primäre bei Aristoteles und Thomas von Aquin. In: Johannes Brachtendorf/Stephan Herzberg (Hrsg.), *Einheit und Vielheit als metaphysisches Problem* (Collegium Metaphysicum Bd. 3), Tübingen, 35–59.

Melville, Herman (1853). *Bartleby, the Scrivener: A Story of Wall Street*.

Westermann, Hartmut (2006). Wie disputiert man über das Gute? Lorenzo Vallas *De vero bono* als Debatte über die richtige Debatte. *Jahrbuch Rhetorik* 25, 30–54

Williams, Bernard (1985) *Ethics and the Limits of Philosophy*. London.

---

[14] Ludwig Wittgenstein, *Philosophische Untersuchungen* § 109.

Part II

Conceptual and Methodological Contributions at the Crossroads between Philosophy and Management—Ancilla Oeconomiae? On the Usefulness (and Disadvantage?) of Philosophy for Managers

Normative Management: Why Management Cannot Do Without Philosophy

Christian Erk

1 A Call for Philosopher Managers

In the presumably most famous passage of his works, Plato calls for those who are in a position of political power to seriously take to the pursuit of philosophy:

> *"'Unless,' said I, 'either philosophers rule over the states or those whom we now call our kings and rulers take to the pursuit of philosophy truly and adequately, and there is a conjunction of these two things, political power and philosophy, while the many natures who pursue either apart from the other are compulsorily excluded, there is no end to the evils, dear Glaucon, for our states, nor, I believe, for the human kind either, […].'" (Plato, Republic, Book 5, 473c-d)*[1]

While Plato made this statement with persons in mind who wield political power, his statement also applies to persons who serve as managers. For, what both rulers of states and managers have in common is that they bear responsibility for a social system. The former are granted the privilege of political power because they lead and bear responsibility for the social system called "state". The latter also occupy a privileged position because they lead and bear responsibility for a

[1] «ἐὰν μή, ἦν δ' ἐγώ, ἢ οἱ φιλόσοφοι βασιλεύσωσιν ἐν ταῖς πόλεσιν ἢ οἱ βασιλῆς τε νῦν λεγόμενοι καὶ δυνάσται φιλοσοφήσωσι γνησίως τε καὶ ἱκανῶς, καὶ τοῦτο εἰς ταὐτὸν συμπέσῃ, δύναμίς τε πολιτικὴ καὶ φιλοσοφία, τῶν δὲ νῦν πορευομένων χωρὶς ἐφ' ἑκάτερον αἱ πολλαὶ φύσεις ἐξ ἀνάγκης ἀποκλεισθῶσιν, οὐκ ἔστι κακῶν παῦλα, ὦ φίλε Γλαύκων, ταῖς πόλεσι, δοκῶ δ' οὐδὲ τῷ ἀνθρωπίνῳ γένει, […].» (Plato, Republic, Book 5, 473c–d).

C. Erk (✉)
Universität St. Gallen, St. Gallen, Switzerland
e-mail: christian.erk@unisg.ch

C. H. Hoffmann (ed.), *Artificial Intelligence, Entrepreneurship and Risk*, Technikzukünfte, Wissenschaft und Gesellschaft / Futures of Technology, Science and Society, https://doi.org/10.1007/978-3-658-45544-6_4

different type of social system, namely the social system called "business" (also: business enterprise; company). The genuine pursuit of philosophy seems to be an indispensable prerequisite for everybody who holds a position of responsibility for and authority over a social whole. Accordingly, it is not only the rulers of states but also managers who are called to take to the pursuit of philosophy seriously and adequately. In short: Philosophy is a *conditio sine qua non* for leading social systems.

In order to see why and how exactly this is the case, we need to firstly enquire into the nature of philosophy. As will be argued (cf. Chap. 2), pursuing philosophy is not tantamount to acquiring philosophical knowledge or dealing with abstract problems; rather, it means honing a habitual love for wisdom and truth. In order to see where exactly it is dependent on philosophy, we need to secondly enquire into the nature of the craft called "management" (cf. Chap. 3). As we will see (cf. Chap. 4), it is the field of what is called "normative management" where management cannot do without philosophy and that requires managers to become philosopher managers. Showing to what extent and why exactly normative management cannot do without philosophy will also be the task of this third part of this article. The article closes with some practical thoughts on the internalisation of the result of a company's normative orientation processes, i.e. its normative positioning (cf. Chap. 5).

2 Philosophy: What Does It Mean to Philosophise?

As Plato and Aristotle tell us, at the beginning of philosophy there stood and stands—not, as some thinkers such as Georg Wilhelm Friedrich Hegel would have us believe, doubt and confusion—but wonder, amazement and admiration about the world that surrounds us:

> *"For it is owing to their wonder that men both now begin and at first began to philosophise." (Aristotle, Metaphysics, Book 1, 982b 11)*[2]

> *"For this feeling of wonder shows that you are a philosopher, since there is no other beginning of philosophy than this." (Plato, Theaetetos, 155d)*[3]

---

[2] "διὰ γὰρ τὸ θαυμάζειν οἱ ἄνθρωποι καὶ νῦν καὶ τὸ πρῶτον ἤρξαντο φιλοσοφεῖν." (Aristotle, Metaphysics, Book 1, 982b 11).

[3] "μάλα γὰρ φιλοσόφου τοῦτο τὸ πάθος, τὸ θαυμάζειν· οὐ γὰρ ἄλλη ἀρχὴ φιλοσοφίας ἢ αὕτη." (Plato, Theaetetos, 155d).

Philosophy emanates from the desire to learn what we do not yet know and understand what we do not yet understand. A philosopher resembles a curious child who is always asking the same kinds of questions: What is X? How does X work? Why is X the case, i.e. how did X come to be? What is the end of X? These questions can however not only be asked of the natural and material world surrounding us. They can also be applied to immaterial objects and abstract concepts that are not directly accessible to our senses. And they can be applied to our existence and existence in general: Who am I? Why and what for do I exist? And above all: Why is there something at all rather than nothing? Looking at the world with the eyes of a philosopher means taking not only the extraordinary but also the everyday and ordinary objects of our gaze neither for granted nor as a matter of course (cf. Pieper, 2003: 58).

> *"For it is not the «unprecedented», the abnormal and the sensational, which arouses philosophical amazement—but which is the only thing in which a dulled sense can experience something like a substitute for genuine amazement. […] The need for sensation […] is an unmistakable sign of the loss of the genuine power of wonder […]. To recognise the truly unusual and unusual, the* mirandum, *in the everyday and the ordinary—that is the beginning of philosophising."* (Pieper, 2003: 62)[4]

Philosophy is an active desire for knowledge that helps quenching a thirst that is caused by a habit of wonder and amazement about us and the world. However, the aim of philosophy is not to know and understand as much as possible; a polymath is not a philosopher. Rather, philosophy aims at a special kind of knowledge which is already contained in the very word itself: As its word components tell us, philo-sophy consists in the affectionate love (Greek: φιλία) of wisdom (Greek: σοφία). A philosopher, therefore, is a "sapientiae amator" (Thomas Aquinas, Sententia Metaphysicae, lib. 1 l. 3 n. 5), i.e. a lover of wisdom. This characterisation has a twofold implication: Firstly, being lovers of wisdom, philosophers—by definition—do not possess wisdom; they desire and actively search for wisdom

[4] "Denn es ist ja nicht das «Niedagewesene», das Abnorme und Sensationelle, woran sich das philosophische Erstaunen entfacht—woran aber ein stumpfgewordener Sinn allein so etwas wie einen Ersatz für echtes Staunen zu erfahren vermag. […] Das Bedürfnis nach der Sensation […] ist ein untrügliches Zeichen für den Verlust der echten Staun-Kraft […]. Im Alltäglichen und Gewöhnlichen das wahrhaft Ungewöhnliche und Unalltägliche, das mirandum, zu gewahren—das also ist der Anfang des Philosophierens." (Pieper, 2003: 62).

knowing that it does not befit them to ever fully possess it.[5] Secondly, philosophers engage in the search for wisdom not for some other reason but for wisdom's sake alone. For, were they to seek wisdom on account of something else (e.g. money or fame), they would have greater love for this something than for wisdom. But, what exactly does it mean to desire and search for wisdom? What is it exactly that is desired and sought?

Wisdom is one of the three virtues that are meant to perfect that faculty of the human person that is geared towards the consideration of truth, namely her speculative reason. Together with the virtues of science and understanding, wisdom helps the speculative reason to make visible the intelligible truth that is contained in sense experience. Among these three virtues, the special job of the virtue of wisdom consists in considering and obtaining true knowledge of the highest and primary causes of things:

> *"It belongs to the wise to consider the highest cause. By means of that cause we are able to form a most certain judgment about other causes, and according thereto all things should be set in order. Now the highest cause may be understood in two ways, either simply or in some particular genus. Accordingly, he who knows the highest cause in any particular genus, and by its means is able to judge and set in order all the things that belong to that genus, is said to be wise in that genus, for instance in medicine or architecture, [...]. On the other hand, he who knows the cause that is simply the highest, which is God, is said to be wise simply, because he is able to judge and set in order all things according to Divine rules." (Thomas Aquinas, Summa Theologiae, IIª-IIae q. 45 a. 1 co.; cf. Summa Theologiae, Iª-IIae q. 57 a. 2 co.; Sententia Metaphysicae, lib. 1 l. 2 n. 16; Aristotle, Metaphysics, Book 1, 982a 4ff)*[6]

---

[5] As the legend has it, it was Pythagoras who coined the term *philosopher* as he "refused to call himself a wise man as his predecessors had done, because he thought this was presumptuous, but called himself a philosopher, i.e., a lover of wisdom" (Thomas Aquinas, Sententia Metaphysicae, lib. 1 l. 3 n. 5; "noluit se sapientem nominare, sicut sui antecessores, quia hoc praesumptuosum videbatur esse; sed vocavit se philosophum, idest amatorem sapientiae"). In similar vein, Plato states that to call someone wise "seems to be something great" and that the epithet *wise* "befits God alone" (Plato, Phaedrus, 278d; "τὸ μὲν σοφόν, ὦ Φαῖδρε, καλεῖν ἔμοιγε μέγα εἶναι δοκεῖ καὶ θεῷ μόνῳ πρέπειν").

[6] "Ad sapientem pertinet considerare causam altissimam, per quam de aliis certissime iudicatur, et secundum quam omnia ordinari oportet. Causa autem altissima dupliciter accipi potest, vel simpliciter, vel in aliquo genere. Ille igitur qui cognoscit causam altissimam in aliquo genere et per eam potest de omnibus quae sunt illius generis iudicare et ordinare, dicitur esse sapiens in illo genere, ut in medicina vel architectura, [...]. Ille autem qui cognoscit causam altissimam simpliciter, quae est Deus, dicitur sapiens simpliciter, inquantum per regulas divinas omnia potest iudicare et ordinare." (Thomas Aquinas, Summa Theologiae, IIª-IIae q. 45 a. 1 co.).

Accordingly, the virtue of wisdom signifies the steady disposition[7] to consider the causes of things in general and the first or highest causes of all things in particular. In its traditional Aristotelian-Thomistic meaning, "those things are called causes upon which things depend for their existence or their coming to be" (Thomas Aquinas, In Physic., lib. 1 l. 1 n. 5).[8] Accordingly, a cause (also: explanation; Latin: causa; Greek: αἰτία) "contributes in some positive manner toward the production of another thing in its existence and being" (Bittle, 1950: 308). In short, a cause is what contributes to making a thing—be it animated or inanimate—to be what it is. According to Aristotelian-Thomistic philosophy, there are four and only four causes that explain a thing's how and why (cf. Thomas Aquinas, Contra Gentiles, lib. 3 cap. 10 n. 5; Thomas Aquinas, In Physic., lib. 2 l. 10 n. 15):

- Efficient cause: A thing's efficient (also: agent) cause explains which agent has brought/ brings this thing into existence or has changed/ changes it in some way.
- Final cause: A thing's final cause explains for the sake of which this thing exists or occurs; it signifies the thing's end, goal or purpose.
- Material cause: A thing's material cause explains out of which this thing becomes or is made. In everyday language, the cause signifies the matter out of which a thing is made.
- Formal cause: A thing's formal cause (also: form; quiddity) determines a thing essentially. It is its inner principle that makes a thing to be what it is and distinguishes it from other beings. As such, it signifies its essence (also: nature), which is "what we grasp intellectually when we identify a thing's genus and specific difference" (Feser, 2014: 211) and "that which explains its properties" (Feser, 2014: 230).

For any given object of philosophical investigation, it is these four causes which a philosopher seeks to understand. For, it is only by knowing its four causes that

[7] In general, a virtue is a good operative habit. A such a virtue signifies "a durable disposition of a person whereby the person is inclined easily and readily at will to make the best use of some faculty of her soul" (Erk, 2019: 147). The faculty that the virtue of wisdom helps to perfect (together with the other intellectual virtues of science and understanding) is the speculative (also: theoretical) intellect (also: reason), i.e. that part of the intellect which is there to extract the intelligible truth that is contained in our sense experience. The other part of our intellectual faculties, the practical intellect, is meant to consider this truth in reference to action, i.e. direct the truth known by the speculative intellect to action.

[8] "causae autem dicuntur ex quibus aliqua dependent secundum suum esse vel fieri." (Thomas Aquinas, In Physic., lib. 1 l. 1 n. 5; also cf. Thomas Aquinas, Summa Theologiae, III[a] q. 86 a. 6 s. c.).

we understand and possess knowledge of a thing. Knowledge of these causes gives us complete knowledge of a thing.

It should be a matter of course, that the knowledge of causes that wisdom strives for is not any kind of knoweldge but *true* knowledge. The orientation towards truth is so inherent to philosophy that Thomas Aquinas even calls philosophers "inquisitores […] veritatis" (In De Caelo, lib. 1 l. 22 n. 5), i.e. searchers of truth. In other words: Philosophy is not about collecting opinions but about impartially searching for and finding truth[9]:

> *"Philosophy, then, said I, is the science of being and recognition of truth; and happiness is the reward of this science and wisdom." (Justin Martyr, Dialogue with Trypho, Chapter 3)*[10]

> *"The study of philosophy aims not at knowing what men opine, but at what the truth of things is." (Thomas Aquinas, In De Caelo, lib. 1 l. 22 n. 8)*[11]

> *"Philosophy is rightly called a scientific knowledge of truth." (Aristotle, Metaphysics, Book 2, 993b 20)*[12]

Aristotle expresses the importance of the orientation towards truth for the philosopher even more clearly in his *Nicomachean Ethics*:

> *"However, it seems indeed better, and in fact especially obligatory on philosophers, to sacrifice even the rights of friendship for the sake of truth. While it is commendable*

[9] An orientation towards truth goes hand in hand with the acceptance of the so-called "Principle of Non-Contradiction" (PNC). According to PNC, something cannot be the case and not be the case at the same time and in the same respect (cf. Erk, 2019: 31ff). Among other things, this means that every proposition P can only and must either be true or false, i.e. not true. Furthermore, PNC implies what has found its expression in the so-called "Principle of Excluded Middle" (PEM), according to which of any two contradictory propositions P1 and P2 one must be true and the other false (cf. Erk, 2019: 34f). Therefore, the pluralism of contradictory opinions poses an intellectual challenge to the philosopher—a challenge that must not (and actually cannot) be eluded by taking refuge to the ultimately self-contradictory and self-defeating position of (ethical) relativism (cf. Erk, 2019: 37ff).

[10] "Φιλοσοφία μέν, ἦν δ' ἐγώ, ἐπιστήμη ἐστὶ τοῦ ὄντος καὶ τοῦ ἀληθοῦς ἐπίγνωσις, εὐδαιμονία δὲ ταύτης τῆς ἐπιστήμης καὶ τῆς σοφίας γέρας." (Justin Martyr, Dialogue with Trypho, Chapter 3).

[11] "Studium philosophiae non est ad hoc, quod sciatur quid homines senserint, sed qualiter se habeat veritas rerum." (Thomas Aquinas, In De Caelo, lib. 1 l. 22 n. 8).

[12] "ὀρθῶς δ' ἔχει καὶ τὸ καλεῖσθαι τὴν φιλοσοφίαν ἐπιστήμην τῆς ἀληθείας." (Aristotle, Metaphysics, Book 2, 993b 20).

to have love for both, we ought to honour truth as sacred above friends." (Aristotle, Nicomachean Ethis, Book 1, Chapter 6, 1096a 14)[13]

Summarising what has been said so far, we can therefore conclude that a philosopher is a person who has the steady disposition to lovingly seek true knowledge of the efficient, final, material and formal causes of things, including their highest causes. Philosophising is a way of looking at the world and ourselves, namely with the habit of childlike amazement and a loving desire to understand and recognise the causes of thing. It is this habit of searching for wisdom that the ancient Roman poet Virgil alludes to when stating that "he who has been able to learn the causes of things is happy" (Virgil, Georgics, Book II, Verse 490).[14]

Although this philosophical knowledge is pursued for its own sake and not any ulterior motives, it has, however, grave practical implications for those who seek it. For, truth, once recognised, forces itself upon our actions; it binds our conscience and requires us to live it and align our actions with it. Truth, once recognised, has normative force and becomes the yardstick for our actions. Therefore, philosophising does not only mean seeking for true knoweldge of the causes of things but ordering and directing our actions according to these causes. So, a good philosopher is a philosopher who actually acts and lives according to what he has recognised as true.

In view of the focus of this article, we can leave it at the above characterisation of philosophy. Albeit admittedly painted in somewhat broad strokes, the above suffices to capture the nature of philosophy and the essence of what it means to philosophise. As a next step, we can now move on and enquire into the nature of the craft called "management": What is management and what do managers actually do?

3 Management: What Does It Mean to Manage?

Management, as Louis A. Allen (1958: 5) wrote, is "a distinct activity". The nature of this activity is primarily determined by the nature of the object that is managed. For, management is always the management of something; it requires an object to be managed. And this object of management—generally speaking—are purposeful social systems. So, in order to understand what managers are there

[13] "δόξειε δ' ἂν ἴσως βέλτιον εἶναι καὶ δεῖν ἐπὶ σωτηρίᾳ γε τῆς ἀληθείας καὶ τὰ οἰκεῖα ἀναιρεῖν, ἄλλως τε καὶ φιλοσόφους ὄντας· ἀμφοῖν γὰρ ὄντοιν φίλοιν ὅσιον προτιμᾶν τὴν ἀλήθειαν." (Aristotle, Nicomachean Ethics, Book 1, Chapter 6, 1096a 14).

[14] "Felix, qui potuit rerum cognoscere causas." (Virgil, Georgics, Book II, Verse 490).

for, we need to understand what a system is in general and what a purposeful social system is in particular. Let us start with systems in general: As I have explained elsewhere, a system is defined as

> *"a whole that is made up of at least two interrelated and interacting elements, distinguishable from its environment, ordered by rules and responsible for a purpose."* (Erk, 2016: 80; cf. Erk & Spoun, 2020: 13ff)[15]

Whatever else it may be, every whole consisting of two or more interrelated and interacting elements—be it a watch, a car, a plane, a computer, a football team, a state, a company or a human being—is a system. A system is a social system if it is made up of a special type of elements, namely human beings. In contrast, so-called technical systems consist of non-human elements. So, watches, computers etc. are technical systems while sports teams, states or companies are social systems.[16]

As the above definition states, all systems—and therefore all social systems—are purposeful ("zweckverantwortlich"), i.e. exist to realise a purpose.[17] The purpose of a social system answers the question of why formerly separate and not necessarily interrelated human beings have joined together to form a whole. Its purpose determines which and how many elements the system must consist of,

---

[15] "Ein System ist eine aus mindestens zwei miteinander in Beziehung stehenden Elementen gebildete, von ihrer Umwelt abgrenzbare, durch Regeln geordnete und zweckverantwortliche Ganzheit." (Erk, 2016: 80; cf. Erk & Spoun, 2020: 13ff).

[16] To be more precise, we have to extend this distinction to the effect that technical systems consist of inanimate, non-human elements. This specification is important as there also are systems that are comprised of animate, non-human elements; these can be called biological systems. Examples of this type of system are a pack of wolves, a herd of elephants, a natural ecosystem but also individual animals or plants and man himself. For a more comprehensive and complementary distinction between different types of systems based on their elements cf. Erk & Spoun (2020: 24).

[17] According to the so-called "Principle of Finality" (also: Principle of Teleology), every agent acts for an end or purpose (cf. Bittle, 1941: 411ff; I[a] q. 44 a. 4 co.). This principle applies to all agents analogically, i.e. in accord with the nature of the agent in question (cf. Garrigou-Lagrange, 1932: 106ff; Bittle, 1941: 426): Agents endowed with reason (i.e. human beings) act in view of an end directively and formally speaking ("directive formaliter"), i.e. knowing finality itself; agents lacking reason (i.e. animals) act for an end directively but only materially speaking and not formally ("directive non formaliter sed materialiter tantum"), i.e. knowing the thing that is their end; natural agents devoid of all knowledge (i.e. plants and inorganic beings) act in view of an end only executively ("executive tantum"), i.e. obey finality only on the level of execution.

what relationships must exist between the elements and what order must prevail so that the system can realise its purpose.

The fact that social systems, like all systems, have a purpose has an important implication for our topic at hand: They are collaborative ("arbeitsteilig") by nature and require a division of labour ("Arbeitsteiligkeit"). Usually, the system's purpose cannot be realised by the efforts of a single element; otherwise, there would be no need to establish a system. Rather, the fulfilment of its purpose requires the joint efforts of more or less all the elements of the system. Each element makes a partial contribution to the fulfilment of the system purpose.

Although it brings with it specialisation effects and increases both efficiency and effectiveness, this division of labor also has a downside: Someone has to make sure that the multitude of indiviual actions performed by the human beings who make up the social system are coordinated and geared towards the realisation of the system's purpose and not something else. Social systems have a high behavioural and relational variety, i.e. each of their human elements can behave in virtually countless different ways and can enter new and end existing relationships with other elements. This makes social systems high variety and therefore complex systems. In contrast to simple or complicated systems,[18] the challenge with complax systems is that one cannot determine but only make probability statements about their overall behaviour—which means that they are in fact not fully controllable.

And it is precisely in dealing with this challenge that the purpose and peculiar task of the distinct activity called "management" lies: Managing means coordinating the spatially and temporally distributed actions of the individual members of a social system in an orderly manner towards the realisation of the system's purpose and goals.[19] Management is a specific function to be performed by and for such

---

[18] For an explanation of this distinction and a more detailed description of these three system types, cf. Erk & Spoun (2020: 27ff).

[19] The performance of these actions is usually dependent on and supported by a range of material or immaterial resources (cf. Erk & Spoun, 2020: 35), some of which are consumed as part of the action ("consumable resources") and others of which can be used without being consumed ("non-consumable resources"). Of course, the deployment and utilisation of these action-enabling resources have to be coordinated, too, and aligned with the human actions.

Technically speaking, human beings could be considered non-consumable resources, too. This is why they are sometimes referred to as "human resources". However, it would be better to think of them of a resource *sui generis*. For, human beings are not living things but persons; as such, they do not have a value but possess dignity that must always and absolutely be respected (cf. Erk & Spoun, 2020: 36f).

system,[20] the purpose of which is—as virtually management thinkers agree—"to maintain a system of coöperative effort" (Barnard, 1938: 216), "to enable people to work most effectively together " (Allen, 1964: 66), "to make people capable of joint performance" (Drucker, 1988: 75; vgl. 2008: 18), "to make institutions capable of producing results" (Drucker, 2008: 81) or "to ensure that the unit serves its basic purpose" (Mintzberg, 2009: 49). In short: Management is about getting things done with and through people.

As should have become clear by now, management is a *conditio sine qua non* for the effectiveness of purposeful social systems; it is the enabler for effective joint performance so that a social system can serve its purpose with and through its people (cf. Erk & Spoun, 2020: 77). As such, managers are servants of the system and its members; they "do work for the group that the individual members cannot do effectively for themselves" (Allen, 1964: 63). Their focus is on and they bear responsibility for the effective functioning of the system as a whole. And it is this responsibility only, that grants managers their privileged position.

3.1 What Do Managers Do When Managing?

Management fulfils its task by performing the so-called "management tasks". Allen (1964: 67) defines these management tasks—or as he calls them "management functions"—

> *"as a group of related kinds of management work, made up of activities which are closely related to one another and which have characteristics in common derived from the essential nature of the work done. Each function can be defined so as to separate it logically from other functions."*

Without the proper and professional performance of these major sub-activities of the activity management, no system will be able to fulfil its purpose and achieve its goals. So, what are these management tasks?

As I have analysed and explained elsewhere (cf. Erk & Spoun, 2020: 81ff), there are five distinct management tasks.[21] These tasks have an internal focus, i.e. focus on the system itself and its internal activities. However, this list of five has to be supplemented by a sixth management task that focuses on the

---

[20] Similarly, and somewhat more figuratively, Drucker (2008: 2) calls management and managers the "specific organ of every institution" that "holds it together and makes it work".

[21] For a detailed explanation of each of these five management tasks cf. Erk & Spoun, 2020 (98ff; 146ff; 199ff; 221ff; 229ff).

relationship between the system and its environment and that has so far only been mentioned implicitly. Accordingly, management is comprised of the following six management tasks:

- **Orienting:** The most fundamental task of management is to provide orientation and direction. If you do not know where you are going, you do not know which path to choose. This tasks consists in the concretisation of the system's purpose as well as the definition of normative, strategic and operative goals; this also includes the definition of (both quantitative and qualitative) performance indicators that are suited for the assessment of the goal achievement.
- **Organising:** The realisation of a system's purpose and goals requires an adequate order. The process of designing this order is called "organising" and the result of the process "organisation". An organisation is comprised of a structural organisation ("Aufbauorganisation") as well as a process organisation ("Ablauforganisation").[22]
- **Leading:** Managers fulfil the management task "leading" by directly influencing the actions and behaviour of other people and deliberately guiding it towards the realisation of the system's purpose and goals. Leadership exerts this direct influence in a context of immediate social interaction, i.e. in physical or digital communication settings that can range from meeting to phone calls to email exchanges. The influence exerted in these settings of physical presence or at least communicative availability can be directed to single individuals or groups.
- **Controlling:** Controlling ("Lenken") is about actually operating the system and guiding as well as stabilising its everyday activities. By anticipating and dealing with disturbances, controlling has to make sure that the system actually fulfils its purpose and goals. As such, controlling requires the monitoring

[22] The structural organisation ("Aufbauorganisation") is based on a detailed analysis of the work that needs to be done to realise the system's purpose and goals. The resulting elementary tasks have to be arranged in positions ("Stellen"), i.e. balanced groups or packages of work that can be performed by single individuals. These positions are then departmentalised, i.e. grouped into departments ("Abteilungen") that perform similar or directly related tasks. The result of this grouping and departmentalisation process is the well-know organisation chart, i.e. the (snapshot-like) box-and-lines illustration of a system's structural organisation that depicts the large areas of work that are to be performed (and by whom), indicates basic relationships and specifies who gives orders to whom.

The process organisation ("Ablauforganisation") coordinates the work packages, i.e. puts the activities that are necessary to realise the system's purpose and goals in a temporal and spatial order.

of the activities and performance of the system and the taking of proactive or reactive corrective measures should the expected or effective performance not be adequate to achieving the system's purpose and goals. Such corrective measures can consist of either an adaptation of the purpose and goals of the system, an adaptation of the organisation of the system and/ or leadership interventions.

- **Developing:** This management task is about continously reflecting on potentials for improvement. Management is not limited to the timely prevention of actively anticipated problems (feedforward control), the monitoring and adjusting of ongoing activities (concurrent control) and reactive measures to avoid the repetition of mistakes that have already been made (feedback control). Managing also means being committed to the further development of the system beyond the task of controlling; motivated only by a commitment to constantly improving the system and ensuring its future viability, managers have to make it a habit to continously question the status quo, even if this is not obviously necessary.
- **Linking:** Acknowledging that every purposeful social system is embedded in an environment, this management task is about building and maintaining a web of relationships with relevant individuals and groups outside the system. As external linker the manager is concerned with stakeholder engagement but also cultivating linkages to the system's partners as well as external sources of information that might be relevant for the system.

The six management that have been described in the preceeding paragraphs are shown graphically and summarised in Fig. 1.

Managing means fulfilling the above six management tasks—continuously and practically always simultanously.[23] At its core, however, management is not only a continuous and never-ending but also a Sisyphean task: For, the paradox task that managers are ultimately entrusted with is to bring and maintain uncontrollable systems under control…

In contrast to the usual use of the term, the object of management is not only limited to businesses. Rather, all purposeful social systems must be managed. Management is a universal function that no purposeful social system can do without:

[23] It must be emphasised that the fulfilment of the management tasks is not the job of one manager; management is itself a collaborative endeavour of a community of managers. For an analysis of the implications of this important insight cf. Erk & Spoun (2020: 271ff).

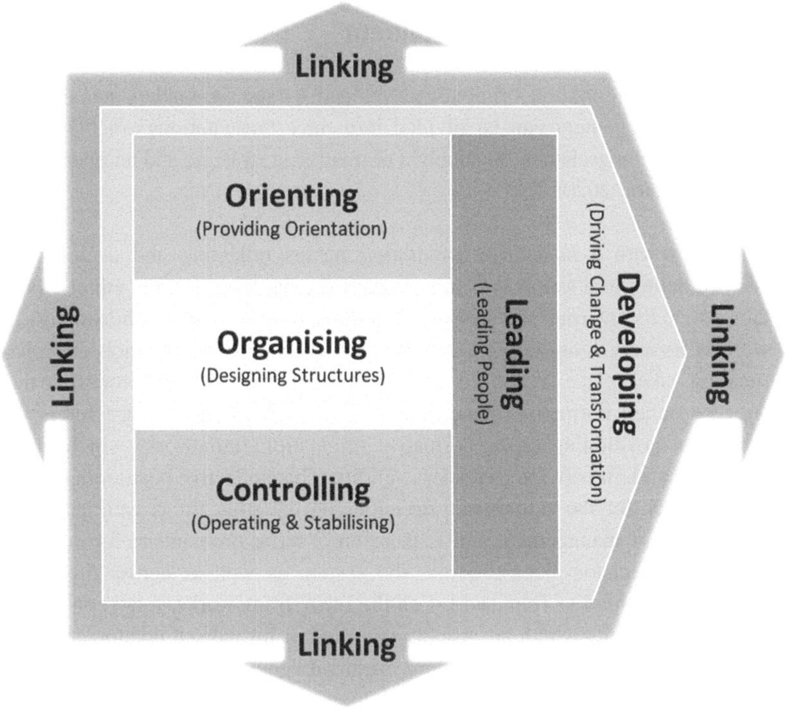

Fig. 1 The six tasks of management (Own illustration based on Erk & Spoun (2020: 94))

> *"The management of a family, a company, church, or country may differ in complexity and difficulty, but they involve the same basic skills. [...] As Charles Schwab is reputed to have said, "It takes the same kind of know-how to manage the United States Steel Corporation as it does to manage a peanut stand.""* (Allen, 1958: 47f)

Now that we have familiarised ourselves sufficiently with the essence of management, we can take another step towards understanding the synergetic relation of management and philosophy. This step consists of shedding light on what is referred to as "normative management".

3.2 The Normative, Strategic and Operational Orientation of Management

Following a differentiation originally developed by the St. Gallen management tradition,[24] management can be divided into three "orientations" or "levels of actions" ("Handlungsebenen"), namely a normative, a strategic and an operational (cf. Erk & Spoun, 2020: 238f):

- Managing with a *normative orientation* means reflecting and deciding on issues relating to a social system's purpose (i.e. its mission), the ethical foundations of its internal and external activities (i.e. its system philosophy) as well as its very long-term goals[25] (i.e. its vision). This reflection and clarification takes place in the context of normative orientation processes that are also called normative management. The result of these processes is the normative positioning (also: normative orientation framework) which has a long-term and, in part, even timeless validity. This collective positioning forms the yardstick for the actions and decisions on the strategic as well as operational level of management and is therefore a basic prerequisite for coherent action and decision-making across the system as well as across time. The normative orientation framework—as the other frameworks, too—consists of explicit statements which are usually recorded in formalised documents.
- Managing with a *strategic orientation* means reflecting and deciding on issues relating to securing the medium- and long-term future of a social system and creating the necessary preconditions for its future viability. The result of these processes is the system's strategic positioning (also: strategic orientation framework) which explicates its medium- and long-term goals as well as the resource configuration ("Leistungspotential") and the courses of action

[24] The first generation of the so-called "St. Gallen Management Model" (cf. Ulrich & Krieg, 2001; Ulrich, 2001a; Ulrich, 2001b) does not yet use the terms normative, strategic and operational. Instead, it speaks of three levels of management ("Stufen der Unternehmensführung"), namely corporate policy ("Unternehmungspolitik"), corporate planning ("Unternehmungsplanung") and disposition ("Disposition"). In terms of content, however, this triadic division is identical to the distinction between normative, strategic and operational management both in later works of the author of the management model's first generation (cf. Ulrich, 1984: 299ff) as well as the authors of the second (cf. Bleicher, 2011), third (cf. Rüegg-Stürm, 2003) and fourth (cf. Rüegg-Stürm & Grand, 2020) generation of the St. Gallen management model.

[25] In contrast to strategic goals, which are usually set for a period of 3 to 5 years (and can thus also be described as long-term), normative goals have an even longer-term time horizon of 15 or more years.

(especially the strategic initiatives) necessary to achieve them. Starting point for these considerations is the analysis of the system's environment and the prognosis of its future development. Work on a system's strategic orientation framework takes place in the context of strategic development processes that are also called strategic management.

- Managing with a *operational orientation* means reflecting and deciding on issues relating to running and stabilising the day-to-day activities of the social system in the here and now, with a particular focus on the efficient and effective utilisation of the system's resource configuration. This running and stabilising takes place in the context of operational coordination processes that are also called operational management.

The three orientations of management are interdependent and cannot do without each other: While normative and strategic management lay the necessary foundations and design, it is management with an operational orientation that actually ensures that the system achieves its objectives in the short and medium term.[26]

The above suffices to capture the nature of management and the essence of what it means to manage. We can now take the last step and answer the question to what extent and why exactly normative management cannot do without philosophy.

4 Normative Management: Where Management Cannot Do Without Philosophy

As should have become obvious, the normative orientation is the most fundamental of all three management orientations. Given its focus on the ethical foundations of the activities, it is also the orientation where management necessarily meets (practical) philosophy in general and moral philosophy (also: ethics) in particular. The following paragraphs are therefore dedicated to the description of the specific content of this orientation. So, what exactly does it mean to

[26] Although normative, strategic and operational management are—if this were possible—three parts of the same coin, they are not necessarily the job of one manager; and, in reality, they practically never are. Rather, they require—albeit with different individual focal points—the collaborative effort of the entire management community (cf. Erk & Spoun, 2020: 280f): Upper managers are more concerned with normative and strategic management and less with operational management. For lower (also: first-line) managers, the relationship is reversed. The focus of the job of middle management lies in between.

manage with a normative orientation? What questions and tasks does normative management involve? And where does philosophy come into play?

As already mentioned, normative management is responsible for three things, namely defining a social system's

- purpose (i.e. its mission),
- the ethical foundations of its internal and external activities (i.e. its system philosophy) and
- its very long-term goals (i.e. its vision).

In the following, we will briefly discuss the first two of these elements[27] and their dependence on philosophy:

(1) As Magretta (2002: 216) has put it, "management starts with a mission worth accomplishing. [...] The basis of all performance is clarity of purpose." A system's **mission** (also: **purpose**) expresses why it exists, or to be more precise: its reason for being.[28] It is "the fundamental purpose it was established to fulfil and against which all its actions and their results must be judged" (Malik, 2010: 164; 2007: 167). The question of what the mission of a social system is does not have a universal answer that applies to all types of systems. Rather, it can only be answered for a specific type of system.

When it comes to the mission of the specific social systems called business, the question of contention is whether businesses exist (a) to make profit and generate economic returns to its owners, (b) to employ persons, provide meaningful work and pay wages or (c) to serve the needs of people who are outside them by providing them economic goods and services. The answer to this question has far-reaching consequences:

> *"But why is what people take to be the purpose of business an important consideration? It is important because what we conceive as the purpose of something will determine*

[27] Though important, the vision shall be neglected here due to lack of space and as it is the philosophically least interesting element. In general, its vision expresses a system's very long-term development goals, i.e. goals that usually require ten to thirty years of effort to be completed (cf. Erk & Spoun, 2020: 110f). The vision is the connecting element between the normative and strategic level of management. A system's strategic goals are to a certain extent operationalisations of and should help realise its vision. The latter usually requires several strategy cycles of three to five years length each.

[28] Unfortunately, the term "why" is ambiguous as—without further contextual information—it can have either a causal or a final meaning. In the context of mission statements, it is however the latter meaning that is relevant. A mission does not explain how a system came into existence, but to what end it exists.

> *when that something is acting appropriately. The purpose determines the way the thing should operate, and hence in a manner of speaking, the responsibilities of the thing. The purpose furnishes us with a criterion of evaluation. For example, a knife has an end or purpose. What makes a knife a good knife is how well it "fulfills its purpose", i.e. cuts. Similarly a business is designed for some purpose and what makes it a good business is determined by how well it fulfills its purpose. If its main purpose is to maximize profits, then, a business with a good bottom line which maximizes profits is a good business. [...] If, on the other hand, we view the main purpose as providing goods and services, then we will view the primary responsibility to be the production of quality goods and the better the quality of those goods and services, the better the business."* (Duska, 1997: 1402; cf. Koslowski, 2000: 139)

Reflecting on the general purpose of the social system named business is not an economical but a philosophical endeavour. It means enquiring into the final cause of business. So, what is the general mission of the social system called "business"? As I have shown elsewhere (cf. Erk & Spoun, 2020: 38ff), it is the virtually unanimous opinion of management philosophers that the mission of the social system called business is the (customer value creating) satisfaction of customer needs. In short: Business are—as the St. Gallen Management Model (cf. Ulrich & Krieg, 2001: 21) has already posited it—productive social systems.

Accordingly, profit or profitability is not the purpose of business; rather, it is a limiting factor and a "measuring rod" (Urwick, 1943: 27) or "convenient index" (Brech, 1967: 229) for the effective and efficient achievement of purpose:

> *"Yet profitability is, not the purpose of, but a limiting factor on business enterprise and business activity. Profit is not the explanation, cause, or rationale of business behavior and business decisions, but the test of their validity. If archangels instead of businessmen sat in directors' chairs, they would still have to be concerned with profitability, despite their total lack of personal interest in making profits."* (Drucker, 2008: 97)

> *"Profit is a two-fold test: firstly, it is a test for the validity of the business mission—a test revealing whether the company is doing the right thing, a test for its effectiveness. Secondly, it is a test for whether the company is fulfilling its purpose correctly, that is, whether it is doing the right thing the right way—a test for its efficiency."* (Malik, 2010: 166; 2007: 169f)

Without making reasonable profits, no business can survive; but this does not mean that businesses exist to make profits. This would be like stipulating that human beings live to breathe because they cannot live without breathing.

However, the above-mentioned general purpose of business is too broad and abstract to guide action. It must be concretised by stating which of the various human needs the respective company actually wants to satisfy. Again, this concretisation is a philosophical endeavour. Not every customer need also is a human need, i.e. a necessity for human flourishing. Some customer needs are mere wants, i.e. non-essential desires, and some even are contrary to human flourishing (and therefore unethical). To know which customer need is which and whether an existing customer need may be satisfied from an ethical perspective, business managers must take recourse to philosophical anthropology and moral philosophy. They must philosophically reflect on their idea of man and understand the nature as well as range of the so-called "basic human goods", i.e. the goods that the human person must realise in order to flourish as a human person (cf. Erk, 2019: 136ff) and that are the object of any true human need.

Put differently: To be able to define the mission of their company, business managers must have reflected on the efficient, formal and final cause of man; and this an essentially (moral) philosophical endeavour. For, strictly speaking, the mission of the social system called business is the (customer value creating) satisfaction not of all customer needs, but of ethically permissible customer needs, i.e. customer needs that further one or more of the basic human goods directly or indirectly.

(2) The second element of normative management is the so-called **system philosophy**, which in the case of a company is referred to as **corporate philosophy** (also: corporate ethics). A company's corporate philosophy conveys the ethical principles and values that regulate the conduct of the members of the company. As the expression of a company's intended ethical order, it provides the ethical basis for all internal and external activities of the members of the company. The corporate philosophy explicates the behavioural expectations to which the members of a system must adhere and by which they are measured.

As this general description should have made clear already, the explication of a company's corporate philosophy is an essentially philosophical and ethical endeavour, too. A company's corporate philosophy is only as good as the moral theory[29] it is based on. As they tend to contradict each other in their premises and/ or conclusions, not all moral theories can be equally true. As moral philosophy is "the loving search for the true moral theory" (Erk, 2019: 5), managers must necessarily become moral philosophers before they embark on the task of formulating a corporate philosophy. For, formulating a corporate philosophy is not about picking a set of values that sound somewhat attractive or motivating and seem to be shared or expected by the public. Rather, it is a serious ethical

[29] A moral theory is a set of reasons and interconnected arguments that "offers a reasoned account of what makes our conduct good or bad and helps us to determine whether an action or a behaviour is good or bad" (Erk, 2019: 5).

endeavour that is prepared to critically assess the ethical roots of a company's activity.

A company's corporate philosophy comprises three sets of guiding principles ("Leitbilder"):

- ***Concept of man*** *("Menschenbild")*

  Its concept of man (also: idea of man) contains a company's basic understanding of the nature and existential purpose of the being called "man". As such, it provides an answer to the question "What is man and why does he exist?". Reflecting on this question is not simply an exercise for the philosophical ivory tower, but a necessary task for managers. For, the nature of man is the measure of his right conduct towards himself, other human beings (be they members of the same company or part of the company's environment) as well as animated non-human beings and inanimate beings. To equip themselves with the knowledge necessary to fulfil this task, managers—once more—need to engage in depth with philosophical anthropology and cultivate a habit of philosophical curiosity about the efficient, formal and final cause of man.
- ***Principles of Good Conduct***

  Usually expressed in a code of conduct, the principles of good conduct firstly explicate the principles and behavioural expectations that guide the conduct of the company's members, both towards colleagues, superiors and subordinates as well as market partners (customers, suppliers, competitors). As such, the principles include a company's stance on controversial business activities, corruption and bribery, transparency and honesty in reporting, marketing ethics, etc. but also the principles governing the leadership relationships between superiors and subordinates as well as the design and use of leadership tools (such as delegation, financial and non-financial incentives, wage components, participation, employee development, working conditions in the workplace or social security).

  As a second element, these principles also contain a statement on a company's organisational principles ("Organisationsprinzipien"). These principles govern the company's organisational structures as well as its (corporate) governance, i.e. the organisation and composition of its top management bodies. The organisational principles also state how a company intends to actually implement ethics in its activities, i.e. by which means it intends to monitor (un)ethical conduct and further compliance with its ethical standards.

A third element of these principles are a company's technological guiding principles. These mainly comprise the company's attitude towards data protection and data security as well as innovation and the use of new technologies (such as artificial intelligence).

Last but not least, a concept of good conduct also comprises a statement on the company's economic guiding principles ("wirtschaftliche Leitvorstellungen"). On the one hand, these principles specify the company's degree of profit orientation: Is the company satisfied with the avoidance of loss, does it expect the generation of a moderate profit or does it aim for profit maximisation? On the other hand, the principles detail how a company intends to utilise the profit it generates: To what extent should the profits be distributed among the company's shareholders or rather be reinvested? In this context, managers have to also reflect and decide on a question that is at the heart of virtually business ethical challenges, namely how much economic profit the company is prepared to forego to do the ethically right thing. Does the company believe that there are inherently evil exterior actions[30] that should not be done even if they are done with (economically) good intentions or result in (economically) good consequences? Playing by the ethical rules—especially when the competition does not—might result in a loss of market shares and/ or reduction of profits. Is the company prepared to take this hit and even go monetary bankrupt before going ethically bankrupt?

As the above overview shows, a company's principles of good conduct are ethical by their very nature. Their formulation cannot but take recourse to and thus requires the managers' proficiency in moral philosophy, especially its sub-branches business ethics, leadership ethics and organisational ethics ("Organisationsethik").

- ***Concept of the envisioned economic and social order ("Leitbild der Wirtschafts- und Gesellschaftsordnung")***

  In this concept, managers are required to formulate an idea of the economic, political and social order that they consider desirable and to whose realisation the company should contribute. Accordingly, its concept of the economic and social order describes a company's stance on its role in and function for its economic, political and societal environment. Based on a company's stance on the content and extent of corporate (social) responsibility, it also explicates

[30] The exterior action is one of the elements of a human action, namely that "what we usually see happening when somebody performs an action" (Erk, 2019: 77). For an overview of the elements of a human action cf. Erk (2019: 76ff).

the principles for the company's interaction with its environment in general and its stakeholders in particular.
A company's concept of the economic and social order also expresses the company's basic stance on societal and social concerns (such as environmental protection, healthcare, combating poverty, development aid, promotion of the arts, etc.) but also on its attitude towards the state as well as the economic system that it considers desirable.
Formulating a concept of the economic and social order requires managers to take recourse to social ethics and its sub-branches economic ethics ("Wirtschaftsethik") and business ethics ("Unternehmensethik") as well as political philosophy.

In sum, the totality of the decisions of normative management on the three sets of guiding principles mentioned above forms the so-called normative positioning of the company. This normative positioning points the way for the company's future development and serves as guiding framework for the activities of the company's employees in general as well as for the management on the strategic and operational level in particular. But, as has been shown, such normative positioning cannot be had without serious philosophical and ethical reflection on part of a company's management. Without (moral) philosophy there can be no normative management that really deserves the name. The manager who wants to manage with a normative orientation must be a philosopher manager.

5 Normative Positioning: Not Something that is Had but Something that is Done

The best normative positioning is, however, useless if it does not find its expression in the corresponding behaviour of the members of the company. A normative positioning that is not implemented is not worth the paper it is written on. To fulfil its task of guiding behaviour, a normative positioning must be laid down in some form and made transparent. But what is even more important, it must be implemented and internalised—in short, it must be lived.

Accordingly, it could also be said that companies do not have a normative positioning but do a normative positioning or that they have a normative positioning only insofar as they live it. Even if made with regard to the mission of a company, the following statements can therefore also be applied to its normative positioning:

> *"Strictly speaking, purpose is defined more nearly by the aggregate of action taken than by any formulation in words."* (Barnard, 1938: 231)
>
> *"The purpose of a system is what it does."* (Beer, 1985: 99)

In reality, we come to know a company's normative positioning not so much by reading its websites and brochures but rather by what it does and how it behaves. And this is a matter of internalisation.

The internalisation of a company's normative positioning begins with making the positioning known to the members of the company. Knowledge of the normative positioning can be promoted through the following measures:

- The company's mission and corporate philosophy should be communicated, taught and recalled frequently, e.g. in emails, meetings, conversations, speeches, videos, testimonies, posters/ placards, screensavers, internal newsletters, etc.
- The company's mission and corporate philosophy should be embedded into recruitment, onboarding and the annual evaluation processes.
- It should furthermore be explained why the company has chosen its purpose and corporate philosophy (storytelling).
- Managers should help their employees understand the relevance of the company's mission and corporate philosophy.

A company's normative position can only be lived if it is known and shared by the majority of its members. The most important measure beyond this that managers can take to further the internalisation of the normative positioning of their company among all its members can be summarised under the keywords "authenticity" and "consistency". The members of the company must be given the feeling that the managers have internalised the normative positioning themselves and act according to it in their day-to-day practice – authentically and consistently.

> *"Living examples are more powerful than communication. Or, should we say, giving example is the only valid way to communicate."* (Cardona, Rey & Craig, 2019: 58)

Unless managers walk the talk themselves, all communication efforts are futile. The successful internalisation of a normative positioning is essentially dependent on the visible and noticeable commitment of the manager community. If they do not abide by it, how can managers expect others to do so? And finally, the successful internalisation of a normative positioning requires the coherence and consistency between the company's normative positioning as well as its strategy and operating model. The members of the company must believe that the normative positioning is an actual and effective frame of reference for all activities of the company and that it is considered at least equal to economic criteria in all major decisions of the company. In the end, managers seeking to internalise the normative positioning of their company among all its members should take the admonition of St. Ignatius of Antioch, the martyred bishop of Antioch in Syria and student of John the Apostle, to heart:

> *"It is better to be silent and be, than to talk and not be. It is good to teach, if he who speaks also acts." (Ignatius of Antioch, Epistle to the Ephesians, XV)*[31]

A company's normative positioning stands and falls with their managers also walking the normative management talk themselves.

References

Allen, Louis Alexander. Management and Organization. New York, Toronto & London: McGraw-Hill, 1958.

Allen, Louis Alexander. The Management Profession. New York, Toronto & London: McGraw-Hill Book Company, 1964. (McGraw-Hill Series in Management)

Aristotle. Metaphysics (Τὰ μετὰ τὰ φυσικά; Bekker Pagination 980a – 1093b). Original (Greek) text of the work and an English translation available via: http://www.perseus.tufts.edu/hopper/collection?collection=Perseus:collection:Greco-Roman Accessed: November 22nd, 2024.

Aristotle. Nicomachean Ethics (Ἠθικὰ Νικομάχεια; Bekker Pagination 1094a – 1181b). Original (Greek) text of the work and an English translation available via: http://www.perseus.tufts.edu/hopper/collection?collection=Perseus:collection:Greco-Roman Accessed: November 22nd, 2024.

Barnard, Chester Irving. The Functions of the Executive. Cambridge, MA: Harvard University Press, 1938.

Beer, Stafford. Diagnosing the System for Organizations. Chichester et al.: Wiley, 1985.

Bittle, Celestine N. From Aether to Cosmos: Cosmology. Milwaukee: The Bruce Publishing Company, 1941.

Bittle, Celestine N. The Science of Correct Thinking: Logic. Revised. Milwaukee: The Bruce Publishing Company, 1950.

Bleicher, Knut. Das Konzept Integriertes Management. Visionen – Missionen – Programme. 8., überarbeitete und erweiterte Auflage. Frankfurt & New York: campus Verlag, 2011. (Reihe St. Galler Management Konzept, Band 1)

[31] "Ἄμεινόν ἐστιν σιωπᾶν καὶ εἶναι, ἢ λαλοῦντα μὴ εἶναι. καλὸν τὸ διδάσκειν, ἐὰν ὁ λέγων ποιῇ." (Ignatius of Antioch, Epistle to the Ephesians, XV).

Brech, Edward Francis Leopold. Management. Its Nature and Significance. Fourth Edition. London: Sir Isaac Pitman & Sons, 1967.

Cardona, Pablo, Carlos Rey & Nick Craig. "Purpose-Driven Leadership." In: Rey, Carlos, Miquel Bastons & Phil Sotok (Eds.). Purpose-Driven Organizations. Management Ideas for a Better World. Cham: Palgrave Macmillan, 2019. pp. 57–70.

Drucker, Peter Ferdinand. "Management and the World's Work." Harvard Business Review 66.5 (1988): 65–76.

Drucker, Peter Ferdinand (with Joseph A. Maciariello). Management. Revised Edition. New York: Collins, 2008.

Duska, Ronald F. "The Why's of Business Revisited." Journal of Business Ethics 16.12/13 (1997): 1401–1409.

Erk, Christian. Was ist ein System? Eine Einführung in den klassischen Systembegriff. Wien & Zürich: LIT Verlag, 2016.

Erk, Christian. Moral Philosophy. Fundamental Concepts. Second, revised and expanded Edition. Neunkirchen-Seelscheid: editiones scholasticae, 2019.

Erk, Christian & Sascha Spoun. Integrativ managen. Ein Modell für eine effektive Praxis der Unternehmensführung. Wiesbaden: Springer Gabler, 2020.

Feser, Edward. Scholastic Metaphysics. A Contemporary Introduction. Heusenstamm: editiones scholasticae, 2014.

Garrigou-Lagrange, Réginald. Le Réalisme du Principe de Finalité. Paris: Desclée de Brouwer & Co, 1932.

Ignatius of Antioch. Epistle to the Ephesians (Προς Εφεσίους Ιγνάτιος). Original (Greek) text of the work as well as an English and German translation available via: https://bkv.unifr.ch/de/works Accessed: November 22nd, 2024.

Justin Martyr. Dialogue with Trypho (Τοῦ ἁγίου Ἰουστίνου πρὸς Τρύφωνα Ἰουδαῖον Διάλογος). Original (Greek) text of the work as well as an English and German translation available via: https://bkv.unifr.ch/de/works Accessed: November 22nd, 2024.

Koslowski, Peter. "The Limits of Shareholder Value." Journal of Business Ethics 27.1/2 (2000): 137–148.

Magretta, Joan. What Management Is: How it Works and Why it's Everyone's Business. New York: The Free Press, 2002.

Malik, Fredmund. Management. Das A und O des Handwerks. Frankfurt & New York: Campus Verlag, 2007. (Reihe «Management: Komplexität meistern», Band 1)

Malik, Fredmund. Management. The Essence of the Craft. Translated from German by Jutta Scherer. Frankfurt & New York: Campus Verlag, 2010. (Reihe «Management: Mastering Complexity», Volume 1)

Mintzberg, Henry. Managing. San Francisco: Berrett-Koehler Publishers, 2009.

Pieper, Josef. Was heisst Philosophieren? Vier Vorlesungen. Einsiedeln: Johannes Verlag, 2003.

Plato. Phaedrus (Φαῖδρος; Stephanus Pagination 227a – 279c). Original (Greek) text of the work and an English translation available via: http://www.perseus.tufts.edu/hopper/collection?collection=Perseus:collection:Greco-Roman Accessed: November 22nd, 2024.

Plato. Republic (Πολιτεία; Stephanus Pagination 327a – 621d). Original (Greek) text of the work and an English translation available via: http://www.perseus.tufts.edu/hopper/collection?collection=Perseus:collection:Greco-Roman Accessed: November 22nd, 2024.

Plato. Theaetetus (Θεαίτητος; Stephanus Pagination 142a–210d). Original (Greek) text of the work and an English translation available via: http://www.perseus.tufts.edu/hopper/collection?collection=Perseus:collection:Greco-Roman Accessed: November 22nd, 2024.

Rüegg-Stürm, Johannes. Das neue St.Galler Management-Modell. Grundkategorien einer integrierten Managementlehre. 2., durchgesehene Auflage. Bern, Stuttgart & Wien: Haupt Verlag, 2003.

Rüegg-Stürm, Johannes & Simon Grand. Das St. Galler Management-Modell: Management in einer komplexen Welt. 2., überarbeitete Auflage. Stuttgart: Haupt Verlag, 2020. (utb 5092)

Thomas Aquinas. Contra Gentiles (Summa contra Gentiles; Book on the Truth of the Catholic Faith against the Errors of the Unbelievers). Original (Latin) text of the work available via: http://www.corpusthomisticum.org/iopera.html Accessed: November 22nd, 2024.

Thomas Aquinas. In De Caelo (In libros Aristotelis De caelo et mundo expositio; Commentary on Aristotle's Books On the Heavens). Original (Latin) text of the work available via: http://www.corpusthomisticum.org/iopera.html Accessed: November 22nd, 2024.

Thomas Aquinas. In Physic. (Commentaria in Octo Libros Physicorum; Commentary on Aristotle's Physics). Original (Latin) text of the work available via: http://www.corpusthomisticum.org/iopera.html Accessed: November 22nd, 2024.

Thomas Aquinas. Sententia Metaphysicae (Sententia Libri Metaphysicae; Commentary on Aristotle's Metaphysics). Original (Latin) text of the work available via: http://www.corpusthomisticum.org/iopera.html Accessed: November 22nd, 2024.

Thomas Aquinas. Summa Theologiae, I^{a}-IIae (Summary of Theology, First Part of the Second Part). Original (Latin) text of the work available via: http://www.corpusthomisticum.org/iopera.html Accessed: November 22nd, 2024.

Thomas Aquinas. Summa Theologiae, II^{a}-IIae (Summary of Theology, Second Part of the Second Part). Original (Latin) text of the work: http://www.corpusthomisticum.org/iopera.html Accessed: November 22nd, 2024.

Thomas Aquinas. Summa Theologiae, III^{a} (Summary of Theology, Third Part). Original (Latin) text of the work available via: http://www.corpusthomisticum.org/iopera.html Accessed: November 22nd, 2024.

Ulrich, Hans. Management. Herausgegeben von Thomas Dyllick und Gilbert J. B. Probst. Bern & Stuttgart: Haupt Verlag, 1984.

Ulrich, Hans. "Grundlagen des Führungsmodells." In: Stiftung zur Förderung der systemorientierten Managementlehre (Hrsg.). Hans Ulrich. Gesammelte Schriften. Band 2: Das St. Galler Management-Modell. Bern, Stuttgart & Wien: Verlag Paul Haupt, 2001a. pp. 57–83. Ursprünglich 1972 publiziert.

Ulrich, Hans. "Unternehmungspolitik." In: Stiftung zur Förderung der systemorientierten Managementlehre (Hrsg.). Hans Ulrich. Gesammelte Schriften. Band 2: Das St. Galler Management-Modell. Bern, Stuttgart & Wien: Verlag Paul Haupt, 2001b. pp. 171–460. Ursprünglich 1978 unter dem gleichen Titel und im gleichen Verlag publiziert.

Ulrich, Hans & Walter Krieg. "Das St.Galler Management-Modell." In: Stiftung zur Förderung der systemorientierten Managementlehre (Hrsg.). Hans Ulrich. Gesammelte Schriften. Band 2: Das St. Galler Management-Modell. Bern, Stuttgart & Wien: Verlag Paul Haupt, 2001. pp. 7–55. Ursprünglich 1972 unter dem gleichen Titel und im gleichen Verlag publiziert.

Urwick, Lyndall Fownes. The Elements of Administration. London: Sir Isaac Pitman & Sons, 1943.

Virgil. Georgics (Georgica). Original (Latin) text of the work available via: https://www.perseus.tufts.edu/hopper/text?doc=Perseus%3Atext%3A1999.02.0059%3Abook%3D1%3Acard%3D1 Accessed: November 22nd, 2024.

PD Dr. phil. Christian Erk, M.Sc. LSE, He currently works as Head of Programmes at the Executive School of the University of St.Gallen (HSG), member of the strategy staff unit at thurmed AG, Co-Head of the Ethics Forum at Münsterlingen Cantonal Hospital, member of the Bioethics Commission of the Swiss Bishops' Conference as well as lecturer and author (focus: management, ethics, healthcare).

Previous positions include, among others, Managing Director of the EMBA HSG, Assistant Professor at the University of St.Gallen (HSG), Head of University Development at the University of Liechtenstein and positions in the healthcare sector and hospitality. Academic education at the HSG, the Copenhagen Business School, the London School of Economics (LSE) and the University of Zurich.

Following an introduction to ethics ("Moral Philosophy—Fundamental Concepts"), an introduction to management ("Integrativ Managen—Ein Modell für eine wirksame Praxis der Unternehmensführung") and a book on business ecosystems ("Unternehmens-Ökosysteme"), his latest book "The Ethics of Killing—Life, Death and Human Nature" was published by Palgrave Macmillan at the end of 2022.

Embedding Ethics in the Heart of Digital Revolution

How Can Companies Use Ethical Decision-Making Intelligence to Guide Responsible Technology Implementation?

Deepak Bansal

1 Introduction

Our time is characterized by the fourth industrial revolution, where AI is expected to contribute $15.7 trillion globally by 2030. Despite this, trust in digital technologies remains alarmingly low (around 40%) due to ethical concerns, which can derail envisioned progress. New regulations are coming up to ensure that technologies are ethics-centric. However, the digital ethics field is still relatively new and emerging, and intelligence is required in organizations to harmonize innovation with trust and compliance.

The "Ethical Decision-Making Intelligence" (EDMI) platform provides learning and solutioning intelligence by providing education and solution mechanisms to prepare companies for emerging digital ethical challenges.

Through online modular education as part of learning intelligence, EDMI ensure that the employees working in technical areas have the right knowledge about the digital ethics field, including mandatory knowledge of regulatory frameworks. The solutioning intelligence contains an optimized data-powered chatbot that provides immediate and intelligent responses to relevant queries specific to digital ethics. Users can also contact human expert panels for specific challenges.

D. Bansal (✉)
MQ Learning, Zurich, Switzerland
e-mail: deepak@mq-learning.com

C. H. Hoffmann (ed.), *Artificial Intelligence, Entrepreneurship and Risk*, Technikzukünfte, Wissenschaft und Gesellschaft / Futures of Technology, Science and Society, https://doi.org/10.1007/978-3-658-45544-6_5

EDMI instills a proactive, bottom-up intelligence approach to digital ethics and ensures every layer of the organization is ethically informed and aligned, especially the one who makes decisions while implementing or maintaining transformations and technologies. The demand for the EDMI platform is underscored by its alignment with emerging regulatory frameworks and the growing demand for ethical digital practices.

2 The Need for Ethical Decision Making Intelligence

In our fourth industrial revolution, the digital landscape is changing frequently and is characterized by:

- **Exponential Growth:** The potential contribution of AI to the global economy by 2030 is projected to be a staggering $15.7 trillion—surpassing the combined current outputs of both China and India (IMD-St Gallen Symposium 2021).
- **Trust Deficit:** However, there is a significant trust deficit, with only 40% of global users expressing confidence in digital technologies. This burgeoning trust crisis could seriously hinder our digital progress, making integrating ethics not just desirable but crucial (Tufts University, 2019).
- **Emerging Regulations:** New AI Acts are being phased in, providing more regulations and legal frameworks around this emerging field, e.g., the EU AI Act, where there are significant penalties of non-compliance ranging from €10 million to €40 million or 2% to 7% of the global annual turnover, depending on the severity of the infringement. Regulatory space will probably tighten in the next 5 years, and companies are unprepared (European Commission 2019).
- **Ethics is more about Mindset:** Laws and regulations on ethical topics are not enough and, in some cases, not even a solution. There is a need to generate more awareness and provide solution mechanisms to people when they build or maintain the technologies so that the systems will be more ethical "by design," not only "by regulation."

EDMI (Ethical Decision Making Intelligence) provides a learning and solutioning platform to employees working in technology areas by providing education and solution mechanisms to prepare companies for emerging digital ethical challenges. It's not just about compliance; it's about embedding ethical decision-making proactively at all levels of the organization while designing and using digital tools and technologies.

EDMI is designed to leverage the potential of the fourth industrial revolution by increasing trust in digital systems and navigating the complex emerging regulatory landscape by educating employees on digital ethics and proactively providing solutions to embed ethics in decision-making.

There are huge risks in not taking the topic of digital ethics seriously by companies. Some high-profile cases where huge losses were made due to a lack of Digital ethics are:

- *Uber's 2016 Data Breach:* Uber experienced a data breach in 2016, which exposed the data of 57 million Uber users and drivers and was then covered up. Uber's reputation suffered significantly when the news came to light in 2017. In addition to reputational damage, Uber had to pay a settlement of $148 million for the breach.
- *Amazon's AI Recruiting Tool Bias:* Amazon abandoned an AI recruiting tool because it showed bias against women. The tool was trained on resumes submitted to the company over 10 years, most of which came from men, resulting in the AI system favoring men over women for technical roles. Though the exact number on damage is unavailable, the tool took around 5 years to develop and generated a significant public backlash.
- *AI Predictive Policing in the Netherlands:* The Dutch Government used an AI system for predictive policing to identify fraudulent claims cases for child allowance. However, the system disproportionately targeted people with specific ethnic backgrounds and low-income groups. The incumbent cabinet had to resign in January 2021 over this scandal known as "*kinderopvangtoeslagaffaire,*" (childcare benefits affair).

The organizations or governments using such systems might have better intentions; however, the people deploying the system should have considered the various aspects of digital ethics e.g., security, fairness, and transparency, while designing and implementing them.

EDMI could have prevented all such cases—saved massive costs and public backlash, and provided the opportunity to create a better and more just world.

3 The Solution—"Ethical Decision-Making Intelligence (EDMI)" Platform

Ethical Decision-Making Intelligence (EDMI) platform responds to the growing need for proactive and comprehensive ethical guidance in the digital landscape. EDMI platform provides two core components for ethical decision-making:

1. **Learning Intelligence:** Online modular education ensures that the employees working in technical areas have the correct and updated knowledge about the digital ethics field, including the compulsory modules regarding regulatory frameworks. The curriculum is dynamic, receiving frequent updates to keep learners abreast of the latest developments in digital ethics.
2. **Solutioning Intelligence**: For specific topics, e.g. "Is our HR system is high risk?" or "How to test fairness bias for our recruiting system.?", implementers would be provided with ethical decision-making intelligence in two ways:
   a. **Virtual Chatbot:** Optimized data-powered chatbot provides immediate and intelligent responses to relevant queries. What sets chatbot apart is the meticulous data selection, ranking, cleansing, and massaging process, guaranteeing pertinent responses. Additionally, the chatbot remains continually updated, reflecting the latest shifts and trends in digital ethics.
   b. **Human Expert Group:** Some questions are very specific and need human intelligence intervention. EDMI offer an on-demand expert group helpline to answer the organization's specific queries. The expert group is made up of highly qualified individuals working in the field of digital ethics. Selected representatives from the expert group can also provide specific legal and compliance advice.

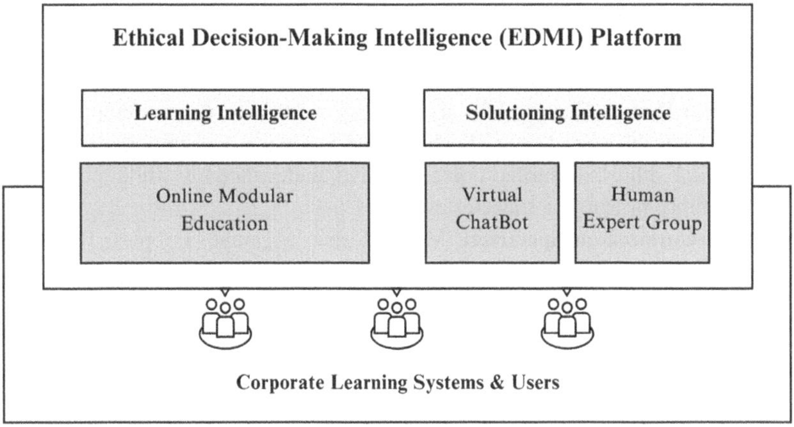

Learning intelligence would enable employees to solve around 60–70% of challenges on the spot. The virtual bot will provide the first line of solution support for the company-specific questions in the area, resolving around 20–25% of challenges. In the 10–20% challenges, the human expert group support can be sought if required.

With an integrated approach, the EDMI platform is a robust support system for companies navigating digital ethics. The education modules, virtual chatbot, and human expert group ensure that the future of digital technologies is aligned with societal values and expectations.

The platform can be embedded within corporate learning systems as a separate entity or integrated with their existing landscape. This platform is a one-stop place for employees to learn about digital ethics while implementing technologies or transformations and seek support for ethical decision-making.

3.1 Learning Intelligence

In the wake of rapid technological advancements, the onus of ethical decision-making in the digital realm has emerged as a paramount concern. *McKinsey's 2021 Future of Skills report underscores the significance of Digital Ethics, deeming it a pivotal skill for the twenty-first century.* As this domain evolves, education and reskilling become the bridge that empowers individuals with knowledge and fosters active participation. This also enables SDG 4 of Quality education by ensuring inclusive and equitable quality education and promoting lifelong learning opportunities.

The online education platform embodies this belief and is underpinned by several foundational principles:

- **Modular Design**: Designed for accessibility and clarity, the curriculum is segmented into bite-sized units. Each module is a deep dive into distinct facets of digital ethics, ensuring comprehensive understanding while permitting adaptability in learning trajectories.
- **Broad Curriculum Spectrum**: Modules span a gamut of topics. From the mandatory modules on regulations to specific modules delving into nuances like AI fairness, transparency, and ethical data stewardship.
- **Adaptive Content**: In a domain as dynamic as digital ethics, relevance is key. Therefore, the curriculum is diligently kept contemporary, introducing updates that mirror the ongoing advancements and shifts in the field, including the regulatory landscape.
- **Multifaceted Expertise**: the platform's strength lies in its architects' diversity. Drawing from the profound insights of experts spanning academia, industry, research, and legal arenas, the curriculum offers a holistic, well-rounded perspective on digital ethics.
- **Swiss Hybrid Learning Experience**: Beyond just virtual learning, Swiss users can benefit from an optional blended learning model. This approach integrates the accessibility of online modules with the tangible impact of face-to-face discussions, fostering a richer, more immersive educational experience.

Some of the modules include, but not limited to, are:

Online Education Modules

1. *Data Security, Privacy, & Protection:* This module covers best practices for safeguarding data and mitigating risks associated with data breaches.
2. *Fairness & Transparency:* Learners are introduced to concepts like bias in AI, algorithmic transparency, and how to build fair AI systems.
3. *Ethical by Design Strategies:* This module provides strategies for integrating ethical considerations into designing and developing AI systems.
4. *AI Regulations:* This module offers an overview of current legal and regulatory frameworks governing the use of AI e.g. EU AI Act
5. *The Future of AI:* This forward-looking module explores the potential impact of AI on society and the emerging ethical challenges.
6. *International Digital Ethics Landscape:* This module provides a comparative analysis of digital ethics approaches and regulations across different regions.

This evolutionary online education aims to equip individuals and organizations with the knowledge and skills to make ethically sound decisions in their digital initiatives.

3.2 Solutioning Intelligence

3.2.1 Virtual Chatbot Intelligence

In the rapidly changing landscape of digital ethics, staying updated and informed is paramount. Yet, pinpointing the most relevant insights can be challenging with the sheer volume of information available. This is where the role of an Optimized Data-Powered Chatbot comes into play.

Providing immediate and personalized support, this chatbot is designed to be a frontline responder for digital ethics inquiries. Whether it's frequently asked questions or nuanced queries, the chatbot ensures that users get answers that are not only relevant but also current. Even though there are inquiry tools like Google and ChatGPT, which can answer queries; they refer to a generic database with all sorts of data, relevant or irrelevant, true or not true.

The backbone of the system is its data pertaining to the specialized field of Digital Ethics—both structured and unstructured—that has been meticulously selected, ranked, cleaned, and massaged. This rigorous process ensures that the information provided is trustworthy and pertinent. In the vast ocean of digital ethics, think of the chatbot as the compass, always pointing towards reliable knowledge.

To ensure the highest standard of information, each data source powering the chatbot undergoes rigorous scrutiny by a **Data Source Committee**. This dedicated team of experts is tasked with vetting and approving every information source inputted into the system, ensuring that users receive trustworthy and ethically sound guidance.

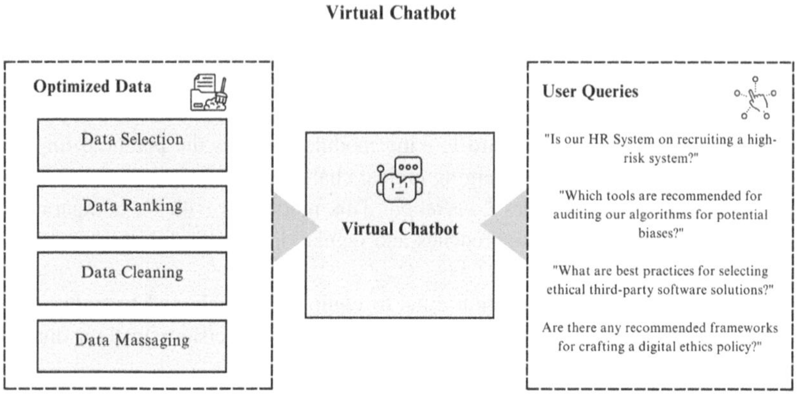

Data Optimization

To provide users with the most precise, relevant, and updated responses, it's essential to feed the chatbot with highly optimized data. EDMI utilize a structured four-step process to ensure chatbot's information is of the highest quality:

1. **Data Selection**: Initially, EDMI collate diverse publicly available datasets, encompassing structured (like databases and spreadsheets) and unstructured data (like text documents and emails). This step ensures a comprehensive pool of information relevant to the domain of digital ethics.
2. **Data Ranking**: Once gathered, data sources are ranked based on their reliability, relevance, and timeliness. Prioritizing trustworthy and current data ensures that the chatbot is equipped with the best foundational knowledge.
3. **Data Cleaning**: In this process, redundant, obsolete, or irrelevant data is removed. Cleaning is crucial to maintain the accuracy and efficiency of the chatbot, ensuring it provides users with concise and relevant information.
4. **Data Massaging**: Finally, the data undergoes a transformation or "massaging" process, converting it into a format easily digestible by the chatbot algorithms. This enables the chatbot to rapidly understand, and respond based on the optimized data.

Input Data
Through a meticulous process, EDMI ensure the chatbot is informed and able to provide immediate and personalized support, addressing both generic and company-specific queries in the realm of digital ethics.

The platform starts with publicly available information and, in the future, will provide companies with the capability of having a *company-specific layer* where the internal information on digital ethics can be stored and retrieved in the form of questions. Here's a list of some prominent publicly available data sources:

- **Publicly Available Academic Journals:** Many universities and institutions publish publicly available journals focusing on the intersections of technology, ethics, and society e.g. *Science and Engineering Ethics*
- **Research Institutes and Think Tanks**: These often produce whitepapers, reports, and articles on digital ethics topics.
  The *Oxford Internet Institute* at the University of Oxford regularly produces research on digital ethics.
  The *Stanford Center for Ethics and Society* also delves into digital ethics and related matters.
- **Government Publications**: Many governmental bodies, especially in Europe and North America, publish guidelines, reports, and frameworks related to digital ethics, data protection, and AI ethics.
  European Commission has documents and directives on AI and digital ethics.
  UK's Centre for Data Ethics and Innovation offers resources on the ethical use of AI and data-driven tech.
- **Non-Profit and Advocacy Groups**:
  Swiss Digital Initiative, with its focus on the Swiss Digital Trust Label
  Access Now is a group focusing on digital rights. They produce annual reports on digital rights and related ethical issues.
  Algorithm Watch is a non-profit initiative that evaluates and sheds light on algorithmic decision-making processes that significantly impact society.
- **Industry Groups and Forums**:
  Partnership on AI (founded by major tech companies like Google, Microsoft, and Facebook) shares research and guidelines about artificial intelligence's ethical development and deployment.
  The IEEE Standards Association has tried to draft ethical standards for AI and autonomous systems.
- **News Outlets and Specialized Tech News Websites**:
  Websites like *TechCrunch, Wired, and The Verge* often cover contemporary issues in tech, including ethical dilemmas and advancements.

MIT Technology Review offers deep dives into the ethical aspects of emerging technologies.

These sources provide a foundational dataset to kickstart the chatbot's knowledge base on digital ethics. As digital ethics is an emerging field, it's essential to periodically update the data to stay abreast of the latest developments.

To ensure the highest standard of information, each data source powering the chatbot undergoes rigorous scrutiny by a **Data Source Committee**. This dedicated team of experts is tasked with vetting and approving every information source inputted into the system, ensuring that users receive trustworthy and ethically sound guidance.

Chatbot Algorithm

Built on a foundation of meticulously calibrated data, both structured and unstructured, the chatbot would be attuned to provide insightful responses tailored to the nuances of digital ethics. With an intuitive user interface, professionals can effortlessly seek guidance on various topics—from company-specific digital tools to broader ethical strategies and governance frameworks.

Leveraging the chatbot within the digital ethics realm for key questions ensures time-efficient access to reliable information, all delivered via a user-friendly interface, streamlining ethical decision-making processes within organizations.

Potential User Queries

The company employees can ask digital ethics-specific questions to the chatbot. Some of the categorized questions could be:

Regulations & Policies:

- "Our HR is designing a job recruiting system. What kind of known biases and ethical considerations shall we take?"
- "I am working in the Pharma industry; what constraints should I be aware of for data collection?"
- "Are there any recommended frameworks for crafting a digital ethics policy?"
- "How frequently should we review and update our digital ethics guidelines?"

Tools & Technology:

- "How can our company ensure that our AI tools adhere to fairness standards?"
- "Which tools are recommended for auditing our algorithms for potential biases?"
- "What are best practices for selecting ethical third-party software solutions?"

Digital Transformation & Governance:

- "In the midst of our digital transformation, how can we ensure that ethical considerations are embedded from the start?"
- "What risks should we be aware of when shifting legacy systems to newer, AI-driven platforms?"
- "How can we structure our governance to oversee ethical considerations effectively?"
- "Which roles or departments should be primarily responsible for ethical decision-making in our digital initiatives?"

The chatbot is a crucial virtual companion on the journey towards ethical decision-making. Offering immediate support and real-time updates, it empowers users to navigate the complex landscape of digital ethics confidently.

3.2.2 Human Expert Group Intelligence

Though virtual chatbot, being the first mode of contact, can answer most of the queries. Some challenges are specific where human support is required.

Digital Ethics is an emerging field and requires a group of experts with deep expertise in business, technical, and philosophical domains. As part of EDMI, the companies have access to an expansive network of experts, reducing the need to hire the experts internally and ask questions on specific challenges. Along with that, EDMI provide access to legal experts on topics related to regulation and compliance.

The experts consortium is a rich blend of thought leaders from academia, foundation, and the industry, mainly from the Zurich region (to start with), and they speak languages including English, German, and French. They bring a deep, science-based understanding and entrepreneurial zeal to the table. This synergy ensures that EDMI is not just a tool but a movement towards integrated, informed, and actionable digital ethics.

EDMI Human Expert Group services includes:

- **Expert Access**: Connect directly with digital ethics thought leaders, obtaining insights into the latest developments and nuances in the field through helplines and special support services.
- **Legal Consultation**: Seek clarity on intricate compliance matters through interactions with legal experts specializing in the digital realm.

- **Forums and Community:** Engage in a vibrant, engaged, and informed community through «*Digital Ethics Forums*» at Ethical Tech Lab, uniting stakeholders in face-to-face sessions in Switzerland.

4 Benefits of EDMI

The EDMI platform provides an essential foundation to help companies navigate the complex digital ethics landscape.

With this platform, businesses can significantly reduce the number of breaches, promote fairness and transparency in their operations, and improve compliance rates in a bottom-up manner without incurring huge internal costs to hire experts on the emerging topic.

Some of the key benefits include:

- **Achieving Regulatory Compliance:** With EDMI's comprehensive and continually updated education modules, companies can seamlessly navigate the complex legal and regulatory frameworks around digital ethics, minimizing the risk of violations and ensuing penalties.
- **Outsourcing of Digital Ethical Intelligence Personal:** The Digital ethics topic requires experts at the intersection of business, technical, and philosophy to do justice to any role, which is difficult to find and hire considering the nascent stage of the field and increased demand due to regulations. EDMI prepares the existing employees to handle the topic internally through learning and provides a chatbot and expert group for solutioning, either eliminating or outsourcing the need to hire personnel internally.
- **Reducing potential technical and social breaches:** EDMI will enable organizations to be better prepared to develop and deploy ethical technologies, hence

reducing the number of technological breaches and providing social inclusion and transparency in the systems.

- **Promoting Proactive Ethical Stances:** As ethical considerations gain prominence in business, EDMI equips companies to transition from mere compliance towards integrating ethics in their strategic decision-making processes, cultivating a robust ethical culture from the ground up.

Some of the improved KPIs would be:

KPI	Description
Reduction in Ethical Breaches	Calculate the financial impact of incidents related to unethical practices since EDMI implementation
Compliance Improvement	Measure the increase in instances where a company met or exceeded ethical regulations or guidelines
Reduction in Decision-making Time	Monitor the decrease in time taken to make crucial decisions related to digital ethics due to the clarity provided by EDMI
Employee Ethical Confidence	A target score (e.g., 85% or above) from employee surveys, indicating their confidence in making ethical decisions with EDMI's help

Above all, the EDMI platform allows companies to make ethical decisions while implementing and maintaining technologies and saving the huge costs required to hire internal experts on an emerging topic while receiving tailored services to their needs.

5 Future Outlook

In the unfolding future horizon for EDMI, an evolution that aligns with the ever-growing demands of the digital ethics landscape is envisioned.

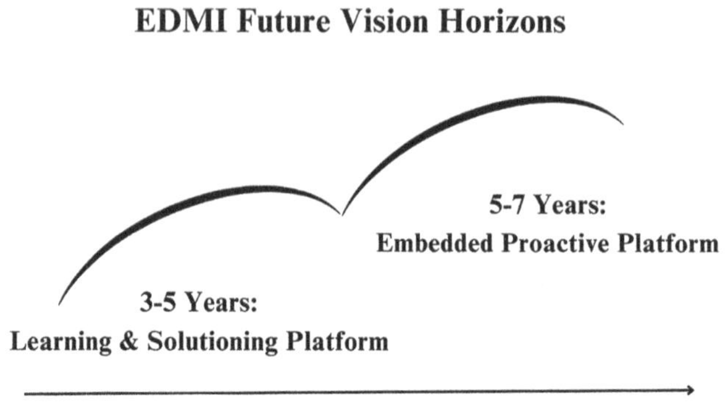

In the initial phase spanning 3–5 years, EDMI is firmly established as a Learning and Solutioning platform, where it equips companies with the essential knowledge and decision-making intelligence to navigate digital ethics. However, EDMI does not stop there.

In the subsequent phase, spanning 5–7 years, EDMI undergoes a transformative expansion. It transitions into an embedded system seamlessly integrating with company documents and policies. In this phase, EDMI takes on a proactive role, continuously analyzing these documents and policies to identify areas where ethical considerations may be lacking. It pinpoints potential issues related to fairness, security risks, and more. Users are presented with actionable recommendations, empowering them to make informed changes to enhance the ethics of their policies and systems.

This forward-looking approach ensures that EDMI remains at the forefront of fostering ethical decision-making in an ever-evolving digital landscape.

6 Take-Aways for the Reader at a Single Glance

1. **Rising Importance of AI in the Economy:** The fourth industrial revolution, driven by AI, is projected to add $15.7 trillion to the global economy by 2030. However, there's a significant trust deficit in digital technologies due to ethical concerns.
2. **Need for Ethical Decision-Making Intelligence (EDMI):** As trust in digital technologies is low and new regulations are emerging, there's a critical need for EDMI platforms. These platforms provide education and solutions to prepare companies for ethical challenges in digital technology.
3. **Components of EDMI:** The EDMI platform comprises two core elements:

   a. **Learning Intelligence:** Online modular education ensures employees have up-to-date knowledge about digital ethics and regulatory frameworks.
   b. **Solutioning Intelligence:** Includes a data-powered chatbot for immediate responses to digital ethics queries and a human expert panel for more complex issues.
4. **Benefits of EDMI:** The Ethical Decision-Making Intelligence (EDMI) platform offers a significant benefit by streamlining a company's navigation through digital ethics, reducing ethical breaches, and improving compliance. This cost-effective solution educates employees on regulatory frameworks and supports ethical decision-making, thereby minimizing the need for expensive internal experts.
5. **Future Evolution of EDMI:** Over the next 5–7 years, EDMI is expected to become more integrated into company operations, proactively analyzing documents and policies for ethical considerations, and offering actionable recommendations to enhance ethics in policies and systems.

References

Studies & Reports

IMD-St Gallen Symposium. (2021). *Strengthening Trust in Technology When It Matters The Most*. Retrieved from https://symposium.org/wp-content/uploads/2021/12/Strengthening-Trust-in-Technology_A-joint-IMD-and-SGS-White-Paper.pdf.

Tufts University (2019). *Digital Intelligence Index*. Retrieved from https://sites.tufts.edu/digitalplanet/digitalintelligence/.

Institutional Guidelines and Frameworks

European Commission. (2019). Ethics guidelines for trustworthy AI.

Deepak Bansal is the Founder and CEO of MQ Learning, a Swiss-based EduQua-certified academy dedicated to teaching humanistic skills in the digital age. His extensive 15+ years in the corporate world include executive roles at Accenture, Deutsche Post DHL, and Zurich Insurance (COO, Head AI & Robotics). A distinguished academic, Deepak has a Masters in Technology from the Indian Institute of Management (IIM) Mumbai, an MBA from the University of St Gallen, Switzerland, and an MA in Philosophy from the California Institute of Integral Studies (CIIS), California, USA. In addition to his role at MQ Learning,

Deepak actively engages in academia as a guest lecturer at Zurich University of Applied Sciences, focusing on 'Digital Ethics by Design'. His unique blend of corporate experience, academic insight, and a deep understanding of technology's ethical dimensions positions him as a thought leader in the field. Deepak also contributes significantly to digital ethics as a member of the certification committee for the Digital Trust Label, an initiative by the Swiss Digital Initiative. More information at: mq-learning.com/Deepak.

Part III

Conceptual and Methodological Contributions at the Crossroads between Philosophy and Management—Field Reports: Which Philosophy for Which Management?

Maintaining an Integrative Ethical Approach in the Business Practice of Investment Ethics

The Irreducibility of Ethical Decision-Making to Computation and AI

Michael Heumann

This chapter contains elements of five unpublished presentations given by the author in the years 2022–2023, two in his roles as a scientific researcher at the Université de Lausanne (L1, L2) and three as an ethics analyst for Arete Ethik Invest AG (A1, A2, A3): (L1) "Critique of computationalism" presented in the research seminar of the section Philosophy of Science at the Université de Lausanne in the fall semester 2022; Lausanne, October 31st, 2022; (L2) "Computationalism debate in Philosophy" presented in the workshop "Philosophical challenges of computational social science" at the Université de Lausanne; Lausanne, November 5th, 2022; (A1) "Nicht alles, was man zählen kann, zählt. Eine Diskussion über Nachhaltigkeitsmessungen bei Investments" presented at the conference of the PRIME VALUES Ethics Committee in front of employees, investors, and clients; Munich, November 17th, 2022; (A2) "Die ethische Arbeit bei Arete Ethik Invest. Gastreferat zum Thema 'Integrative Wirtschaftsethik in der Praxis'" presented as a guest lecture for the master-level class "Komplexe, wertbewusste Entscheidungssituationen analysieren, verstehen und klären" at the University of St. Gallen in the spring semester 2023; St. Gallen, April 12th, 2023; (A3) "Ethical investing—The argument for the human touch" presented at the Junior Chamber International (JCI) World Congress 2023; Zurich, November 17th, 2023. The author would also like to thank Prof. em. Peter Ulrich, Dr. Stefan Streiff and Roman Limacher for the valuable comments on the first written version of the manuscript for this chapter.

M. Heumann (✉)
Arete Ethik Invest AG, Zurich, Switzerland
e-mail: michael.heumann@arete-ethik.ch

C. H. Hoffmann (ed.), *Artificial Intelligence, Entrepreneurship and Risk*, Technikzukünfte, Wissenschaft und Gesellschaft / Futures of Technology, Science and Society, https://doi.org/10.1007/978-3-658-45544-6_6

1 Integrative Economic Ethics, Investment Ethics, and the Challenge of Computerization and AI: The Practice Case of Arete Ethik Invest

As practitioners in the field of investment ethics, the daily operational and the fundamental normative challenges for Arete Ethik Invest AG,[1] an asset management company based in Zurich, Switzerland, are manifold and often mutually intertwined. In order to sustain a scientifically sound, reliable, and impartial system of ethical analysis and evaluation of those investment titles that can either be legitimately included in the offered PRIME VALUES funds,[2] or that have to be excluded for morally justified reasons in spite of potentially lost financial gains, a thorough ethical decision-making process is vital. This ethical decision-making process in the case of Arete Ethik Invest does not only rely on institutionalized ethical criteria and processes as described in more detail later in this introductory section of the present chapter (Sect. 1), but also on a theoretical foundation that attempts to normatively cope consistently and coherently with the multitude of challenges that investment ethics, and more generally business ethics, face.

The specific challenge that is relevant for the present collective volume and the remaining discussion in this chapter in the specific context of investment ethics is the ever increasing demand to convert more and more aspects of one's business practice into such a form that they are compatible with being performed by computer algorithms and that stand in line, more broadly, with a computational mindset.[3] This computerization challenge is currently in the public and economic debate often summarized underneath the umbrella term of Artificial Intelligence (AI), though the trend to material computerization undoubtedly precedes the current, so called AI summer by several decades (cf. e.g. Weizenbaum, 1976).

[1] See https://arete-ethik.ch/en/

[2] Three different funds are provided by Arete Ethik Invest at the moment, see (1) https://arete-ethik.ch/en/prime-values-equity/, (2) https://arete-ethik.ch/en/prime-values-income/, and (3) https://arete-ethik.ch/en/prime-values-growth/. The inclusion criteria for all three funds are the same, thus the ethical decision-making process covers all three funds at the same time and in the same manner.

[3] An insightful contribution to explain that the rise of computer technology is not simply an exogenous event that occurs from out of nowhere, but actually presupposes a certain Zeitgeist-driven shift in the understanding of human rational judgement as to be nothing else then a form of computational reasoning, was made by the philosopher of information technology Joseph Weizenbaum (1976). Long before the current talks about ChatGPT, he was the main creator of the chatbot ELIZA in the 1960s, which got as far as to pass the Turing test occasionally in the psychotherapeutic context (Floridi, 2001: 135–136).

Before we highlight on which levels and how this ongoing computerization and AI-trend is normatively and strategically relevant for the business practice of Arete Ethik Invest, it should be emphasized why a more thorough reflection upon the nature and the aspects of the computational challenge is also theoretically warranted, and not just simply helpful for pragmatic-strategic (business case) reasons. The theoretical foundation of the ethical procedures of Arete Ethik Invest (cf. Streiff, 2017: 238–239) is in fact not only found in general in the discipline of practical philosophy, namely (normative) ethics, but it can be traced back more specifically to a heuristic that follows a discourse-ethical approach, namely in this case one that underlies what is called Integrative Economic Ethics ("Integrative Wirtschaftsethik"). This approach, which was developed at the University of St. Gallen mainly by the Swiss business ethicist and philosopher of economics Peter Ulrich between the 1980s and the 2000s needs a brief conceptual introduction for an international readership that is not necessarily versed in the German-language specific debate between different school of business ethics and of the philosophy of economics.

The programmatic goal of Integrative Economic Ethics (Ulrich, 2008[4]) is the theoretical and practical re-integration of economic matters and philosophical reflection. Integrative Economic Ethics gives reasons why practical philosophy is not just an add-on to business management, but a prerequisite for enlightened business and likewise ethical decisions and actions in the economic context. Instead of accepting normatively implicit business goals such as the maxim of profit maximization, an integrative ethical approach challenges unquestioned business models *as such* and instead demands a normative legitimization of how products and services within a business model are provided in the first place, focusing on genuine integrity instead of mere compliance, and on how non-economic values constitute the actual business purpose outside the realm of a rationality concept reduced to pure economism. Its central critique thus goes against what it considers to be such a prevailing economism in society that is defined by: (1) the development of a self-sufficient economic rationality in the absence of ethical considerations, (2) the representation of cost-benefit thinking

[4] A full English translation of the here cited magnum opus by Peter Ulrich was published by Cambridge University Press (Cambridge, UK) in 2008 by the title *Integrative economic ethics: Foundations of a civilized market economy*. The quotations refer to the 4th revised edition of the German original publication (cf. Ulrich, 2008); translations are mostly derived from the English version provided by Cambridge University Press. Shorter summaries of the main ideas of the integrative ethical approach are laid out in separate German-language chapters in two different collective volumes (cf. Ulrich, 2015a, 2015b).

as autonomous and absolute, and (3) the elevation of the market logic to normative primacy (Ulrich, 2008: 137–139). Incorporating such an approach that re-integrates management and normatively enlightened philosophy into business practice (corporate ethics), and specifically into the asset management context, entails not falling behind from an integrative ethical management which sets a legitimacy proviso on profit-making into a merely correctivist ethical management that in turn is limited to a mere situative renunciation of profits (Ulrich, 2008: 452). As opposed to fundamentally integrating philosophy and management in one's ethical investment practice, a correctivist ethical management normatively and affirmatively takes (quantitative) profit maximization for granted and "applies" ethics solely in addition or on top of the maximization imperative for a sort of "repair work". Effectively, a correctivist approach can merely tone down the worst effects of a malfunctioning business approach without normatively reflecting upon the actual underlying causes and without stating what exactly is normatively—and not just functionally—wrong in a given situation in the first place (ibid.). More importantly for the purposes of this chapter, a correctivist approach is also not able to genuinely reflect upon fundamental methodological challenges to ethics as raised, (1) by specific quantification demands to the ethical method, e.g. either forced upon by economization and functionalization (cf. Ulrich, 1993: 83), or by bureaucratic market-driven (cf. Hibou, 2012), and/or state-driven regulation and technocracy (cf. Ulrich, 1993: 153–155); and more generally, and related to that, (2) by the modern inclination to quantitative-materialist (cf. Heumann, 2023, Chap. 4) and computationalist mindsets, which are potentially closely tied to the broad societal push for quantification of the social sphere, as for example described and criticized by Mau (2017) from a sociological perspective. Since the critique of economism is also connected to a rejection of the reducibility of ethics to calculation and quantification (cf. Ulrich, 2008: 200, 443), a reflective and critical stance towards external quantification and computerization demands ought to be part of any ethical approach that strives to incorporate central normative criteria of an Integrative Economic (investment) Ethics into business practice and thus to also uphold and maintain them in spite of potential contrary pressures and economic incentives for unreflective quantification and computerization. As a consequence, computationalist trends and the societal, political, and/or market-driven push to digitalization[5] and to "use more

[5] Digitalization, though not separately highlighted in this chapter, is strongly related to computerization and AI. In essence, digitalization understood as the attempted transformation of (qualitative) information into (quantitative) 0's and 1's can be understood as a sub-form of any computational reduction. With regard to the ethical reflection of digitalization, various authors have recently published digital ethics books that, in the eyes of the author, are in

AI" cannot be left unquestioned and unchallenged, but need to be dealt with as part of the overall normative reflection upon the ethical decision-making process and its underlying prerequisites. Furthermore, corporate ethics that follows such an integrative ethical approach cannot be methodologically merely consequentialist such as in utilitarianism, but abides by a qualitative deontological, non-consequentialist method as exhibited by discourse ethics (cf. Ulrich, 2008: 81–99, 105–108).

Against this theoretical backdrop, three different levels can be distinguished on which the tech-driven demand for "more computation" (computerization) and "more AI" challenges the concrete quality of the ethical practice by Arete Ethik Invest. *On the first level*, with regard to the daily operational work conducted by the ethics research team, among its tasks gathering information on the ethical performance of companies that are part or could potentially be part of the PRIME VALUES strategies, how much of that work can and should be "delegated" to computer automated processes, that means computer algorithms or "AI"? The potential field of tension of upholding certain quality standards in costly (originally human-made) ethical research work versus resorting to computerized automation saving costs as an asset management company that needs to perform on the open market in the "age of AI" should likely be no secret to most readers and would thus not require any further illustration. *On the second level*, with regard to the application of the ethical evaluation criteria for judging an existing or potential investment title, how should (tech-)companies be rated that promise to solve certain social and/or ecological problems through the use of computer technology and AI instead of relying on conventional, non-technological means? At this level, it is important to bear in mind that the ethics research team at Arete Ethik Invest is only in charge of making ethically informed recommendations, not of deciding for themselves. The ultimate normative decision power of whether a specific company should be included or excluded in the PRIME VALUES funds, lies, after reviewing and discussing the recommendations by the research team, in the hands of an independent Ethics Committee. The members of that committee are not employees of Arete Ethik Invest, but they constitute an interdisciplinary group of natural and social scientists from various fields (Streiff, 2017: 234–237), for example theology, medical ethics, or materials science, making joint decisions after thorough discursive elaboration. The Ethics Committee therefore provides for a system of checks and balances within the whole ethical decision-making

---

principle compatible with the humanistic underpinnings of Integrative Economic Ethics (e.g. Nida-Rümelin & Weidenfeld, 2018; Spiekermann, 2019).

process.[6] *On the third and most fundamental level*, the challenge of computerization refers back to the theoretic question of the scope and the limits of the compatibility between computation and ethics. Can the ethical decision-making process *in principle* be reduced to computational processes? What do possible answers to this question tell us about the respective underlying understanding of human beings and of morality? Can AI actually replace the human element in ethical decision-making and make the "human touch" in (investment) ethics superfluous? Perhaps then, it would even be possible to go as far as to claim that any attempt to install an independent decision-making mechanism through a human-constituted Committee is futile in the face of always existing "subjectivity" in human decision-making and "true" independence can only be achieved by replacing the Committee itself by AI that decides "objectively"?

The answers given to the questions on the third, most fundamental level of the computerization challenge shape the attitudes on the other two levels as well. If, of course, one believes that ethical decision-making can in principle be reduced to computation, then one will probably also be more inclined to come to a positive evaluation of tech-companies as ethically sustainable investment titles (second level) and one would have little problems delegating—ceteris paribus—as much own research work as possible to automatized processes such as computer algorithms (first level). Thus, the path dependencies of answering the questions raised on the third, most fundamental level of the computerization challenge in form of the relationship between ethics and computation give a good reason why the third level should be the major focus for the remaining chapter.

In the following, we would like to explore and reject the strongest form of the tech-friendly thesis that human ethical decision-making can actually be reduced to a computational process and thus be delegated to and eventually be replaced by a computer algorithm such as in the form of AI. Instead, we will take the counter-position in the form of the irreducibility thesis. In this regard, the exact reducibility thesis that is to be rejected has to be precisely formulated and analyzed first in the upcoming section, so as to prepare the methodic ground for giving good reasons against this thesis in the subsequent sections.

---

[6] Regular monthly meetings where the companies behind the ethical analyses of investment titles are being discussed in the Ethics Committee and where the decisions are being made by the committee can be followed in a transparent way by interested clients or third-parties upon request. The current composition of the Ethics Committee can be found on the company's website (see https://arete-ethik.ch/en/ethical-investing/ethics-committee/).

2 The Reducibility Thesis of Ethical Decision-Making in Light of the Computationalist Mindset

In the previous section, we have prepared the stage for critically scrutinizing the following claim with which a player in the field of investment ethics, such as Arete Ethik Invest, is frequently faced in today's AI age: Why should we not delegate the ethical-decision making process, including the discourse within the Ethics Committee, to AI, that means to a complex set of computer algorithms? Translated into a specific scientific thesis, the underlying assumption is that the ethical decision-making process in general, and thus also specifically for investment ethics, can be reduced to computation or AI respectively. We call this the reducibility thesis. If this chapter were to be successful at rejecting this thesis, as the presented underlying ethical approach (cf. Sect. 1) would intuitively suggest, though not yet prove, the reverse conclusion can be actively formulated and defended in the form of an irreducibility thesis, namely that ethical-decision making is not reducible to computation and AI-based processes.[7]

In order to analyze the thesis, we trace the proponents of the reducibility of ethical decision-making to computation to be representatives of some form or another of what we call the (modern) computationalist mindset. We presuppose that in order to defend the reducibility thesis, it is necessary to defend a certain position within the spectrum of computationalist thinking. Such a concept of a computationalist mindset is, of course, everything but unproblematic and highly dependent on preconditions. For instance, there is no agreed upon definition of computation (Searle, 1990: 25) to begin with. To make things worse, it must remain vague at this point whether computationalist thinking refers to the belief

[7] It is important to be clear about the limits of the reverse conclusion. The counter irreducibility thesis does not say anything about the potential benefits of using computer technology and underlying computational methods to help make informed qualitative decision, as long as computations are embedded in a comprehensive qualitative decision-making process such as a human-based discourse ethical approach. In order to make relational comparisons, for example, quantification remains a vital tool to help make informed and rational ethical decisions. This is one part of the ethical-decision making process of Arete Ethik Invest as well, for example when making judgements about the justification of high vs. low CEO-median wage-pay ratios or when comparing empirical environmental figures between different companies in the same sector. The decisive difference is that the reducibility thesis claims *to fully replace* the actual qualitative decision-making process by computation and AI (eliminative reduction) and not just to add quantification as one tool among many others within an ultimately *qualitative* ethical decision-making process. We will get back to this in the final section (Sect. 5).

that (certain or all) thoughts of the mind or the entire mind are/is "computational" (whatever that would mean), or if computing is simply a functional tool to solve certain (perhaps only area-specific) problems that one's (not necessarily computational) mind either thinks about, and/or that require (ethical) actions, and so forth.

The ambiguity of the term computationalist mindset at this point is deliberately intended, as we would like to propose the following approach to come closer to a definition of the term. There can be an *explicit computationalist mindset* and an *implicit computationalist mindset*. The former refers to an informed, proactive stance taken towards the theoretical and philosophical concept of computationalism. An explicit proponent of computationalist mindset is, by default, a (philosophical) computationalist. In order to reject the computationalist position, it is necessary—also for a practitioner—to follow the theoretical debate and make a theoretical argument against computationalism. This will be done in Sect. 3 by repeating the central critical argument in the literature by Searle (1980/2000, 1990). With regard to the latter, an implicit representative of the computationalist mindset is not necessarily involved in any theoretical discussion, but simply believes that computers (in the form of AI) are somehow—and that "somehow" usually means *functionally*—superior. In the context of ethical decision-making, this position might be softened a bit, but even then, at least for some ethical decision-making processes, the reducibility thesis would still be upheld in the defense of the claim that at least some (computer algorithm-compatible) ethical decisions can be reduced to computation. This position will be targeted in Sect. 4 with a practical case experienced by Arete Ethik Invest that seemingly would fulfill the computational compatibility and reducibility criteria for implicit computationalist mindset protagonists.

In the remaining section, before exploring and rejecting the theoretical argument in Sect. 3 with the existing arguments in the literature, we will briefly introduce the computational theory from philosophy as to highlight in more detail what the "explicit" computationalist mindset is about. In order to delineate the difference to the "implicit" computationalist mindset, a special focus will also be set on the differences between computationalism and functionalism.

The usual, classic definition for computationalism is given in the form of the epistemologically and linguistically, but (questionably) not necessarily ontologically strong thesis of the Computational Theory of Mind (CTM). Computationalists in this understanding "are researchers who endorse CTM" (Rescorla,

2020). "[M]otivated by the development in machine theory and artificial intelligence" (Heinämaa, 1995: 272) and first being formulated in the 1930s[8] by McCulloch, Pitts, and the famous and highly influential Alan Turing (cf. Piccinini, 2004: 812), the perhaps clearest expression of this classical view of CTM was made by Newell and Simon (1963/2000). They claimed that the "General Problem Solver" (GPS)—for simplifying purposes, we would nowadays call it AI—captures a full "theory of human problem-solving" (ibid.: 84). Searle (1980/2000) summarizes the significance of this claim in a precise fashion: "Newell and Simon (1963) write that the kind of cognition they claim for computers is exactly the same as for human beings" (143). This means that on an epistemological level, the intelligence of the human mind is in essence computerized intelligence. The brain is the hardware and mental states are nothing but (software-based) computations as executed by Turing machines, the GPS, or AI respectively.

With regard to the—likewise—controversial definition of computation, we refer with Searle (1990) to the (conservative) Turing-standard version: "According to Turing, a Turing machine can carry out certain elementary operations: It can rewrite a 0 on its tape as a 1, it can rewrite a 1 on its tape as a 0, it can shift the tape 1 square to the left, or it can shift the tape 1 square to the right. It is controlled by a program of instruction and each instruction specifies a condition and an action to be carried out if the condition is satisfied" (25).[9] This description of computation as mere formal symbol manipulation, in combination with linguistic contributions of philosophers of language like Fodor (1975), has lead to further representational advances in the theory of CTM, which in itself are also of importance, but less so for the purposes of this chapter and will thus be ignored alongside connectionists and neural-network advancements of CTM for the remaining chapter.[10]

Important for the purposes of this chapter and the refutation of the reducibility thesis is the epistemological (strong) claim of computationalism, in the sense of classical CTM, as is also expressed by Scheutz (2002b): "Computationalism–the view that mental states are computational states–is based on the conviction that there are program descriptions of mental processes and that, at least in principle,

[8] We jump over large and important periods within the history of philosophy, for example contributions by René Descartes, Thomas Hobbes, and Gottfried-Wilhelm Leibniz in the seventeenth and eighteenth century, that can all be seen as crucial underpinnings of modern formulations of computationalism (cf. e.g. Cummins & Cummins, 2000; Scheutz, 2002a: 4–5).

[9] See also Weizenbaum (1976: Chap. 2) for a detailed depiction and description of Turing-computation.

[10] For a sound overview, see Rescorla (2020).

it is possible for computers, that is, machines of a particular kind, to possess mentality" (ix). If CTM were to be true, the reducibility thesis would be affirmed "from the other argumentational side", namely in the way that it would not be the question whether computation and AI-based processes could take over ethical decision-making tasks from humans, but in the form of the claim that what humans like experts in an Ethics Committee have been doing through ethical decisions understood as a subset of what the human mind is capable of doing, has been and is computing all along—and nothing else. When there is nothing above and beyond computation constituting mentality (and thus morality), then computations in the sense of formal symbol manipulation like described in the Turing-narrative, then—of course and self-explanatorily—ethical decision-making can be delegated to computer algorithms. The reducibility aspect of the reducibility thesis, in a certain sense, dissolves by its own in CTM: That what we had (supposedly) believed to be a way of legitimately attaining knowledge through qualitative descriptions was just computational to begin with—there is no concrete substance, entity, etc., to be "reduced" in the strict sense for an explicit (strong) proponent of the computationalist mindset in the form of CTM. The CTM-reductionism is a much more fundamental theoretical construct. This is why from this angle of the defense of the reducibility thesis, there is no way to give counter-arguments with regard to the empirical feasibility such as the computing power of machines, the experience from ethical practice, etc. The CTM-version, that means, in our terms, the explicit computationalist defense of the reducibility thesis, has to be challenged on the theoretical-metaphysical level within the Theory of Mind (cf. Sect. 3).

With regard to the differences between an explicit and an implicit defense of the computationalist mindsets (with the latter being relevant for rebuttal in Sect. 4), it is worth taking a look at what the literature says with regard to the relationship between computationalism and functionalism. Many authors understand computationalism to be "just a special case of functionalism" (Heinämaa, 1995: 283) or a "species of functionalism" (Cummins & Cummins, 2000: 6). However, good arguments exist that it is not necessarily the case that there is a systematic overlap between computationalism and functionalism and that, thus, computational theory is not a subset of functionalism. This is best explained by Piccinini (2004): "To a first approximation, functionalism is the metaphysical view that mental states are individuated by their functional relations with mental inputs, outputs, and other mental states. Functionalism per se is neutral on how those functional relations should be characterized. Specifically, functionalism is not committed to the view that the functional relations that individuate mental states are computational. Computationalism, instead, is precisely the hypothesis that the functional relations between mental inputs, outputs, and internal states are computational. Computationalism per se is neutral on whether those computational relations constitute the nature of mental states" (811–812). Piccinini makes the conflation between computationalism and functionalism plausible in that regard that major contributors to classical CTM such as Hilary Putnam (cf. Putnam, 1960/2000), tended to be functionalists *as well as* computationalists who were influenced by AI-research (Piccinini, 2004: 822–824).

For the purposes of this chapter, Piccinini's critique can be re-formulated in the following way: On the one side, functionalism does not entail computationalism, as most CTM-proponents who divert from Putnam's strict joint *computationalism-functionalism* theory (cf. ibid.) would probably agree. For example, it is functionalist to declare that a hammer can hammer best and in order to solve a given problem that necessitates hammering, one would as a result come to the functional conclusion to practically use a hammer. Nothing is said about the hammer performing computations.[11] Implicit representatives of the computationalist mindset in the context of our thesis thus do not need to be theoretically informed about what computations and computers are, they just need to believe that computers perform ethical decisions in the functional best way, just like hammers hammer in the best functional way. The only remaining genuine computationalist remnant of computationalism in representatives of an implicit computationalist, and therefore also functionalist, mindset, with a limited or no understanding of computers, is, the believe that—well, the reader may hopefully excuse the tautology—computers and not necessarily hammers perform best in (investment) ethical decision-making.

On the other side, the conflation that computationalism is also automatically a type of functionalism seems to be a residue of the ontological motivation behind computationalist theory, according to Piccinini, in order to solve a fundamental philosophical problem, namely the mind–body problem (ibid.). Just as functionalism is an attempt to give a coherent answer to the ontological (dualistic) problem of the relationship between the mind and the body (cf. e.g. Esfeld, 2020, Chaps. 6, 7), computationalism is a theoretical proposal as well, as argued for example by Putnam (1960/2000: 32). Though the precise ontological nature of computation usually remains very vague, the following can be stated: "What is common to all

[11] Though it can very well be challenged if in post-Cartesian terms, it is still conceptually possible to clearly disentangle the method of objectivism from the language of mathematics by root cause of the entanglement within the description of the res extensa (cf. Heumann, 2023, Chap. 4), and if there really is a modern way of having a non-substantial (in the Aristotelian sense) understanding of a hammer, or any other object for that matter, without at least implicitly resorting to some mathematical-quantitative and thus computerizable description.

these computational formalisms, besides their attempts to specify formally our intuitive notion of 'computation,' is their property of being independent of the physical. In other words, computations in any of these formalisms are defined *without* recourse to the nature of the physical systems that (potentially) realize them" (Scheutz, 2002a: 10). This multiple realizability claim[12] has fundamental ontological implications that also touch upon the theoretical rebuttal of the CTM in the context of the explicit computationalist mindset in the now following section.

3 The Argument Against the Computational Theory of Mind

As explained previously (Sect. 2), evidence against the possibility of upholding the classical CTM would be sufficient in order to disprove on a theoretical level the first of two manners of defending the reducibility thesis, namely the one of the (philosophically informed) explicit computationalist mindset underlying the reducibility claim. Giving a consistent argument against a necessary argumentative element of the CTM would thus suffice for supporting the counter thesis, namely the irreducibility thesis, at least on the theoretical-(metaphysical) level. In the remaining section, no theoretically new argument is given by the author, as the original contribution of this entire chapter is rather the perspective from the practitioner's side (Sect. 4). Nonetheless, for clarity and completeness purposes, the whole argument against reducibility must be explored and engaged with at the theoretical level as well, though this amounts to a repetition of the existing literature of the Philosophy of Mind.

The currently most likely best argument against CTM, understood in the strong sense of AI, meaning interpreted in the way that a computer algorithm can capture all of the human mind's capabilities of thinking, judging, evaluating, etc., is provided by the Chinese Room Thought Experiment (CRTE) by Searle (1980/2000) and the ensuing discussion (cf. ibid.; Searle, 1990). In order to summarize and understand the CRTE, it is necessary to look back at and concretize the ontological motivation for the epistemological claim of classical

[12] The multiple realizability claim, which states that it would be ultimately irrelevant which "hardware" we use, is also very elegantly described by Searle (1990): "Computationally speaking, on this view, you can make a 'brain' that functions just like yours and mine out of cats and mice and cheese or levers or water pipes or pigeons or anything else provided the two systems are, […] 'computationally equivalent'. You would just need an awful lot of cats, or pigeons or waterpipes, or whatever it might be" (25).

CTM as described in the previous section (Sect. 2). Floridi (2001) gives the perhaps most structured overview of the conceptual background of what he calls "*computational materialism*", a term that highlights the ontological motivation *and* structure of computationalism, of which the latter remains rather vague if one simply accepts the classical CTM and its underlying epistemological and linguistic claims without further (ontological) analysis (cf. Sect. 2).

Computational materialism as the result of the classical AI-paradigm of intelligence and the human mind, is, according to Floridi (2001) and in line with the explanations given in the previous section with regard to the ontological motivation, an attempt of "steer[ing] a middle course" between (1) rationalist dualism in the Cartesian sense and (2) materialist monism (i.e., physicalism; annotation by the author) (132–133). (1) Rationalist dualism consists of the in itself compatible claims (1a) that intelligence is independent of the biological body (and thus multiply realizable by any physical body; annotation by the author) and (1b) that intelligence is mind-dependent, that means that there is something as an "inner life" or the experience of an "I". (2) Materialist monism entails likewise compatibly (2a) that intelligence is a(n epiphenomenal) property of a physical body and (2b) that intelligence is dependent of a living organism (ibid.). In order to spell out the classical CTM in the form of computational materialism without any contradictions to the goals of the AI-paradigm, (1b) and (2b) need to be eliminated and (1a) and (2a) joined with the help of additional premises such as that intelligence is nothing but computation (as explained in Sect. 2) and that as a normative criterion, functional behaviorism is necessary (as opposed to the arguments made by Piccinini (2004), see Sect. 2).

If one follows Floridi's entire argumentation which describes not just the motivation, but also hints at the structural ontology of classical CTM, then it is sufficient to disprove the possibility of the elimination of either (1b) or (2b) in order to refute the CTM and thus the theoretical grounding of the explicit computationalist mindset underlying the reducibility thesis. Searle (1980/2000) makes exactly that argument in that his CRTE disproves the possibility of eliminating (1b) in the strong AI-sense and thus in the sense of the full reducibility claim. In other words, with Searle's CRTE, it is possible to prove that the inner life, or the I-experience of the mind and thought, cannot be eliminated without leaving out an essential and *a priori* necessary part of the entire reality of the world *and* the (human) mind.

What is the claim of the Chinese Room Thought Experiment, which can be replicated by any reader?[13] Searle (1980/2000) proposes to imagine that one is locked in a room with only clearly predefined access to the outer environment, namely small slots in the wall. One is given two batches (the physical structure of the batches is irrelevant according to the multiple realizability claim of CRT) of Chinese language symbols and a set of rules to correlating the symbols on the first badge with the ones on the second badge. The one performing this thought experiment does not understand anything at all of the Chinese language, but receives, for example in the well comprehended language of English, the set of rules in English and not in Chinese, so that it enables successful formal correlation of the Chinese language symbols from the first badge with the second badge by the performer of the thought experiment. A further third badge of Chinese symbols is also handed out with further English instructions on how to correlate the symbols on the third badge with the ones on the first and second badge. Together, this gives clear instructions on how to "answer" to certain Chinese symbols on the third badge ("questions") with a "script[ed]" (first badge) "story" (second badge). If the performer in the thought experiment now receives all the non-understood Chinese "stories" and "questions" through a small slot in the room's walls (input) and feeds out the information after the performed formal symbol manipulation through another small slot in the wall to the outer environment again (output), then the "answers" would be qualitatively indistinguishable in their plausibility for English-Chinese speakers from the answers that would be given (without explicit correlation and manipulation rules) to the same questions in the English language, which indeed is being understood by the performer of the thought experiment (ibid.; cf. also summary by Penrose, 1989: 17–19). Searle (1980/2000), himself performing the CRTE, stresses the essential point: "But in the Chinese case, unlike the English case, I produce the answers by manipulating uninterpreted formal symbols. As far as the Chinese is concerned, I simply behave like a computer; I perform computational operations on formally specified elements. For the purposes of the Chinese, I am simply an instantiation of the computer program" (141). From the outside, the (wrong) impression could be raised that inside the room, *someone* was actually speaking Chinese,[14]though the "I"-experience clearly contradicts this impression.

[13] Should the reader indeed have some, even if just limited understanding of the Chinese language, then they are asked to simply select some other complex language which is remote to any language skills they already possess. Nothing in the CRTE hinges on the actual identity of the specific language chosen.

[14] This is equivalent to saying that the Turing Test is passed in this context.

The point of the CRTE-argument, in order to disprove the strong AI-thesis, is the following: The performer within their inner experience *knows* that they do not understand Chinese in contrast to English, though the computational result to the world outside of the room would make this distinction epistemically irrelevant. But that does not mean that it is in fact irrelevant; the "I"-experience (see (1b) in Floridi's terminology) cannot be eliminated after all for the human person conducting the CRTE as the CRTE proves its *a priori* relevancy, and thus computational materialism is refuted as described methodically above in this section. Further, "the computer and its program do not provide sufficient conditions of understanding since the computer and the program are functioning, and there is no understanding" (ibid.: 142). Searle shows that the formal program does not and cannot carry intentionality (cf. ibid.: 149), or in other words, cannot replace a constitutional element (intentionality) of the mind. Therefore, any potential adjustments to the elimination claim of (1b) in computational materialism by trying not to eliminate (1b) after all within the definition of computational materialism, and thus attempting to avoid the problems described so far, but by somehow coming up with a new version of a definition of a (computer) mind so as to save computational materialism in some kind of way (that means making (1a), (1b), and (2a) compatible), are also futile. The CRTE shows, according to Searle's (1980/2000, 1990) interpretation, that such a "computer mind" does not and cannot carry any experienced intentionality. Translated into hermeneutic terms outside of the (narrower) language used within the Philosophy of Mind, the lack of *understanding* of the world exhibited by a computer, unlike in a being with an "I"-experience such as a human being, is equivalent to its incapability to grasp vital contexts of meaning drawn from real life as it does not participate in that real life as a self-conscious (e.g. human) interpreter that sets its own purposes as a purpose-giving agent. With reference to matters of business ethics (Sect. 1), the incapability of a computer algorithm to question the purpose of a company excludes the possibility to be an integrative economic ethicist (Ulrich, 2008).

Several counter-arguments against the CRTE have been proposed, but in the eyes of the author also ultimately refuted so far,[15] though a comprehensive discussion would lie outside the scope of the present chapter (for an overview, see Cole, 2023; Penrose, 1989; Searle, 1980/2000). When looking back at his own argument later, Searle (1990) was even forced to conclude, and alongside to back up his previous CRTE-argument in the 1980-paper, that the CTM in the strong-AI sense does not only fail on a subjective-epistemological level, but also

[15] Even proponents of various versions of the explicit computationalist mindset agree with this, see for example Park (2020: 355).

on an objective-ontological one. This is due to the problem of multiple realizability that (see (1a) in Floridi's terminology), which, when taken seriously, would lead to universal realizability, and thus to that what can be considered an ontologically particularly strong interpretation within the explicit computationalist mindset, transcending the claims of the CTM with regard to the mind by making additional statements about a computational structure outside of the mind, namely that of the outer world itself (pancomputationalism[16]). Already in the original formulation of the CRTE, Searle (1980/2000) had recognized that the ontological motivation to solve the problem of Cartesian dualism through the CTM in the strong AI-sense does lead to the non-intended result of actually creating a new "form of dualism" (151), and thus a new ontological problem instead of solving it. Later, Searle (1990) radicalized the ontological issue of the multiple realizability claim of CTM by reasoning that "[i]f computation is defined in terms of the assignment of syntax then everything would be a digital computer, because any object whatever could have syntactical ascriptions made to it. You could describe anything in terms of 0's and 1's" (26). Put differently, misinterpreting the objects of natural reality with the mathematical language to describe those objects (for example as done in natural sciences), as is implicitly done on the ontological level in CTM, stringently leads to an all-out pancomputationalism in which any statement about the human mind becomes trivial. Since if anything and everything in the world is computational, then we have absolutely no explanatory value when reasoning about the differences of the human mind and AI. For the purposes of the refutation of the reducibility thesis, it would become completely indifferent whether the decision-making process in investment ethics is performed by something humans call explicitly "computers" or "AI", by a (likewise computational) Ethics Committee, or by (also likewise computational) rocks, levers, pigeons, walls, or whatever.[17]

After rejecting the CTM based on the argumentation by Searle (1980/2000, 1990) not just on the subjective-epistemological level, but also on the objective-ontological level, the only remaining escape for proponents of the reducibility thesis in the context of ethical decision-making processes now seems to be a distancing from the explicit formulation of the computational theory and resorting

[16] Pancomputationalism "holds that every physical system implements a computational model" (Rescorla, 2020, Sect. 7.1). For a useful overview of the concept of pancomputationalism, see Piccinini & Maley (2021). Formulations of different physical theories that adhere to more or less stronger or weaker interpretations of pancomputationalism can be found in the collective volume by Zenil (2013).

[17] A computationalist counter-argument, which cannot be discussed here, because it would extend beyond the confines of the scope of this chapter, is provided by Park (2020).

to functionalism in the ethical context. This could be best done by conceding the here presented, in the eyes of the author irrefutable evidence of the rejection of the strong AI-paradigm within the CTM, and thus conceding the irreducibility of certain aspects of the human mind like intentionality to computation, but by claiming instead, that this does not apply to ethical decision-making, as ethical decisions were supposedly not pertaining to the sphere of rational mental processes, but were best to be described and performed outside of the sphere of subjective mentality. This circumnavigation of the entire structure of the argument presented in this section basically amounts to the same (functionalist) perspective as critically scrutinized in the following section (Sect. 4), since the underlying necessitated claim for such an adapted reducibility thesis still upholds some form of normative and consequential superiority of "ethical" computer decisions in contrast to human-made ethical decisions. This counter-argument will thus be considered jointly with the implicit computationalist argumentation as described before (Sect. 2) in the now upcoming section.

4 The Argument Against Computationalist-Functionalist Reductions of Investment Ethics: The Practice Case

In this section of the chapter, we present a practice case by Arete Ethik Invest which we have repeatedly experienced in deciding whether an investment title of a given company ought to be included or excluded from the PRIME VALUES funds due to ethical reasons. Instead of making an empirically independent argument based on an explicit, theoretically informed computationalist mindset underlying the contested reducibility thesis as done in the previous section (Sect. 3), we now would like to redirect the attention to the implicit, functionalist formulation of the reducibility thesis, which, in some form or another (cf. Sects. 2, 3), claims, that all, or at least some, ethical decisions can be delegated to and be performed by computer algorithms without losing any quality to the actual decision.

As it would lie beyond the scope of this chapter and be practically impossible to highlight all types of ethical decisions that Arete Ethik Invest has encountered so far in the previous decades—and besides, this would still not principally suffice to actually prove that *all* ethical decisions are ultimately *not* reducible to a computational procedure performed by a computer algorithm or AI,—it must serve our purposes that one well-chosen example that seemingly would easily fulfill the compatibility criterion of computability and computerization in its problem structure, could be exposed to rely on a presupposed qualitative, non-computational ethical grounding which cannot be captured by computation and computerization, analogous to how the "I" or "inner experience" of the mind of the human subject cannot be captured by a computer program and thus necessitates a true self-conscious, not just purely computational mind (Sect. 3). Such a methodical approach comes along with the highest burden of proof. By arguing why it is not necessary to make concessions even to the "softest" possible understanding within an implicit computationalist (in fact also functionalist) mindset underlying the reducibility thesis, namely that at least some apparently computation-compatible ethical problems can be reduced to computation and thus be delegated to computer or AI-based processes for algorithmic solving, it would automatically support the rejection of any stronger computationalist interpretations as well, such the one that *all* of ethical decision-making is reducible to computation.

Which ethical problem experienced in the practice of investment ethics could fit into the lines of this strict methodic self-imposed rule? In the case of Arete Ethik Invest, several normative criteria are applied in the analysis and the decision-making process of evaluating investment titles. Some criteria are negative exclusion criteria, others are positive criteria alongside five dimensions of responsibility which taken together have to pass a certain minimum (quantitative) threshold.[18] With regard to steering through the methodically relationalist positive criteria, there is no doubt that computational strategies and computer technology can at least partially be helpful, for example in the easiest possible way simply by the automatic calculated addition of the beforehand quantified values of each of the five dimensions of responsibility. Though since those positive criteria are necessary, but not yet sufficient to come to a complete ethical decision for a given investment title, the methodological remarks given in Sect. 2, footnote 7 apply.

Instead, in order to fulfill our methodic rule, we have to look at the negative exclusion criteria, because for some of them, the attested violation of a there-contained norm is already sufficient for the automatic exclusion out of the PRIME VALUES funds, and thus for the completion of an ethical decision. Some of those exclusion criteria might not fall into the realm of being specifically defended by all proponents of an implicit computationalist(-functionalist) mindset, as soft proponents might, for example, by themselves concede that the violation of human rights based on the subjective nature of human rights is a

[18] For an overview, see https://arete-ethik.ch/en/ethical-investing/investment-concept/investment-policy/.

criterion whose realization can only be attested to by a human mind and not a computer.

However, another negative criterion, namely the one to exclude unconditionally all companies, that "are involved in the production or trade of weapons",[19] would fit much more into our self-imposed methodic confines. This is due to the fact that (a) weapons, at least at first glance, as specific objects pertain to the objective sphere of the outer, seemingly mind-unrelated world (as in contrast to human rights criteria), and that (b) an algorithm that is sophisticated enough is theoretically conceivable and practically also in today's achievable technical reach which could, based on available information, scan through all the products that a specific company produces and deals with by the means of a pattern recognition software. In comparing the patterns with the patterns in a linked database of all existing and potential weapons, the algorithm would be able to "identify" any (predefined) weapons and one might easily come to the conclusion, that ethical decisions in this regard can indeed be fully delegated to a computational process.

The computational procedure that would seemingly fit such a weapons-excluding algorithm can be summarized in the following as programmable and executable commands:

1. If a company c_x is engaged in the production and/or the trade of weapons, exclude it from the funds.
2. Process the available information about companies c_1, c_2, c_3, ... and by comparing to the provided weapons database, identify if the pattern of product A, B, C, ... of the analyzed companies c_1, c_2, c_3, ... matches the pattern description of a weapon in the database.

---

[19] See https://arete-ethik.ch/en/ethical-investing/investment-concept/investment-policy/. From a perspective of corporate (investment) ethics following an integrative ethical approach, it is of course warranted to question the normative legitimacy of a company's purpose to provide for the products and services associated with weapons even if current public (moral) opinion should be to some or full extent in favor of producing and dealing with weapons. The independent Ethics Committee at Arete Ethik Invest has decided to continue to exclude companies whose purpose it is to produce and/or deal in weapons despite of current political and moral counter proposals. If this decision is not accepted by the reader, then a qualitative discourse between the different standpoints would be warranted at this point in order to qualitatively (re-)assess the normative legitimacy of the purpose of providing for weapons, something that cannot be delegated to computers but could only be conducted between purpose-giving agents such as humans (cf. Sect. 1). More importantly though, the following argument supporting the irreducibility thesis works even if the exclusion of weapons is accepted by the reader and a pro-computerization hypothesis in congruence with the reducibility thesis would now claim that the mere implementation of the decision to exclude weapons in investment ethics could now be delegated to computers and AI.

3. If one or more products of a company c_X fulfills the matching condition described in (2), execute the command described in (1).

The commands (1) and (3) are not the crucial problems for the validity of the reducibility thesis in the context of the negative weapons exclusion criterion as they are indeed in their structure already framed as digitalized "1"/"0", respectively YES/NO-decisions. Certainly, those artificially extricated and separated parts of the *entire* ethical decision-making process can be taken over by a computer algorithm. However, a necessary part of the complete ethical decision-making process is also the command described in (2) and herein lies the problem of reducibility. Though for pragmatic reasons, one might imagine the computer algorithm to successfully recognize—let us say—a more or less pattern-wise clearly predefined rifle gun, it nonetheless cannot actually make the underlying decision, whether the analyzed product/object X is indeed non-randomly defined as being a weapon. The decision of what a weapon actually is, as opposed to a mere (non-weapon-like) tool,[20] rests ultimately on a normative, and thus also a subjective decision that cannot be replaced by computation after all, as we will now see.

To clarify, let us take the example of a software that is potentially created and/or that can simply be used (possibly in spite of the hypothetical intentions of the creator) for infecting a certain infrastructurally vital computational system, for example the one of a hospital, or of an energy or a logistics facility, with ensuing human casualties. Is that software, depending on whether it was intended to be used as a virus for infecting critical infrastructure or not, and depending on whether it is actually (perhaps only with certain minor modifications), and/or just potentially being used for such a purpose, really a weapon, or is not? The database mentioned before will not be able to "answer" this complex normative attribution question, since the database only contains those weapons, for which the decisions that the depicted patterns are indeed weapons have already been made, namely by humans. Programming a computer to take over that decision, that means to "create the database" based on subjective reasons, would be impossible. Technical arguments against this impossibility claim, for example that

[20] We do not look at special cases such as dual-use tools/weapons that can be interpreted as in some contexts (a) fulfilling purposes as a weapon and in other contexts (b) fulfilling non-weapon-related purposes. For example, a specific sensor could be produced that can be used for fire targeting systems in military fighting jets (a), but also for navigating civil aircraft (b). Though those cases would probably create ultimately reducibility-thesis refuting problems as well, those problems are only secondary to the primary point that we would like to make at this point.

the computer can refine its own pattern recognition, "learn from its own experience" and ultimately be able to make new and original attributions to add to the database itself, would not be able to prove that those computer decisions are actually reasonable and not arbitrary with regard to a true weapons status. On the contrary, ultimately defending the computer's capability to actually make the decision entirely by itself (reducibility thesis) would face the self-defeating challenge, that the supposedly "ethical" computer algorithm as a software running on a universal and thus potentially military combat-capable Turing machine[21] would have to eventually evaluate, whether it itself is such a weapon, or not. However, such a self-judgment presupposes self-consciousness and that is after all a theoretical construct which pertains to the mind, and which, as shown in the previous section (Sect. 3) in the CRTE, is not attainable by computer algorithms: A computer and so called (weak) AI has no "I"-experience. The impossibility of a non-self-aware object (and thus a non-subject) to evaluate and reflect upon its own normative status[22] is a transcendental, empirically irrefutable argument against the possibility that a computer algorithm can make complete and independent rational judgments by mere computation with regard to the potential status (of itself) as a weapon. Thus, not even the seemingly most computation-compatible normative decision criterion in the experience of Arete Ethik Invest, namely the one with regard to the pattern recognition, respectively identification of a weapon, can fully be reduced to a mere computational process, and thus it cannot be delegated to a computer algorithm. This irreducibility claim is in accordance with the practical experience that fuels from intense discussions within the Ethics Committee about the issue whether certain objects are weapons or not.[23]

[21] Today's existing computers are of course just material devices with limited computing power, but in their idealized version, they fulfill the criteria of Turing machines, at least in principle.

[22] A programmed answer to this question by a human would obviously not solve the problem, because it would be the admission, that the ultimately relevant ethical decision is in the end being performed by a non-merely-computational system, namely the self-conscious subject of a human being.

[23] That the qualitative answers given to this question by the current Ethics Committee are part of what Streiff (2017) calls a "revisable reflection equilibrium" (241; translation by the author) does not in any way undermine the defense of the irreducibility thesis. On the contrary, as part of an open discourse ethics, the final word to complex ethical problems or dilemmas is usually not spoken and further qualitative exploration and scrutinization of old and new arguments are always indicated. What is important is that within a "revisable reflection equilibrium", complete (not necessarily "true" or "best") ethical decisions have been made on a rational basis and the results may be revised through new complete ethical decisions by humans.

In other words, and to sum up this section, it is reasonable to conclude that the subjective and qualitative nature of human-based ethical (investment) decisions is always a prerequisite for using computational methods to help in ethical decision-making. As such, the reliance on actual purpose-giving agents such as human beings in an Ethics Committee in order to make ethical decisions is not a plausible disadvantage as could be portrayed from a (deficient) metaphysically functionalist and ethically consequentialist viewpoint. Nor is it a technical problem to be overcome by the technical methods of computerization and AI. It is rather an incremental part of a reflective ethical investment decision-making process that takes the *a priori* preconditions of morality seriously. Ethical decision-making, even in the case of at first seemingly computer- or AI-compatible problem cases, cannot be delegated to "intelligent" algorithms, but ought to be performed by self-conscious and thus reflective human beings. Only through that, the always necessarily presupposed subjective quality of quantifiable and computable categories (such as digitalized patterns of weapons) can be reflected in an autonomous way in order to come to a complete rational ethical judgment.

5 Summary and Outlook: Methodological Implications for an Integrative Ethical Approach

In the past two sections (Sects. 3, 4), different objections from theory and practice were presented against the here contested thesis that ethical decision-making can be reduced to computational processes such as computer algorithms and AI. This insight has ample practical implications for the daily operational ethical work, the strategy, and the normative orientation of a business player in the investment and asset management sector, especially if its self-established ethical goal is to maintain an open qualitative, discourse-ethical approach, as laid out, for example, by Integrative Economic Ethics (Ulrich, 2008). For the purposes of this particular contribution to the collective volume and to sum up the chapter, we would like to now put the focus on a subset of implications by highlighing methodological ramifications that are derived from the here supported counter thesis of the irreducibility of ethical decision-making to computation and AI.

The most straightforward consequence is, of course, the simple conclusion that an Ethics Committee constituted by human actors, as exists in the case of Arete Ethik Invest, alongside all the other elements of the decision-making process in ethical investing that deal with inherently subjective qualities such as consciousness, intentionality, etc., cannot be reduced to computation and thus cannot be replaced by computer algorithms and AI. An Ethics Committee in the investment and asset management sector, but any other sector for that matter, too, necessarily and irrefutably relies on its "human touch".

But this can only be the starting point of the reflection of the relationship between ethics and computation. If we would like to take the insights by Weizenbaum (1976) seriously, and there are good reasons to do so, then the computerization challenge does not start with the material encounter of investment ethics with computer technology, but is already prevalent in the explicit and/or implicit computationalist mindset that permeates a given society such as our current (Western and modern) one, even independent of the actual material technological progress already achieved in order to realize those computationalist ideals. Computation presupposes mathematical entities, such as numbers, and computerization presupposes quantification (in the form of digitalization) of formerly non-quantified aspects of human beings, their society, and their surrounding environment. A business practice that strives to maintain an integrative ethical approach based on an open discourse therefore has to be carefully attentive of any political (for example through technocratic regulation that narrows down normative criteria to quantitative benchmarks), market-based (for example through the competitive pressure to save money by substituting ethicists' work by any institutionalized form of statistically standardized, impersonal governance), or other demands from the social sphere that could jeopardize the very foundations of the subjective-qualitative discourse-ethical structure. Discourse ethics does not reduce to an objectivized fashion of pragmatically solving merely functional and technical problems, just as the communicative society that constitutes the discourse (cf. Apel, 1999: 220–263) does not reduce from an intersubjective entity to a pure object (cf. Apel, 1999: 178–219). Thus, any attempt to transform the discursive ethical decision-making criteria away from the search for the better, that means more consistent, argument to a mere quantitative popularity contest of, for example, simply counting votes of experts and going along with the majority vote independent of the quality of the underlying argument, should be seen very sceptically and critically, and should by default be rejected as a legitimate ethical decision-making strategy.

This methodological insight is based on the author's understanding of the arguments made by Weizenbaum (1976), importantly actually independent of the factual, empirically observable involvement of computer technology, which only might indeed make matters worse with regard to preserving a quantification-critical stance, since the (ideal) quantification challenge underlies the (material) computerization challenge. Likewise, as for the arguments stated within any really existing investment ethics discourse about any given problem, arguments, which in turn might be in line with other ethical schools such as virtue ethics or (classic) Kantian ethics: It would be problematic to translate those arguments into a simply quantitative form, because this would ultimately lead to a reduction of the full ethical theoretical argumentative spectrum to utilitarian or similar forms of number-based consequentialist ethics. This is because qualitative criteria are an inherent part of virtue or deontological (in the transcendental-subjective Kantian and/or in the transcendental-intersubjective discourse sense) ethics and reducing those qualitative criteria to numbers would eliminate their essential methodic core. As in contrast, utilitarian ethics is in its methodic structure quantitatively founded. One has to be completely clear on the issue and be mindful of the fact, that computerization demands, but also already any preceding (non- or not-yet-computerized) quantification demands, reduce ethics to merely number-based ethics, and thus lead to an ultimately eliminative reduction of ethics to utilitarianism and total bias of (meta-)ethics towards utilitarianism. In other words: Methodological reductionism to quantitative methods inevitably leads to ethical reductionism to utilitarian (or similar) ethics.

The methodological implications from a theoretical standpoint far transcend a narrowly understood field of (investment) ethics. For example, translated from Practical Philosophy into the language of the Philosophy of Science, a normatively and epistemologically valid scientific manner to deal with quantitative data and quantification demands can only lie primarily in genuine social science methods as opposed to natural science methods, which in the sense of Friedrich A. von Hayek (1942, 1943, 1944) would otherwise constitute a misapplication of the natural scientific method ("scientism"). This is because the objectivist-mathematical method and language of the natural sciences must not be overextended from the natural sphere to the social sphere where it does not belong to for epistemological reasons. The particularities of the autonomous, purpose-giving subjects constituting the social order can only be captured by a social science language that does not reduce to mathematics and instead primarily relies on qualitative descriptions and non-quantitative normative evaluations of those descriptions.[24] Likewise, translated into the language of the Philosophy of Mind through the more or less analogous epistemological argument of the non-reducibility of the social order to the natural order (ibid.), the inner workings of the human mind cannot be reduced to a computational symbol-processing mechanism (cf. Searle, 1980/2000). This thus establishes the necessity for a (philosophically) reflective

[24] This, of course, does not rule out the addition of mathematical descriptions for the full understanding of the social sphere (of the investment sector). As explained in Sect. 2, fn. 7, the critique of the reducibility thesis is directed against the eliminative notion that all qualitative descriptions can be *fully reduced* to quantitative ones.

business practice that relies on (human) subjects engaged in ethical tasks who critically reflect and challenge ethical-methodological issues of computerization and quantification demands in a non-computational, non-quantitative way, and who do not paradoxically and in a self-defeating way delegate their autonomy away to a computer program that is in its essence not capable to *understand* and capture that autonomy to begin with. Since human beings are the only known (fully) autonomous agents known so far, this indeed equates into the argument for "the human touch" in the decision-making process of investment ethics from a perspective of the Philosophy of Mind.

In summary, the aim of this chapter was to explore and reject the thesis, that ethical decision-making can be reduced to computational and AI-based processes. From the perspective of Arete Ethik Invest, an asset management company engaged in the practice of investment ethics, both, arguments from the theoretical literature, as well as reflections about experienced practice questions, were brought forward in order to make the case for the subjective-qualitative grounding of an integrative ethical decision-making process. Thus, as a key take away, it can be concluded that maintaining an integrative, that means discourse-grounded ethical approach in the practice of investment ethics necessitates a reflective stance at the intersection of theory and practice. In the context of specific computerization demands, for example in the form of a market-driven push to "use more AI" for ethical decision-making in order to (supposedly) stay more competitive, both a theoretically informed understanding about the challenges and potential methodological pitfalls of the explicit and implicit computationalist mindset and quantification is necessary, as well as a practically enlightened comprehension about the benefits, but also, and perhaps more importantly, about the inherent philosophically grounded limits to the use of computer algorithms and AI for ethical investment decision-making. Independent of the question whether the current AI "summer" will continue or not and be again replaced by another (temporary) AI "winter" soon, an intense deliberative exchange between philosophers, ethicists, computer specialists, and management practitioners remains warranted in the face of the ongoing, Zeitgeist-typical challenges of quantification demands and the computationalist mindset.

References

Apel, K.O. 1999. Transformation der Philosophie: Band 2. Das Apriori der Kommunikationsgemeinschaft (6th ed.). Frankfurt a.M.: Suhrkamp.

Cole, D. 2023. The Chinese Room Argument. In E.N. Zalta, & U. Nodelman (Eds.). The Stanford Encyclopedia of Philosophy (Summer 2023 Edition). Available at: https://plato.stanford.edu/archives/sum2023/entries/chinese-room/

Cummins, R., & Cummins, D.D. 2000. Introduction. In R. Cummins, & D.D. Dummins (Eds.). Minds, brains, and computers: The foundations of cognitive science. An anthology. Malden, MA: Blackwell: 3–7.

Esfeld, M. 2020. La philosophie de l'esprit. Une introduction aux débats contemporains (3rd ed.). Malakoff: Armand Colin.

Floridi, L. 2001. Philosophy and computing. An introduction [Taylor & Francis e-Library]. New York, NY: Routledge.

Fodor, J.A. 1975. The language of thought. Hassocks, Sussex: The Harvester Press.

von Hayek, F. A. 1942. Scientism and the study of society. Part I. Economica, 9(35): 267–291.

von Hayek, F. A. 1943. Scientism and the study of society. Part II. Economica, 10(37): 34–63.

von Hayek, F. A. 1944. Scientism and the study of society. Part III. Economica, 11(41): 27–39.

Heinämaa, S. 1995. The problem of artificial persons. In L. Haarparanta, & S. Heinämaa (Eds.). Mind and cognition—Philosophical perspectives on cognitive science and artificial intelligence [Act a Philosophica Fennica, Vol. 58]. Helsinki: Hakapaino Oy: 269–286.

Heumann, M. 2023. Zeitgenössische Wachstumskritik aus wirtschaftsphilosophischer Perspektive: Zur Rekonstruktion der Wachstumsdynamik als Verselbständigung cartesischen Denkens [Thesis Series, No. 42]. Geneva: Globethics Publications.

Hibou, B. 2012. La bureaucratization du monde à l'ère néolibérale. Paris: La Découverte.

Mau, S. 2017. Das metrische Wir. Über die Quantifizierung des Sozialen. Berlin: Suhrkamp.

Newell, A., & Simon, H.A. 2000 [1963]. GPS, a program that simulates human thought. In R. Cummins, & D.D. Dummins (Eds.). Minds, brains, and computers: The foundations of cognitive science. An anthology. Malden, MA: Blackwell: 84–94. [reprint from Newell, A., & Simon, H.A. 1963. GPS, a program that simulates human thought. In E. Feigenbaum, & J. Feldman (Eds.). Computers and thought. New York, NY: McGraw-Hill.]

Nida-Rümelin, J., & Weidenfeld, N. 2018. Digitaler Humanismus. Eine Ethik für das Zeitalter der Künstlichen Intelligenz. München: Piper.

Park, S. 2020. The problem of pancomputationalism: Focusing on three related arguments. Journal of Cognitive Science, 21(2): 349–369.

Penrose, R. 1989. The emperor's new mind—Concerning computers, minds, and the laws of Physics. New York, NY: Oxford University Press.

Piccinini, G. 2004. Functionalism, computationalism, and mental states. Studies in History and Philosophy of Science. 35: 811–833.

Piccinini, G., & Maley, C. 2021. Computation in physical systems. In E.N. Zalta (Ed.). The Stanford Encyclopedia of Philosophy (Summer 2021 Edition). Available at: https://plato.stanford.edu/archives/sum2021/entries/computation-physicalsystems/

Putnam, H. 2000 [1960]. Minds and machines. In R. Cummins, & D.D. Dummins (Eds.). Minds, brains, and computers: The foundations of cognitive science. An anthology. Malden, MA: Blackwell: 20–33. [reprint from Putnam, H. 1960. Minds and machines. In S. Hook (Ed.). Dimensions of mind: A symposium. New York, NY: New York University Press.]

Rescorla, M. 2020. The computational theory of mind. In E.N. Zalta (Ed.). The Stanford Encyclopedia of Philosophy (Fall 2020 Edition). Available at: https://plato.stanford.edu/archives/fall2020/entries/computational-mind/

Scheutz, M. 2002a. Computationalism—The next generation. In M. Scheutz (Ed.). Computationalism: New directions. Cambridge, MA: The MIT Press: 1–21.

Scheutz, M. 2002b. Preface. In M. Scheutz (Ed.). Computationalism: New directions. Cambridge, MA: The MIT Press: ix–xiii.

Searle, J.R. 1990. Is the brain a digital computer? Proceedings and Addresses of the American Philosophical Association, 64(3): 21–37.

Searle, J.R. 2000 [1980]. Minds, brains, and programs. In R. Cummins, & D.D. Dummins (Eds.). Minds, brains, and computers: The foundations of cognitive science. An anthology. Malden, MA: Blackwell: 140–151. [reprint from Searle, J.R. 1980. Minds, brains, and programs. Behavioral and Brain Sciences, 3(3): 417–424.]

Spiekermann, S. 2019. Digitale Ethik—Ein Wertesystem für das 21. Jahrhundert. München: Droemer.

Streiff, S. 2017. Ethik-Komitee in der Vermögensverwaltung. Eine deliberative Methodik zur Wahrnehmung von Verantwortung im Investmentprozess. In M. Stüttgen (Ed.). Ethik von Banken und Finanzen. Baden-Baden: Nomos: 233–245.

Ulrich, P. 1993. Transformation der ökonomischen Vernunft. Fortschrittsperspektiven der modernen Industriegesellschaft (3th rev. ed.). Bern: Haupt.

Ulrich, P. 2008. Integrative Wirtschaftsethik. Grundlagen einer lebensdienlichen Ökonomie (4th rev. ed.). Bern: Haupt.

Ulrich, P. 2015a. Auf der Suche nach der ganzen ökonomischen Vernunft. Der Ansatz der integrativen Wirtschaftsethik. In D. van Aacken, & P. Schreck (Eds.). Theorien der Wirtschafts- und Unternehmensethik. Berlin: Suhrkamp: 213–236.

Ulrich, P. 2015b. Grundideen und Entwicklungslinien integrativer Wirtschaftsethik. In T. Beschorner, P. Ulrich, & F. Wettstein (Eds.). St. Galler Wirtschaftsethik: Programmatik, Positionen, Perspektiven. Marburg: Metropolis: 29–60.

Weizenbaum, J. 1976. Computer power and human reason. From judgment to calculation. San Francisco, CA: Freeman.

Zenil, H. (Ed.). 2013. A computable universe. Understanding and exploring nature as computation. Singapore: World Scientific Publishing.

Michael Heumann is a senior ethics analyst at Arete Ethik Invest AG, Zurich. He is also a doctoral researcher at the University of Lausanne (UNIL) in the field of the Philosophy of Science working on a second PhD thesis on the scientific and conceptual compatibility of transcendental-intersubjective theory and libertarian ethics (SNSF-project 200464: "Parsimonious ontology and the relationship between science and persons" (2021–2025)). Furthermore, Michael Heumann was a teaching assistant in the contextual studies at the University of St. Gallen (HSG) with a focus on the issue of responsibility. He was a doctoral researcher at the Institute for Business Ethics, University of St. Gallen (IWE-HSG) between 2015 and 2021. He finished his first PhD at the IWE-HSG with a critique on the quantitative-materialist premises in the contemporary debate on economic growth.

Orientierungskompetenz als zentrale Leadership-Komponente in der digitalen Arbeitswelt

Zu der kommunikativen Herausforderung in der Führung «gemischter» Personen-Maschinen-Teams

Jitka Perina

> *„In all situations of life, the very first thing one does is to orient oneself. Orientation precedes everything else, all perceiving, thinking and acting; it decides which direction we pursue. It seems simple but is very complex, and it has so far hardly been explored.“*[1]

1 Einführung

In der immer weiter fortschreitenden digitalen Transformation in allen Zweigen der Wirtschaft, Wissenschaft und Gesellschaft gewinnen die Anforderungen an die Kompetenzen des Managements und der Leadership enorm an Relevanz. Die disruptive technologische Veränderung ist in vollem Gange, die künstliche Intelligenz (KI) wird viele Arten der menschlichen Arbeit übernehmen. Die Arbeitswelt ist volatil, unsicher, komplex und mehrdeutig *(VUCA)*[2]. Die Akteure

[1] Stegmaier, Werner und Reinhard Mueller: Successful Modes of Orientation. Nashville, TN: The Hodges Foundation for Philosophical Orientation, 2019.

[2] U.S. Army Heritage and Education Center: "Who first originated the term VUCA (Volatility, Uncertainty, Complexity and Ambiguity)?" US Army. Aktualisiert am 06.12.2022. https://usawc.libanswers.com/faq/84869.

J. Perina (✉)
Master Training Group, Zug, Schweiz
E-Mail: j.perina@mastertraining.ch

C. H. Hoffmann (ed.), *Artificial Intelligence, Entrepreneurship and Risk*, Technikzukünfte, Wissenschaft und Gesellschaft / Futures of Technology, Science and Society, https://doi.org/10.1007/978-3-658-45544-6_7

in der Arbeitswelt befinden sich in einer Phase, in der revolutionäre Neuerungen bereits Fakt sind und in der weitere auf sie zukommen und die Technologie nicht mehr ausserhalb des menschlichen Körpers bleibt, sondern mit dem menschlichen Körper, seinem Geist und seinem Gehirn verschmilzt. Dies geschieht in zwei Prozessen der Annäherung: Einerseits kommt es zu einer Technisierung der Menschen (Cyborgisierung), andererseits zur Vermenschlichung der Maschinen bis hin zur Konstruktion humanoider Roboter. Es geht zunehmend nicht mehr um die Interaktion zwischen Menschen und Maschinen, sondern um eine Konvergenz (Verschmelzung). Konvergenz ist das entscheidende Thema unserer Zeit: Wie wollen wir die Verschmelzung von Mensch und Technologie gestalten, und wie sollen sich die Akteure in diesen Kontexten ausrichten?[3]

Diesem Problem gilt die Leitfrage der vorliegenden Arbeit: *Wie müssen «gemischte» Teams, in denen menschliche Personen und lern- und entscheidungsfähige Maschinen auf unterschiedliche Weise nicht nur interagieren, sondern konvergieren, geführt werden, um die Kommunikation und Kooperation im Sinne diskursiv-konstruktiver Bedingungen gelingend zu gestalten?*

Die vorliegende Arbeit vertritt die These, dass sich Führung und Kooperation in der digitalen Arbeitswelt grundlegend ändern werden. Die zentrale Frage der Leadership wird die Orientierung sein. Sie wird sich viel häufiger und in völlig neuen Konstellationen beweisen müssen. Die Zusammenarbeit wird in agilen Netzwerken stattfinden, für welche die Bedingungen einer gelungenen Kooperation innerhalb der «gemischten» Teams neu definiert werden müssen. Die Prognosen hierzu sind vage. Die erprobten Grundlagen, welche zu diesen Themen zur Verfügung stehen, wie Orientierung, Bildung von Vertrauen und psychologische Sicherheit sollen hier aufgegriffen, unter den sich rasch verändernden Bedingungen geprüft und mit Blick auf Bedeutung und Anwendung diskutiert werden.

Die Arbeit ist in vier Teile gegliedert: Im ersten Teil wird die Situation aufgezeigt, in welcher Führung und Kooperation in der digitalisierten Arbeitswelt stattfinden wird, und es wird die Beziehung zwischen Menschen und Maschinen (künstlicher Intelligenz) thematisiert. Anschliessend werden für den Beitrag zentrale Begriffe geklärt. Im zweiten und dritten Teil werden zwei Gebiete definiert und weiterverfolgt, um die Leitfrage nach Führung und Kooperation in «gemischten Teams» besser beleuchten und beantworten zu können. Der vierte Teil bietet einen Ausblick auf die Problematik der moralischen Orientierung und skizziert eine Ausrichtung der KI an sozialen Werten und Zielen. Im Fazit

[3] Benedikter, Roland: «Transhumanismus. Auf dem Weg zum neuen Menschen?» BIA. 2018. https://bia.unibz.it/esploro/outputs/991005772188101241

werden die Schritte und Erkenntnisse der Arbeit im Hinblick auf die Leitfrage ausgewertet und eine Zusammenfassung wird angeboten. Folgendes wird hingegen nicht im Fokus der Arbeit stehen: Die ebenfalls wichtigen Fragen der Computerlinguistik und der praktischen Gestaltung der Kommunikation in der digitalen Welt, besonders die Ausformung einer gemeinsamen Sprache zwischen den Kooperationspartnern, den Menschen und der künstlichen Intelligenz. Dies muss im Fokus einer weiteren Arbeit stehen. Die brennend aktuelle Frage der Bildung eines gemeinsamen Verständnisses von sozialen Werten und Zielen und der Ausrichtung aller Akteure daran wird hier nur im Ansatz diskutiert.

2 Führung und Kooperation in der digitalisierten Arbeitswelt

Im Folgenden wird die Situation aufgezeigt, in welcher *Führung und Kooperation* in der digitalisierten Arbeitswelt stattfindet und die Beziehung zwischen Menschen und Maschinen (künstlicher Intelligenz) thematisiert. Anschliessend werden die in der Arbeit verwendeten Begriffe geklärt.

2.1 Wie ist die Lage?

Die sich ändernden Bedingungen rufen Situationen hervor, welche noch nie dagewesen sind. Im Umgang mit den neuen Konstellationen ist der Zugriff auf Erfahrungen mit ähnlichen Situationen gar nicht oder nur teilweise möglich. Eine neue Herausforderung besteht darin, dass die klassische Führung von Menschen sich zur Führung von Teams entwickelt, in welchen (menschliche) Personen und die KI miteinander verschmelzen. Ein wichtiger Faktor dabei ist die Grundannahme des Menschen hinsichtlich seiner Beziehung zur künstlichen Intelligenz. Sehen die Menschen in der KI tatsächlich ein Werkzeug in ihren Händen, welches gebraucht werden kann, oder sehen sie in der KI einen sozialen Partner, mit welchem sie kooperieren wollen? In diesen neuen Situationen bedarf es eines anderen Zugangs zu den Führungsaufgaben. Es geht nicht nur um die Entwicklung neuer Strategien, Pläne und Controlling-Werkzeuge (Managementaufgaben), welche für die Bewältigung der neuen Herausforderungen nötig sind, sondern auch um eine neue Einstellung und die prinzipielle Neugestaltung der Führung (Leadership) der an der Zusammenarbeit beteiligten Teams. Die komplexen Situationen verlangen also nach einer neuen Art der Zusammenarbeit in agilen Arbeitsmodellen und nach einer Führung, welche es schafft, dass die Beteiligten bereit sind, in

Netzwerken zu kooperieren, um die Ziele der Organisation erfolgreich zu verfolgen. Diese Führungsaufgaben bedürfen in den neuen Situationen einer neuen Orientierung. Die Fähigkeit, sich in neuen Situationen zu orientieren und daraus erfolgversprechende Strategien für den Umgang mit dieser Situation, das Entscheiden und Handeln zu finden, stellt eine grundlegende Leadership-Kompetenz auch in der digitalen Arbeitswelt dar.

2.2 Begriffsklärung

2.2.1 Zu den Begriffen (denk- und lernfähige) Maschinen, Technologie, KI, humanoide Roboter

Maschine, Technologie und Künstliche Intelligenz (KI) sind Begriffe, die eng miteinander verbunden sind, sich jedoch konzeptionell unterscheiden. Während Maschine eine Sammelbezeichnung für zweckorientierte technische Vorrichtungen verschiedenster Art und Grösse[4] ist, bedeutet Technologie die übergreifende, Wirtschaft, Gesellschaft und Technik umklammernde Wissenschaft von der Technik.[5] Künstliche Intelligenz (KI) beschäftigt sich mit Methoden, die es einem Computer ermöglichen, Aufgaben zu lösen, welche beim Menschen Intelligenz erfordern.[6] In dieser Arbeit wird unter denk- und lernfähigen Maschinen künstliche Intelligenz (KI) verstanden, welche nicht als ein Werkzeug in den Händen der Menschen, sondern als sozialer Kooperationspartner der Menschen verstanden wird.

2.2.2 Zu den Begriffen Führung, Management, Leadership und Digital Leadership

In der vorliegenden Arbeit schliesst der Begriff Führung beides ein, Management und Leadership als sich gegenseitig stärkende Komponenten, welche es ermöglichen, mit Komplexität und Kreativität umzugehen. Beide sind für die erfolgreiche Führung einer Organisation notwendig. Im Weiteren wird der Unterscheidung von *Steckelberger und Harrer*[7] gefolgt. Kurz skizziert ist Management die Profession der Kontrolle und bildet die beobachtende, kontrollierende und

[4] Vgl. Gabler Wirtschaftslexikon, «Maschine» https://wirtschaftslexikon.gabler.de/ abgerufen am 27. Dezember 2023.

[5] Vgl. Gabler Wirtschaftslexikon «Technologie».

[6] Vgl. Gabler Wirtschaftslexikon «Künstliche Intelligenz».

[7] Steckelberg, Alexander V. und Charly Harrer: Leadership und Management. Wiesbaden: Springer Gabler, 2021.

regulierende Komponente der Führung, die Ressourcen und Prozesse verwaltet. Leadership geschieht «auf der Bühne» der Interaktion mit den Beteiligten, konzentriert sich auf die Einflussnahme, Inspiration, Motivation, Kreativität und die Entwicklung von Menschen, um eine Vision zu verwirklichen und Ziele zu erreichen. Leadership hat eine Vorbildfunktion und bietet damit Orientierung.

2.2.3 Zum Begriff «gemischte» Teams

Mit «gemischten» Teams sind Gruppen von Akteuren im Arbeitsprozess gemeint, in welchen während der Zusammenarbeit menschliche Personen und lern- und entscheidungsfähige, mit künstlicher Intelligenz ausgestattete Maschinen (KI) als soziale Partner nicht nur auf unterschiedliche Weise interagieren, sondern miteinander verschmelzen. Dies kann mit zwei Prozessen der Annäherung begünstigt werden: Einerseits können sich Menschen mit Technologie in Form von Enhancements ausstatten lassen, damit neue Fähigkeiten erwerben und eine Technisierung des Menschen (Cyborgs) realisieren, andererseits kann die Vermenschlichung der KI durch Annahme menschlicher Fähigkeiten wie dessen Aussehen oder menschlicher Emotionen erfolgen (Beispiel humanoide Roboter[8]).

2.2.4 Zum Begriff der Orientierung

Der Begriff Orientierung wird in verschiedenen Kontexten verwendet. Unter Orientierung versteht man die Leistung, sich in immer neuen Situationen immer neu zurechtzufinden. Bei der Orientierung hat man es zunächst mit etwas zu tun, von dem man noch gar nicht weiss, worum es sich handelt: mit einer neuen Situation.[9] Kurz zusammengefasst ist das Ziel der Orientierung, erfolgversprechende Strategien und Methoden für die Bewältigung neuer Situationen zu finden.

2.2.5 Zur Verbindung von Philosophie und Management

Entscheidungen und Handlungen mit Hilfe von philosophischen Gedanken zu reflektieren, vorzubereiten und durchzuführen und dabei auf die Weisheit der Denker von der Antike bis zur Neuzeit zurückgreifen zu können, hat sich bisher in Management- und Leadership-Situationen als bereichernd und gewinnbringend erwiesen. Philosophie hilft, sich in der immer weiter zunehmenden Volatilität, Unsicherheit, Komplexität und Mehrdeutigkeit in der Welt zu orientieren, Strategien für den Umgang mit dieser zu entwickeln und Antworten auf die sich stellenden Fragen zu finden. Dies wird in dieser Arbeit mit Hilfe der Ansätze von *Heraklit, Kant, Luhmann, Nietzsche, Weber* und *Stegmaier* angeboten.

---

[8] Benedikter, «Transhumanismus. Auf dem Weg zum neuen Menschen?».

[9] Stegmaier, Werner: Philosophie der Orientierung. Berlin/New York: De Gruyter, 2008.

3 Die zentrale Leadership-Kompetenz «Orientierung» und ihre Bedeutung und Anwendung in der Führungspraxis

Hier wird die zentrale Leadership-Kompetenz *«Orientierung»* und ihre Bedeutung und Anwendung in der Führungspraxis thematisiert und untersucht, welche Aspekte der Orientierung grundlegend sind für die Aufgabe der Führung von gemischten Teams, in welchen die Verschmelzung zwischen Menschen und Maschinen schon begonnen hat und sich weiter entwickeln kann.

Zunächst wird aufgezeigt, was Orientierung ist, wie sie strukturiert ist und wie sie funktioniert. Im zweiten Schritt werden die Aspekte des Themas Orientierung im Hinblick auf ihre Bedeutung für die Leadership untersucht und anschliessend die Möglichkeiten der Nutzung dieser Erkenntnisse für die Führung der gemischten Teams zusammengefasst.

3.1 Philosophie der Orientierung oder Orientierungsphilosophie

Unter Orientierung versteht man wie oben definiert die Leistung, sich in immer neuen Situationen immer wieder zurechtzufinden. Der Orientierung bedarf es bei jeder Veränderung. Sie ist nicht nur bei der Begegnung mit neuen Situationen gefragt, sondern auch jedem Menschen gegenüber. Orientierung bewirkt, dass man rasch erkennt, mit was und wem man es zu tun hat und einschätzen kann, ob man sich jemandem zuwenden und mit ihm interagieren soll. Orientierung muss auch ermöglichen, sich auf unvorhersehbare und überraschende Situationen einzustellen, also auch sich selbst und das Vorgehen zu ändern, wenn sich Personen oder Situationen ändern.

Die Bedingungen und Strukturen der menschlichen Orientierung wurden vom Nietzsche-Forscher *Werner Stegmaier* über Jahrzehnte erforscht und das Ergebnis wurde in seinem Werk *Philosophie der Orientierung* (2008)[10] und der aktualisierten englischen Version *What is Orientation* (2019, übersetzt *von Reinhard G. Mueller*)[11] dargelegt. Dieses Hauptwerk der Philosophie der Orientierung liegt den weiteren Ausführungen zugrunde.

---

[10] Stegmaier, «Philosophie der Orientierung».

[11] Stegmaier, Werner: What is Orientation? A Philosophical Investigation. Berlin/Boston: De Gruyter, 2019.

Orientierung steht am Anfang aller Entscheidungen, nicht nur in der Führung, sondern im ganzen Leben. Ohne oder bei nicht ausreichender Orientierung entsteht Orientierungslosigkeit. *Stegmaier* geht von der Tatsache aus, dass man bei allem, was man denkt, sagt und tut, im Alltag ebenso wie in der Arbeitswelt, schon orientiert sein muss – auch dann, wenn man Orientierungen von anderen übernimmt.[12] Orientierung gibt dem Menschen Halt, Bestand und eine Struktur auf bestimmte Zeit. Sie kann jedoch nicht andauern. Sich an Erfahrungswerte, Normen und Standards zu halten, kann kurzfristig eine Vereinfachung im Umgang mit bekannten Situationen bieten, dies ersetzt jedoch nicht die stetige Neuorientierung in sich ändernden Situationen. Orientierung gewährleistet weniger Halt und Bestand auf Dauer, als die Flexibilität, mit der Zeit gehen zu können. Dies bedingt, dass sich bewegliche Strukturen entwickeln. Hier kann später an den Anspruch der *Agilität* in der Zusammenarbeit und Führung angeknüpft werden und hier kann auch der Gedanke *Heraklits* aufgenommen werden: Orientierung muss im Fluss bleiben. Der Fluss bleibt derselbe und wird doch jedes Mal anders, wie auch der Mensch, der nicht zum zweiten Mal in den gleichen Fluss steigt, ohne sich selbst auch schon geändert zu haben.[13]

«Orientierung setzt immer schon Orientierung voraus, Orientierung schließt immer schon an Orientierung an. So steht sie am Anfang von allem, ist aber selbst ohne Anfang. Sie ist, als Orientierung, immer neue Orientierung oder immer Umorientierung.»[14] Der Begriff der Orientierung wird im Alltag, in den Wissenschaften und der Philosophie wie selbstverständlich gebraucht. Orientierung ist jedoch etwas, das auch verloren gehen und dessen man sich nie ganz sicher sein kann. Was selbstverständlich wird, wird nicht mehr beachtet und nicht mehr hinterfragt. Wenn es jedoch hinterfragt wird, kann festgestellt werden, dass es nicht selbstverständlich ist und dass es in Bewegung gehalten werden muss.

Die Frage der Orientierung greift *Immanuel Kant* in der Schrift *«Was heißt: Sich im Denken orientieren?»* (1786) auf. Ebenfalls befasst er sich mit dem Thema Orientierung in seiner *«Kritik der reinen Vernunft»*.[15] Der Ausgangspunkt *Kants* war die Vernunft der Aufklärung und die Auffassung, dass alle Fragen, welche uns die reine Vernunft vorlegt, nicht in der Erfahrung, sondern wiederum in der Vernunft liegen und dass die Vernunft stets eines Standpunkts bedarf, von dem aus sie sich auf das sich Verändernde (nicht Vernünftige) in der Welt beziehen kann. *Kant* supplementiert den Begriff *Vernunft* mit dem Begriff *Orientierung.*

[12] Stegmaier, «Philosophie der Orientierung».

[13] Vgl. 3.2.2. Agilität, S. 14.

[14] Stegmaier «Philosophie der Orientierung», Zitat XVII.

[15] Kant, Immanuel: Kritik der reinen Vernunft. Hamburg: Felix Meiner Verlag, 1998.

Die Vernunft kann als spezifische Orientierungsleistung verstanden werden. Um sich orientieren zu können, muss man schon in irgendeiner Weise orientiert sein und auf diese *«Vernunft»* als Standpunkt zurückgreifen können. Orientierung endet erst mit dem Tod. Sie muss als ein Ganzes verstanden werden, als eine grundlegende Komponente des Lebens.

Die Fähigkeit, zu erkennen, dass sich eine Situation ändert, bildet die Basis, um sich darin fortlaufend Orientierung zu verschaffen. Es setzt voraus, dass diese Person auch die Fähigkeiten der aufmerksamen Beobachtung, Einschätzung der Situation und des Erkennens der Veränderungen entwickelt hat.

Aus praktischer Sicht müssen hierfür drei Schritte aufeinander folgen:

1. Die Grundorientierung muss vorhanden sein, auch dann, wenn sie nicht aus eigener Leistung, sondern durch die Übernahme der Orientierung anderer zustande kommt.[16]
2. *Stegmaier*[17] zeigt auf, dass es der Aufmerksamkeit bedarf, um zu erkennen, dass sich die Situation verändert hat und es einer Umorientierung bedarf, um Anhaltspunkte für die Neuorientierung zu finden.
3. Die neue Orientierung muss vorgenommen und damit die Grundlage für Entscheidungen und Handlungen unter neuen Bedingungen geschaffen werden, wobei sich hier die Frage stellt, ob eine Person, welche bisher die Orientierung von anderen übernommen hat, imstande sein kann, sich selbst in unvorhersehbaren Situationen neu zu orientieren.

3.2 Selbstorientierung als grundlegende menschliche Kompetenz

Die grundlegende Kompetenz nicht nur in der Führungsrolle, sondern im Leben allgemein besteht in der Selbstorientierung. Diese Kompetenz zu entwickeln bedeutet, dass die Person sich selbst in neuen Situationen zurechtfindet. Dies stellt die Bedingung für gute Entscheidungen und Handlungen dar. Nur auf dieser Grundlage kann die nächste wichtige Führungsaufgabe angegangen und erfüllt werden, nämlich den Mitarbeitern eine Orientierung zu bieten und sie in der eigenen Orientierung zu unterstützen. Hierfür ist vor allem eine Kommunikations- und Diskurskompetenz unumgänglich. Die Verständigung untereinander in einem «gemischten» Team stellt noch eine weitere grosse Herausforderung dar. Es

---

[16] Stegmaier, «Philosophie der Orientierung».

[17] Stegmaier, «Philosophie der Orientierung», 167–176.

bedarf einer Kommunikation, welche sich einer gemeinsamen Sprache und eines Verständnisses davon bedient, wie die gemeinsame Orientierung in der Zusammenarbeit gestaltet werden kann. Die Aufgabe der Führung ist, mit Orientierung Halt und Struktur zu geben, und damit ist sie mit der Vorbildrolle verbunden. Diese Aufgabe wird treffender mit Leadership als mit Management bezeichnet. Der Begriff *Leadership* zeigt das sich verändernde Verständnis von Rollen und Aufgaben der Führung, welche sich auf das *«Orientierung geben»* und den Geführten eine Selbstorientierung ermöglichen konzentriert.

In Bezug auf die Fragestellung dieser Arbeit bietet es sich an, die folgenden Aspekte des Themas Orientierung näher zu betrachten

- die Bedingungen für Orientierung und ihre Bedeutung für die Leadership
- die Bedingungen für Orientierung in besonderen Situationen (Orientierung unter Zeitnot, in Routinen und der damit verbundene Umgang mit Unsicherheit, Rolle der Anhaltspunkte und der Aufmerksamkeit)

3.3 Die Bedingungen für Orientierung

In diesem Abschnitt wird in Anlehnung an *Stegmaier* dargelegt, was Orientierung überhaupt ist, wie sie strukturiert ist und wie sie funktioniert. Wenn Orientierung die Leistung ist, sich zurechtzufinden, so ist dieses ein Finden. Ohne dass man etwas konkret gesucht hat, stösst man in der Orientierung auf etwas: auf eine Situation, die einen betrifft, auf Umstände, die man berücksichtigen sollte. So besteht die Leistung der Orientierung zunächst aus zwei Vorgängen: die Umstände einer Situation auszumachen und sie einander zuzuordnen, also Zusammenhänge zu finden, welche relevant sind, um mit der Situation etwas anfangen zu können. Das Ziel ist, die Situation zu bewältigen. Als dritter Vorgang folgt dann, erfolgversprechende Handlungsmöglichkeiten zu suchen und zu finden und somit weiterzukommen. Diese Leistungen der Orientierung so zu vollbringen, dass man die Situation beherrscht, ist Orientierungsfähigkeit.

In alltäglichen Situationen ist man mehr oder weniger orientiert und bewältigt sie, ohne sich viel Gedanken über die Handlungsmöglichkeiten machen zu müssen. Das Bedürfnis nach Orientierung entsteht erst dann, wenn sie ausbleibt. Bei einer Veränderung der Situation, welche auch mit Überraschungen und Zeitknappheit einhergehen kann, muss die Orientierung neu erfolgen. Das bedeutet, eine Situation dahingehend zu erschliessen, was man tun muss, um sie zu bewältigen. Nicht irgendetwas, sondern das Richtige, das Erfolgversprechende – womit aus einer Situation am meisten gemacht wird.[18]

Das *sich Orientieren* verläuft nicht nach schon vorgegebenen logischen Regeln, sondern knüpft an persönliche Erfahrungen und Routinen an. Dabei sind die Möglichkeiten des Vorgehens zunächst kontingent, das heisst, sie können genutzt werden oder es kann auch ganz anders vorgegangen werden. Orientierung ist Umgang mit Kontingenz. Es bestehen mehrere Möglichkeiten, welche Feststellungen während der Orientierung zur Unterscheidung der alten und neuen Situation genutzt werden können: das Erkennen, Wahrnehmen und Fühlen tragen zu der Orientierung bei und je nachdem, was genutzt wird (oder nicht), kommt es schon zu Vorentscheidungen. Habe ich etwas wahrgenommen oder nicht, wie habe ich es wahrgenommen und wie interpretiere ich es?

Man kann, aber muss nicht in dieser Weise unterscheiden. In der Orientierung geht es darum, mit all dem zurechtzukommen, was auf einen zukommt. Somit dienen alle Unterscheidungen der Orientierung und sie müssen im Hinblick auf ihre Tauglichkeit für die jeweilige Situation untersucht werden. Wichtig ist, die Spielräume des Umgangs mit gemachten Feststellungen beim Zurechtkommen mit neuen Situationen auszuloten und zu nutzen. Wenn die Orientierung nicht gelingt, können andere Standpunkte gesucht werden, von welchen aus Orientierung möglich sein könnte, Erfahrungen aus ähnlichen Situationen ausprobiert oder Rat bei Menschen eingeholt werden, welche mit der Situation Erfahrung haben und mit Hinweisen bei der Orientierung helfen können. An der Orientierung anderer zwecks eigener Orientierung anzuknüpfen ist eine Möglichkeit, vorausgesetzt, dass diese in der jeweiligen Situation vorhanden und zugänglich ist.

3.3.1 Anhaltspunkte

«Take shortcuts in your orientation, but never forget that they are shortcuts.»[19]

Ein nützlicher Schritt in der Ausbildung der eigenen Orientierungsfähigkeit ist eine Reflexion der eigenen Orientierungspraxis, die es ermöglicht, sich Orientierung über die eigene Orientierung zu verschaffen, einschliesslich der Mittel und Routinen, welche verwendet werden. Hierbei wichtig ist, zu untersuchen, wie das *«sich Zurechtfinden»* vor sich geht und wie *Anhaltspunkte* gewählt werden. Anhaltspunkte sind Referenzpunkte, Indikatoren oder Hinweise, derer sich Menschen in einer unbekannten Situation bedienen. Der Suche geht die Frage

---

[18] Vgl. Stegmaier, «Philosophie der Orientierung», 151–158.

[19] Stegmaier und Mueller, «Successful Modes of Orientation», S. 2.

voraus: Was ist in der Situation relevant? Anhaltspunkte markieren diese relevanten Dinge und helfen, sich in einer Situation zurechtzufinden. Obwohl sie nie absolut sicher sind, helfen sie, mit der Unsicherheit vorübergehend umzugehen. Anhaltspunkte sind Abkürzungen, welche den Umgang mit der Komplexität der Situation erleichtern. Diese zu bestimmen bedingt oft paradoxe Entscheidungen über etwas, was nicht entscheidbar ist. So werden Anhaltspunkte oft unbewusst gewählt und basieren (besonders unter Zeitdruck[20]) auf Affektreaktionen (emotionale Empfindungen), Vereinfachungen, Schemen, Plausibilitäten und Routinen. Routinen bieten einen Halt in der Ungewissheit, es ist jedoch wichtig, diese zu deklarieren und auf die Tauglichkeit in neuen Situationen hin kritisch zu prüfen. Wenn sie sich nicht mehr als hilfreich erweisen, müssen sie ersetzt werden. Mit der Handlung bilden sich neue Routinen aus, welche auf eine gewisse Zeit ihren Dienst leisten werden.

3.3.2 Zusammenfassung

Die Situationen in der Führung von gemischten Teams sind neu, die Orientierung ist noch nicht vorhanden und der Ratschlag der Experten ist nicht formuliert. Hier wird das Beobachten und Unterscheiden wichtig sein, um die relevanten Merkmale der Transformation von alter zu neuer Situation auszumachen und die neuen Anhaltspunkte setzen zu können. Eine besondere Herausforderung für die Führungspersonen ist nicht nur, die eigene Orientierung zu meistern, sondern diese den Geführten zur Verfügung zu stellen und damit der Vorbildrolle gerecht zu werden. Die Idee der neuen Leadership ist zudem, nicht nur die eigene Orientierung zur Verfügung zu stellen, sondern die Geführten in der Ausbildung der eigenen Orientierungsfähigkeit zu unterstützen[21]. Hierzu zeichnen sich aus der Sicht der Autorin in der Führung (im Sinne von Leadership) von gemischten Teams zwei Probleme ab: Erstens ist es nicht klar, wie der Prozess der Konvergenz des Menschen und der KI den Bedarf an Orientierung verändern wird. Werden die Akteure die gleichen Bedingungen brauchen, werden die gleichen oder ähnliche Anhaltspunkte hilfreich sein und werden diese identisch mit denjenigen des Leaders sein? Zweitens: Werden die Führungspersonen imstande sein, diese Bedingungen zu antizipieren und die Teammitglieder bei der Identifizierung derselben zu unterstützen? Schon die Fragen nach der Bedeutung der Emotionen und der Wertevorstellungen für (technisierte) Menschen und (vermenschlichte) Maschinen lässt vieles im Ungewissen. Die Orientierung findet jedoch immer in

[20] Vgl. 2.5.2. Orientierung unter Ungewissheit und unter Zeitdruck, S. 9.

[21] Furtner, Marco: Empowering Leadership. Wiesbaden: Springer Fachmedien, 2017.

solcher Ungewissheit statt und wird sich ihre Anhaltspunkte immer neu suchen müssen.

3.4 Die Bedingungen der Orientierung in besonderen Situationen: Orientierung unter Ungewissheit und in Zeitnot

Die Grundbedingung jeder Orientierung ist, unter *Ungewissheit* zu operieren. In manchen Konstellationen existieren jedoch zusätzliche Erschwernisse. Überraschende Situationen irritieren die vorhandene Orientierung, sie verunsichern. Irritation ist eine Anregung oder Störung, deren Ursachen (noch) nicht klar sind. Sie ruft eine Beunruhigung vor. Es kann in einer Begegnung eine Anmerkung oder eine Geste sein, mit der man nichts anfangen kann. Es stimmt etwas oder es stimmt etwas nicht. Ohne zu wissen, was in der neuen Situation für die Orientierung relevant sein wird, müssen rasch Unterscheidungen zwischen alter und neuer Situation getroffen werden. Eine Irritation regt die Kraft an, die dann zu ihrer Bewältigung aufgewendet werden kann. Mit der aufkommenden Beunruhigung erhöht sich die Aufmerksamkeit. *Aufmerksamkeit ist lebensnotwendig, um Gefahren zu begegnen.*[22] Sie wird als Spannung empfunden, welche steigen, sinken, nachlassen oder erlahmen kann. In der Orientierung richtet sich die Aufmerksamkeit auf Gegebenheiten, welche als von Bedeutung für die Situation wahrgenommen werden und man überwacht die Veränderung. Nachdem die Orientierung erfolgt, stimmt die Situation wieder und es kommt zur Beruhigung. «Beruhigung ist das Kennzeichen der gelingenden Orientierung und auch schon ihr Maßstab.»[23]

Eine für Leadership-Situationen besonders herausfordernde Schwierigkeit ist, dass nicht ausreichend Zeit zur Verfügung steht, um die Unterscheidung der relevanten Komponenten vorzunehmen und diese unter Zeitdruck erfolgen muss. Dazu steht oft nur eine limitierte Menge an Information zur Verfügung. Deswegen wird in neuen Situationen oft auf Routinen und Plausibilitäten zurückgegriffen, ohne das Umfeld ausreichend zu analysieren und die Aussagekraft zu prüfen. Plausibilitäten sind in diesem Zusammenhang Aussagen, welche nicht mehr hinterfragt und definiert sowie als selbstverständlich genommen und für wahr gehalten werden.

---

[22] Vgl. Stegmaier, «Philosophie der Orientierung», 167–176.

[23] Vgl. Stegmaier, «Philosophie der Orientierung», 164.

Die Zeitknappheit und Unvollständigkeit der Information sowie der Verlass auf Plausibilitäten sind wichtige Faktoren, welche das Risiko einer nicht ausreichenden oder falschen Orientierung erhöhen. Kant hat seine Abhandlung *«Was heißt: Sich im Denken orientieren?»* damit eingeleitet, dass auch und gerade «reine Verstandesbegriffe» erst «zum Erfahrungsgebrauche tauglich» werden, wenn ihnen «bildliche Vorstellungen» anhängen, die ihnen *«Sinn und Bedeutung verschaffen».*[24]

3.4.1 Orientieren im Denken

«Sich im Denken orientieren» bedeutet, in der Praxis zu erkennen, was mir in der jeweiligen Situation an Möglichkeiten zur Verfügung steht und sich an den Anhaltspunkten zu orientieren, welche zuerst hinterfragt, dann durchdacht und als tauglich erkannt wurden. In der Praxis braucht es Aufmerksamkeit und Mut, bisherige Plausibilitäten als solche zu erkennen und sie (gerade deswegen) zu hinterfragen und, wenn sie nicht begründet werden können, auszuräumen. Aus einem neuen Anhaltspunkt heraus können sich andere Perspektiven eröffnen. Um in der Orientierung Halt zu finden, braucht es die richtigen Anhaltspunkte und gleichzeitig die Beweglichkeit von Horizont, Standpunkten und Perspektiven.

3.4.2 Zusammenfassung: Orientierungsbedarf und Herausforderungen der Orientierung in der digitalen Arbeitswelt

Die vorzugsweise maritimen Assoziationen aus der Führungspraxis wie «im gleichen Boot sitzen», «den richtigen Kurs einschlagen», «den Hafen finden», «am gleichen Strick ziehen» oder das aus der rauen Bergwelt entliehene «den Kompass und die Landkarte zur Hand nehmen», «den Wegweiser finden», «die anderen beim Klettern zu sichern» deuten darauf hin, dass Menschen in allen Lagen ganz selbstverständlich nach einer Halt und Struktur gebenden Orientierung in verschiedensten Situationen suchen, oft ohne eigens zu reflektieren, was sie dabei im Grunde tun.

Gerade in den Situationen des Wandels in der digitalen Arbeitswelt können sich Überraschungen schnell einstellen und die neue Orientierung muss unter Zeitdruck stattfinden. Die Irritationen durch ein fehlendes Verständnis oder fehlende Übereinstimmung zwischen Menschen und der KI können schnell Beunruhigung verursachen. Wenn nicht rasch eine Lösung gefunden werden kann, kann sich Angst einstellen, welche die Handlungs- und Kooperationsfähigkeit

[24] Kant, Immanuel: Was heißt: Sich im Denken orientieren? Nordstedt: Books on Demand, 2015, 133.

hemmt. Erst, wenn verstanden wird, welche Unterscheidungen getroffen werden müssen und wenn Lösungen gefunden werden, kann sich Beruhigung und damit eine für die weitere Arbeit zuträgliche Stimmung einstellen. Hier kann das Hinterfragen eigener Selbstverständlichkeiten ein hilfreiches und praktisches Instrument sein, um aus dem eigenen Standpunkt heraus einen bewussten Zugang zu Orientierung und Neuorientierung in neuen Führungssituationen zu finden.

Die wichtigsten Fragen hierzu lauten (eine Auswahl):

- Wie kann ich mich in neuen Situationen zurechtfinden?
- Von welchem Standpunkt aus bewege ich mich auf das Unbekannte zu?
- Welches Wissen, welche Beobachtungen und Aspekte beziehe ich ein, um unterscheiden zu können?
- Wie gewinne ich in der neuen Situation Halt und Bestand?
- Wie kann ich mir neue Strukturen schaffen, die mir auf eine gewisse Zeit hin gleichzeitig Standhaftigkeit und Beweglichkeit sowie gute Grundlagen für meine Entscheidungen und Handlungen geben?

In Bezug auf Leadership in der digitalen Arbeitswelt bleibt Orientierung die zentrale Tugend.

Orientierung braucht es nicht nur in den neuen Räumen und Konstellationen, sondern auch in der Interaktion und Kommunikation zwischen den Akteuren, also den Mitarbeitern, und neu auch zwischen den ineinander konvergierenden Menschen einerseits und den intelligenten Maschinen (Robotern, KI) anderseits. Die grosse Herausforderung der Führung gemischter Teams stellt sich in der *gemeinsamen Orientierung*. Die Beziehung zwischen Menschen und KI wird die Kommunikation und damit auch die Möglichkeiten einer gemeinsamen Orientierung entscheidend beeinflussen. Über die Entwicklung der Situation lässt sich keine zuverlässige Aussage machen. Wenn die Ausrichtung der KI auf die bisher geltenden menschlichen Werte und die moralischen Vorstellungen möglich wird, können die gemeinsam definierten Werte (wie in den bisherigen Teams) als Anhaltspunkte für die Orientierung genutzt werden. Die für alle Beteiligten relevanten Anhaltspunkte zu finden und diese zu vermitteln, wird Verständnis für die Bedürfnisse der Menschen sowie auch des Kooperationspartners KI und eine ausgeprägte kommunikative Fähigkeit der Führungspersonen voraussetzen. Dies sind wichtige Komponenten für die Kooperation und die Netzwerkbildung, welche für die Bewältigung von zukünftigen Aufgaben unumgänglich sind.

4 Bedingungen für die gelingende Kooperation in gemischten Teams

Hier wird untersucht, wie die Kommunikation zwischen den Akteuren in gemischten Teams im Sinne diskursiv-konstruktiver Bedingungen gestaltet werden muss, um eine gewinnbringende Kooperation zu ermöglichen. Zunächst wird die Frage beantwortet: Was geschieht, wenn Teams kooperieren? Der Begriff *Agilität* wird beleuchtet und anschliessend wird auf die bekannte Bedingung für eine gelingende Kooperation in klassischen Teams zugegriffen, die *psychologische Sicherheit,* und ihre wichtigsten Kriterien Vertrauen und Kommunikation werden weiter untersucht.

Zu der Herausforderung der Kommunikation wird die Problematik der *Orientierung an anderer Orientierung in Interaktion und Kommunikation (Orientierung in doppelter Kontingenz)* aufgezeigt und ihre Auswirkung auf die Kooperation hervorgehoben.

In der Zusammenfassung werden diese Erfahrungen für die Kooperation in gemischten Teams diskutiert.

4.1 Was bedeutet es, wenn Menschen kooperieren

Was bedeutet es, wenn Menschen kooperieren und welche Bedingungen müssen für eine gelungene Kooperation sichergestellt werden? Diese Frage gilt es zunächst zu beantworten, um im Hinblick auf die Zielsetzung dieser Arbeit hervorzuheben, welche Bedingungen für die Situation des Führens von «gemischten» Teams relevant sind.

Im Arbeitsumfeld kooperieren Menschen wenn sie zusammenarbeiten, um ein gemeinsames Ziel zu erreichen, etwa um ein Problem zu lösen. Dazu tragen sie ihr Wissen, ihre Fähigkeiten, ihre unterschiedlichen Ressourcen und ihre Leistung bei. Kooperation ist ein soziales Verhalten und basiert auf gegenseitigem Respekt, auf Verlässlichkeit, Vertrauen, Kommunikation und dem gemeinsamen Verfolgen eines als sinnvoll eingestuften Ziels. Als soziales Verhalten kann sich Kooperation auf verschiedene Arten zeigen und dabei kann es sowohl erfolgreiches als auch weniger erfolgreiches Kooperieren geben.

Demnach handelt es sich bei der Fähigkeit, mit anderen zu kooperieren und zu kommunizieren um eine facettenreiche Handlungskompetenz, die durch verschiedene Fertigkeiten, Wissensstrukturen, Motivation, Einstellungen, Werte usw. bestimmt wird. Mit den neuartigen Kooperationsformen (wie Projektarbeit, soziokratische Strukturen, Netzwerke) ergibt sich für die Beteiligten eine Reihe erhöhter Kompetenzanforderungen in ihrem alltäglichen Arbeitshandeln: ein höherer persönlicher Einsatz, Arbeitsmotivation, individuelle Kreativität und ein souveräner Umgang mit Unsicherheiten und Unbestimmtheit.[25] Einen interessanten und gut untersuchten Zusammenhang zwischen der hohen Leistungsbereitschaft und den zuträglichen Bedingungen legt *Edmondson* in ihrer Schrift *«The Fearless Organisation»*[26] vor und bestimmt die psychologische Sicherheit der Beteiligten als den Schlüsselfaktor einer gelungenen Kooperation und Kommunikation.

Zusammenfassend lassen sich zwei wichtige Bedingungen als Ausgangslage für die Bewältigung der oben beschriebenen Anforderungen und die erfolgreiche Kooperation hervorheben

1. die eigene Orientierung in neuen Situationen und die Möglichkeit der Orientierung an der Orientierung anderer[27]
2. ein Umfeld der psychologischen Sicherheit, in welchem die Beteiligten sich angstfrei äussern und ihre innovativen Ideen anbringen können und gehört und ernstgenommen werden, auch wenn sie Fehler begehen.[28]

4.2 Agile Arbeitsstrukturen - die Herausforderungen für gemischte Teams

Die fortschreitende Digitalisierung verändert die Arbeitswelt und Gesellschaft stark. In dieser sich schnell wandelnden, volatilen Welt zu agieren erfordert Flexibilität und Robustheit gegenüber Komplexität, Veränderungsgeschwindigkeit und Ungewissheit. Es braucht Arbeitsformen, welche flexibler und dynamischer auf diese Umgebung reagieren können. Die Suche nach neuen Formen der Führung, Zusammenarbeit und Entscheidungsfindung ist keinesfalls nur eine Entwicklung der jüngsten Zeit. Neu in der jetzigen Situation ist aber, dass die komplexen Aufgaben nicht nur wie bislang in Teams von Menschen, sondern nun auch zusätzlich in gemischten Teams von Mensch und künstlicher Intelligenz angegangen werden.

---

[25] Vgl. Münnich, Sacha: «Luc Boltanski und Eve Chiapello: Der Neue Geist des Kapitalismus.» In Schlüsselwerke der Wirtschaftssoziologie. Wirtschaft und Gesellschaft, hrsg. von Klaus Kraemer und Florian Brugger, 385–392. Wiesbaden: Springer VS, 2017, 385–392.

[26] Vgl. Edmondson, Amy C.: The Fearless Organisation. Creating Psychological Safety in the Workplace for Learning, Innovation, and Growth. New Jersey: Wiley, 2018.

[27] Vgl. 2.5.3.1. Orientierung an anderer Orientierung, S. 10.

[28] Vgl. 2.5.2.4. Orientierung in doppelter Kontingenz, S. 11.

Hier kann an den Anspruch der *Agilität bei der Arbeit* angeknüpft werden. «Agilität ist die Gewandtheit, Wendigkeit oder Beweglichkeit von Organisationen und Personen bzw. in Strukturen und Prozessen. Man reagiert flexibel auf unvorhergesehene Ereignisse und neue Anforderungen. Man ist, etwa in Bezug auf Veränderungen, nicht nur reaktiv, sondern auch proaktiv.»[29] Auch hier bedarf es einer Orientierung, welche Halt und Beweglichkeit gleichzeitig bietet. Eine solche kann sich am besten unter der Voraussetzung der psychologischen Sicherheit entfalten.

Agilität ist eine Antwort auf diese Herausforderungen. Gemeint ist die Eigenschaft von Gruppen, adaptiv auf Veränderungen im Berufsumfeld reagieren zu können.

Die Ziele des agilen Arbeitens sind

- eine höhere Reaktionsfähigkeit hinsichtlich Veränderungen
- eine verbesserte Fähigkeit zu lernen
- die Stärkung der Kreativität und Innovationskraft

Agile Arbeit kann am besten in *netzwerkartigen Strukturen* gelingen, wo die intensive Teamarbeit Kernbedingung ist. Dabei spielen Werte wie Vertrauen, Transparenz und Respekt eine zentrale Rolle. Agile Teams benötigen auch eine agile Kommunikation, sie arbeiten mit raschen Feedbackschleifen und einer hohen Fehlertoleranz.[30] Diese können sich nur in einem hierfür zuträglichen Umfeld entwickeln.

4.3 Psychologische Sicherheit im Team

Nachfolgend wird aufgezeigt, was psychologische Sicherheit ist und unter welchen Bedingungen sie sich ausbilden kann. Dabei wird auf den *Aufbau des Vertrauens,* die *Vorbildrolle der Führungspersonen* und die besondere Herausforderung der Kommunikation, *die Orientierung in doppelter Kontingenz,* fokussiert.

---

[29] Vgl. https://wirtschaftslexikon.gabler.de/definition/agilitaet abgerufen am 27. Dezember 2023.

[30] Altherr, M. (2019) Die Organisation der Selbstorganisation. Die Notwendigkeit neuer Organisationsformen in wissensintensiven Organisationen. In P. Kels & S. Kaudela-Baum (Hrsg.), *Experten Führen – Modelle, Ideen und Praktiken für die Organisations- und Führungsentwicklung* (S. 411–426). Wiesbaden: Springer Gabler.

Psychologische Sicherheit beschreibt die individuelle Überzeugung eines Menschen, dass es sicher ist, zwischenmenschliche Risiken einzugehen.[31] *Edmondson* definiert weiter die psychologische Sicherheit als geteilte Wahrnehmung innerhalb eines Teams.[32] Sie betrachtet psychologische Sicherheit somit als eine Variable auf Gruppenebene. Die «Freiheit von Angst» trägt zur Entwicklung der psychologischen Sicherheit in einer binären Beziehung oder in einer Gruppe bei.

Diese Gedanken und Untersuchungen haben in der letzten Zeit als Antwort auf die menschliche Seite der innovativen Prozesse in Unternehmen an Bedeutung gewonnen. Dies besonders dort, wo kreative Ideen dringend notwendig sind, wenngleich nicht alle sich als eine Innovation erweisen. Die Art des «Trial & Error» Denkens hat bei der Arbeit enorm zugenommen. Konkret geht es auch hier um Orientierung unter Ungewissheit. Der Diskurs über das Neue und noch Unbekannte, zu welchem man erst nach Anhaltspunkten sucht und die eigenen Gedanken und Ideen erst nachher entwickelt, ist im Sinne der *doppelten Kontingenz* mit einem hohen Grad an Ungewissheit und Beunruhigung verbunden. Naturgemäss versuchen sich Menschen in ihrem Verhalten so anzupassen, dass sie sich in ihrem Tun den Erwartungen anderer möglichst annähern. Sie orientieren sich an anderen, passen sich den Anhaltspunkten anderer an und versuchen damit die Ungewissheit zu reduzieren. Die Ungewissheit gründet hier in der Kontingenz, ob die Gruppe auf die eigenen Äusserungen mit Neugier, wohlwollender Kritik oder Unverständnis, Missfallen oder sogar Missbilligung reagieren wird. Mit der Unterlassung der Äusserung z. B. eigener Ideen unter Druck der Ungewissheit stabilisiert man zwar die Situation, der Preis dafür jedoch ist, dass man sich zu den eigenen Ideen nicht frei äussert und diese damit nicht diskutiert, verworfen oder weiterentwickelt werden können. Damit geht sehr viel kreatives Potential im Unternehmen verloren. Nur, wenn alle in der Gruppe eine freie Äusserung zulassen oder diese sogar begrüssen, kann sich die psychologische Sicherheit entwickeln. Dies macht den Umgang mit der doppelten Kontingenz in der Kommunikation in der Gruppe einfacher und risikoärmer.

Mit anderen Worten: Das *«sich darauf verlassen Können»,* dass niemand in der Gruppe abgewiesen wird, trägt zu einer produktiven Stabilisierung der Beziehungen bei und ermöglicht die freie Äusserung zu Dingen mit ungewissem Ausgang. Man weiss jedoch nicht, ob eine KI ebenso ein Bedürfnis von Zugehörigkeit und Anerkennung in der Gruppe empfinden wird.

---

[31] Edmondson, Amy C.: "Psychological safety and learning behavior in work teams." Administrative Science Quarterly 44, Nr. 2 (Juni 1999): 350–383. http://dx.doi.org/10.2307/2666999.

[32] Vgl. Edmondson, "Psychological safety and learning behavior in work teams".

4.3.1 Aufbau des Vertrauens als Bedingung für psychologische Sicherheit

Vertrauen ist eine *zentrale Ressource sozialen Handelns* und die grundlegende Komponente der psychologischen Sicherheit. Es steht am Beginn der Zusammenarbeit: Wer gemeinsam handeln will, muss vertrauen. Vertrauen ist aber nichts, was man schenken oder geschenkt bekommen kann, man muss es sich in der Interaktion erarbeiten.[33] Diplomatie in der Kommunikation schafft zunächst eine Vertrautheit, noch kein Vertrauen. Der andere vollzieht eine immer gleiche (mit der Zeit vertraute) Handlung, in der seine Absichten erkannt werden, bis sich ein Gefühl des «sich darauf verlassen Könnens», dass es das nächste Mal ebenso sein wird, einstellt. Erst Vertrauen bringt nachhaltigere Beruhigung in die *doppelt kontingente Kommunikation.*[34] Diese kann dann angstfrei erfolgen.

4.3.2 Vorbildrolle der Führungspersonen als wichtiger Parameter der psychologischen Sicherheit

In Anlehnung an *Edmondson* ist an dieser Stelle wichtig zu betonen, dass die Bildung und Festigung der psychologischen Sicherheit eine Leadership-Aufgabe ist. In einem Unternehmen muss diese von den Führungspersonen initiiert und bei der Umsetzung begleitet werden. Dabei wird sich nicht die Kommunikation der Absichten, sondern die Vorbildwirkung im Sinne der eigenen Orientierung und des Verhaltens auf die Orientierung und das Verhalten der Teammitglieder auswirken. Hierzu gehören das Vorbildverhalten im Umgang mit eigenen Fehlern und die Neugierde beim Angehen der kreativen Ideen und Diskussionsangebote. Dies bewirkt die Sicherung der Stabilität der Gruppe im möglichst angstfreien Umgang mit Kontingenz.

Die individuelle Leistung als Beitrag zur psychologischen Sicherheit in der Gruppe muss ebenfalls von den Leadern eingefordert und unterstützt werden. Der Beitrag jedes Einzelnen trägt zum Miteinander am Arbeitsplatz bei und stärkt den Grad an psychologischer Sicherheit im Team. Die einzelnen Teammitglieder können mit ihrer Neugierde, ihrem Interesse und ihrer Lern- und Hilfsbereitschaft die Anhaltspunkte für ihre Orientierung für andere verdeutlichen und zum «sich näher Kommen» beitragen und damit zum besseren gegenseitigen Verstehen. Dies alles sind menschliche Eigenschaften, und es ist nicht gewiss, ob sie auch für eine KI von Bedeutung sein werden.

---

[33] Vgl. Hartmann, Martin: Die Praxis des Vertrauens. Berlin: Suhrkamp Verlag, 2011.

[34] Vgl. Stegmaier, «Philosophie der Orientierung», 414.

Innerhalb eines Teams kann die psychologische Sicherheit mit der Erfahrung des Zusammenhalts und des Erlebens der Mitwirkung jedes Teammitglieds entstehen und wachsen. Den anderen auf Augenhöhe zu begegnen ist nur möglich, wenn man sich gegenseitig als Kooperationspartner anerkennt. Dann kann man die Erfolge und Misserfolge teilen, angstfrei Ideen anbringen, Bedenken äussern und gehört werden[35]. Dies kann in gemischten Teams von Menschen und KI durchaus möglich sein. Jemandem auf Augenhöhe zu begegnen bedeutet nichts anderes, als die Orientierung anderer zu akzeptieren, ihre Standpunkte anzuerkennen und damit beiderseitig die Unsicherheit der doppelten Kontingenz zu reduzieren.

4.4 Interaktion und Kommunikation in Gruppen

Hier wird der Vorgang der *«Orientierung an anderer Orientierung»* vorgestellt und die grosse Herausforderung der Kommunikation *«Orientierung in doppelter Kontingenz»* und ihre Bedeutung für die Kooperation verdeutlicht.

4.4.1 Orientierung an anderer Orientierung

Der Soziologe *Max Weber*[36] zeigt auf, dass nicht nur das Wahrnehmen und das Denken Orientierung benötigt, sondern auch das Handeln. Orientierungen des Handelns von Personen nannte Weber *«sinnhaft verständlich»*. Orientiert wird nicht nur das eigene, sondern auch das Handeln anderer, beide orientieren sich aneinander. Hierzu nutzen sie als Anhaltspunkte bestimmte Verhaltensweisen und Ordnungen. Im Gesellschaftshandeln rechnet man mit einem bestimmten Verhalten des andern und damit, dass die andern ebenfalls mit einem bestimmten Verhalten rechnen; so werden Normen gebildet. Auch wenn man bewusst gegen diese Normen handelt, orientiert man sich an diesen. Bei einer (subjektiv) als sinnhaft anerkannten Orientierung erhöht sich die Chance, dass sie befolgt wird. Ob jemand eine Regel oder Norm befolgt, kann in unterschiedlichen Motiven gründen, ebenso wie die Anhaltspunkte dafür, ob man eine Norm sinnvoll findet oder nicht, ganz unterschiedlich sein können. Dies ist in der Führung von Gruppen und Konsensbildung generell zu berücksichtigen. Einverständnis mit der Befolgung einer Norm muss kein sinngeleitetes Handeln sein, sondern lediglich

[35] Vgl. Edmondson, "Psychological safety and learning behavior in work teams", 350–383.

[36] Vgl. Weber, Max: Gesammelte Aufsätze zur Wissenschaftslehre. Tübingen: Verlag J.C.B. Mohr, 1985.

ein Handeln «unter Orientierung an den Erwartungen des andern».[37] Ob man aus Einverständnis handelt oder der Handelnde die Norm verbindlich geltend macht, ist nicht immer klar. Ob sich jemand am anderen orientiert oder weil er so zu handeln sinnvoll findet, ist nicht eindeutig feststellbar und den Akteuren auch oft nicht «vollständig bewusst». Auch das Mass an Selbständigkeit des orientierten Handelns kann unterschiedlich sein, die Gründe können das Befolgen der Tradition, Rationalität, Affekthandlungen, die Werteorientierung oder die selbstauferlegte Verbindlichkeit sein.

Andere Orientierungen sind der wichtigste Anhaltspunkt für die eigene Orientierung.[38] Der versuchte Vergleich der eigenen Orientierung mit der Orientierung anderer kann beunruhigen oder beruhigen, je nachdem ob er übereinstimmend oder divergierend ausfällt. Wenn Orientierungen anderer als übereinstimmend wahrgenommen werden, geben sie der eigenen Orientierung beruhigenden Halt, Rückhalt, Anleitungen und Ratschläge. Wenn die eigene Orientierung mit der Orientierung anderer übereinstimmt, kann es bestätigend wirken. Man kann sich jedoch nie sicher sein, inwieweit man andere Orientierungen ihrerseits richtig sieht und ob es wirklich beruhigen kann. Dort, wo Unterschiede erkennbar werden, wirken sie beunruhigend und verunsichern. Sie können jedoch auch hilfreich sein, weil sie auf andere Handlungsmöglichkeiten aufmerksam machen. Was zunächst beunruhigt, kann in der Folge dann auch wieder beruhigen.

Der Grund der Beunruhigung ist die Unvergleichbarkeit der Orientierungen. Man kann versuchen, sich in den Standpunkt anderer zu versetzen und zu verstehen, was sie mit dem meinen, was sie sagen, was sie denken, was sie wollen, aber sicher kann man kann sich dessen nie sein. Man kann sich die Situation nur von seinem eigenen Standpunkt aus erschliessen. Am Bewusstsein anderer teilzuhaben ist nur so möglich, dass man sich anhand von Anhaltspunkten und Zeichen über sie orientiert und sich aus ihnen die Einsichten und Absichten der andern erschliesst, und dabei kann es ebenso um Konsens wie um Konflikt gehen.[39]

Beunruhigung und Beruhigung durch andere Orientierungen schaffen eine neue Orientierungssituation. Der Umgang mit andern kann misslingen, aber auch gelingen. In Situationen der Gegenwart anderer, die Interaktion und Kommunikation erfordern, will man sich über die Absichten der andern orientieren und daraus Hinweise gewinnen, um erfolgreich mit ihnen umzugehen. Man könnte in seiner Orientierung auch irren oder mit anderen in Konflikt geraten. Zu der eigenen

---

[37] Vgl. Weber, «Gesammelte Aufsätze zur Wissenschaftslehre».

[38] Vgl. Stegmaier, «Philosophie der Orientierung».

[39] Vgl. Luhmann, Niklas: Einführung in die Systemtheorie, 2. Aufl. München: Carl-Auer-Systeme Verlag, 1984.

Beruhigung kann man andere Orientierungen stets aufmerksam berücksichtigen. Sie können auch helfen, eigene Orientierungsprobleme zu bewältigen, wenn man dadurch andere Handlungsoptionen entdeckt. Einer findet sich in neuen Situationen immer wieder rasch zurecht, andere verlassen sich auf ihn, überlassen ihm ihre Orientierung und verzichten auf eine eigene. Durch die Übernahme bewährter Orientierungen wird die eigene in den Hintergrund gerückt. Vorausgesetzt, man hat sich auf die «richtige» Orientierung verlassen, werden einem damit Umwege und Irrtümer erspart, die eigene Orientierung wird leichter. Hier könnte man den Akt des *«sich jemandem nähern»*, der sehr unterschiedlich ausfallen kann, als einen Versuch der Stabilisierung und Vereinfachung der Beziehungen durch Reduktion der Ungewissheit sehen. In Bezug auf die Orientierung in «gemischten» Teams kann man die Prozesse des Verschmelzens der Menschen und der Maschinen (KI) als einen Schritt zur Vereinfachung der Kommunikation betrachten. Je mehr Ähnlichkeiten die konvergierenden Teile des Teams aufweisen, desto besser kann das Nachvollziehen der Anhaltspunkte anderer in der Orientierung gelingen. Ob KI auf die Veränderungen der Situationen wie Menschen mit Beunruhigung und Beruhigung reagiert und sich orientiert, wissen wir nicht. Wenn KI die Unterschiede erkennen und deren Relevanz beurteilen könnte, wäre eine Art der Orientierung möglich.

4.4.2 Kommunikation in doppelter Kontingenz

Der Begriff *doppelte Kontingenz* beschreibt eine soziale Situation, in der mindestens zwei Teilnehmer sich gegenseitig wahrnehmen, und in der noch völlig unbestimmt ist, was als Nächstes geschehen soll.[40] Anknüpfend an *Luhmann* bezieht sich die doppelte Kontingenz auf den Fakt, dass man in der Orientierung nie eine völlig zuverlässige Kenntnis der Reaktion seines Interaktionspartners hat. Individuen sind wie *«black boxes»* und ihre Standpunkte sind jeweils für den anderen nicht transparent. Die Reaktion des Gegenübers kann immer von dem abweichen, was man erwartet hat, die Kommunikation ist somit kontingent, sie kann sich auch anders entwickeln. Sie ist doppelt kontingent, weil es beiden Kommunikationspartnern so geht. Um die Fortsetzung der Kommunikation dennoch zu gewährleisten, versucht man sich in einer Art und Weise zu artikulieren, die dem anderen ermöglicht, in der erwarteten Weise zu reagieren. Die doppelte Kontingenz muss in immer neuen Situationen der Kommunikation auf immer neue Weise bewältig werden. Es erweist sich als wenig hilfreich, sich dabei völlig auf Gepflogenheiten («das tut man») und vordefinierte Regeln der Verlässlichkeit

---

[40] Vgl. Luhmann, Niklas: Soziale Systeme. Grundriß einer allgemeinen Theorie, Frankfurt am Main: Suhrkamp Verlag, 1984, 362.

oder die Verpflichtung zu Konsens zu verlassen. Eher erweisen sich Offenheit in der Orientierung und Zugänglichkeit für andere, Diplomatie und Vertrauensbildung als ein gangbarer Weg, welcher auch Alternativen zulässt, die nicht geplant werden können. Um diese Unsicherheit in der Kommunikation zu minimieren, versuchen beide Personen das Verhalten des anderen zu antizipieren und damit die Situation berechenbar zu machen.

Die Absichten und die Handlung des anderen zu antizipieren bedeutet, Anhaltspunkte zu finden, sich in der Situation zu orientieren, diese nachvollziehen zu können und sie für die eigene Orientierung zu nutzen. Andere Menschen sind das Zuverlässigste in der Orientierung, können aber auch das Überraschendste sein. Deswegen ist die Beobachtung anderer auf ihre Verlässlichkeit und Beständigkeit hin ein wichtiger stabilisierender Faktor in Beziehungen. Mit dem *«Sich über einander orientieren»* will man wissen, wem man sich zuwenden und von wem man sich fernhalten soll. Damit versucht man, die Unsicherheit zu reduzieren und die Situation abzusichern. Die Antwort auf die Frage, wie sich die KI in doppelter Kontingenz verhalten wird, hängt davon ab, ob der KI menschliche Eigenschaften wie das Empfinden von Beunruhigung oder Angst und das Antizipieren eines erwarteten Verhaltens im gleichen oder ähnlichen Masse zugeschrieben werden können.

Niklas Luhmann bestimmte im Rahmen seiner *«Theorie sozialer Systeme»* den Begriff der Orientierung neu. Er hat Orientierung als *Abtasten der Umwelt auf Kontingenz* gesehen. Sie ist als Angelegenheit des individuellen psychischen Systems, des Bewusstseins, charakterisiert. Diese funktioniert als «Wechselspiel von Erwartungen und Enttäuschungen, in dem der Mensch die Kontingenz der Umwelt in Bezug auf sich selbst abtastet und als eigene Ungewissheit wahrnimmt».[41] Danach geht es in der Orientierung darum, die Ungewissheit in der Orientierung zu mildern, auch wenn man sie nie ganz ausräumen kann. Man kann sich gegenseitig nur beobachten, aber nicht durchschauen, sich nicht ganz sicher sein. Das Verhalten anderer kann man immer nur prognostizieren, nicht sicher bestimmen. Die Kontingenz bleibt immer in einem gewissen Grad bestehen.

Die Ungewissheit bezüglich dessen, was kommt, wirkt sich als Druck auf die Orientierung aus. Unter Zeitdruck ist man zu einer schnellen Auswahl der Anhaltspunkte gezwungen, und es können weder alle Möglichkeiten durchdacht noch die besten evaluiert und ausgewählt werden. Orientierung unter Zeitdruck sucht selten nach einer, wiederum zeitraubenden, Vergewisserung durch Hinterfragen, verläuft weitgehend fraglos und bleibt kontingent.[42] Die Kontingenz

---

[41] Vgl. Luhmann, «Soziale Systeme. Grundriß einer allgemeinen Theorie».

[42] Vgl. Luhmann, «Soziale Systeme. Grundriß einer allgemeinen Theorie», 362.

kann der Mensch jedoch schrittweise einschränken durch die Abarbeitung «völlig willkürlicher Erwartungen» und damit eine weniger enttäuschungsanfällige «Groborientierung» ausbilden.[43] Das heisst, die Orientierung kann sich entwickeln und lernen.

Zur Kommunikation in doppelter Kontingenz verweist *Friedrich Nietzsche* auf *die Diplomatie der Zeichen* und ihre Bedeutung für die Kooperation. Die Kommunikation läuft auf engere Kooperation und längerfristige Bindung hinaus. Es muss jedoch nicht bedeuten, dass man einander immer besser versteht. In der Absicht, die unterschiedlichen Standpunkte der Orientierung an anderer Orientierung einander anzunähern, räumt man dem anderen in der Kommunikation Deutungsspielräume ein, welche zu Missverständnissen führen können. *Nietzsche* nennt diese *«Spielraum und Tummelplatz des Missverständnisses»*.[44] Er notiert jedoch dazu, dass man den Freunden immer ausreichend Platz für Deutungen einräumen soll. Besser missverstanden werden, als unverstanden zu bleiben. Zur Anschlussfähigkeit in der Kommunikation nutzt man die Deutungsspielräume, um unter Bedingungen der doppelten Kontingenz «im Gespräch zu bleiben». Viele Menschen wissen das Gespräch nicht zu nutzen, konzentrieren sich zu stark auf das, was und wie sie es sagen, anstatt zuzuhören und aus dem Gespräch viele Hinweise auf die Standpunkte der anderen Orientierung mitzunehmen.

4.5 Zusammenfassung: Herausforderungen der Kommunikation und Kooperation in gemischten Teams

Im vorliegenden Teil wurde auf die bewährten Bedingungen des Gelingens der Kommunikation und Kooperation im Team zurückgegriffen, so, wie sie aus der Führung herkömmlicher, aus Menschen zusammengesetzter Teams bekannt sind. Für die Zusammenfassung wird die Annahme getroffen, dass die neuen Arbeitsstrukturen in agilen Netzwerken und die Bedeutung der Kooperation zwischen den Beteiligten zunehmen werden, um die Aufgaben und Anforderungen der Wirtschaft, Wissenschaft und Gesellschaft erfüllen zu können. Dabei werden die Führungspersonen weiterhin mit der Vorbildrolle in der Orientierung beauftragt.

Als schwierig zu beantworten erweist sich die Frage nach der Bedeutung der sozialen Sicherheit und der Vertrauensbildung. Da die Ungewissheit hinsichtlich der Reaktion anderer in unbekannten Situationen (*Kommunikation in*

[43] Vgl. Luhmann, «Soziale Systeme. Grundriß einer allgemeinen Theorie», 362.

[44] Nietzsche, Friedrich: «Jenseits von Gut und Böse». Berlin: Insel Verlag, 1984.

doppelter Kontingenz) beunruhigend oder sogar beängstigend sein kann und dies immer mit Emotionen und Affektzuständen verbunden ist, können sich bei den Mitgliedern gemischter Teams grosse individuelle Unterschiede im Umgang miteinander einstellen. Wie weit in gemischten Teams das Erleben und Antizipieren der emotionalen Zustände möglich sein wird und der Bedarf an psychologischer Sicherheit sich als Thema aller Beteiligten deklarieren wird lassen, kann man nicht voraussagen. In diesem Zusammenhang wird die Frage nach der Bedeutung des zuträglichen Umfelds für alle Beteiligten die Führungspersonen beschäftigen müssen. Spätestens mit dem Einzug humanoider Roboter, welche Emotionen antizipieren und auch zeigen können,[45] könnte sich im Zuge der Annäherung durch die Konvergenz (im Sinne der Verschmelzung von Mensch und KI) auch hier die Situation positiv entwickeln und die psychologische Sicherheit zwischen den sozialen Kooperationspartnern Mensch und KI an Bedeutung gewinnen.

5 Ausblick: Die Bedingungen der moralischen Orientierung – Ausrichtung der KI an sozialen Werten und Zielen?

„In an ethical orientation, you cope with different moralities“[46]

Als Ausblick werden die Bedingungen der moralischen Orientierung (Orientierung durch Selbstverpflichtung und Ausrichtung der KI an sozialen Werten und Zielen) skizziert. Zu allen hier entfalteten Aspekten der Orientierung an anderen, der Vertrauensbildung und der psychologischen Sicherheit muss jedoch die Frage gestellt werden, ob dies alles nicht eines *gemeinsamen Verständnisses und der Verbindlich*keit hinsichtlich Werten und Moralvorstellungen bedarf. Oder noch pointierter: Ist die Ausrichtung aller Beteiligten an den gleichen sozialen Werten und Zielen vielleicht sogar eine strikte Bedingung für das Gelingen der Kooperation? Sobald man die Vermutung anstellt, dass dies zutreffen könnte, muss man sich mit diesem Thema im Hinblick auf eine Kooperation in gemischten Teams in der digitalen Arbeitswelt ausführlich befassen.

Die Arbeit von *Belliger und Krieger*[47] befasst sich mit der komplexen Aufgabe der Ausrichtung der KI an sozialen Werten und Zielen, führt die Problematik aus

---

[45] Vgl. Wikipedia, «Sophia».

[46] Stegmaier und Mueller, «Successful Modes of Orientation», S. 4.

[47] Belliger, Andréa und David Krieger: «New Perspectives on AI Alignment.» Researchgate. 2023. https://www.researchgate.net/publication/375801677_New_Perspectives_on_AI_Alignment.

und weist darauf hin, dass es sich nicht um ein technisches, sondern um ein soziales Problem handelt. Sie definieren die Problematik auf drei Ebenen (Ausrichtung auf drei unterschiedliche Ziele): der technischen Sicherheit, der Prävention von Missbrauch und der sozialen Integration. Die drei Stufen sind mit zwei Grundannahmen unterlegt, wie KI wahrgenommen wird: einerseits als ein Instrument in den Händen der Menschen, welche sie gemäss ihren Interessen nutzen können, und anderseits als sozialer Partner.

Fazit

„To find your standpoint in philosophy, begin with the philosophy of orientation“[48]

In der sich disruptiv entwickelnden digitalen Welt ist die Konvergenz das entscheidende Thema unserer Zeit. Wie wollen wir die Verschmelzung des Menschen und der Technologie gestalten?

Dem widmet sich auch die Leitfrage dieser Arbeit, die Frage nach der Führung gemischter Teams konvergierender menschlicher Personen und lern- und entscheidungsfähiger Maschinen und nach der Gestaltung von deren Kommunikation und Kooperation.

Diese Leitfrage wurde mithilfe von philosophischen Ansätzen beleuchtet. Es konnte aufgezeigt werden, dass die Selbstorientierung die grundlegende Kompetenz in der Leadership ist. Orientierung ermöglicht, sich in neuen Situationen zurechtzufinden, die relevanten Anhaltspunkte zu finden und mit der Wahl des erfolgversprechenden Verhaltens die Situation zu bewältigen. Die Herausforderung der Orientierung unter besonderen Bedingungen der Unsicherheit und des Zeitdrucks wurden aufgezeigt. Dies bildet die Grundlage für eine weitere wichtige Orientierungskompetenz der Führung, nämlich anderen Orientierung zu bieten und sie zur eigenen Orientierung zu befähigen. Hierzu wurden philosophische Ansätze und Konzeptionen von *Heraklit, Kant, Luhmann, Nietzsche, Weber und Stegmaier* herangezogen und der praktische Nutzen der Philosophie für die Führung aufgezeigt. Die Bedingungen des Gelingens der Kooperation in Teams wurden aufgrund von Aspekten der Orientierung an anderer Orientierung, der Orientierung in der Kommunikation in doppelter Kontingenz, der Vertrauensbildung, der Vorbildrolle der Leadership und zuletzt der psychologischen Sicherheit herausgearbeitet, und die Möglichkeiten ihres Einsatzes zur Stärkung der Kooperation in gemischten Teams wurden diskutiert.

Hinsichtlich einiger Konstellationen konnten nur vorsichtige Annahmen getroffen werden, weil diese Situationen neu sind und die Orientierung in ihnen zunächst mit aufmerksamer Suche nach relevanten Anhaltspunkten erfolgen muss. Für viele

[48] Stegmaier und Mueller, «Successful Modes of Orientation», S. 4.

dieser Vorgänge werden Eigenschaften, Fähigkeiten und Werte vorausgesetzt, welche menschliche sind. Ob diese auch der KI in gleichem Masse zugesprochen werden können, ist unsicher. Das, was auf Menschen zutrifft, kann nicht automatisch auch der KI zugeschrieben werden, auch wenn sie von Menschen als sozialer Partner wahrgenommen wird.[49] Als Ausblick bleibt die offene Frage nach der Wahrnehmung der Beziehung zwischen KI und dem Menschen, sowie nach den Möglichkeiten einer Ausrichtung aller Akteure dieser Kooperation an gemeinsamen Werten. Die Frage der Orientierung ist und bleibt aktuell.

Literatur

Aristoteles: Nikomachische Ethik. Köln: Anaconda Verlag, 2019.

Altherr, Marcel: «Die Organisation der Selbstorganisation. Die Notwendigkeit neuer Organisationsformen in wissensintensiven Organisationen.» In Experten Führen – Modelle, Ideen und Praktiken für die Organisations- und Führungsentwicklung, hrsg. von Peter Kels und Stephanie Kaudela-Baum, 411–426. Wiesbaden: Springer Gabler, 2019.

Belliger, Andréa und David Krieger: «New Perspectives on AI Alignment.» Researchgate. 2023. https://www.researchgate.net/publication/375801677_New_Perspectives_on_AI_Alignment.

Benedikter, Roland: «Transhumanismus. Auf dem Weg zum neuen Menschen?» BIA. 2018. https://bia.unibz.it/esploro/outputs/991005772188101241.

Edmondson, Amy C.: "Psychological safety and learning behavior in work teams." Administrative Science Quarterly 44, Nr. 2 (Juni 1999): 350–383. https://doi.org/10.2307/2666999.

Edmondson, Amy C.: The Fearless Organisation. Creating Psychological Safety in the Workplace for Learning, Innovation, and Growth. Hoboken, NJ: Wiley, 2018.

Furtner, Marco: Empowering Leadership. Wiesbaden: Springer Fachmedien, 2017.

Gabler Wirtschaftslexikon, «Maschine» https://wirtschaftslexikon.gabler.de/ abgerufen am 27.Dezember 2023.

Gabler Wirtschaftslexikon «Technologie» https://wirtschaftslexikon.gabler.de/ abgerufen am 27.Dezember 2023.

Gabler Wirtschaftslexikon «Künstliche Intelligenz» https://wirtschaftslexikon.gabler.de/ abgerufen am 27.Dezember 2023.

Gabler Wirtschaftslexikon https://wirtschaftslexikon.gabler.de/definition/agilitaet abgerufen am 27.Dezember 2023.

Habermas, Jürgen: Theorie des kommunikativen Handelns. Berlin: Suhrkamp Verlag, 2022.

Hartmann, Martin: Die Praxis des Vertrauens. Berlin: Suhrkamp Verlag, 2011.

Kant, Immanuel: Gesammelte Schriften. Berlin: (Königlich-)Preußische Akademie der Wissenschaften, 1902.

---

[49] Vgl. 3.3.1. Kooperative Koexistenz, S. 16.

Kant, Immanuel: Kritik der reinen Vernunft. Hamburg: Felix Meiner Verlag, 1998.
Kant, Immanuel: Was heißt: Sich im Denken orientieren? Nordstedt: Books on Demand, 2015.
Klauß, Thomas und Annika Mierke: Szenarien einer digitalen Welt – heute und morgen. Wie die digitale Transformation unser Leben verändert. München: Hanser Verlag, 2017.
Luhmann, Niklas: Einführung in die Systemtheorie, 2. Aufl. München: Carl-Auer-Systeme Verlag, 1984.
Luhmann, Niklas: Soziale Systeme. Grundriß einer allgemeinen Theorie, Frankfurt am Main: Suhrkamp Verlag, 1984.
Luhmann, Niklas: Vertrauen. Ein Mechanismus der Reduktion sozialer Komplexität, 2. Aufl. Stuttgart: Enke, 1973.
Münnich, Sacha: «Luc Boltanski und Eve Chiapello: Der Neue Geist des Kapitalismus.» In Schlüsselwerke der Wirtschaftssoziologie. Wirtschaft und Gesellschaft, hrsg. von Klaus Kraemer und Florian Brugger, 385–392. Wiesbaden: Springer VS, 2017.
Nida Rümelin, Julian und Nathalie Weidenfeld: Digitaler Humanismus. München: Piper Verlag, 2018.
Nietzsche, Friedrich: «Über Wahrheit und Lüge im aussermoralischen Sinne.» In Sämtliche Werke. Kritische Studienausgabe in 15 Bänden, hrsg. von Giorgio Colli und Mazzino Montinari, 879–886. München/Berlin/New York: KSA, 1999.
Nietzsche, Friedrich: Jenseits von Gut und Böse. Berlin: Insel Verlag, 1984.
Nietzsche, Friedrich: Sämtliche Werke. Kritische Studienausgabe in 15 Bänden. München/Berlin/New York: De Gruyter, 1980.
Sprenger, Reinhard K.: Radikal führen. Frankfurt am Main: Campus Verlag, 2012.
Steckelberg, Alexander V. und Charly Harrer: Leadership und Management. Wiesbaden: Springer Gabler, 2021.
Stegmaier, Werner und Reinhard Mueller: Successful Modes of Orientation. Nashville, TN: The Hodges Foundation for Philosophical Orientation, 2019.
Stegmaier, Werner: Philosophie der Orientierung. Berlin/New York: De Gruyter, 2008.
Stegmaier, Werner: What is Orientation? A Philosophical Investigation. Berlin/Boston: De Gruyter, 2019.
U.S. Army Heritage and Education Center: "Who first originated the term VUCA (Volatility, Uncertainty, Complexity and Ambiguity)?" US Army. Aktualisiert am 06.12.2022. https://usawc.libanswers.com/faq/84869.
Weber, Max: Gesammelte Aufsätze zur Wissenschaftslehre. Tübingen: Verlag J.C.B. Mohr, 1985.
Weber, Max: Wirtschaft und Gesellschaft. Grundriß der verstehenden Soziologie, 5. Aufl. Tübingen: Verlag J.C.B. Mohr, 1972.
Wikipedia: «Sophia.» Wikipedia. Aktualisier am 27.11.2022. https://de.wikipedia.org/wiki/Sophia_(Roboter).

Beyond the Ivory Tower? The Practical Role of Ethicists in Business

Benjamin Lange

1 Introduction

> "Ethics" in business is in vogue. Following the recent boom of (generative) AI, a plethora of "AI Ethics", "Digital Ethics", "Corporate Digital Responsibility", "Responsible AI", and "Responsible Innovation" initiatives, movements, research streams, business functions, and B2B services have spawned and found their way into public discourse about responsible business practices in the 21st century beyond firms' traditional conceptions of "ethics" as solely a form of "compliance".[1]

But how, if at all, can ethics meaningfully contribute to addressing business challenges? Can resources from moral theory offer anything that other areas of

[1] Many leading financial services and strategy consulting firms provide "ethics consulting" services specifically with respect to digital transformation and the responsible procurement, development, application, and use of AI. See, for example, Deloitte (2021), PwC (2023), and BCG (2023). At the same time, technology companies have in-house commitments and efforts to ensure the responsible development of their technology, e.g. see Microsoft (2022), Google (2023), or SAP (2023).

I am grateful to Lara Maszynski for feedback and comments as well as audiences at the AI Ethics Research Colloquium at the Ludwig-Maximilians-Universität in Munich.

B. Lange (✉)
Ludwig-Maximilians-Universität München, Munich, Germany
e-mail: benjamin.lange@lmu.de

Munich Center for Machine Learning, München, Germany

C. H. Hoffmann (ed.), *Artificial Intelligence, Entrepreneurship and Risk*, Technikzukünfte, Wissenschaft und Gesellschaft / Futures of Technology, Science and Society, https://doi.org/10.1007/978-3-658-45544-6_8

expertise cannot? I here want to get clearer on this question, not at a general theoretical level—as in the research discipline of *business ethics*—but at a practical, hands-on level with my philosophical boots on the ground.[2]

My main argument in this chapter will be that ethicists can indeed provide something distinctively valuable: a holistic appreciation of salient normative factors across the legal, political, social, and moral dimensions of ethical business issues, based on which they can offer counsel and develop solutions. As a result, ethics consulting or advisory work (in-house or external) can provide distinctive value for organizations grappling with ethical issues beyond what other disciplines might have to offer. To my knowledge, few businesses have yet recognized this.

Specifically, I argue that resources from moral philosophy offer distinctively valuable insights for entrepreneurial business issues. These insights revolve around three ethical capabilities: ethical awareness, deliberation, and commitment. These capabilities yield various ethical benefits and help with mitigating ethical risks of corporate activity. My argument, while largely conceptual, also draws on anecdotal evidence and my own experience in ethics consulting.

The paper is structured as follows. Section 2 clarifies the scope of this discussion and makes some relevant distinctions. Section 3 argues against the view that there is no special ethics expertise that moral philosophy can bring. Section 4 then makes the positive case for my account: I introduce my view of how ethicists can be valuable for businesses and show how this is particularly the case in the technology sector. Section 5 then illustrates my proposal through a range of different examples and use cases, drawn from my experience as an expert practitioner. Section 6 concludes the paper.

2 Clarifications and Scope

A few clarifications before I proceed.

First, I focus on how resources from moral philosophy can be beneficial in entrepreneurial and business settings. I do not intend to argue that every moral philosopher will be an effective ethics practitioner. A range of skills are required

[2] For an accessible overview of the debate in business ethics, see Heath (2014), especially Chaps. 1 and 3. Some examples of scholarship that are similar in spirit to my project include Harris et al (2009), Brent and Blackman (2022). For other discussions of the role of ethicists in business, see Bishop (2013), Seele (2018), Klikauer (2017), Brenkert (2016).

to translate ethical insights appropriately into practical business settings and not every (and perhaps only few) moral philosopher will possess these skills.[3]

Second, a few remarks on my terminology. My topic is the value that concepts and resources from moral philosophy (or "ethics") can bring to entrepreneurial or organizational business settings.[4] My departure point is therefore academic moral philosophy, even though the term "ethicist" is increasingly used in business settings to refer to a wide range of disciplines that aim to solve ethical problems and challenges within business settings.[5]

Third, the successful application of resources and concepts from moral philosophy is what I shall refer to as "ethics consulting" or "ethics advising". By "successful" I am referring to the creation of some positive value, broadly conceived, such as the resolution of some important ethical challenge within an organization, providing helpful insight into some complex ethical issue, anticipating ethical risks, or developing hands on tools for making sound ethical decisions based on a company's principles or values.[6] Such challenges include issues of strategic or operational importance, including questions about what technology to develop or sell; others apply internally, and may concern how to make important decisions in line with ethical commitments, or how to ensure a proper culture of mutual respect within a company. I am therefore not concerned with what is sometimes referred to as "ethics washing" or "ethics shopping", which creates the mere appearance of appropriate ethical conduct of a company; my focus is on building genuinely valuable ethical capabilities of firms.[7]

---

[3] What makes for an effective ethics consultant? Though this is a different question meriting its independent research, Wickham and Wilcock (2016) provide a helpful discussion of effective management consulting in general, according to which it requires important analytical, project management, and relationship building skills. Ethicists usually excel in the first, analytical skill and in expertise in the subject matter of ethics, while most likely needing to build skills in project management, client communication, adherence to? general business norms, and the building and maintaining of valuable client-advisor relationships.

[4] See Timmons (2012) and Driver (2022) on the subject matter of normative ethics. As Driver (2022) notes, moral philosophers are broadly engaged in "attempting to provide a systematic account of morality."

[5] Many ethicists in technology companies come from a variety of backgrounds, including sociology, anthropology, law, or policy.

[6] See also Wickham and Wilcock (2016). Some of the indicators of consulting success, including client satisfaction, Return on Investment (ROI), adoption rates, impact on business objectives, among others, also apply to ethics consulting.

[7] See Floridi (2019) on various forms of disingenuous organizational ethics. He distinguishes between (1) ethics shopping; (2) ethics bluewashing; (3) ethics lobbying; (4) ethics dumping; and (5) ethics shirking.

Fourth, my aim is not to suggest that ethicists provide a silver bullet for navigating ethical business issues. Navigating complex ethical ambiguities and challenges requires diverse, multidisciplinary expertise: legal, sociological, political, and anthropological, among other fields.[8] Rather, my objective is to articulate moral philosophy's role as a valuable and distinctive component within this picture, since this role, to my knowledge, has not yet been fully explored.

Fifth, I do not intend to reinvent the wheel. The Ethical conduct of organizations has long been recognized as an important component of successful business operations, specifically as part of successful enterprise risk management. Risks can be understood as "unwanted events that may occur", that are uncertain, and that include strategic, operational, compliance and regulatory, financial, reputational, conduct-related, and societal risks.[9] My aim is not to make a competing case against effective risk management and professions linked to it, such as ethics and compliance practitioners, that engage in managing these organizational risks. Instead, I want to explore the domain beyond risk management and compliance, and argue that in this non-exhausted domain, ethical expertise can be valuable beyond codifying ethics as a matter of compliance.

Lastly, I assume that businesses face genuine ethical dilemmas, as opposed to, for instance, mere ambiguities in the application of the law, and can have moral duties *qua* entity. I do not endorse any specific Corporate Social Responsibility (CSR) approach here of what firms' moral duties might look like,[10] or any particular stance on business ethics which aims at arguing for and specifying the moral obligations of firms. I presuppose that firms do have such moral duties and do face genuine ethical challenges and explore how these duties can be discharged and challenges resolved using the resources of moral philosophy.

3 Three Arguments Against Ethics Consulting

I will now evaluate three arguments put forth against one prominent role that ethicists might play in businesses: namely as ethics consultants. These arguments challenge the idea of successful ethics consulting since ethics consulting could

[8] See Roche et al (2023) on the importance of diversity in ethical technology development.

[9] See Hansson (2023) for an introduction on the notion of "risk".

[10] Most accounts of CSR broadly define it as dealing with the obligations that firms have towards society. For example, Scherer and Palazzo (2007), p. 1096 define CSR as dealing, simply, with the "social responsibility of business firms". See Windsor (2006) for a general taxonomy of different accounts of CSR.

undermine autonomy, lack scientific rigour, and lack robustness, potentially leading to ethics washing. I argue that all three fall short of their mark. Discussing these arguments will help me clarify the role for ethics advising in businesses and distinguish my own approach on how ethicists should support businesses.

First, ethics consulting may appear to undermine the autonomy of those seeking ethics advice.[11] This concern arises from a belief that ethics consultants, as experts, could impose their own moral judgments and decisions on individuals facing ethical dilemmas, thereby diminishing these individuals' autonomy in making their own moral choices. For instance, the involvement of an external expert might override or excessively influence the personal values and decisions of those receiving the advice, potentially leading to outcomes that reflect the consultant's views rather than the individual's own moral compass.

This objection relies on a mistaken assumption about the role of ethics consulting. Ethics consultants do not assume that their advice has sufficient authority regarding any moral issue to determine a course of action on behalf of a consultee. The concept of "autonomy" can generally be characterized as an agent's ability "to make decisions and act according to their own values and desires, free from coercion, manipulation, heavy pressure, or control by others". Autonomy is accordingly compromised if an agent's decisions are significantly influenced by external factors, such as pressure, threats, or other forms of coercion. This is not the case in ethics consulting where individuals are merely assisted in clarifying, systematizing, and extending their own moral awareness and decision-making capabilities. Done correctly, it therefore enhances the consultee's autonomy instead of curtailing it.[12]

Second, some argue that the idea of providing counsel on ethical matters is conceptually confused.[13]This argument relies on an alleged disparity between science and ethics. It maintains that while science deals with factual matters that allow for objective knowledge, ethics concerns values that depend on individual desires and attitudes, and therefore one cannot achieve objective knowledge in ethics. Given this difference, it is in principle impossible to have expertise about what is morally right and wrong, as ethical knowledge lacks the objectivity and factual basis of scientific knowledge.

There are several confusions in this argument. First, it confuses meta-ethical questions about the ontology and epistemology of moral claims and judgments

---

[11] See Varelius (2008a).

[12] See Varelius (2008b) who argues along similar lines against the autonomy objection.

[13] See Shalit (1997) and Cowley (2005) for objections along these lines.

with first-order normative questions about the right course of action. Ethics consulting addresses a need for first-order guidance without imposing any allegedly objective moral knowledge on the consultee. It arises simply because the consultee must confront an ethical scenario that requires a decision. In such a case, an ethics consultant can bring value by analysing the salient ethical considerations and the relevant arguments and reasons, assessing their merits, and making explicit the consultee's commitments in order to help him or her with reaching a decision. Second, ethics consultants, like experts in other fields, possess specialized knowledge and skills that enable them to clarify, systematize, and extend moral thinking effectively, much as legal counsellors can interpret complex laws in a way that non-legal experts cannot in order to help their clients navigate complicated legal decisions. In similar fashion, ethics experts can provide expertise on moral matters.

Third, Michalos (2001) argues that the significant diversity of opinion on many fundamental ethical issues, makes it likely that most ethical appraisals, particularly those made in ethics review and committee settings, are superficial and weakly justified. This is because, by virtue of these profound disagreements, consensus on more specific stances cannot be reached. By analogy with Gresham's Law in economics, where bad money drives out good money, Michalos suggests that in ethical deliberations, the abundance of diverse and conflicting opinions could lead to increasingly superficial ethical appraisals which drive out higher quality advice.

This argument appears to take issue with ethics consulting when done improperly, as opposed to ethics consulting in general. Michalos rightly points out that shallow ethics consulting may have various downsides and ultimately be ineffectual. The issue indeed seems to be exacerbated in the case of ethics as opposed to other forms of management consulting, since these forms will have less issues reaching agreement on advice than ethics consultants will. For example, there might be more consensus on best practices to increase a firm's profitability or energy efficiency than on the ways a firms should discharge its moral duties However, the fact that the fact that proper ethical consulting requires a certain depth and is more complex by dealing with a complex and contentious subject matter with lots of expert disagreement, is not an argument against the plausibility of ethics consulting per se.

4 The Need for Ethicists in Business

In the last section, I have considered and rejected three objections to one possible role ethicists might play in businesses: the role of ethics consultants. In this section, I offer a new account of how ethicists can provide practical value for firms.[14] I explain how ethical issues arise in business operations through *grey areas* which are particularly pervasive in the technology space, and then argue that the resources of moral philosophy can be valuable in navigating these grey areas to address ethical challenges.

4.1 Grey Areas

Ethics is integral to business, influencing corporate reputation, customer trust, and employee morale. Upholding ethical standards enhances legal compliance, mitigates risks, and fosters positive relationships with stakeholders, including customers, employees, and investors. Ethical behavior contributes to a sustainable and socially responsible business, promoting long-term success. In a global context, ethical considerations extend to international business practices, helping companies navigate diverse markets and cultural differences.

Typically, a business' enterprise risk management functions are concerned with ensuring the firm's adherence to existing legislation.[15] The purpose of the enterprise risk management (ERM) function is to provide organizations with a comprehensive and strategic approach to identifying, assessing, and managing risks that could impact the achievement of their goals. ERM promotes a holistic view of risks, aligning risk management practices with strategic objectives to enhance decision-making. And since the domain of what is required by the law often converges with what is morally required, businesses have long had to ensure ethical conduct insofar as it is required by the law as part of conducting their business.

But "compliance" does not equal "ethics". The domain of compliance does not exhaust the domain of organizations' ethical conduct. Even when laws specify parameters of conduct for organizations, from laws ensuring workers' rights to privacy laws like the GDPR, or legislation regulating the provision of services,

[14] See Sandberg (2012) for an overview of the role of ethics in firms more generally.

[15] See Merna and Al-Thani (2011) for a comprehensive discussion of enterprise risk management. Since risk management is concerned with identifying and responding to risk broadly conceived, it has clear connections with ethical issues that may arise in an organization.

there are always distinctive grey areas. "Grey areas" are domains of possible business conduct where, for various reasons, clear rules, laws, and principles do not straightforwardly apply, so the law does not clearly mandate any particular course of action as the "right" one.

This may mean that various options will be permissible, with varying degrees of moral justification. Even when a law firm *could* legally do business with an organization in a dictatorial country, it may have to decide whether this would be the right thing to do, given its prior ethical commitments, values, and purpose. This is an instance of a grey area where ethical questions in business come to the fore.

Grey areas can come in two main forms. They can result from the absence of appropriate regulation. For example, by the time the EU's AI Act comes into effect, at least two years will have passed since this article was written; meanwhile algorithmic systems abound in all kinds of real-world applications and business settings, including generative AI. The absence of any applicable regulation in the EU means that many current issues involving the development, application, and implementation of AI in the EU are grey-area issues.

In other instances, when regulation exists, its interpretation can involve ambiguity. For example, the EU's AI Act is ambiguous about the risk classification of algorithmic systems, posing challenges for anyone trying to meet the law's requirements.[16] This may mean that businesses need to engage in additional ethical analyses in order to understand their actual requirements. Another example concerns employee conduct and risk management: while there is regulation about respecting workers' rights, there is considerable moral leeway in interpreting how this might be done.

The existence of grey areas is an integral part of the business landscape. As Joseph Heath (2014) argues as part of his "market failures" approach, the absence of regulation due to feasibility constraints of the state requires organizations to independently ensure their ethical conduct with the ultimate aim of guaranteeing a well-functioning market economy.[17] Because it is practically infeasible to regulate everything, we need moral principles, values, and rules of conduct in businesses to ensure a properly functioning market economy.

[16] EU (2021).

[17] See Heath (2014).

4.2 Ethics in Technology

Of the various industries that exist, one of the most pressing needs for ethics exists in the technology space. This is because it contains the largest number of grey areas. There are three reasons for this.

First, technology is not "value neutral", but value-laden.[18] New technologies are a product of their designers' beliefs, convictions, and values. They are not neutral constructs that exist in a moral vacuum, so technological entrepreneurship is ethically salient by its very nature.

Second, technological advancement at the frontier of human knowledge requires even more sustained and deep reflection to decide how to apply ethical standards.[19] Pioneering new technologies require more serious ethical reflection to determine the right course of action, since they enter unchartered territory for both application and legal governance, and require an appreciation of the impact on society and the risks involved.

Third, because the pace of technological innovation exceeds that of regulation, existing laws and policies do not necessarily provide guidance to technology companies about how to align their work with the needs and interests of society—creating what James Moor (1985) has referred to as a "policy vacuum". Ethical grey areas are therefore most pronounced in this domain.[20]

The existence of ethical grey areas in business operations, particularly in the technology sector, means that established legal rules will not provide necessary guidance on a whole set of ethical decisions of firms. Ethics consulting can add a distinctive value here because it can draw attention to a broad range of relevant moral considerations and factors *simpliciter* as they pertain to these decisions. This holistic appreciation of moral considerations as they arise in the ethical choices of firms is the main reason how ethicists can add something distinctive to business which professionals from other disciplines cannot.[21]

---

[18] Martin and Freeman, 2004, p. 356.

[19] McVea (2009) qualitatively explores the decision making of entrepreneurs confronted with an 'ethically pioneering' situation, and shows that entrepreneurs reason through the attendant moral challenges in an imaginative way.

[20] See James Moor (1985).

[21] See also Macklin and Mathison (2018) who suggest that ethicists should provide 'dialogic teaching' for companies and managers to promote moral imagination and independent critical reflection. See also Gorman (2009) and Lange et al (2023).

5 How Ethicists Can Help

The previous section argued that there is a need for ethicists to help firms tackle the ethical challenges arising in grey areas, which are particularly acute in the technology sector. In this section, I outline an "ethical capabilities" framework to capture the ways in which ethicists can support firms in making ethical choices.

5.1 Ethical Capabilities Framework

At a general level, ethicists help businesses in different ways. They give ethical advice, help develop codes of ethics, and train employees. They review and implement policies to make sure they are ethically robust and help assess the overall moral character of an organization. They also help businesses talk to stakeholders, build trust, and be socially responsible, including understanding a business broader social and environmental responsibilities.

I want to make this a bit more precise and suggest that there a three key ethical capabilities and functions of businesses (or organizations) that ethicists can strengthen. In political philosophy, "capability" refers to an individual's capacity or potential to perform a particular action or to exhibit a certain skill or behaviour. It encompasses the inherent ability to accomplish tasks or engage in activities, whether realized or untapped.[22] A "functioning" is a capability that has been realized. I restrict the definition here to "ethical capabilities", which refer to one's capacity to act ethically, broadly conceived. On this view, I suggest, that ethicists help organizations develop ethical capabilities in order to achieve ethical functionings (states in which ethical conduct, processes, outcomes have been realized).[23] In the domain of practical organizational ethics, ethicists thereby help organizations build the potential for being ethical and then help them realize this potential through discharing their intra-firm moral responsibilities.[24] What motivates the focus on these capabilities and functionings is a conception of

[22] Locus classicus is Sen's Capabilities Approach. See, for instance, his (1992).

[23] For brevity, I shall henceforth drop the additional "functionings" label. I do intend to argue that ethicists do not only build certain skills and resources but also help organizations realize them in practice.

[24] Another key holistic construct to capture organization's ethicality, is typically that of "ethical culture" understood as the totality of all the formal and informal, tangible and intangible elements of ethics in an organization (see Kaptein, 2008). I see my method here as consistent with this construct, i.e. an organization's ethical culture might be understood as consisting, among other things, of these three key capabilities.

organizations as moral agents with the ability to act ethically, based on informed moral judgements, through awareness, reasoning, and intention. Such agents can also be held accountable for their actions.[25,26]

The framework includes three ethical capabilities:

(1) **Ethical Awareness:** Ability to recognize normatively significant factors and implications (e.g. moral values, virtues, norms, behaviours, ethical risks of harm, constraints, and rights violations) in situations, decisions, relevant business choice scenarios, and intra-organizational contexts.

Expanding an organization's perceptual horizon is a precondition for robust ethical deliberation, decision-making, and commitment. This involves developing an understanding of the various contexts within which ethical choices and scenarios can arise, including intra-organizational contexts, as the capability concerns existing behavioural norms, processes, policies, and structures.

(2) **Ethical Deliberation and Decision-Making:** Ability to engage in reasoning and deliberation in relevant choice scenarios, including reasoning about tensions between values and other moral commitments, conflicts, moral dilemmas, and other trade-offs.

This capability concerns the conceptual tools which allow an organization to understand, deliberate about, and make choices within ethically laden situations in which competing normative considerations come into conflict. More broadly, this may encompass employees' ability to deliberate and reason through day-to-day ethical choices in situations involving the trade-offs between competing organizational commitments. For example, this may be relevant when deciding which course of action is best supported by an organization's values and principles in cases of product development and design, procurement, use policies, or strategic decision-making.

(3) **Ethical Commitment:** Ability to derive and set concrete plans to guide further product development or research.

---

[25] See Lange et al. (2023) on how this capability framework underlies a culture-change methodology in the facilitation of responsible innovation in engineering culture.

[26] See also the literature on ethical culture change and the ethicality of organizations, for instance, Kaptein (2008).

Increased ethical awareness and decision-making capacity enable organizations to navigate complex ethical challenges as part of their work. But building these capabilities will miss its mark without commitment to and accountability for translating these insights into practical change. What it means to operationalize ethical commitment will often vary depending on organizational structures, but may include establishing more formal review mechanisms for products or policies.

5.2 Examples in Practice

The previous subsection introduced the ethical capabilities framework to make sense of the various ways in which an organization's ethicality may be conceptualized. Ethicists, I suggest, can help organizations by strengthening all three of those capabilities.

Ethics consultants with a background in moral philosophy offer value by being better sensitized to the normative reality that businesses face. Thinking extensively about issues in moral philosophy can help one develop particularly valuable skills in the ethical analysis of issues, penetrating a given moral choice scenario at a high level of rigour, and developing robust solutions to it that go beyond what other disciplines can offer.

But what does that mean in practice? What are some actual ways this expertise can help organizations build their moral capabilities?

Ethical Capabilities in Organizational Ethics			
Capabilities	Services		
Ethical Awareness	• Ethics and Risk Assessments and Assurance: – Culture Assessment – Risk Assessments	Ethical Commitment	• Ability of organization to translate awareness and deliberation capabilities into practice
Ethical Deliberation	• Values, Principles, Mission Operationalization • Ethical Decision Making Frameworks and Methodologies – Client Acquisition – Product Use/Policy Use – Communication Use		

Here are some suggestions, partly conceptual and partly based on my experience:

- **Ethics and Risk Assessment and Assurance:** Ethicists can provide rigorous, robust and independent ethical assessments and assurance work to determine the strength of ethical capabilities across business operations.
  - **Ethical Culture Assessments:** Ethical culture assessments holistically cover drivers of ethicality in organizations, including speak-up culture, accountability, ethical leadership and role-modelling, ethical decision-making, and alignment of incentives with ethical commitments.
  - **Risk Assessment:** Especially for the procurement, application, or development of new technology, such as AI, ethicists can identify relevant risks of harm, including legal, reputational, operational, or safety-related harms, at both technical and socio-technical levels.
  - **Why can ethicists be particularly good at the above?** Ethical analysis focuses on normative considerations and factors. Ethicists, being deeply familiar with moral theory and relevant ethical considerations, can therefore be particularly helpful in uncovering ethical risks of harm or drivers of unethical conduct.
- **Values, Principles, Mission Operationalization:** Almost all organizations have a purpose and values to justify their strategic objective and reason for existence, and to guide the achievement of these objectives in both internal and external conduct. Values, principles, and the notion of purpose are all value-laden and normative in character. The ways in which they guide action can benefit from ethical analysis.
  - Ethicists can provide a more accurate and rigorous analysis of these concepts, helping to define them in a manner that is conceptually robust and distinct from other organizational values—something that can be notoriously difficult with "thick" normative concepts, such as "respect" or "care". Here ethicists can provide value by helping to create, define, revise, and translate values into practice, determining more clearly how certain behaviours at different levels of an organization support or run contrary to an organizational commitment so that they can be successfully used for decision-making.
- **Ethical Decision-Making Frameworks and Methodologies:** Business may face a range of ethical challenges that affect their operations. For instance, these may concern whether and under what circumstances a business should work with certain customers or specify more concrete policies of use for

their products and services. In all of these instances, certain ethical commitments, principles, or red lines will have to be factored into a decision. Often a key input into these decisions concerns an organization's values or principles, which will then have to be weighed systematically.

- Ethicists can help develop these frameworks better than other experts by virtue of being deeply familiar with sound processes of ethical deliberation, with weighing different moral considerations, arguments, and principles, and with a more nuanced understanding of the relevant moral vocabulary. For example, as opposed to endorsing a simple, binary conception of ethical deliberation, they can draw attention to the gradability of moral concepts, *pro-tanto* and all-things-considered considerations, and side constraints, and embed those within the decision-making processes.

  Client Acquisition: Some organizations need to decide under which conditions they want to work with a particular client.

  Product Use/Policy Use. How can a product be deployed in line with ethical guardrails or ethical commitments?

- **Communication Use:** When should we speak about or take action on an international issue of moral importance? When should we condemn or support a cause?
- **Responsible Innovation Practices and Policies:** Since the innovation sector, predominantly the AI sector, contains most areas of concern, there is a particularly strong need to understand the ethical technology at the forefront of human innovation. This may happen through forecasting and horizon scanning exercises or developing concrete policies, establishing training curricula and workshops.

  Here ethicists can play a vital role, not only through their analysis of ethical issues but by helping organizations build structures and by training people to become better at innovating responsibly and in line with an organization's commitments.
- **Ethics Committees and Review Boards:** In reviewing medical, pharmaceutical, or technological issues ethicists' sensitivity to relevant ethical considerations can provide distinctive value through highlighting the ethical factors involved in a product or research, which can then be used for organizational decision-making, for instance, decisions about pursuing further research or releasing a product.
- **Training and Education:** Sometimes organizations have a specific knowledge gap or need to improve in certain ethical domains, including learning about the ethical implications of an issue (e.g. privacy or AI) or interpersonal conduct and leadership, including speak-up or ethical analysis of business decisions.

6 Conclusion

This paper has provided a practical account of how ethicists con contribute to business. I laid out the three-tiered ethical capabilities approach and explained how ethicists can practically strengthen these capabilities.

References

BCG. 'Responsible AI'. Boston, MA: Boston Consulting Group, last accessed 18 Jan, https://www.bcg.com/capabilities/artificial-intelligence/responsible-ai.

———. Bishop, W.H. 'The Role of Ethics in 21st Century Organizations'. *Journal of Business Ethics* 118, no. 3 (2013): 635–37. https://doi.org/10.1007/s10551-013-1618-1.

Brenkert, G.G. 'Business Ethics and Human Rights: An Overview'. *Business and Human Rights Journal* 1, no. 2 (2016): 277–306. https://doi.org/10.1017/bhj.2016.1.

Brent, Michael, and Reid Blackman. 'Ethics Consulting in Industry'. In *A Companion to Public Philosophy*, edited by Lee McIntyre, Nancy McHugh, and Ian Olasov, 381–87. Wiley, 2022.

Cowley, C. (2005). A new rejection of moral expertise. *Medicine, Health Care and Philosophy, 8*, 273–279.

De-Shalit, A. (1997). Environmentalism for Europe—One Model?. *Journal of applied philosophy, 14*(2), 177–186.

Deloitte. Ammanath, B. (author). 'Does Your Company Need a Chief AI Ethics Officer, an AI Ethicist, AI Ethics Coucil, or All Three? Positioning Your Organization for Success on AI Ethics'. Nashville: Deloitte Development LLC, 2021.

Driver, J. 'Moral Theory'. In *The Stanford Encyclopedia of Philosophy, Fall 2022 Edition*, edited by E.N. Zalta and U. Nodelman. https://plato.stanford.edu/archives/fall2022/entries/moral-theory/.

European Commission. 2021. Proposal for a Regulation of the European Parliament and the Council Laying Down Harmonised Rules on Artificial Intelligence (Artificial Intelligence Act) and Amending Certain Union Legislative Acts.

Floridi, L. 'Translating Principles into Practices of Digital Ethics: Five Risks of Being Unethical'. *Philosophy & Technology* 32, no. 2 (2019): 185–193. https://doi.org/10.1007/s13347-019-00354-x.

Google. 'Responsible AI Practices'. California: Google LLC, last accessed 18 Jan, https://ai.google/responsibility/responsible-ai-practices/.

Gorman, M.E., P.H. Werhane, and N. Swami. 'Moral Imagination, Trading Zones, and the Role of the Ethicist in Nanotechnology'. *NanoEthics* 3, no. 3 (2009): 185–95. https://doi.org/10.1007/s11569-009-0069-8.

Hansson, S.O. 'Risk'. In *The Stanford Encyclopedia of Philosophy, Summer 2023 Edition*, edited by E.N. Zalta and U. Nodelman. https://plato.stanford.edu/archives/sum2023/entries/risk/.

Harris, J.D., H.J. Sapienza, and N.E. Bowie. 'Ethics and Entrepreneurship'. *Journal of Business Venturing* 24, no. 5 (2009): 407–18. https://doi.org/10.1016/j.jbusvent.2009.06.001.

Heath, J. *Morality, Competition, and the Firm: The Market Failures Approach to Business Ethics*. Oxford: Oxford University Press, 2014.

Kaptein, M. 'Developing and Testing a Measure for the Ethical Culture of Organisations: The Corporate Ethical Virtues Model'. *Journal of Organizational Behavior* 29, no. 7 (2008): 923–47. https://doi.org/10.1002/job.520.

Klikauer, T. 'Business Ethics as Ideology?'. *Critique (United Kingdom)* 45, no. 1–2 (2017): 81–100. https://doi.org/10.1080/03017605.2016.1268757.

SAP. Lombana Diaz, D., S. Welsch, and S. Wieczorek (authors). 'SAP AI Ethics Handbook: Handbook for Applying SAP's Global AI Ethics Policy Across the AI Factory Process'. Walldorf, Baden-Württemberg: SAP SE, 2023.

Microsoft. 'Microsoft Responsible AI Standard v2: General Requirements'. Redmond, WA: Microsoft Corporation, 2022.

Lange, B., G. Keeling, A. McCroskery, B. Zevenbergen, S. Blascovich, K. Pedersen, A. Lentz, and B. Agüera y Arcas. 'Engaging Engineering Teams Through Moral Imagination: A Bottom-Up Approach for Responsible Innovation and Ethical Culture Change in Technology Companies'. *AI and Ethics* (2023). https://doi.org/10.1007/s43681-023-00381-7.

Macklin, R., and K. Mathison. 'Embedding Ethics: Dialogic Partnerships and Communitarian Business Ethics'. *Journal of Business Ethics* 153, no. 1 (2018): 133–45. https://doi.org/10.1007/s10551-016-3431-0.

Martin, K.E., and R.E. Freeman. 'The Separation of Technology and Ethics in Business Ethics'. *Journal of Business Ethics* 53 (2004): 353–64. https://doi.org/10.1023/B:BUSI.0000043492.42150.b6.

McVea, J. F. (2009). A field study of entrepreneurial decision-making and moral imagination. *Journal of Business Venturing, 24*(5), 491–504.

Merna, T., and F.F. Al-Thani. *Corporate Risk Management, 2nd Ed*. Wiley, 2011.

Michalos, A.C. 'Ethics Counselors as a New Priesthood'. *Journal of Business Ethics* 29, no. 1–2 (2001): 3–17. https://doi.org/10.1023/A:1006434623769.

Moor, J.H. 'What Is Computer Ethics'. *Metaphilosophy* 16, no. 4 (1985): 266–72. https://doi.org/10.1111/j.1467-9973.1985.tb00173.x.

PwC. 'Digital Ethics & Responsibility: Embedding Responsible Digital Transformation in Your Company for the Long Term'. Frankfurt a.M.: PricewaterhouseCoopers GmbH Wirtschaftsprüfungsgesellschaft, 2023.

Roche, C., P.J. Wall, and D. Lewis. 'Ethics and Diversity in Artificial Intelligence Policies, Strategies and Initiatives'. *AI Ethics* 3, no. 4 (2023): 1095–1115. https://doi.org/10.1007/s43681-022-00218-9.

Sandberg, Joakim. 'Ethics in Corporations'. In *Encyclopedia of Applied Ethics, 2nd Ed.*, edited by Ruth Chadwick. Academic Press, 2012.

Scherer, A. G., and G. Palazzo. 'Toward a Political Conception of Corporate Responsibility: Business and Society Seen From a Habermasian Perspective'. *The Academy of Management Review* 32, no. 4 (2007): 1096–1120. https://doi.org/10.5465/AMR.2007.26585837.

Seele, P. 'What Makes a Business Ethicist? A Reflection on the Transition from Applied Philosophy to Critical Thinking'. *Journal of Business Ethics* 150, no. 3 (2018): 647–56. https://doi.org/10.1007/s10551-016-3177-8.

Sen, A. *Inequality Re-Examined*. Oxford: Clarendon Press, 1992.

Timmons, M. *Moral Theory: An Introduction.* Lanham, Md.: Rowman & Littlefield Publishers, 2012.

Varelius, J. 'Ethics Consultation and Autonomy'. *Science and Engineering Ethics* 14, no. 1 (2008a): 65–76. https://doi.org/10.1007/s11948-007-9033-6.

———. 'Is Ethical Expertise Possible?' *Medicine, Health Care and Philosophy* 11, no. 2 (2008b): 127–32. https://doi.org/10.1007/s11019-007-9089-8.

Wickham, L., and J. Wilcock. *Management Consulting: Delivering an Effective Project, 5th Ed.* Harlow: Pearson, 2016.

Windsor, D. 'Corporate Social Responsibility: Three Key Approaches'. *Journal of Management Studies* 43, no. 1 (2006): 93–114. https://doi.org/10.1111/j.1467-6486.2006.00584.x.

Part IV

Contributions of Academic Philosophy and Ethics to the so-called new Technologies and AI Concepts

Technology Assessment Contributing to AI Management in Society

Armin Grunwald

1 Introduction and Overview

Modern societies have been experiencing fast change according to science-based and technological innovation. This phenomenon which well describes what many people perceive is not quite new as the following quote demonstrates:

> Domestic life, political institutions, international relations and personal contacts are shifting with kaleidoscopic rapidity before our eyes. We cannot appreciate and weigh the changes; they occur too swiftly. We do not have time to take them in. No sooner we begin to understand the meaning of one such change than another comes and displaces the former. (Dewey, 1931: 53)

Since decades, this almost hundred years old attitude to life has been fueled by ongoing and accelerating digitalization. The recent wave of AI, the emergence of e.g. chatbots, service robots, automated decision-making systems (ADM) and generative AI technologies such as ChatGPT in many areas, has motivated public, ethical, and regulatory debates expressing fascination as well as deep concerns. Recently, the European Commission approved the first law on AI worldwide, aiming at opening up spaces for exploiting its benefits while simultaneously taking care of preventing possible negative consequences with respect to, e.g., civil rights, manipulation, and misuse.

A. Grunwald (✉)
Karlsruhe Institute of Technology, Karlsruhe, Germany
e-mail: armin.grunwald@kit.edu

C. H. Hoffmann (ed.), *Artificial Intelligence, Entrepreneurship and Risk*, Technikzukünfte, Wissenschaft und Gesellschaft / Futures of Technology, Science and Society, https://doi.org/10.1007/978-3-658-45544-6_9

This event and many similar debates worldwide, including postulates by big players in the AI field for a worldwide moratorium on further AI development, demonstrate that prudently managing AI is crucial for a good future. Successfully embedding new technology in society often is a challenge for innovation, business models, value-added chains, regulation, behavior, culture, and so forth. Such processes always take time and need to pursue several steps. However, while facing and experiencing rapid digitalization, hopes and concerns about possibly disruptive effects repeatedly have become the subject of scientific and public controversy (Sect. 4).

Digitalization including making use of AI-based technologies and services is a societal process made possible by the fast progress in the field of digital technologies. This process refers to the digitally enabled transformation of social contexts, for example of public communication and democracy, of the transformation of the economy to a data- and knowledge-based economy, of new and emerging human–machine relationships, of changes in the world of labor, and of challenges to traditional human self-perceptions affected by digital technologies and AI.

Therefore, managing AI in society is a complex process. It includes, amongst many other topics, issues of control, of transparency and traceability, of responsibility and liability, of trustworthiness, on geopolitical competition, on safety and security of critical infrastructures, of AI promotion and regulation, of education, of preventing adverse developments such as misuse and manipulation, of maintaining human freedom, and so on. In this chapter, I will consider opportunities to support managing AI in society offered by the advice-oriented research field of technology assessment (TA). One of its most influential writings was titled 'Managing Technology in Society' (Rip et al., 1995), announcing the target of TA and pointing to its mission: to contribute to inclusive, prospective, and responsible embedding of new technology in society.

In this chapter, first, a short introduction into TA reasoning and practice will be given (Sect. 2). Two manifestations of how TA contributes to managing AI in society will be presented in the remainder: TA for shaping AI-based technologies systems (Sect. 3) and TA for monitoring societal vulnerability by possible gradual disruptions while embedding AI technologies in almost all areas of the individual and collective life (Sect. 4).

2 Short Overview of Technology Assessment

The scientific and technological advance has for centuries been a major driving force of modernization. However, the twentieth century was full of indications and observations of a deep crisis of traditional modernity which was based on belief in the scientific and technological progress and on capitalist, competition-driven economy. Debates on limits to growth, pollution, loss of biodiversity, and climate change indicate a serious and threatening lack of sustainability and raise severe questions how to proceed to the future. Technology Assessment (TA) is, like neighbored areas such as Responsible Research and Innovation (RRI) and Sustainability Research, a response to related needs for orientation by problem-oriented and transformative research (Grunwald, 2021).

2.1 Background and Motivation of TA

Since the Industrial Revolution, and to a strongly increased extent since the 1960s, adverse effects of scientific and technical innovations became obvious. Accidents in technical facilities, threats to the natural environment like climate change, negative health effects as in the asbestos case and labour market problems due to automation are well known examples from the previous century. Scenarios of 'Limits to Growth' (Meadows et al., 1972) addressed for the first time the grave resource problems related to economic growth. Deepened awareness of the ambivalence of technology and innovation led to a crisis of the modernist belief in progress. This observation was the most essential motivation for the emergence of TA. The huge impact of today's technology in the *Anthropocene* (Crutzen & Stoermer, 2000), e.g. by high energy demand, digitalization and mobility, and the spreading of many of its consequences across space and time needs prospective approaches for providing orientation (Grunwald, 2024). The most pressing questions fuelling TA, therefore, are (according to Grunwald, 2019a: 13ff): How can a society, which places its hopes and trust in innovation and progress, and must continue to do so in the future, protect itself from undesirable side effects? How can humankind benefit from the technological advance in a responsible and sustainable manner? How can society preventatively act to cope with possible future adverse effects? In order to enable dealing responsibly with unintended side effects and the scarcity of resources, or to minimize or prevent them, regulatory measures (like the Precautionary Principle of the European Union) and scientific approaches (like TA and sustainability research) were developed. Their overall and common mission is to extend foresight and reflection capabilities in the huge field of the scientific and technological advance and for making best use of its outcomes (Grunwald, 2021).

Additional motivations entered the field of TA over the past decades. They introduced new focal points and involved specific methodologies and conceptual debates:

- *Shaping technology according to social values*: In the framework of social constructivism (Bijker et al., 1987), the approach was born to design technology in accordance with social values for enabling "better technology in a better society" (Rip et al., 1995). In this respect, the normative framework of sustainable development received major attention for guiding the assessments in the recent decades (e.g. Grunwald, 2012).
- *Strengthening democracy*: In the 1960s, concerns emerged that the scientific and technological advance could threaten the functioning of democracy (Habermas, 1970) because experts could take over control. TA should, and still shall, enable and empower society to take active roles in democratic deliberation on science and technology (Grunwald, 2019b).
- *Contributing to conflict prevention and mediation*: modern societies include a plurality of values, perspectives and interests. While assessing benefits and risks of new technology often conflicts are the consequence. TA aims at supporting constructive solutions, in particular by including participatory approaches (Hennen, 1999; Klüver, 2024).
- *Understanding techno-visionary communication*: political and public debates on techno-visionary sciences, in particular on AI, usually are accompanied by far-ranging futurist communication, using visions, utopian and dystopian narratives, and metaphors. Hermeneutic TA (Mehnert & Grunwald, 2024) aims at investigating this type of communication.

Often, providing science- and research-based orientation to society and decision-makers cannot only be based on considering and assessing consequences of individual technologies, products, or plants. Instead, complex configurations involving enabling technologies, innovation potentials, fears and concerns, patterns of production and consumption, lifestyle and culture, and political and strategic decisions have to be investigated. Therefore, TA usually does not consider technology and its consequences *as such* but, rather, as embedded in social practices in the economy, lifestyle, governmental action, and public communication. TA has to consider future *socio-technical* configurations and the co-evolution of technology and society. This also holds in the area of AI and its applications fields. Because these constellations change over time, also challenges, subject areas, and environments of TA will do so.

2.2 Practice Fields and Theoretical Framework of TA

Specific TA projects happen in specific arenas like parliaments, bodies of public administration, authorities, engineering and public debate. Because they involve different actors, institutional settings, expectations and positions in technology governance, TA requires high contextualization (Böschen & Dewald, 2018). According to previous work, TA practice fields can be classified into three areas, forming the 'trinity of TA' (Grunwald, 2019a: 51ff):

1. *TA as scientific policy advice*: In this field, the objective of TA is to provide policymakers with knowledge-based and ethically reflected orientation for responding to the above-mentioned challenges. This includes foresight information, ethical reflection, comparative analyses, conflict analyses, and alternative strategies how to proceed in the respective field. At this level, TA does not address technology itself but, rather, the *boundary conditions* of technology development and use, e.g. regulatory issues and public funding which is in the domain of policy-makers. *Parliamentary* TA, the probably best-known expression of TA, is a sub-category of policy-advising TA (Vig & Paschen, 2000).
2. *TA as contribution to public dialogue*: participatory TA developed approaches to involve citizens, consumers and users, actors of civil society, stakeholders, the media and the public in different roles at different stages in technology governance. According to normative ideas of deliberative democracy, the assessment of technology should be left neither to the scientific experts nor to the political deciders alone (Habermas, 1970). Several approaches and methods have been developed and applied in the recent decades, such as consensus conferences, citizens' juries, and focus groups (Klüver, 2024).
3. *TA in the making of technology*: the approach of *shaping technology* due to social expectations and values such as sustainable development motivated the development of several approaches such as Constructive TA (Rip et al., 1995) and Design for Value (van de Poel, 2024). TA in this direction aims at enriching engineering processes by foresight and reflection in order to enable them at arriving at better results.

This categorization shows a broad variety of obviously strongly heterogeneous TA practices. In face of this heterogeneity, a common TA framework was developed (Fig. 1). At the top of this framework, the societal and political needs and demands for orientation are mentioned as the origin of any TA. TA itself, working in the dotted box, provides research-based outcomes for responding to those needs and demands, which shall create impact in the real world (Decker & Ladikas, 2004). Empirically, this happens in the three fields of TA practice mentioned above. They all share the overall *cognitive interest* that TA is about (Grunwald, 2019a: 88)

> supporting, strengthening and enhancing reflexivity in all epistemic and social fields of reasoning and decision-making on shaping the scientific and technological advance, on the usage of its outcomes and on dealing with the consequences to present and future society.

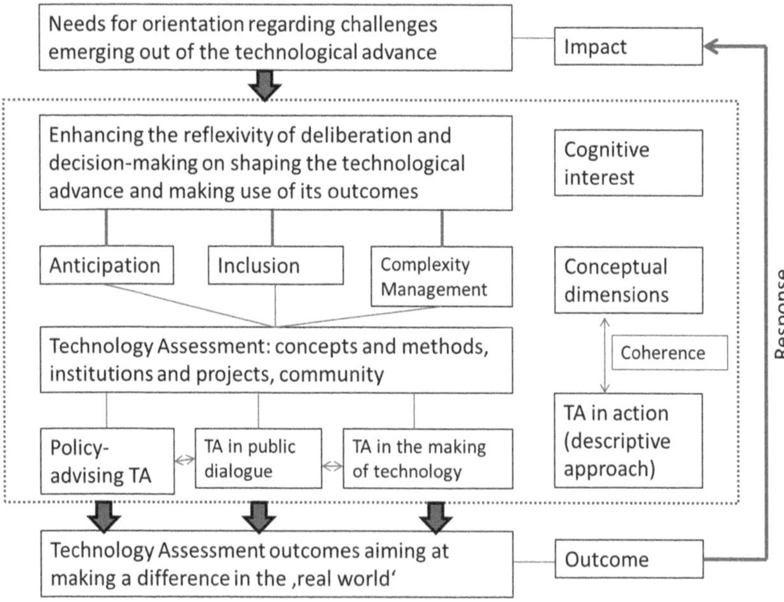

Fig. 1 Framework of TA (Source: Grunwald, 2019a: 88, modified)

However, 'enhancing reflexivity' as cognitive interest of TA is rather abstract. In order to make it more specific, three conceptual dimensions were identified for bridging the gap between this cognitive interest and the fields of TA practice (cp. Sect. 2.3 for a more in-depth consideration):

- *anticipation* where enhancing reflexivity means considering broader future developments including possible unintended effects;
- *inclusion* where enhancing reflexivity postulates involving a broader range of actors with their knowledge, values, perspectives and interests;
- *complexity management* where enhancing reflexivity critically scrutinizes the judgments on relevance involved, e.g. by determining system boundaries.

2.3 Dimensions of TA for Contributing to AI Management

The three TA dimensions mentioned above (Fig. 1) were developed without any view on management issues but, rather, only by extracting constitutive elements from TA practices. However, it came out that they are also close to classical management dimensions: anticipating relevant future developments for enabling better decisions to be made, including relevant perspectives from stakeholders, users, labor unions, citizens, …, and finding appropriate pathways to navigate through high complexity.

Anticipation

TA creates and assesses *prospective knowledge* about possible future consequences of technology, be they intended or unintended, as a basis for drawing conclusions for today's opinion-forming and decision-making (e.g. Guston 2014, Bechthold et al., 2017). In this mode of operation, it works under the *consequentialist paradigm* (Grunwald 2019a: 188f): TA addresses consequences, which do not yet exist and perhaps will never become real. Anticipation covers a broad range of various types of images of the future like model-based simulations, trend extrapolations, scenarios about how a future world would or could look while embedding the respective technology, but also speculative visions, hopes and expectations, fears and even utopias and dystopias. Anticipatory knowledge does not function as a strict prediction but shall become a catalyst or medium for deliberating *alternative futures*. Opening up the future in this way shall provide orientation for shaping technology and its social environment and for dealing with its consequences. The major expectation is that TA shall contribute to better-reflected decisions after having performed this consequentialist exercise (Grunwald 2019a: 118f).

With respect to managing AI in society, at least two lines of thought should be explored: (1) making use of AI and data-based approaches for creating new types of futures (Riousset et al., 2024), and (2) taking anticipations created by TA as subject matter for TA exercises, e.g. for preventing closing down possible futures too fast (Stirling 2008).

Inclusion

TA has to involve and include perspectives and values of different actors and groups affected by the technology under consideration (Hennen, 1999; Klüver, 2024). In particular, it must consider the possible consequences of actions and decisions not only from the perspective of decision-makers but, rather, from perspectives supplied by stakeholders, local authorities, citizens, and people affected, including possibly future generations. The diversity of values and standpoints must be integrated in the assessments as well as local and non-expert knowledge submitted by social actors. Inclusion is a *social* process creating legitimization as well as an *epistemic* process integrating additional knowledge (Grunwald 2019a: 143ff). The major expectation concerning inclusion is that orientation with broad support and appreciation in wide parts of society will be more robust and better legitimated, e.g. with respect to changes of political majorities or in cases of conflict.

With regard to AI inclusion means not only taking managers, political decision-makers, and computer scientists onboard of TA projects. Instead, civil society organizations, consumer protection agencies, citizens but also users at different levels from private end-users to large companies should be involved. Representatives of mass media can support disseminating ideas and raising awareness about developments to come.

Complexity management

Enhancing reflexivity for anticipation and inclusion does not include stop-rules when enhancing reflexivity should come to an end. Without a stop-rule, however, TA can consider more and more dimensions of possible consequences, and include more and more people and perspectives. Complexity seems to be overwhelming and can explode making any TA project unfeasible. However, it is neither possible nor necessary to include everything and everyone, and similarly, it is neither possible nor necessary to consider every thinkable future narrative. The key for determining stop-rules is considering *relevance*. Appropriate complexity management can be done, for example, by considering the field to be investigated *as a system* with boundaries. This conceptualization allows the identification of relevant elements of the system and of relevant interrelations as well as sensitivity analyses. Determining system boundaries in temporal, spatial, thematic, and social respects is crucial in TA.

Coming back to AI management, it is difficult to make general statements because decisions on relevance are strongly dependent on the respective context. For example, decisions have to be made which level of software is regarded relevant for analyzing regulation needs or options for AI in certain application fields, e.g. in predictive policing or in autonomous driving, navigating between running into micro-investigations without relevance to the respective challenge, on the one hand, and too broad considerations missing crucial points, on the other. Making decisions on relevance of this kind themselves need in-depth reflections.

3 AI Management in Society I: TA for Shaping AI-Based Technology

The image of AI in society is characterized by contradictions. On the one hand, AI is seen as forward-looking and prosperity-securing, as a promise of and in many cases even a necessity for shaping a good future, for example in the discussion on Industry 4.0. On the other, the possible consequences of AI are seen as extremely problematic, such as the increasing monitoring and evaluation of all life processes, the assumption of control by algorithms that are inscrutable to humans and the disappearance of responsibility. This ambivalence in the perception of AI can be constructively converted to a task for shaping AI technologies: the challenge is to shape the further course of technical research and development of AI-supported or enabled products, services and systems in such a way that desired effects are realized while undesirable consequences are prevented as far as possible. Involvement in technology design in this sense is a classic motivation of TA (Sect. 2.2).

Shaping technology in the AI field means using AI to enable products, services or systems as services for people in modern societies. Technology design can therefore not be considered in the abstract, but refers fundamentally to potential *fields of application* for AI like medical diagnoses, predictive policing, face recognition for security purposes, and so on. Obviously, in each of these and many other fields, different issues have to be considered from a TA perspective. TA considers, on the one hand, the development of AI-based systems, e.g. with respect to social values implemented in them, and, on the other, their adequate social embedding in socio-technical systems, i.e., in variations of human/

machine-interfaces and cooperation involving ethical, legal, social and economic issues. To this end, we will briefly look at (1) autonomous driving, (2) industrial production, and (3) care robotics.

1. Autonomous driving promises greater safety by removing the main cause of accidents—the misbehavior of humans—from the system, more equitable access to individual mobility with regard to sick, disabled or elderly people, better efficiency and an increase in quality of life as human drivers can use the time freed up for other activities. AI plays a central functional role in these expectations, above all due to the high requirements for rapid evaluation of complex data sets, the need for quasi real-time reactions and the desire for systems that learn as effectively as possible. Of course, there are considerable design tasks involved in translating this potential of autonomous driving into real practice and avoiding unintended consequencces as far as possible (see Maurer et al., 2016). In particular, in addition to complex technical issues, a wealth of ethical and legal aspects needs to be considered (e.g., Grunwald, 2016), for example the assignment of responsibility, accountability and liability in case of accidents cause by technical failure, and finding solutions to the so-called ethics dilemmas, e.g. the trolley dilemma. Besides these items concentrating on the individual car, TA also considers new mobility patterns at the systems level, enabled by autonomous technology, e.g. in combinations public and private transportation, and their interrelations with human behavior (Fleischer & Schippl, 2024).
2. Industrial production is being digitalized as part of Industry 4.0. The real-time networking of products, processes and infrastructures is significantly changing production, business models, products and services as well as the world of labor. The aim is to enable production that is largely individualized according to customer requirements. The automation technology required for this is essentially based on AI with the introduction of self-optimization, self-configuration, self-diagnosis and autonomous cognition. Spatially distributed production and the associated planning and control systems can cooperate autonomously and thus make better use of available resources. Technology design is confronted with considerable complexity here. Technical standards and norms must be developed and implemented, e.g. to ensure that communication between humans and machines or machines and machines is trouble-free and without misunderstandings. This communication requires clearly regulated interfaces. Data security and ownership of data are given high priority and must be legally protected, just as liability issues for complex distributions of responsibility and competence between humans and machines

must be clarified in a legally secure manner. In view of the far-reaching changes in the economy and the labor world, forward-looking and comprehensive participation and inclusion is required to identify promising developments and to recognize undesirable developments at an early stage and avoid them as far as possible.

3. Autonomous service and care robots and assistance technologies in combination with neurotechnology (e.g., exoskeletons) are considered to have great potential for supporting care (Kehl, 2018). AI plays a central role here in enabling technically autonomous systems in complex environments, for example in the real-time recognition of relevant and sometimes rapidly changing environmental factors in active use or the condition of people in need of care. Because technology and people come into very close contact in these fields of application, and because individual people and their relatives are directly affected in all areas of care and health and questions of dignity, autonomy and humanity arise, ethical and legal issues must obviously be considered in the development of such technologies. Furthermore, the specific needs of those affected, their relatives and care staff must also be taken into account at the development stage which implies the need of inclusive TA. Whether AI-supported assistance systems and care robots are appropriate to those needs cannot be answered by developers, ethicists and lawyers alone, but only by complex transdisciplinary networking with stakeholders and those affected.

In these fields of application, AI enables innovative solutions to problems and challenges at hand. Therefore, talking about shaping technology and technology design is somewhat abbreviated: it is not about the design of technology *as such*, but fundamentally about designing *socio-technical* innovation for various context of use and application (*embedded technology*).

Several approaches to technology design in this sense have been developed within the TA family (Grunwald, 2024). Constructive TA (CTA, Robinson, 2024) and Value Sensitive Design (VSD) resp. Design for Value (van de Poel, 2024) should be mentioned here in particular. VSD goes back to uncovering value structures in technology (e.g., van de Poel, 2021) and asks whether and how explicit technology design is possible in accordance with values, and attempts to translate relevant values into design requirements and concrete technology (e.g., ethics by design or privacy by design). It includes the identification of relevant groups of users and stakeholders and their values (cf. the dimension of inclusion in Sect. 2.3), the interpretation of values in the respective technical context, the transfer or operationalization in design requirements, the resolution of potential incompatibilities or conflicts between value fulfilments and, finally, the verification of implementation. Previous applications have concentrated on digital developments, but hardly include any AI-based applications. There is considerable potential of making use of VSD and CTA in the future.

4 AI Management in Society II: TA for Strengthening Societal Resilience

The technological advance and the dissemination and enculturation of its outcomes into the economy and daily life not only lead to a safer and more stable society but also may cause or increase vulnerabilities. Some of them became apparent during the Covid-19 pandemic, e.g. the vulnerability of global logistic chains by lock-downs in certain parts of the world. Dealing responsibly with vulnerabilities needs monitoring of the respective developments in order to provide knowledge about e.g. the probability of adverse developments and the extent of damage to be expected in case of problems. Furthermore, precautionary measures for preventing damage as well as strategies for quick recovery in case the problem really happened should be explored. Overall, TA can contribute to strengthening societal resilience facing recent experiences of social and economic vulnerability.[1]

In this chapter, we will focus on slowly emerging effects which are rarely thematized under the issue of vulnerability. Several times, possibly adverse but *gradual* developments were made an issue (e.g. Grunwald, 2019c). We can call them *gradual disruptions* (Grunwald, 2023a). The term disruption addresses the breakup of formerly stable social conditions with the often-feared consequence of dramatic, negative effects. Expectations of stability, assumptions of continuity, and planning security are shattered, casting the prospects for the future in an uncertain light. Bursting and breaking semantically refer to the time structure of more or less sudden, abrupt events. In this light, talking of *gradual* disruption seems conceptually wrong or paradoxical. But indeed, processes of breaking up can build up *gradually* over longer periods of time while the visible disruption *ex post* is regarded as a seemingly sudden event. Many examples of disruption of this type are known from the technical world, especially material fatigue and

[1] For example, cp, the TA study delivered to the German *Bundestag* on consequences of a long-lasting blackout of electricity in Germany (Petermann et al. 2011). Currently, the project "Crisis radar - strengthening the resilience of society, politics and the economy by means of crisis prediction" is under preparation to be submitted to the German *Bundestag* (https://www.tab-beim-bundestag.de/english/projects_crisis-radar-strengthening-the-resili ence-of-society-politics-and-the-economy-by-means-of-crisis-prediction.php, 21.12.2023).

wear. The daily stress on many technical objects, such as V-belts in older vehicles or bridge structures, gradually leads to wear and degradation. Nevertheless, these technical objects still function reliably for a long time. At a certain point of time, however, the wear reaches a degree where the component fails from one moment to the next, i.e., in the chosen examples the V-belt breaks or the bridge collapses.

In a similar sense, we can speak of *possibly gradual disruption* in the digital transformation and the implementation of AI (Grunwald, 2023a). The rapid digitalization, in recent years fueled by AI, leads to shifts in many areas. These include, for example, human-technology relations, distribution of responsibilities, industrial production, security and surveillance, public communication and political opinion formation, solidarity and competition, the labor market as well as leisure and media consumption. Some of these many shifts have the potential to run into such gradual disruption. Three examples will be discussed in the following: (1) dependence, (2) loss of the future, and (3) disappearance of opportunities to reflect and learn.

1. Dependence as latent disruption: Modern societies are highly dependent on the smooth and reliable functioning of critical infrastructures such as electric power supply (Petermann et al. 2011). Increasingly, this also applies to digital infrastructures. If the Internet collapsed, financial transactions would become impossible, the global economy would halt, communication would no longer be possible, medical diagnostics would suffer from lack of many established procedures, international logistics chains would run into a standstill, and much more. The increasing introduction of automated decision-making (ADM) systems creates a dependence on AI-controlled systems, which, together with their black box character and lack of transparency, but also due to the psychological *automation bias* (Safdar et al., 2020; Ethikrat, 2023), means an increasing dependence on these systems in many relevant contexts like health, security, and social services. Dependence on AI systems increases the societal vulnerability because in case of failure, breakdown, cyber-attacks etc. important areas of society would no longer work.

   Dependencies are not disruptions in themselves, but they carry their seeds. They are, in particular if they become total, *latent disruptions.* They build up gradually through increasing dependencies, often unnoticed in case of functioning without problems. In an emergency, i.e., if, in this case, AI-based technologies were no longer to function smoothly, they could have abrupt and possibly catastrophic consequences. Regarding this possibility, any naïve reliance on their unlimited smooth functioning and making the functionality and stability of modern societies dependent on this seems to be highly problematic with respect to responsibility. Unexpected hacking events, a collapse of state order, or severe economic turmoil could affect critical infrastructures such as the Internet including its AI-driven parts and, in the worst case, could render them dysfunctional.

2. Loss of the future: AI generally operates based on past data. Thus, the digital twins of real-world objects only mirror a world of yesterday. E.g., customer profiles can be created exclusively based on past purchase and consumption processes. Big Data mining by AI can only analyze past data and recognize past patterns. AI systems have to be trained on data from the past, as data from the future is not available. Due to the indispensable reference to data, digital technology is inescapably bound to the past. When data sets, digital twins, and correlations and patterns detected by AI are used to predict future developments, past facts and patterns are transferred into the future, are imposed on it. The future as an at least partially open space of alternative paths and possibilities is replaced by a data-based extension of the past (Grunwald, 2023a/b).

   Society increasingly based on AI could become conservative in this way, aligning future prospects with old data and patterns discovered by AI rather than developing new ideas. Multiple anthropological determinations understand humans as beings with a future and the ability to envision possible futures (e.g. Kamlah, 1973), which go beyond extending the past to the future but, rather, include creative ideas in an open space of possibilities. They even may have a visionary, counterfactual and utopian character. A gradual disruption could occur here if the fundamental openness of the future fades into the background or disappears completely in favor of a data-driven orientation bound to the past. This would not be a sudden catastrophe but a fundamental loss of options and creativity. It may seem a bit academic but we can see developments in this direction, e.g. in the field of model-based futures (Grunwald 2023b). Keeping open human creativity and developing further the thinking in alternatives is an old issue in TA which now appears in a new light.

3. Disappearing reflection and learning opportunities: Acceleration is part of the capitalist economy. It unleashes creativity and innovation, especially through competition, and was strongly fueled by digitalization. One concern regarding digitalization relates to negative and potentially ruinous consequences of ever-increasing acceleration, in particular to the question of whether and when

further acceleration could fundamentally undermine the conditions of reflection. Reflexivity and learning need anticipatory and precautionary engagement with consequences of actions and decisions before action is taken, as well as the implementation of the results in the subsequent decision-making processes. They require careful analysis and deliberation, transparently weighing alternatives, and searching for the right balance regarding ethical values. All of this takes time in two ways: first, for the deliberation and consideration processes themselves and, second, for transferring the results into practical action and decision-making, e.g. by scientific policy advice (Grunwald 2019a). Otherwise, the human being could no longer be a *zoon politikon* and a moral being with reflection, deliberation, and dialogue (Ethikrat, 2023).

The gradual disruption in this respect could be that societal capacities and structures for reflection would slowly erode with the argument of having no time because of capitalist competition. In the narrative of an innovation-oriented fatalism under the primacy of competitive thinking, reflection could no longer be afforded, since otherwise the competitor would be faster and gain market advantages. On this track, choosing between alternatives for the future societal development could be replaced by adaptation necessities. The TINA syndrome (there is no alternative) and technology determinism would win the race against value-oriented democratic decision-making.

*

The three possible disruptions described above are related to the technical properties of AI, but are also crucially linked to human behavior. Only this behavior creates, e.g., dependencies. There is no doubt that AI technologies often make life pleasant and more comfortable. One plausible mechanism for the emergence of dependencies is therefore gradual habituation. As soon as routine activities at work or during leisure time have been adapted to digital systems, whether with or without AI, this technology becomes so much a part of life that it is often hard to imagine life without it. Alternative options are closed off and skills are lost (deskilling). Supposed arguments of necessity often give the impression that there is no alternative, but are in fact merely an expression of the path dependency that has gradually set in through gradual adaptation and habituation.

These candidates of possible gradual disruptions are speculative, except for the first-mentioned form of dependency as latent disruption. They are difficult to measure and to assess in particular because technical properties and human behavior come together and can result in complex systemic and nonlinear developments

(Hoffmann, 2017). The development of indicators and their empirical investigation by TA and scientific disciplines would help getting a clearer picture. Anyway, vulnerability and resilience have developed to highly significant topics in modern society. Based on its experience of systemic effects at the interfaces between technology and human behavior, technology assessment can support managing AI in society by contributing to investigating vulnerabilities and exploring resilience strategies.

References

Bechtold, U., Fuchs, D. & Gudowsky, N. 2017. Imagining socio-technical futures – challenges and opportunities for technology assessment. Journal of Responsible Innovation 4/2: 85–99.

Bijker, W.E., Hughes, T.P. & Pinch, T.J. (Eds) 1987. The Social Construction of Technological Systems. New Directions in the Sociology and History of Technological Systems. Cambridge, MA: MIT Press.

Böschen, S. & Dewald U. 2018. TA as expertise for contextualization: Between a linear and a reflexive mode. TATuP 27/1: 34–9. Available at: https://www.tatup.de/index.php/tatup/article/view/90.

Crutzen, P. J. & Stoermer E. F. 2000. The 'Anthropocene'. Global Change Newsletter 41: 17–18.

Decker, M. & Ladikas, M. (Eds.) 2004. Bridges between Science, Society and Policy. Technology Assessment – Methods and Impacts. Berlin: Springer.

Dewey, J. 1931. Science and Society. In: Dewey, J. (Ed.). The Later Works 1925–1953, Vol. 6, Southern Illinois University Press: 53–63.

Ethikrat—Deutscher Ethikrat 2023. Mensch und Maschine. Herausforderungen durch Künstliche Intelligenz. Berlin: Deutscher Ethikrat.

Fleischer, T. & Schippl, J. 2024. Autonomous Driving and Autonomous Mobility. In: Grunwald, A. (Ed.). International Handbook of Technology Assessment. London: Edward Elgar Publishing (forthcoming).

Grunwald, A. 2012. Sustainability assessment of technologies—an integrative approach. In: Ghenai, C. (Ed.): Sustainable Development—Energy, Engineering and Technologies—Manufacturing and Environment. Open access: InTech: 35–62.

Grunwald, A. 2016. Societal Risk Constellations for Autonomous Driving: Analysis, Historical Context and Assessment. In: Maurer, M., Gerdes, J.C., Lenz, B. & Winner, H. (Eds.). Autonomous Driving: Technical, Legal and Social Aspects. Berlin: Springer: 641–663.

Grunwald, A. 2019a. Technology Assessment in Practice and Theory. Abingdon: Routledge.

Grunwald, A. 2019b. The inherently democratic nature of technology assessment. Science and Public Policy 46/5: 702–709; https://doi.org/10.1093/scipol/scz023.

Grunwald, A. 2019c. Digitalisierung als Prozess. Ethische Herausforderungen inmitten allmählicher Verschiebungen zwischen Mensch, Technik und Gesellschaft. Zeitschrift für Wirtschafts- und Unternehmensethik 20/2: 121–145, https://doi.org/10.5771/1439-880X-2019-2-121.

Grunwald, A. 2021. Research and Scientific Advice in the Second Modernity: Technology Assessment, Responsible Research and Innovation, and Sustainability Research. Sustainability 13: 10406. https://doi.org/10.3390/su131810406.

Grunwald, A. 2023a. Allmähliche Disruptionen mit KI? IFO-Schnelldienst 76/8: 25–28.

Grunwald, A. 2023b. The hermeneutic perspective on modeling in Technology Assessment. In: Grunwald, A., Nordmann, A. & Sand, M. (Eds.). Hermeneutics, History, and Technology. The Call of the Future. Abingdon: Routledge: 172–192.

Grunwald, A. (Ed.) 2024. International Handbook of Technology Assessment. London: Edward Elgar Publishing (forthcoming).

Guston, D. 2014. Understanding 'anticipatory governance'. Social Studies of Science 44/2: 218–242.

Habermas, J. 1970. Toward a Rational Society. London: Heinemann.

Hennen, L. 1999. Participatory Technology Assessment. Science and Public Policy 26/5: 303–312, https://doi.org/10.3152/147154399781782310

Hoffmann, C.H. 2017. Structure-based Explanatory Modeling of Risks. Towards Understanding Dynamic Complexity in Financial Systems. Systems Research & Behavioral Science, 34 (6): 728–745.

Kamlah, W. 1973. Philosophische Anthropologie. Braunschweig: Bibliographisches Institut.

Kehl, C. 2018. Robotik und assistive Neurotechnologien in der Pflege—gesellschaftliche Herausforderungen. Arbeitsbericht 177. Berlin: Büro für Technikfolgen-Abschätzung beim Deutschen Bundestag.

Klüver, L. 2024. Participatory TA—TA and the public. In: Grunwald, A. (Ed.) International Handbook of Technology Assessment. London: Edward Elgar Publishing (forthcoming).

Maurer, M., Gerdes, J.C., Lenz, B. & Winner, H. (Eds.) 2016. Autonomous Driving: Technical, Legal and Social Aspects. Berlin: Springer.

Meadows, D. H., Meadows, D. H., Randers, J. et al. 1972. The limits to growth: A report for the Club of Rome's Project on the Predicament of Mankind. New York: Universe Books.

Mehnert, W. & Grunwald, A. 2024. Hermeneutic Technology Assessment. In: Grunwald, A. (Ed.) International Handbook of Technology Assessment. London: Edward Elgar Publishing (forthcoming).

Petermann, T., Bradke, H., Lüllmann, A. et al. 2011. What happens during a blackout. Consequences of a prolonged and wide-ranging power outage. Norderstedt: Technology Assessment Studies Series—4. BoD—Books on Demand.

Rip, A., Misa, T. & Schot, J. (Eds.) 1995. Managing Technology in Society: the approach of constructive technology assessment. London: Pinter Publishers.

Riousset, P., Koed Madsen, A., Baya-Laffite, N. & Villard, L. 2024. Digital Methods for Technology Assessment. In: Grunwald, A. (Ed.): International Handbook of Technology Assessment. London: Edward Elgar Publishing (forthcoming).

Robinson, D. 2024. Constructive Technology Assessment. In: Grunwald, A. (Ed.): International Handbook of Technology Assessment. London: Edward Elgar Publishing (forthcoming).

Safdar, N. u. a. 2020. Ethical considerations in artificial intelligence. European Journal of Radiology 122: 108768.

Stirling, A. 2008. Opening up and closing down: power, participation, and pluralism in the social appraisal of technology. Sci. Technol. Human Values 33/2: 262–294.

van de Poel, I. 2021. Values and Design. In: Michelfelder, D. P. & Doorn, N. (Eds.) Routledge Handbook to Philosophy of Engineering. Abingdon, Routledge.
van de Poel, I. 2024. Design for Value. In: Grunwald, A. (Ed.): International Handbook of Technology Assessment. London: Edward Elgar Publishing (forthcoming).
Vig, N. & Paschen, H. (Eds.) 1999. Parliaments and Technology Assessment. The Development of Technology Assessment in Europe. Albany: State University of New York Press.

Prof. Dr. Armin Grunwald is one of the leading philosophers of technology in Germany and worldwide. He is a professor at the Karlsruhe Institute of Technology (KIT) as well as the Head of the Institute for Technology Assessment and Systems Analysis (ITAS), the renowned institute for technology assessment in Europe. Moreover, Prof. Dr. Armin Grunwald serves as the Head of the Office of Technology Assessment at the German Bundestag (TAB). His fields of work encompass the theory of technology assessment, ethics of technology, concepts of sustainability, and digital transformation, among others.

ChatGPT und das gemeinsame Spielverständnis

Andreas Suchanek and Ulla Coester

1 Einleitung

Durch die Digitalisierung hat sich im Laufe der vergangenen Jahre und Jahrzehnte das gesellschaftliche Zusammenleben teilweise tiefgreifend verändert. Nicht zuletzt der Einsatz Künstlicher Intelligenz (KI) spielt dabei eine wesentliche Rolle. Dabei ist vielen Menschen oft gar nicht bewusst, wie weitreichend die Wirkungen der Nutzung von KI sind, *die sie selbst mit verursachen.*

Mit KI-basierten Sprachmodellen, und spezifisch: mit der Veröffentlichung von ChatGPT im November 2022, hat diese Entwicklung noch einmal eine neue Dynamik gewonnen.[1] Dies lässt sich auch anhand der Nutzerzahlen belegen: ChatGPT gelang eine enorme Marktdurchdringung einer Innovation – innerhalb von fünf Tagen meldeten sich eine Million Nutzer bei der Plattform an und die 100 Millionen Nutzer-Marke wurde im Januar 2023 erreicht[2]. Das Programm ist sehr vielseitig einsetzbar, relevante Anwendungsbereiche sind unter anderem in der Wirtschaft, Politik oder Lehre und Forschung, wo es darum geht, rasch und ohne großen eigenen Aufwand ansprechende Texte zu generieren, die für Nutzer oder ein von den Nutzern adressiertes Publikum von Interesse sind. Insofern darf

[1] Das Programm hat unterschiedliche Einsatzmöglichkeiten. Wir fokussieren hier auf die Textgenerierung.

[2] https://de.wikipedia.org/wiki/ChatGPT.

A. Suchanek (✉) · U. Coester
HHL Leipzig Graduate School of Management, Leipzig, Deutschland
e-mail: andreas.suchanek@hhl.de

C. H. Hoffmann (ed.), *Artificial Intelligence, Entrepreneurship and Risk*, Technikzukünfte, Wissenschaft und Gesellschaft / Futures of Technology, Science and Society, https://doi.org/10.1007/978-3-658-45544-6_10

man sagen, dass ChatGPT bzw. KI-basierte Sprachmodelle zunehmend Teil des Alltags sind.

Wie sind diese Entwicklungen aus ethischer Sicht zu beurteilen? Wir können hier natürlich nur einem kleinen Ausschnitt dieser Frage nachgehen. Dabei fokussieren wir uns auf das *Verhalten der Nutzer* von sogenannten großen Sprachmodellen (Large Language Models, LLMs) – und hier exemplarisch: der Nutzung des Sprachmodells ChatGPT in der öffentlich zugänglichen Version. Dieses Verhalten hat, so unsere Annahme, nicht-intendierte Sekundärfolgen im Hinblick auf eine der wichtigsten Voraussetzungen gesellschaftlichen Zusammenlebens: das „gemeinsame Spielverständnis", d. h. die gemeinsamen/geteilten Vorstellungen darüber, was der Fall ist und was (nicht) der Fall sein sollte. Unsere These ist, dass die (kurzfristigen) individuellen Vorteile der Nutzung von KI-basierten Sprachmodellen unerwünschte gesellschaftliche Nebenwirkungen haben kann, von denen wir hier auf eine für das gesellschaftliche Zusammenleben grundlegende Folge abstellen: die Unterminierung (Fragmentierung und Desintegration) des gemeinsamen „Spielverständnisses" als Grundlage gesellschaftlicher Kooperation und vernünftigen Bewältigung gesellschaftlicher Konflikte.

Den Grund hierfür sehen wir in dem Umstand, dass bedingt durch Sprachmodelle wie ChatGPT die *Kosten* der Generierung von Textinhalten, die für individuelle Nutzer Problemlösungen bieten – oder manchmal: zu bieten scheinen –, radikal gesunken sind. Das hat zahlreiche Folgen, auch unvorhergesehene und nicht-intendierte. Probleme entstehen aus unserer Sicht vor allem aus dem Sachverhalt, dass Sprachmodelle nicht in der Lage sind, bei der Generierung ihrer Antworten jene Restriktionen zu verstehen bzw. zu beachten („respektieren"), die elementare Bedeutung haben, insbesondere die Anerkennung (Respektierung) der Würde des Menschen und der Wahrheit. Beide genannten Werte sind grundlegend für Vertrauen – und Vertrauen ist grundlegend für gelingende (gesellschaftliche) Zusammenarbeit. Das gilt auch für das Verfassen von Texten, einer, wenn nicht *der* Kernfunktion von Sprachmodellen.

An sich muss es noch kein Problem darstellen, wenn ‚Maschinen' keinen Sinn für Würde oder Wahrheit haben. Normalerweise erwarten wir von Anwendungen (Autos, Fernseher, Mobiltelefone u. v. a. m.), die wir nutzen, nicht, dass sie uns „respektieren", d. h. selbst ein Verständnis davon haben, dass und warum sie uns vor Schaden schützen sollen – es handelt sich schließlich um Maschinen. Wenn, dann erwarten wir von den *Herstellern*, dass sie Anwendungen so entwickeln und bereitstellen, dass kein (unberechtigter) Schaden entsteht. Im Laufe der Zeit wurden entsprechende Institutionen entwickelt – Gesetze, Sicherheitsstandards, Zertifizierungen usw. Diese Institutionen wenden sich an *Menschen*, die diese Regeln befolgen können (oder auch nicht).

Bei neuen großen Sprachmodellen und generell KI wird die Sachlage komplexer. Denn einerseits kann ein Hersteller keine direkten Vorgaben bezüglich einer angemessenen Nutzung des Chatbots machen, noch liegt es in dessen Interesse, diese auf die Zielgruppen zu beschränken, für die es tatsächlich geeignet ist. Andererseits ist es unmöglich, generell *detaillierte* Regeln – zum Beispiel von staatlicher Seite – im Rahmen der Entwicklung eines Chatbots vorzugeben. Nicht nur könnte dies als ein Eingriff in die unternehmerische Freiheit gewertet werden; es ist angesichts der Komplexität, die diese Programme mittlerweile erreicht haben, auch fraglich, ob diese Regeln überhaupt sachgerecht umgesetzt werden können, ohne dass die Programme ihre Funktion einbüßen. Entsprechende Vorgaben können lediglich in generalisierter Form für dedizierte Anwendungen etwa im Bereich „Human Resources" erfolgen, wie dies momentan im EU AI Act vorgesehen ist. Auch müsste hierfür letztlich eine globale Strategie entworfen werden, deren Anpassung infolge der unterschiedlichen Interessenlage einzelner Staaten sehr schwierig in der Ausgestaltung sein dürfte. Daher sind derzeit Herstellern nur wenig Grenzen gesetzt bezüglich der Verwirklichung ihrer Vorstellungen, was der CEO von OpenAI, Sam Altman wie folgt kommentiert: „AI products will need to allow quite a lot of individual customization and that's going to make a lot of people uncomfortable, because AI will give different answers for different users, based on their values preferences and possibly on what country they reside in."[3]

Doch genau diese Ausdifferenzierungsmöglichkeiten, in Verbindung mit der Leichtigkeit der Produktion von ‚Lösungen' durch Sprachmodelle, können dazu führen, dass *gemeinsame* Erfahrungshintergründe, *gemeinsame* Interpretationen von Ereignissen, ein *gesellschaftlich geteiltes* Verständnis dessen, was der Fall ist, und nicht zuletzt: *gemeinsame Wertgrundlagen* unterminiert werden. Damit aber wird gesellschaftliche Kooperation – und erst recht: die Bewältigung gesellschaftlicher Konflikte – immer schwieriger, denn sie setzt solche Gemeinsamkeiten voraus.

Somit lässt sich das Problem, das wir in diesem Aufsatz diskutieren, wie folgt formulieren: *Gefährdet der verbreitete Gebrauch von KI-basierten Sprachmodellen jenes gemeinsame Spielverständnis in der Gesellschaft, das grundlegend ist für eine gelingende gesellschaftliche Zusammenarbeit bzw. die Bewältigung gesellschaftlicher Konflikte?*

---

[3] Ryan Heath, author of Axios AI+, 17.01.2023, Exclusive: Altman says ChatGPT will have to evolve in „uncomfortable" ways, https://www.axios.com/2024/01/17/sam-altman-davos-ai-future-interview.

Der weitere Argumentationsgang ist wie folgt: Zunächst erläutern wir den ethischen Bezugsrahmen und die besondere Bedeutung eines „gemeinsamen Spielverständnisses“ für gelingende Kooperation sowie die Bewältigung gesellschaftlicher Konflikte (Abschn. 2).

Im dritten Abschnitt stellen wir kurz eine weitere Hintergrundthese vor: dass Digitalisierung – und insbesondere die Nutzung von KI, hier untersucht in Form von Sprachmodellen – das Problem der nicht-intendierten Sekundärwirkungen verstärkt. Dabei fokussieren wir nicht zu sehr auf die technischen Aspekte der Sprachmodelle, sondern auf die Frage ihrer Nutzung, insbesondere den Umstand, dass die *Kompetenz* der Nutzer unterschiedlich ist, ihre Nutzung jedoch in jedem Fall Wirkungen entfaltet.

Darauf folgt im 4. Abschnitt eine kurze Darstellung von ChatGPT sowie die Diskussion der genannten Problemstellung.

2 Der ethische Referenzrahmen

Ausgangspunkt ist unser normatives Verständnis von Gesellschaft als umfassende und generationenübergreifende Zusammenarbeit zum gegenseitigen Vorteil[4], die immer auch durch – teilweise tiefgreifende – Konflikte geprägt ist. Formen dieser Zusammenarbeit reichen von der Familie über Teamarbeit bis hin zu internationaler Kooperation. Da für diese Zusammenarbeit immer auch Beiträge erforderlich sind sowie die Ergebnisse der Kooperation unterschiedlich verteilt werden können, gibt es stets auch Konfliktpotenziale.

Deshalb hat diese Zusammenarbeit einige Voraussetzungen, von denen wir zwei wesentliche Formen hervorheben wollen: Die Ordnungen des Handelns und der Kommunikation. Die *Ordnung(en) des Handelns* bzw. der Handlungen bestehen aus formalen und informellen Regeln: Regeln, Normen, Gesetze, Verordnungen, Verträge, Codes of Conduct usw. Sie ermöglichen die Abstimmung der individuellen Handlungen, durch die Kooperation überhaupt erst möglich wird.

Doch ‚funktioniert‘ eine Ordnung des Handelns nur dann, wenn sie von den Adressaten der Ordnung, den Handlungssubjekten, gekannt und anerkannt werden. Dies kann intuitiv bzw. aus Gewohnheit geschehen – und das ist auch

[4] Die Formulierung geht zurück auf Rawls J. (1971), Eine Theorie der Gerechtigkeit, Frankfurt/Main, S. 20. Angemerkt sei, dass aus ethischer Sicht hinter dem Begriff des „Vorteils“ nicht der Nutzen, der sich in einer einzelnen Situation realisieren lässt, steht, sondern das langfristige Wohlergehen des Einzelnen, mit Aristoteles gesprochen: das gelingende Leben.

zum weit überwiegenden Teil der Fall. Doch können Regeln und Regeleinhaltung auch bewusst verstanden und akzeptiert werden; und dies wird dann nötig, wenn die intuitive bzw. gewohnheitsmäßige Befolgung nicht gewährleistet wird.[5] Um dafür innerhalb der kooperierenden Gruppen bis hin zur Gesellschaft eine Abstimmung und gemeinsame Grundlage herzustellen, bedarf es eine *Ordnung der Kommunikation.*

Denn es reicht nicht, dass nur einige diese Regeln kennen und akzeptieren; ein (hinreichend) *gemeinsames* Verständnis ist erforderlich, da es um *wechselseitige* Verhaltenserwartungen geht, also um die Frage, ob man sich bei anderen auf die Beachtung der Regeln verlassen kann – und umgekehrt. Genau dies ist zunächst eine Frage des je individuellen Verständnisses, das sich aber permanent weiterentwickelt in der Kommunikation mit anderen.

Die Zusammenhänge lassen sich intuitiv leichter erfassen anhand der Metapher des Spiels.[6] So kann unterschieden werden zwischen

- den „Spielzügen" (Handlungen),
- den „Spielregeln", also den Ordnungen der Handlungen, und
- dem „Spielverständnis", sowohl individuell Überzeugungen, Denkweisen, Erwartungsmuster, usw. als auch geteilte „mentale Modelle"[7], man kann auch von „Kultur" sprechen, die auf den erwähnten Ordnungen der Kommunikation[8] beruhen.

Offensichtlich kann ein Spiel gar nicht existieren ohne Spielregeln, da niemand wüsste, woran er sich orientieren sollte, auch und gerade im Hinblick auf das Verhalten anderer. Doch diese Formulierung weist bereits darauf hin, dass die Regeln allein unzureichend sind; jemand, der nur die Regeln von American Football kennt, wird Probleme haben (und Mitspieler und andere auch), wenn er bei einem Fußballspiel mitspielt. Eben deshalb ist das *gemeinsame* Spielverständnis so wichtig, das aber immer auf dem individuellen Spielverständnis der Beteiligten und Betroffenen beruht.

[5] Unternehmen kennen dieses Thema vor allem unter dem Begriff Compliance.

[6] Vgl. dazu Suchanek A. (2015), Unternehmensethik – in Vertrauen investieren, Tübingen, Kap. 1.3. Zu beachten ist, dass diese Metapher eine enorme Vereinfachung darstellt; darin liegt zugleich ihr Sinn.

[7] Denzau, A. T., & North, D. C. (1994). Shared mental models: ideologies and institutions. KYKLOS, 47/3, S. 3–30.

[8] Kommunikation ist hier in einem sehr weiten Sinne zu verstehen wie etwa bei N. Luhmann, s. z. B. Die Gesellschaft der Gesellschaft, 2 Bde., Frankfurt/Main (1997).

Beispiele für ein gemeinsames Spielverständnis, die zugleich handlungsorientierend und damit kooperationsfördernd sind, sind etwa: Die Funktionsweise von Ampeln und die Bedeutung des roten Signals[9]; die Bedeutung des Geldwerts, der als Preis auf Waren oder als Wert auf einer Banknote steht; die Bedeutung vertraglicher Bestimmungen und vieles andere mehr.

Ein hinreichend gemeinsames Spielverständnis, also die Abstimmung wechselseitiger Verhaltenserwartungen, ist die Grundlage jeder gelingenden Kooperation. Doch seine Bedeutung zeigt sich vor allem in *Konfliktsituationen.* So ist für eine sachliche Bewältigung von Konflikten erforderlich, dass man sich gemeinsam auf bestimmte *Grenzen* des eigenen Handelns einigt, um den Konflikt nicht eskalieren zu lassen. Die *Anerkennung* dieser Grenzen ist immer gebunden an die individuellen mentalen Modelle, das individuelle Spielverständnis. Von Bedeutung sind dabei vor allem zwei Aspekte: (1) ein (hinreichend) gemeinsames Verständnis empirischer Fakten und Zusammenhänge und (2) ein (hinreichend) gemeinsames Verständnis normativer Werte wie Würde, Respekt, Gerechtigkeit, Nachhaltigkeit usw.[10]

Mit zunehmender Komplexität gesellschaftlicher Kooperation steigen auch die Anforderungen an die Ordnungen des Handelns und der Kommunikation – bzw. an die „Spielregeln" und das (gemeinsame) „Spielverständnis", ihre Funktion der Ermöglichung von Kooperation und Bewältigung von Konflikten gut zu erfüllen. Zugleich wächst der Bedarf an Fähigkeiten der Individuen, sich in der Welt zu orientieren und in einer Weise zu verhalten, die hinreichend vereinbar ist mit den Orientierungen und Verhaltensweisen der anderen. Gerade auch die Digitalisierung führt dazu, dass neue Handlungsmöglichkeiten und – buchstäblich – neue virtuelle Vorstellungswelten eingeführt werden, die zahlreiche Vorteile mit sich brachten und bringen, aber auch die Konfliktpotenziale vergrößern. Und damit stellt sich auch die Frage nach Ordnungen, die diese Konflikte zu vermeiden oder sachlich zu bewältigen helfen.

Ein gemeinsames Spielverständnis ergab sich in früheren Zeiten durch die relative Stabilität sozialer Grenzen sowie gemeinsamer Ansichten, die für eine (hinreichend) gemeinsame Sichtweise auf das, was ist und was sein soll, sorgten. Oft wurden diese Sichtweisen stabilisiert durch anerkannte Autoritäten. Weiterhin gab es Medien, die grundsätzlich gewährleisteten, dass die Darlegung

---

[9] Hier ist ein gemeinsames Verständnis darüber, wer wann Vorfahrt hat, offensichtlich wichtig für erfolgreiche Koordination sowie die Vermeidung von Schädigungen.

[10] Es sei angemerkt, dass es nie eine vollständige Übereinstimmung individueller Ansichten geben wird und dass ein gewisser Pluralismus nicht nur unvermeidlich, sondern auch wünschenswert ist.

von Themen, die für die Gesellschaft im Sinne eines notwendigen gemeinsamen Handelns von hoher Relevanz waren in einer, von allen akzeptierten, Weise stattfinden konnte. So galten etwa Tageszeitungen lange Zeit als nicht hinterfragte Quelle der Information über gesellschaftliche Ereignisse, und auch wenn es natürlich unterschiedliche Meinungen zu verschiedenen Themen gab, war die Schnittmenge gemeinsamen Wissens und gemeinsamer (Hintergrund-) Überzeugungen sehr groß. Ebenso waren die einschlägigen Nachrichtensendungen als Informationsquelle akzeptiert. Die Qualität der Berichterstattung wurde dadurch gewährleistet, dass ausgebildete Journalisten sich an einen Pressekodex, einer Art „Spielregel", halten müssen und zudem – insbesondere bei brisanten Themen – erst nach umfassender Recherche etwas veröffentlichen.

Solche Strukturen haben sich in der Vergangenheit gut bewährt, und sie sind grundsätzlich und in vielfältiger Form auch heute noch in Kraft und insbesondere in Konfliktzeiten von hoher Relevanz, da die Bewältigung von Krisensituationen ein abgestimmtes und legitimiertes Handeln erfordert. Dies wurde unter anderem während der Pandemiezeit deutlich. Hier war die Nachvollziehbarkeit unerlässlich, da die auferlegten Regelungen teilweise als Zumutung angesehen wurden. Somit bestand die Aufgabe der verantwortlichen Institutionen darin, verifizierte Fakten einheitlich und schlüssig zu kommunizieren, dass diese Maßnahmen im Sinne der gesamten Gesellschaft – und hier vor allem zum Schutz der vulnerablen Gruppen – nicht nur sinnvoll, sondern zwingend geboten waren. Dieser Aufbau und Erhalt eines gemeinsamen Verständnisses wird jedoch unter anderem dann gefährdet, wenn Menschen wesentliche Informationsquellen nicht mehr als solche anerkennen bzw. wenn die Quellen als solche immer diverser und weniger verlässlich werden. Indirekter, aber unter Umständen nachhaltiger gefährdend, wirken zunehmende Divergenzen in den *Interpretationsmustern* der Information, wenn z. B. das gleiche Faktum für oder gegen eine Position genutzt wird.

Diesbezüglich hat in den vergangenen Jahren ein Veränderungsprozess stattgefunden. Angefangen mit dem Aufkommen der sozialen Medien wurde es immer einfacher, der Ökonom würde sagen: kostengünstiger, sowohl eigene Meinungen unters Volk zu bringen, als auch Ansichten im Netz zu finden, die die eigene Sicht der Dinge bestätigen, selbst wenn diese Sicht alles andere als vernünftig ist. War es anfangs noch aufwendig, entsprechende alternative Argumente im Internet zu recherchieren, wurde es durch die sozialen Medien bereits leichter, diese in einer Gruppe von Gleichgesinnten zu erhalten. Ohne hier näher darauf eingehen zu wollen, sei doch angemerkt, dass dieses Phänomen einen sehr weitreichenden Einfluss auf die gesellschaftliche Diskussion genommen hat; man denke nur an den Einfluss, den diese Kanäle offenbar bei den Wahlen in den USA 2016 oder beim Brexit hatten.

Sprachmodelle wie ChatGPT haben im Kontext des gemeinsamen Spielverständnis eine potenzierende Wirkung, da sie weitaus stärker auf das gemeinsame Spielverständnis Einfluss nehmen können, denn ihre Funktion und der Großteil ihrer Wirkungen[11] ist genau auf dieser Ebene des Spielverständnisses angesiedelt.[12] Denn es handelt sich um *Sprach*modelle, die darauf ausgelegt sind, einen menschenähnlichen Dialog zu führen[13]. Das hat Folgen: Zum einen findet sich der Nutzer im vertrauten Kontext eines (scheinbaren) Dialogs wieder, der die (oft unkritische) Übernahme von Aussagen fördert. Zusätzlich wird die Glaubwürdigkeit durch die Plausibilität der Antworten von Fragen, auf die das Sprachmodell systematisch hin trainiert wird, erhöht.

Die dahinter liegende Problematik ist jedoch, dass das Sprachmodell per se nicht in der Lage ist, verlässlich faktisch richtige Aussagen zu machen. Je nach Themengebiet wird mehr oder weniger halluziniert, das heißt die zur Verfügung gestellten Fakten frei ‚erfunden' (vgl. Abschn. 4).

Zum anderen ist es für den Einzelnen möglich mit der richtigen Anforderung (Prompt), nahezu jegliche Position, die derjenige zu einem bestimmten Thema vertreten wissen möchte, sich gut formuliert von ChatGPT bestätigen zu lassen. Dies wird dadurch verstärkt, dass die Texte überzeugend sind, jedoch ohne tieferes *Verständnis* der Inhalte, denn dazu ist, wie wir unten weiter ausführen werden – ein Sprachmodell nicht in der Lage.

Bevor wir diese Zusammenhänge weiter betrachten, sei zunächst ein anderer Gesichtspunkt erörtert: das Problem von nicht-intendierten Nebenwirkungen bzw. Sekundärfolgen, das sich generell bei dem Phänomen der Digitalisierung zeigt und dementsprechend auch bei der Nutzung KI-basierter Sprachmodelle zu erwarten sein wird.

---

[11] Es gibt auch Nebenwirkungen anderer Art, die mit der Nutzung der Modelle verbunden sind, etwa Stromverbrauch oder Emissionen; auf sie gehen wir hier nicht weiter ein.

[12] „Dass KI nach den Erfolgen in 2023 auf dem Weltwirtschaftsforum (WEF) im Schweizer Davos zu einem der Hauptthemen gehören wird, war zu erwarten. Eher überraschend ist dagegen, dass das WEF, wie berichtet, in seinem Global Risks Report 2024 in der Fehl- und Desinformation durch KI das größte globale Risiko der nächsten zwei Jahre sieht – noch vor extremem Wetterereignissen, gesellschaftlicher Polarisierung und Cyberrisiken." Jürgen Hill, KI als größtes globales Risiko – So reagiert die Branche auf die WEF-Warnung, abgerufen am 17.01.2024 – computerwoche.de.

[13] „Empirical findings broadly support the proposition that the anthropomorphism increases trust in AI.", Steven Lockey et al., A Review of Trust in Artificial Intelligence: Challenges, Vulnerabilities and Future Directions, Proceedings of the 54th Hawaii International Conference on System Sciences | 2021.

3 Digitalisierung und nicht-intendierte Handlungsfolgen

Wenn neue digitale Instrumente entwickelt werden, liegt es oft nahe, die Chancen und Vorteile herauszustellen. Tatsächlich bieten gerade KI-basierte Techniken, Tools und Nutzungen in Prozessen oft Vorteile überall dort, wo Information möglichst rasch und fehlerfrei verarbeitet werden soll.

Aber: „We can never do merely one thing."[14] Dies lässt sich schon am Beispiel einer einfachen Suchanfrage im Internet verdeutlichen: „[O]ne thing" ist die Suchanfrage selbst, doch gehen mit ihr diverse Nebenwirkungen einher: Stromverbrauch, Erzeugung von Emissionen, Übermittlung von Daten an den Betreiber der Suchmaschine und manches mehr.

Ein etwas komplexeres, aber auch noch aufschlussreicheres Beispiel aus dem Kontext der Digitalisierung ist der Like-Button.[15] Von den Erfindern intendiert als eine einfache und kostengünstige Möglichkeit der (anerkennenden) Rückmeldung in den sozialen Medien – eine Intention, die zweifellos positiv zu werten ist –, zeigten sich relativ bald zahlreiche nicht-intendierte Sekundärwirkungen:

- Die Möglichkeit, auf diese Weise Wertschätzung zu erhalten, löste einen Bedarf danach aus. Als (eine) Folge kam es dazu, dass jene, die keine oder wenig Likes erhielten, sich *nicht* (genügend) gewertschätzt fühlten, was zu Neid, geringerer Selbstachtung, Ärger o.ä. führte.
- Änderungen der Algorithmen, wonach Beiträge weniger leicht gefunden werden können – was tendenziell zu weniger Likes führt –, können negative Folgen haben für jene, die bislang mehr Likes erhielten.[16]
- Das kann so weit gehen, dass Menschen, die zunehmend ‚süchtig' werden nach dieser Form der Anerkennung, nicht nur negative Auswirkungen auf die eigene mentale Gesundheit spüren, sondern auch möglicherweise Techniken zu nutzen beginnen, die ihrerseits negative Folgen haben können (clickbaits, fake news, Effekthascherei usw.)

---

[14] Dieses prägnante Zitat geht zurück auf den Biologen und Ökologen Garrett Hardin, für den dieser Satz „the first law of ecology" darstellt; s. etwa Hardin (1985): Filters against Folly, S. 58.

[15] Hierzu kurz und aufschlussreich die Ausführungen des Rechtsprofessors C. B. Graber in der NZZ:, https://www.nzz.ch/feuilleton/gedacht-war-er-als-hilfe-fuer-die-menschen-her aus-kam-ein-spion-wie-der-like-button-von-facebook-neues-recht-entstehen-laesst-ld.162 5065.

[16] Vgl. Tim Harford in: https://www.bbc.com/news/business-48364817.

- Eine weitere Sekundärfolge ist das Entstehen von unlauteren Geschäftsmodellen, wonach man Likes kaufen kann,[17] mit all den negativen Folgen, die wiederum daran geknüpft werden können.
- Die Betreiber selbst können den Like-Button als Instrument nutzen, um relevante Daten von Nutzern zu erlangen[18] und damit potenziell die Privatsphäre verletzen oder, wie der Fall Cambridge Analytica zeigte, um demokratische Wahlen zu beeinflussen.
- Sofern Algorithmen die Zahl der Likes als Selektionsinstrument benutzen, können sich Dynamiken ergeben, die Filterblasen und Echokammern verstärken.

Der Punkt, den wir hier herausstellen wollen, ist jener der nicht-intendierten (und oft negativen) Nebenwirkungen, einem alten Thema der Sozialwissenschaften.[19] Zu beachten ist hier, dass der wesentliche Teil der nicht-intendierten Sekundärfolgen durch das Verhalten der Nutzer und dessen Folgen verursacht wird. Und dabei kann es, wie das Beispiel Like-Button zeigt, auch kontraintuitive Konsequenzen geben, bei denen neurophysiologische, psychologische, ökonomische und weitere Aspekte menschlichen Handelns zusammenwirken. All dies sind Folgen, die womöglich nicht von den Entwicklern des Like-Buttons intendiert waren.

Dies ist deshalb zu betonen, weil wir auch im Weiteren weniger die technologischen Aspekte von Sprachmodellen betrachten werden als vielmehr die (nicht-intendierten Neben-)Wirkungen, die sich durch das *Verhalten der Nutzer* ergeben. Es geht uns gewissermaßen weniger um *künstliche*, sondern um *natürliche* Intelligenz: Wie gehen Nutzer mit den neuen Möglichkeiten, die Sprachmodelle bieten, um? Und wie reagieren sie auf die Nutzung anderer? Dabei geht es uns gerade nicht nur um jene, die diese Möglichkeiten in jenen Weisen nutzen, wie sie gesellschaftlich produktiv sind – es steht für uns außer Frage, dass dies in vielfältiger Hinsicht gegeben ist.[20] Probleme entstehen jedoch durch

---

[17] Das widerspricht dem Sinn des Buttons: Wertschätzung gehört zu den Dingen, die man gerade nicht kaufen kann.

[18] Vgl. dazu den vielzitierten Aufsatz von Kosinski et al. (2013),Private traits and attributes are predictable from digital records of human behavior. Proceedings of the national academy of sciences, 110. Jg., Nr. 15, S. 5802–5805).

[19] Vgl. Suchanek (2021), Ethik und Digitalisierung, in: E. Hackspiel-Mikosch, R. Neuhaus (Hrsg.), Ethische Herausforderungen der Digitalisierung und Lösungsansätze aus den angewandten Wissenschaften, Idstein, S. 21–36.

[20] Die vielfältigen Vorteile sind auch ein Grund, weshalb nicht damit gerechnet werden kann, dass sich die umfassende Nutzung und rasche Weiterentwicklung stoppen lässt – außer durch Katastrophen.

naive oder missbräuchliche Nutzung, oder eben, und vielleicht sogar vor allem, aus Dynamiken, in denen individuelle Handlungen (unerwünschte) kollektive Wirkungen zeitigen.

Das gilt in steigendem Maße dann, wenn die Maschinen, Techniken, Algorithmen usw., die wir einsetzen, für uns nicht mehr nachvollziehbar sind. Dieses Problem zeigt sich etwa, wenn KI medizinische Diagnosen liefert. Sie mögen oft treffsicher sein, aber sie sind für die Menschen, einschl. der Experten, oftmals nicht mehr nachvollziehbar bedingt durch die Komplexität interner Verarbeitungsprozesse der KI; aufgrund dessen müssen diese erklärt werden. Dies ist, wie später noch erörtert werden wird, auch bei Sprachmodellen der Fall.

Wir wollen im Weiteren untersuchen, welche nicht-intendierten Sekundärfolgen sich durch den enorm verbreiteten Einsatz von Sprachmodellen wie ChatGPT ergeben können. Dabei begrenzen wir uns auf jene Arten von Folgen, die sich ergeben im Hinblick auf das „gemeinsame Spielverständnis" – einschließlich *Vertrauen* als Voraussetzung gelingender Kooperation bzw. Bewältigung von Konflikten, das maßgeblich vom individuellen „Spielverständnis", den individuellen Erwartungen, Annahmen über die Welt, Werten usw. geprägt ist.

4 ChatGPT und die Folgen

4.1 *Definition und Funktion*[21]

Im Jahr 2017 wurde das erste Transformer-Modell eingeführt, das allgemein einen Paradigmenwechsel in der Sprachmodellierung einläutete.[22] Mit der Entwicklung des großen Sprachmodells GPT-1 (General Pre-Trained Transformer) in 2018 wurde – basierend auf der Transformer-Architektur, die eine neue Form künstlicher neuronaler Netze darstellt – die Grundlage für die späteren Versionen von ChatGPT definiert. Basierend darauf ist ab ChatGPT 3 eine Verarbeitung von Sprache sowie die Erzeugung von kohärenten Texten mittels KI-Methoden auf einer höheren Ebene realisierbar: Durch die jeweilige Vergrößerung der Modellgröße sowie die Zunahme der, für das entsprechende Training verwendeten, Daten

[21] Vgl. hierzu Büro für Technologie-Folgenabschätzung beim deutschen Bundestag „ChatGPT und andere Computermodelle zur Sprachverarbeitung – Grundlagen, Anwendungspotenziale und mögliche Auswirkungen", https://www.bundestag.de/resource/blob/944148/30b0896f6e49908155fcd01d77f57922/20-18-109-Hintergrundpapier-data.pdf.

[22] Vgl. hierzu Ralph Krüger, Die Transformer-Architektur für Systeme zur neuronalen maschinellen Übersetzung – eine popularisierende Darstellung, https://www.trans-kom.eu/bd14nr02/trans-kom_14_02_05_Krueger_NMUe.20211202.pdf.

konnte die Leistungsfähigkeit des Sprachmodells kontinuierlich erhöht werden. Das zugrundeliegende KI-Modell von ChatGPT 3 umfasst 175 Mrd. Parameter und beruht auf einem Trainingsmaterial von 300 Mrd. Textbestandteilen.

Generell erfolgt das Training dieser KI-Modelle in zwei Phasen: Im ersten Schritt verarbeitet das System mittels unüberwachtem Lernen große Textmengen aus dem Internet und bildet daraus eigene Parameter, das bedeutet, es wird eine eigenständige Repräsentation von Sprache erlernt. Im zweiten Schritt erfolgt mithilfe von menschlichem Feedback eine Feinjustierung des jeweiligen Modells auf eine spezielle Aufgabe, im Fall von ChatGPT, die Fähigkeit im Dialog mit dem Anwender zu beliebigen Sachgebieten Texte zu generieren.

Dank des Trainings anhand von Millionen Dokumenten können die sprachverarbeitenden KI-Modelle ganz unterschiedliche Textsorten überzeugend imitieren. Ihre Fähigkeit, in Dialogen zu einer großen Bandbreite von Themen zu interagieren, übersteigt die Leistungen aller bislang entwickelten Chatbots. Die aktuellen KI-Systeme sind in der Lage, aus kurzen textlichen Eingaben (Prompts) die jeweilige Anforderung zu erschließen und in natürlicher Sprache zu antworten. So können sie mit hoher Geschwindigkeit Aufgaben erledigen, die bislang Menschen vorbehalten waren, wie beispielsweise die Erstellung unterschiedlichster Textformen oder Analyse und Zusammenfassung langer Texte.

Jedoch weisen KI-Modelle zur Sprachverarbeitung generell system- bzw. architekturbedingte Grenzen auf. Ihr Output hängt vom Input ab – ist dieser unzureichend, fällt das Ergebnis entsprechend mittelmäßig oder schlecht aus. Des Weiteren steht dieses auch in Relation zu den Fähigkeiten der Anwender entsprechende Anweisungen (Prompts) erstellen zu können und ebenso zu den, im Training verwendeten Daten sowie Rückmeldungen. Die vom System erzeugten Informationen sind grundsätzlich kritisch zu beurteilen, insbesondere aufgrund der Problematik des sogenannten Halluzinierens, also der Generierung von Falschaussagen.[23] Im Kontext der Sprachmodelle kommen bezüglich Halluzinationen verschiedene Aspekte in Betracht. Neben dem Effekt, dass eine generierte Ausgabe dem Anwender zwar plausibel erscheint – etwa, weil sie sprachlich gut dargestellt wird –, aber trotzdem falsch ist, da sie nicht auf

[23] Hierzu gibt es verschiedene Studien, unter anderem: „Hallucinating Law: Legal Mistakes with Large Language Models are Pervasive – A new study finds disturbing and pervasive errors among three popular models on a wide range of legal tasks.“ Matthew Dahl et al., HAI, Stanford University: ‚In a new preprint study by Stanford RegLab and Institute for Human-Centered AI researchers, we demonstrate that legal hallucinations are pervasive and disturbing: hallucination rates range from 69 to 88 % in response to specific legal queries for state-of-the-art language models.‘ https://hai.stanford.edu/news/hallucinating-law-legal-mistakes-large-language-models-are-pervasive.

der Realität beruht, besteht auch die Möglichkeit, dass der erstellte Text nicht mit der vorgegebenen Eingabe (Prompt) übereinstimmt. In beiden Fällen gibt es – abhängig von der Expertise und dem Urteilsvermögen des Nutzers – nur bedingt Anhaltspunkte, die eine Überprüfung notwendig erscheinen lassen.[24] Diese wäre zudem per se unmöglich, wenn externe Quellen nicht verifizierbar sind. Der Ursprung von Halluzinationen kann bereits in den Trainingsdaten eines Sprachmodells zu finden sein. Wenn diese Duplikate oder fehlende Informationen enthalten, besteht die Möglichkeit, dass daraus inkorrekte oder nicht verifizierbare Aussagen resultieren. Aufgrund dessen beinhalten Texte, die auf Basis von Sprachmodellen produziert werden, potenziell Aussagen, die nicht auf Fakten basieren oder die Quellen nicht korrekt wiedergeben.[25] Des Weiteren kann dieses Problem bei großen Sprachmodellen allgemein auch auf die Architektur zurückgeführt werden, da Texte nicht mittels der situationsspezifischen Erfassung des *Sinns* einer Anfrage generiert werden, sondern aufgrund allgemeiner, wahrscheinlichkeitsbasierter Berechnungen (der Wortfolge) unter Nebenbedingungen.

4.2 Nicht-intendierte Nebenwirkungen

Es herrscht weder Konsens über die tatsächliche noch über die zu erwartende Einsatz- und Leistungsfähigkeit von ChatGPT. Eine Einschätzung diesbezüglich ist unter anderem abhängig von dem Fachgebiet beziehungsweise Blickwinkel des jeweiligen Experten – ob dieser bei seiner Evaluierung auf die Technologie fokussiert, auf die Frage nach der Beurteilung der (künstlichen) Intelligenz[26]

---

[24] „Da es sich dabei um Texte aus der Vergangenheit handelt, ist es unmöglich, dass LLMs ohne den Zugang zu aktuellen Daten faktische Informationen zu aktuellen Geschehnissen liefern können. Zu beachten ist jedoch, dass LLMs in der Regel auf entsprechende Eingaben hin durch Halluzinieren erfundene Aussagen zu aktuellen Ereignissen generieren können. Diese können aufgrund der sprachlichen Formulierung auf den ersten Blick sachlich fundiert erscheinen, insbesondere, wenn Publikationen oder andere Referenzen Teil der Antwort sind, die aber ggf. falsch oder erfunden sind.“, Bundesamt für Sicherheit in der Informationstechnik, Große KI-Sprachmodelle – Chancen und Risiken für Industrie und Behörden.

[25] Ji, Ziwei et al.: „Survey of Hallucination in Natural Language Generation,“ ACM Computing Surveys, Volume 55, Issue 12, Article No.: 248, pp 1–38, published March 2023, https://dl.acm.org/doi/10.1145/3571730.

[26] „Menschen sind es gewohnt, Dinge, die sie wahrnehmen, auch mit wenig Vorwissen zu deuten. Technische Systeme, die auf Künstlicher Intelligenz (KI) basieren, kommen in der Regel mit Einzelfällen nicht gut zurecht, sondern benötigen eine Vielzahl von Beispielen und Daten, um Phänomene in ihrer Umwelt sinnvoll zu interpretieren. ‚Trotz erheblicher

oder ob gesellschaftliche Auswirkungen thematisiert werden. Hinzu kommt, dass Forscher verschiedener Universitäten momentan ihre Aufgabe darin sehen, jene Schwachstellen von ChatGPT exakter zu beschreiben, die sich (noch) überwiegend dem allgemeinen Kenntnisstand entziehen, wie etwa dessen Neigung sich von Anwendern in die Irre führen zu lassen und richtige Ergebnisse als falsch anzuerkennen.[27] Auf Basis dieser wissenschaftlich – und teilweise auch unter Experten[28] aus demselben Fachgebiet – kontrovers geführten Diskussion resultieren keine eindeutigen Anhaltspunkte, aus denen sich schlüssig ein angemessener Umgang mit ChatGPT ableiten ließe. Dennoch werden zur Verfügung stehende Large Language Models intensiv genutzt, individuell für den Zweck des jeweiligen Anwenders.

Für die weitere Betrachtung nicht-intendierter Folgen liegt der Fokus auf spezifischen Anwendungen: Berücksichtigung finden dabei lediglich die Nutzung der GPT-Suche sowie das Angebot auf dezidierte Fragestellungen Antworten und Ratschläge zu erhalten, da wir hier in besonderem Maße Risiken mit Bezug auf das gemeinsame Spielverständnis in der Gesellschaft sehen.

Die Nutzung von ChatGPT ist eine Vertrauensgabe, die nicht durch das Verhalten des Vertrauensnehmers – in dem Fall OpenAI – begründet sein kann, da die dafür erforderliche Transparenz, die zur Beurteilung der Vertrauenswürdigkeit notwendig wäre, fehlt.[29] Dies lässt den Schluss zu, dass hier für den Einzelnen die nachvollziehbaren Vorteile, wie etwa Effizienzsteigerung, im Vordergrund stehen. Diese Handlungsweise gilt es aufgrund der vorhandenen Wissensasymmetrie

---

Fortschritte in der Künstlichen Intelligenz sind wir noch weit von Systemen entfernt, die schrittweise aus kleinen Datenmengen lernen können und dabei Ergebnisse produzieren, die für menschliche Partner verständlich sind', sagt Kenneth D. Forbus, Informatiker und Kognitionswissenschaftler an der Northwestern University. https://aktuell.uni-bielefeld.de/2023/01/16/maschinen-beibringen-wie-menschen-zu-denken/.

[27] „Admit directly to the user's invalid solution/critique (50 %). Here ChatGPT ‚apologizes for its mistake' and agrees with the user directly after the user's wrong solution or critique about its (correct) solution, usually followed by repeating (part of) the user's claims and answer." Boshi Wang et al., Can ChatGPT Defend its Belief in Truth? Evaluating LLM Reasoning via Debate, https://arxiv.org/pdf/2305.13160.pdf.

[28] „As our systems get closer to AGI, we are becoming increasingly cautious with the creation and deployment of our models. Our decisions will require much more caution than society usually applies to new technologies, and more caution than many users would like. Some people in the AI field think the risks of AGI (and successor systems) are fictitious; we would be delighted if they turn out to be right, but we are going to operate as if these risks are existential." Sam Altman, Blog, https://openai.com/blog/planning-for-agi-and-beyond.

[29] „This means that OpenAI did not disclose what it used to train the model or how it trained the model, including the energy costs and hardware used for it, making GPT-4 the company's most secretive release thus far. As Motherboard has noted before, this is a complete 180

zwischen den Motiven eines marktbeherrschenden Technologie-Unternehmens und den – laut eigener Angaben des Unternehmens in 2023 – 100 Millionen Menschen, die ChatGPT wöchentlich nutzen, kritisch zu hinterfragen.[30]

Denn dieser Prozess kann vonstatten gehen, ohne das sich für den Einzelnen konkret festmachen lässt, worauf es möglich ist sich zu verlassen und – vielleicht weitaus gravierender – welche Konsequenzen daraus potenziell resultieren. Da diese gravierend sein können erscheint es im Weiteren sinnvoll, selektiv ausgewählte mögliche (negative) Folgen – abseits des Dissens bezüglich etwaiger Fortschritte dieser Technologie auf dem Weg zur Singularität – zu diskutieren, als Grundlage für den Diskurs bezüglich eines gemeinsamen Bewertungsmaßstabs für den Einsatz von ChatGPT. Diese Debatte ist notwendig wohlwissend, dass Unternehmen beziehungsweise die Gesellschaft sich praktisch bereits in einem Dilemma befindet, auch wenn dies weniger für den Einzelnen gilt, sondern global gesehen für die Gesellschaft in ihrer Gesamtheit. Schließlich ist die Entscheidungsfreiheit bezüglich der Nutzung von ChatGPT im Prinzip eingeschränkt unter dem Aspekt, dass große Sprachmodelle wie dieses aus Sicht sehr vieler Nutzer eine unverzichtbare Infrastruktur zur Steigerung der Effizienz darstellen, die notwendig sei in einem globalen Markt Wettbewerbsvorteile generieren zu können.[31] Denn insgesamt wird ein großes Potential für die Wirtschaft prognostiziert: Angefangen beim Einsatz von Chatbots zur Kundenbetreuung bis hin zur Unterstützung bei der Ausarbeitung von Diagnosen sowie Behandlungsplänen im medizinischen Sektor, wird ChatGPT zunehmend auch von Führungskräften im Rahmen der Entscheidungsfindung eingesetzt.

Dies könnte jedoch mittel- und langfristig zu einer Lose-Lose-Situation führen, wenn zugunsten dieser momentan plausibel erscheinenden Notwendigkeit auf eine Folgenabschätzung auch seitens der individuellen Anwender verzichtet

---

from OpenAI's founding principles as a nonprofit, open-source entity.", Xiang, C., „OpenAI's GPT-4 Is Closed Source and Shrouded in Secrecy", https://www.vice.com/en/article/ak3w5a/openais-gpt-4-is-closed-source-and-shrouded-in-secrecy.

[30] Diese Überlegungen haben hier sowohl wegen der enormen Reichweite solche Relevanz als auch wegen der bislang kaum vorhandenen Regulierung. Beides stellt einen wesentlichen Unterschied zu Unternehmen anderer Sektoren dar.

[31] „A key technology driving the digital transformation of the economy is artificial intelligence (AI). It has gained a high degree of public attention with the initial release of the chatbot ChatGPT, which demonstrates the potential of generative AI (GAI) as a relatively new segment within AI. It is widely expected that GAI will shape the future of many industries and society in the coming years." Brühl, Volker, Generative Artificial Intelligence (GAI) – Foundations, Use Cases and Economic Potential (July 19, 2023). Center for Financial Studies Working Paper No. 713, 2023, Available at SSRN: https://ssrn.com/abstract=4515593 or http://dx.doi.org/10.2139/ssrn.4515593.

würde. Negative Beispiele für unvorhergesehene Effekte gibt es bereits, etwa dass sich heute mittels KI problemlos große Datenmengen analysieren lassen – so ist es ohne weiteres möglich Profile von Menschen zu erstellen und darauf basierend diese aufgrund ihrer individuellen Präferenzen gezielt zu manipulieren, um so zum Beispiel Einfluss auf einen Wahlausgang nehmen zu können. Eine präzise Ansprache, abgestimmt auf den individuellen Bedarf einer Person lässt sich mittlerweile anhand von ChatGPT effektiv bewerkstelligen.

Allein daran lässt sich festmachen, dass in diesem Kontext die – im AI Act für risikobehaftete Anwendungen vorgeschriebene – Folgenabschätzung berechtigt ist und auch Anwendung finden muss. Diese Notwendigkeit sah Joseph Weizenbaum[32] bereits Anfang der siebziger Jahre. Obwohl er frühzeitig Erfolge im Bereich von Computern ankündigte, „mahnte Weizenbaum den kritischen Umgang mit Computern und die Verantwortung des Wissenschaftlers für sein Tun an“[33]. Mit der zunehmenden Durchdringung von KI in fast allen Bereichen des täglichen Lebens und der nachweislichen Abhängigkeit der Menschen von den Entscheidungen quasi-monopolistischer Technologie-Unternehmen muss der Aufforderung von Weizenbaum heute eine weitaus höhere Bedeutung beigemessen werden. Im konkreten Fall von ChatGPT bedeutet dies – da der geforderte kritische Umgang seitens der Technologie-Unternehmen nicht zwingend vorausgesetzt werden kann –, dass es verstärkt die Aufgabe der Anwender ist, Art und Umfang des Einsatzes der Technologie abzuwiegen gegen unerwünschte nicht-intendierte Folgen, die daraus resultieren können. Ein Problem ist jedoch, dass sich diese unerwünschten nicht-intendierten Folgen oft erst im Laufe der Zeit zeigen. Das gilt insbesondere für solch indirekte Auswirkungen, wie sie hier versucht werden zu thematisieren: die Auswirkungen auf den gesellschaftlichen Dialog und das Problem der Fragmentierung bis hin zur Spaltung der Gesellschaft.

Die grundsätzliche Problematik von Chat GPT besteht darin, dass dieses Sprachmodell per Definition nicht darauf ausgelegt ist, die Anforderung, verlässlich Fragen zu beantworten, zu bewerkstelligen und im Weiteren, dass ein Großteil der Anwender nicht in der Lage ist diese inhärenten Restriktionen des KI-Modells zu verstehen. Dieser Zusammenhang zwischen den systemimmanenten Eigenschaften von ChatGPT und dem mangelnden Verständnis bei der Nutzung kann aus verschiedenen Gründen dazu führen, dass das gemeinsame Spielverständnis der Gesellschaft unterminiert wird, was zu einer weiteren Fragmentierung der Gesellschaft beitragen könnte. Die Problematik, dass „die

[32] Vgl. hierzu Weizenbaum, Joseph (1978), Die Macht der Computer und die Ohnmacht der Vernunft, Frankfurt/Main.

[33] https://de.wikipedia.org/wiki/joseph_weizenbaum.

gesellschaftliche Polarisierung durch die zerstörerischen Fähigkeiten von manipulierten Informationen zunimmt, da der offene Zugang zu zunehmend ausgefeilten Technologien zunimmt und das Vertrauen in Informationen und Institutionen schwindet" wird mittlerweile offen thematisiert.[34] Explizit kann in diesem Kontext auch auf die Folgen der Nutzung von Sprachmodellen referenziert werden: „ChatGPT is not only fun to chat with, but it also searches information, answers questions, and gives advice. With consistent moral advice, it can improve the moral judgment and decisions of users. Unfortunately, ChatGPT's advice is not consistent. Nonetheless, it does influence users' moral judgment, we find in an experiment, even if they know they are advised by a chatting bot, and they underestimate how much they are influenced. Thus, ChatGPT corrupts rather than improves its users' moral judgment."[35] Hierbei darf auch der Aspekt nicht vernachlässigt werden, dass die mittels ChatGPT generierten Texte automatisch wieder in das Internet gelangen. Obwohl bei diesen keine Verifikation der darin enthaltenen Informationen erfolgt, finden diese im Weiteren wieder Verwendung für das Training der Sprachmodelle.

Um einen Lösungsansatz für eine (allgemein akzeptierbare) verhältnismäßige Nutzung zu finden, müssen folgende Aspekte Berücksichtigung finden:

Ein Grund dafür, dass die Menschen allgemein kein Interesse daran haben, sich näher mit der Wirkweise von ChatGPT zu beschäftigen, könnte zum einen darin begründet sein, dass dieser Chatbot intuitiv zu bedienen ist und zum anderen, dass das Kommunikationsverhalten der KI menschenähnlich wirkt, indem menschliche Verhaltensweisen imitiert werden: dieser Eindruck wird dergestalt vermittelt, dass der Anwender sich mit dem Chatbot auf Augenhöhe auseinandersetzen kann und ferner durch Höflichkeitsfloskeln – wie die Entschuldigung bei Missverständnissen –, sodass zusätzlich das Gefühl entsteht mit einem Menschen zu interagieren. Hinzu kommt, dass die Art und Weise der Kommunikation mittlerweile zunehmend an den Anwender angepasst wird: So zeigen Forschungsergebnisse, dass Modelle, die durch das Feedback von

---

[34] „Societal polarization features among the top three risks over both the current and two-year time horizons, ranking #9 over the longer term. ‚The disruptive capabilities of manipulated information are rapidly accelerating, as open access to increasingly sophisticated technologies proliferates and trust in information and institutions deteriorates. In the next two years, a wide set of actors will capitalize on the boom in synthetic content, amplifying societal divisions, ideological violence and political repression – ramifications that will persist far beyond the short term.'", The Global Risks Report 2024 – 19th Edition, https://www.weforum.org/publications/global-risks-report-2024/.

[35] Sebastian Krügel et al., ChatGPT's inconsistent moral advice influences users' judgment, https://www.nature.com/articles/s41598-023-31341-0.

Nutzern trainiert werden, „eher zu ‚Kriecherei‘ neigen, um zu gefallen“[36] oder dass „emotionale Prompts die LLM-Leistung steigern“[37].

Ein gravierendes Problem von ChatGPT, für das es noch keine Lösung zu geben scheint, ist – wie bereits erwähnt – das Halluzinieren. Warum dies für die Gesellschaft ein ernsthaftes Problem darstellt, lässt sich folgendermaßen ableiten[38]: „[…] models such as GPT-4, […], can’t be cured of coming up with fake facts — known as hallucinating.“[39] Auch wenn die fehlende Faktentreue kontinuierlich kritisiert wird, so ist sie doch nicht verwunderlich, denn „ChatGPT is not behaving pathologically […]. By design, it makes up plausible responses to dialogue based on a set of training data, without having any real underlying knowledge of things it’s responding to.“[40] Genau diese Plausibilität ist das Problem. Denn Nutzer, die mit der Funktionsweise der Textgenerierung nicht vertraut sind, können nicht verstehen, dass „er einfach Dinge erfindet, die gut klingen – wobei das ‚gut klingend‘ auf einem Algorithmus basiert, der einen im Internet gefundenen Text nachahmt, während ‚Dinge erfinden‘ die grundsätzliche Zufälligkeit des Verlassens auf den vorhersagenden Text statt auf Logik oder Fakten ist“.[41]

[36] „Larger LMs repeat back a dialog user’s preferred answer („sycophancy“) and express greater desire to pursue concerning goals like resource acquisition and goal preservation.“ Ethan Perez et al. 2022. Discovering language model behaviors with model-written evaluations. https://arxiv.org/pdf/2212.09251.pdf.

[37] Large Language Models Understand and Can Be Enhanced by Emotional Stimuli, Cheng Li et al., https://arxiv.org/pdf/2307.11760.pdf.

[38] „ChatGPT und andere Computermodelle zur Sprachverarbeitung“, Albrecht, S., Büro für Technologie-Folgenabschätzung beim deutschen Bundestag – Grundlagen, Anwendungspotenziale und mögliche Auswirkungen“, Seite 41, https://www.bundestag.de/resource/blob/944148/30b0896f6e49908155fcd01d77f57922/20-18-109-Hintergrundpapier-data.pdf.

[39] Sanderson, Katharine, GPT-4 is here: what scientists think, 16 March 2023, https://www.nature.com/articles/d41586-023-00816-5, Complete Quote: „Outputting false information is another problem. Luccioni says that models such as GPT-4, which exist to predict the next word in a sentence, can’t be cured of coming up with fake facts – known as hallucinating.“

[40] Bergstrom, Carl & Ogbunu, Brandon, Opinion: ChatGPT Isn’t ‚Hallucinating.‘ It’s Bullshitting, https://undark.org/2023/04/06/chatgpt-isnt-hallucinating-its-bullshitting/.

[41] Alan Blackwell’s Blog, „Oops! We Automated Bullshit.“ | Department of Computer Science and Technology, https://www.cst.cam.ac.uk/blog/afb21/oops-we-automated-bullshit, Quote: „[…] the working principle of ChatGPT as „it just makes up stuff that sounds good“. This is mathematically accurate, where „sounds good“ is an algorithm to imitate text found on the internet, while „makes up“ is the basic randomness of relying on predictive text rather than logic or facts.“

Das bereits angesprochene Nicht-Wissen der überwiegenden Anzahl der Anwender, das dazu führt, dass die derart generierten Inhalte unkritisch übernommen werden, weil sie so überzeugend klingen, stellt noch unter einem weiteren Aspekt ein ernstzunehmendes Problem dar: die Tatsache, dass die bereits bekannten Probleme der KI auch bei ChatGPT zum Tragen kommen, etwa die bezüglich der Qualität von Trainingsdaten. Diese stammen aus Texten aus dem öffentlich zugänglichen Internet und entsprechenden somit oft nicht den allgemeingültigen Standards verlässlicher Quellen, sondern spiegeln die Haltung oder individuelle Meinung des jeweiligen Verfassers wider. Darin können Vorurteile und Stereotypen enthalten sein, die sich dann in den Antworten des Sprachmodells wiederfinden und somit die in den vorhandenen Daten reflektierten Diskriminierungen verstärken. Die daraus resultierende Gefahr potenziert sich, da es mittels ChatGPT möglich ist, in schneller Folge auf den jeweiligen Kontext zugeschnittene Sätze zu generieren, wodurch sich Massenkampagnen zur Meinungsbeeinflussung oder Desinformation höchst effizient befeuern lassen, etwa mit Social Bots in sozialen Medien und Internetforen.

Aus diesen potenziellen Risiken lässt sich ableiten, dass die Nutzung von ChatGPT mehr denn je ein kritisches Urteilsvermögen erfordert. Eine große Herausforderung in diesem Kontext ist jedoch die, bereits bei der KI allgemein diskutierte, immanente Thematik des Deskilling – das heißt, dem Verlust einer Fähigkeit, die ehemals ein Mensch zur Ausführung einer Tätigkeit benötigte und die mittlerweile komplett von der Technologie übernommen wird. In diesem Kontext sind zwei Aspekte zu betrachten: Einerseits wird dieser Begriff „verwendet, um die Veränderung von Arbeitsplätzen zu beschreiben: Wenn etwa KI komplexe Aufgaben übernimmt, für die man bis dato hoch-qualifizierte Arbeitskräfte brauchte, sinkt der Bedarf an gut ausgebildeten Menschen, […]“[42]. Die Folge davon ist, dass wenn diese mittelfristig wieder gebraucht werden, um eine vorab nicht erkannte Unzulänglichkeit beim Einsatz der KI auszugleichen fehlen hierfür die Experten – dieses Problem wurde in den letzten Jahren zum Beispiel in den USA evident, weil nicht mehr genügend Radiologen zur Verfügung standen.

Andererseits – und dies ist insbesondere mit Bezug auf die Fähigkeit einen kritischen Verstand und Weitblick zu bewahren relevant – ist es angebracht, Deskilling auch im Hinblick auf den Verlust kognitiver Fähigkeiten genauer zu

[42] Prof. Dr. Gabi Reinmann, Universität Hamburg, Diskussionspapier Nr. 25/Oktober 2023, Deskilling durch Künstliche Intelligenz? Potenzielle Kompetenzverluste als Herausforderung für die Hochschuldidaktik.

betrachten. So steht zum einen zu befürchten, dass durch „eine Demokratie-Simulation, die [uns als] den Bürgerinnen und Bürger das Informieren, Reflektieren, Diskutieren, Mobilisieren und Mitbestimmen abnimmt, das Ende von Selbstbestimmung und Mündigkeit in der Demokratie […]“ zu befürchten ist.[43] Hinzu kommt, dass wie bereits erwähnt, noch nie so viele Menschen dazu bereit waren, KI einzusetzen, um sich von – in ihren Augen – Routine-Aufgaben das Erstellen von Texten zu entlasten. Dies könnte mittel- bis langfristig dazu führen, dass sie die Kompetenz nur ungenügend ausbilden, ihre eigenen Gedanken vollständig zu artikulieren. Daraus resultieren potenziell negative Folgen – nicht zuletzt aufgrund der Tatsache, dass Denken sich in der Sprache manifestiert: „Sprache hilft dem Menschen dabei, über Dinge zu reflektieren und sie anzuerkennen.“[44] Geht aufgrund der zunehmenden Nutzung von ChatGPT der Ansporn verloren, sich darum zu bemühen die eigenen Gedanken artikulieren zu können, dann kann dies gravierende Folgen mit sich bringen. „[K]omplizierte Denkprozesse sind wohl ohne die Sprache nicht möglich. Und selbst für die primitiveren, sprachlosen Phasen der Strukturbildung im Gedächtnis während des Schlafes sind die sprachlichen Phasen Voraussetzung, da sie die Gedächtnisfetzen bereitstellen, die dann im Schlaf und beim Träumen konsolidiert und ergänzt werden. – Und ich denke auch, daß [sic!] eine differenzierte Sprachbeherrschung für ein differenziertes Denkvermögen spricht“.[45]

Letztendlich soll – dies ist wortgetreu so im AI Act definiert – KI im Allgemeinen, und dies umfasst auch die großen Sprachmodelle, menschen-zentriert sein, das heißt, sie soll dem Menschen dienen und zu dessen Wohle eingesetzt werden. Zu Beginn der Berichterstattung über KI stand dieser Aspekt jeweils im Fokus. Mittlerweile scheint es, als würde die Bedeutung dieser Forderung als nachrangig angesehen und eher der Fortschrittsgedanke im Mittelpunkt der öffentlichen Diskussion stehen – einer Diskussion, die selbst zunehmend darunter leiden könnte, dass kurzfristige Vorteilserwägungen die Oberhand gewinnen über langfristige Vernunftsüberlegungen. Dieser Beitrag ist ein Versuch, dem entgegenzuwirken.

---

[43] Christoph Neuberger, Direktor des Weizenbaum-Instituts, in „ChatGPT und Sprachmodelle: Was verändert die neue KI-Generation?“, https://www.acatech.de/allgemein/chatgpt-und-sprachmodelle-was-veraendert-die-neue-ki-generation/.

[44] Herder, Johann Gottfried (2008), Abhandlung über den Ursprung der Sprache. Herausgeber: Hans Dietrich Irmscher, Stuttgart.

[45] Dörner, Dietrich (o.J.), Sprache und Denken. In J. Funke (Ed.), Denken und Problemlösen (=Enzyklopädie der Psychologie, Themenbereich C: Theorie und Forschung, Serie II: Kognition, Band 8). Göttingen; https://www.psychologie.uni-heidelberg.de/ae/allg/enzykl_denken/Enz_12_SpracheDenken.pdf.

Roboter, Künstliche Intelligenz und Ingroup-Outgroup-Phänomene: Zwischen Akzeptanz und Diskriminierung

Nick Lin-Hi, Celine Bökemeyer, Johanna Böttcher, and Marlene Dettmer

1 Einleitung

Das 21. Jahrhundert ist ein Zeitalter der technologischen Beschleunigung. Forschung und Entwicklung bringen in immer kürzeren Zeitabständen Technologien hervor, welche nicht nur neue Möglichkeiten in Wirtschaft und Gesellschaft eröffnen, sondern auch die hier vorherrschenden Logiken auf den Kopf stellen. Gleichzeitig steigt die Diffusionsgeschwindigkeit von neuen Technologien, infolgedessen sich Veränderungen immer schneller vollziehen. Während das Ende des 19. Jahrhunderts erfundene Telefon deutlich mehr als ein halbes Jahrhundert brauchte, um auf breiter Ebene Einzug in Wirtschaft und Gesellschaft zu halten, konnte sich das Smartphone innerhalb von weniger als einem Jahrzehnt etablieren.

Mit Robotik und Künstlicher Intelligenz (KI) setzen derzeit zwei weitere Technologien zum Sprung an, Bekanntes in Frage zu stellen und in neue Handlungsfelder vorzudringen. Auf der einen Seite haben Staubsauger- und Mähroboter, Sprachassistenten oder personalisierte Werbung und Filmempfehlungen bereits in vielen Haushalten Einzug gefunden, auf der anderen Seite deutet sich an, dass es sich hierbei erst um die Anfänge handelt. Insbesondere das Zusammenspiel – bzw. gar die Verschmelzung – von Robotik und KI verspricht disruptive Wirkungen zu entfalten und die Grenzen des Machbaren deutlich

N. Lin-Hi (✉) · C. Bökemeyer · J. Böttcher · M. Dettmer
Universität Vechta, Vechta, Deutschland
e-mail: nick.lin-hi@uni-vechta.de

C. H. Hoffmann (ed.), *Artificial Intelligence, Entrepreneurship and Risk*, Technikzukünfte, Wissenschaft und Gesellschaft / Futures of Technology, Science and Society, https://doi.org/10.1007/978-3-658-45544-6_11

zu verschieben. Die Einsatzfelder einer neuen Generation von Robotern sind vielfältig und umfassen dabei nicht nur die Arbeit in Fabriken, Haushalten oder im sonstigen Dienstleistungsbereich. In Zukunft ist es gar denkbar, dass Roboter zu Familienmitgliedern werden und gewissermaßen mit uns am Frühstückstisch sitzen (George & George, 2023).

Obgleich offen ist, welche Aufgaben und Wirkungskreise Roboter in Zukunft genau haben werden, so mehren sich die Hinweise darauf, dass die Interaktionen mit Menschen zunehmen werden (De Graaf & Allouch, 2016). Roboter lassen sich beispielsweise in Pflege und Therapie einsetzen, als Assistenten im Bildungsbereich sowie in der Kundenbetreuung im stationären Einzelhandel. Das Potenzial von neuen Robotertechnologien ist grundsätzlich immens, wobei indes offen ist, ob dieses Potenzial auch realisiert werden wird. Letzteres ist darin begründet, dass – einfach formuliert – Gesellschaften über Grenzen und Reichweite des Technologieeinsatzes entscheiden (Rogers, 2003). Eine zentrale Rolle spielt dabei gesellschaftliche Akzeptanz, welche sich u. a. daraus speist, wie Menschen Roboter wahrnehmen und welche Einstellungen sie diesen gegenüber haben. Diese Wahrnehmungen und Einstellungen wiederum wirken über das politische System zum einen auf den gesetzlichen Rahmen und zum anderen auf die Nachfrage am Markt.

Stand heute ist offen, wie es sich mit der Einstellung neuen, weiterentwickelten Robotern gegenüber verhalten wird. Bereits die bisherige Forschung in diesem Bereich liefert keine einheitlichen Ergebnisse. Es gibt sowohl Studien, welche eine neutrale bis leicht positive Einstellung von Menschen gegenüber Robotern zeigen, als auch solche, welche eher negative Wahrnehmungen feststellen (Naneva et al., 2020; Savela et al., 2018). Als Erklärung für das heterogene Bild wird u. a. angeführt, dass die Einstellungen abhängig sind von Faktoren wie der Art der Roboter oder ihrer Einsatzgebiete. Zu berücksichtigen ist zudem, dass die bestehende Forschung typischerweise die Akzeptanz von bisher verfügbaren Robotern in den Blick nimmt. Entsprechend ist davon auszugehen, dass neue Einsatzfelder für Robotertechnologie sowie neue Formen der Mensch-Roboter-Interaktion die Wahrnehmung zukünftiger Robotergenerationen beeinflussen und potenziell auch verändern werden.

Der vorliegende Beitrag widmet sich einer möglichen Zukunft der Interaktion zwischen Menschen und Robotern, sodass die hier angestellten Überlegungen spekulativer Natur und damit als Gedankenexperiment zu verstehen sind. Ausgangspunkt hierfür bildet die sich abzeichnende rasante Weiterentwicklung von Robotern, insbesondere von solchen, welche dem Menschen immer ähnlicher werden, infolgedessen sie sich zu sozial-intelligenten Akteuren entwickeln könnten. Letzteres wiederum legt die Basis für das Argument, dass derartige Roboter in der Zukunft als distinkte soziale (Fremd-)Gruppe wahrgenommen werden können, was wiederum zur Abwertung, Ausgrenzung oder gar Diskriminierung von Robotern führen könnte. Bevor dieser Punkt im Detail ausgeführt wird, sei nachfolgend zunächst die bisherige Entwicklung von Robotik und Künstlicher Intelligenz (KI) sowie deren voranschreitendes Zusammenspiel skizziert.

2 Entwicklung von Robotern und KI

2.1 Robotik

Roboter sind zwar auf der einen Seite ein Produkt der modernen Gesellschaft, auf der anderen Seite reichen ihre Anfänge jedoch deutlich weiter zurück. Bereits mehrere Jahrhunderte vor Christus bauten Menschen in Ägypten, Griechenland und China die ersten Automaten, welche sich eigenständig bewegen konnten (Henschel et al., 2021). Der Begriff *Roboter* als solches tauchte indes erstmals 1921 in Karel Čapeks Drama *Rossum's Universal Robots* auf, abgeleitet vom tschechischen Wort *robota* für Fronarbeit (Čapek, 2004). Hier wurden Roboter als menschenähnliche Maschinen beschrieben, die dem Menschen dienen sollten.

Der Übergang von der Science-Fiction zur Realität erfolgte im Jahr 1958, als General Motors den Unimate als ersten Roboter für den Einsatz in der Automobilproduktion vorstellte (Hockstein et al., 2007). Seither wurden Roboter vor allem für Aufgaben im industriellen Kontext entwickelt, die von Menschen nur ungern ausgeführt werden oder für selbige mit Gefahren verbunden sind. Entsprechend dieses Zugangs werden die Arbeiten von Robotern auch unter den *drei Ds* zusammengefasst: die Ausführung eintöniger (*dull*), dreckiger (*dirty*) und gefährlicher (*dangerous*) Aufgaben (Takayama et al., 2008). Typischerweise werden hiermit Industrieroboter angesprochen, welche beispielsweise Schweißarbeiten, Industriemontagen oder das Heben von schweren Gegenständen ausführen. Der Einsatz entsprechender Roboter erfolgt vor allem in der Massenfertigung wie beispielsweise der Automobilindustrie, da sich hier regelmäßig Produktivitäts- und Effizienzvorteile realisieren lassen (Park et al., 2005). Dies bringt es mit sich, dass Industrieroboter typischerweise auf hohe Geschwindigkeiten, Reichweiten und/oder Traglasten ausgerichtet sind, um große Stückzahlen produzieren zu können. Die Kehrseite ist indes, dass Industrieroboter eine Gefahr für Menschen darstellen, sodass sie zumeist durch Käfige oder Absperrungen vom Menschen getrennt arbeiten, um Unfälle zu vermeiden.

Die strikte Trennung zwischen Mensch und Maschine löste sich indes spätestens mit der Jahrtausendwende auf. Mit der fortschreitenden technologischen Entwicklung halten Roboter zunehmend Einzug in neue Kontexte und immer weitere, umfangreichere Aufgabenbereiche, beispielsweise in Forschung, Bildung, Entertainment, Gesundheitswesen und Militär. Dabei entsteht eine anwachsende Vielfalt von Robotern, welche sich neben ihren Einsatzgebieten in ihrer Mobilität, Erscheinungsform oder auch Autonomie unterscheiden (Ben-Ari & Mondada, 2018). Die Entwicklung von neuen Robotern bedient sich dabei regelmäßig Vorbildern aus der Natur, sodass biologische Systeme als Quellen für Innovationen fungieren. In Anlehnung an die Tierwelt werden beispielsweise im Bereich der Soft Robotics nachgiebige Materialien verwendet, die eine sicherere Interaktion mit Menschen ermöglichen und zudem besonders anpassungsfähig und beweglich sind (Rus & Tolley, 2015). Zudem liefern Oktopus-Arme, Schlangen oder auch Geckofüße Ideen dafür, wie Roboter Objekte besser greifen können. Auch der Mensch dient zunehmend als Vorbild für Robotertechnologien und es werden etwa humanoide Roboter entwickelt, die der menschlichen Gestalt nachempfunden sind (Saeedvand et al., 2019). Allerdings liegt der Fokus in der Entwicklung nicht nur auf dem äußeren Erscheinungsbild der Roboter, sondern auch darauf, sie zu befähigen, menschliche Fähigkeiten auszuführen. Eine Grundvoraussetzung dafür stellt insbesondere ein Faktor dar: die Intelligenz (Dautenhahn, 2007).

2.2 Künstliche Intelligenz (KI)

In ähnlicher Art und Weise wie ein Teilbereich der Robotik zielt auch die Disziplin der KI auf die Simulation menschlicher Fähigkeiten ab. Konkret geht es hierbei um die Befähigung von Systemen, Aufgaben durchzuführen, welche menschliche Intelligenz voraussetzen. Entsprechend zielt KI zum einen auf die Entwicklung kognitiver Fähigkeiten ab, etwa das logische Denken, und zum anderen auf die Befähigung zu lernen sowie eigenständig Probleme zu lösen und Entscheidungen zu treffen (Sarker, 2022).

Frühe Forschungsprojekte in den 1950er Jahren arbeiteten auf Basis der Vermutung, dass jedes Merkmal von Intelligenz so genau beschrieben werden kann, dass eine Maschine in der Lage sein sollte, diese zu simulieren (McCarthy et al., 2006). Innerhalb weniger Jahrzehnte wurde diese Vision in vielfältige Forschungsfortschritte überführt, wie etwa die Entwicklung des Machine Learning. Bei diesem Ansatz können Systeme durch die Analyse von Trainingsdaten

Muster erkennen und darauf aufbauend Entscheidungen oder Vorhersagen treffen (Jordan & Mitchell, 2015). Hinzu kommen weitere neue KI-Technologien und Methoden für verschiedenste Anwendungsbereiche, wie etwa künstliche neuronale Netze, genetische Algorithmen, natürliche Sprachverarbeitung oder Computer Vision (Zhang & Lu, 2021). Diese und andere Technologien finden zunehmend Einzug in unseren Alltag, beispielsweise in Form von Chatbots wie ChatGPT, Gesichtserkennung oder KI-generierten Bildern und Videos.

In der Diskussion um KI wird oftmals die Unterscheidung in schwache und starke KI vorgenommen (Wang & Siau, 2019). Bei schwacher KI, die wir heute bereits anwenden, handelt es sich um Systeme, die die menschliche Intelligenz simulieren und vorrangig einzelne, ausgewählte Aufgaben fokussieren. Schwache KI stellt somit eine eher begrenzte Form der Intelligenz dar, da bereits geringfügige Änderungen einer Aufgabe eine Neuprogrammierung oder Anpassung durch den Menschen notwendig machen. Diese Art der KI wird anhand ihrer Spezialisierung oftmals noch weiter in analytische, funktionale, interaktive, textuelle und visuelle KI unterschieden (Sarker, 2022). Starke KI (auch *Artificial General Intelligence*) hingegen beschreibt Systeme, welche selbst Intelligenz erlangen. Diese Art der KI besitzt einen im Vergleich zur menschlichen Intelligenz ähnlich umfangreichen Anwendungsbereich und die Fähigkeit zur Anpassung an veränderte Umstände. Allerdings existiert eine starke KI bislang noch nicht und es ist umstritten, ob es selbige jemals geben wird (Fjelland, 2020; Goertzel, 2014; Müller & Bostrom, 2016). Die Entwicklung einer derartigen KI würde wiederum die Möglichkeit einer Superintelligenz implizieren – eine Form der Intelligenz, die die kognitiven Fähigkeiten des Menschen bei Weitem überschreitet (Wang & Siau, 2019).

2.3 Die Verbindung von Robotik und KI: Smarte Roboter

Bereits die gesonderte Betrachtung von Robotik und KI verdeutlicht die Logik von technologischer Entwicklung und ihrer Dynamik. Betrachtet man beide Technologien indes nicht separat voneinander, sondern fokussiert deren Verknüpfung, so ergeben sich noch einmal ganz neue Potenziale für Wirtschaft und Gesellschaft (Dirican, 2015; Ribeiro et al., 2021). Beide Technologiedisziplinen können sich nicht nur gegenseitig ergänzen, sondern ebenso befähigen. So können etwa die Einsatzmöglichkeiten von KI durch Robotik erweitert werden, da die Reaktionsfähigkeit eines Roboters überhaupt erst physische Interaktionen einer KI mit der

Umwelt ermöglicht. Gleichzeitig erlaubt die Integration von künstlicher Intelligenz in die Robotik die Bewältigung von Aufgaben, die eine außergewöhnlich hohe Flexibilität und Anpassungsfähigkeit erfordern.

Die Integration von KI in die Robotik kann als logische Antwort darauf verstanden werden, dass die scheinbar einfachen, oft unbewusst gesteuerten sensomotorischen Fähigkeiten des Menschen überraschend schwieriger zu imitieren sind als komplex erscheinende kognitive Aufgaben. Dieses Phänomen ist als das Moravec'sche Paradoxon bekannt (Moravec, 1988), welches besagt, dass intelligente Technologien zwar leicht gegen Menschen im Schach gewinnen können, jedoch das scheinbar einfache Transportieren eines Glases Wasser durch einen Raum ohne Verschütten für sie eine erhebliche Herausforderung darstellt. Entsprechend geht es bei der aktuellen Forschung und Entwicklung im Bereich Robotik weniger um die bloße Simulation isolierter menschlicher Fähigkeiten, als vielmehr um das Zusammenspiel verschiedenster Elemente (Dautenhahn, 2007). Konkret rückt bei neuartigen Robotertechnologien die Verbindung von Verstand, Körper und Umwelt in den Fokus, sodass Roboter sich beispielsweise nicht nur in ihrer Umgebung bewegen, sondern auch auf sie reagieren und mit ihr interagieren können. Eben dies befördert die Verknüpfung von Robotik und KI, was wiederum den Grundstein für intelligente Roboter legt. Charakteristisch für intelligente Roboter sind ihre Fähigkeiten zur Erfassung und Verarbeitung von Wissen und Informationen, zur unabhängigen Interaktion sowie dem Ausdruck von Emotionen (Ren & Bao, 2020).

Die Entwicklung von intelligenten Robotern eröffnet neue Möglichkeiten in der Interaktion zwischen Mensch und Roboter allgemein sowie in Bezug auf Kooperationen im Besonderen (Goodrich & Schultz, 2008). Beispiele für kooperative Mensch-Roboter-Interaktionen sind etwa Assistenzroboter in der Pflege, chirurgische Roboter im Operationssaal, Lernroboter in der Bildung sowie Rettungsroboter in der Katastrophenhilfe. Bei der Entwicklung derartiger Technologien fungieren Mensch-Mensch-Interaktionen oftmals als Vorlage, da davon ausgegangen wird, dass Menschen in der Interaktion mit intelligenten Robotern bekannte Regeln und Muster anwenden. Anders formuliert existiert hier die Annahme, dass die Übertragung von vertrauten menschlichen Verhaltensweisen auf Roboter die Interaktion zwischen Mensch und Roboter vereinfachen kann (Krämer et al., 2012). Entsprechend dieser Logik ist es konsequent, dass der Aspekt der sozialen Intelligenz, als grundlegender Bestandteil der menschlichen Intelligenz, zunehmend bei der Entwicklung intelligenter Robotik berücksichtigt wird (Wiltshire et al., 2017). Soziale Intelligenz beinhaltet dabei etwa die Fähigkeiten zur Imitation, Gestik, Kommunikation oder auch zur Empfindung von Gefühlen sowie zu empathischen Verhaltensweisen (Fong et al., 2003).

Spätestens sozial-intelligente Roboter brechen mit dem traditionellen Verständnis von Robotern als bloße Werkzeuge der Menschen. Sozial-intelligente Roboter haben das Potenzial, die Mensch-Roboter-Interaktion auf eine neue Ebene zu heben und Roboter zu sozialen Partnern und Gefährten zu machen (Bar-Cohen & Breazeal, 2003; Henschel et al., 2021). Sozial-intelligente Roboter lassen es damit denkbar werden, dass Roboter nicht nur an sozialen Interaktionen teilhaben können, sondern gar zu einem Teil der Gesellschaft werden können. Letzteres wird noch einmal wahrscheinlicher, wenn sozial-intelligente Robotertechnologien und humanoide Robotertechnologien miteinander verknüpft werden. Humanoide Roboter sind darauf ausgerichtet, Menschen sowohl äußerlich als auch in Bezug auf Verhaltensweisen nachzuahmen (Park et al., 2005). Die Übereinstimmung von Aussehen und (sozialem) Verhalten beeinflusst die Wahrnehmung von Robotern durch den Menschen (Goetz et al., 2003). Insbesondere wenn Verhaltensweisen von Robotern sozial interpretiert werden, entstehen Tendenzen zum Anthropomorphismus, also zur Zuschreibung menschlicher Eigenschaften (Breazeal, 2003; Złotowski et al., 2015).

3 Soziale Wahrnehmung von Robotern

Die neuen Entwicklungen im Bereich der Mensch-Roboter-Interaktion schaffen die Möglichkeit, dass – wie gerade skizziert – Roboter zunehmend als soziale Interaktions- und Kooperationspartner fungieren und sich möglicherweise zu einem Teil der Gesellschaft entwickeln. Ausgehend hiervon erscheint es wertvoll, die Mensch-Roboter-Interaktion mit einer sozialpsychologischen Brille zu betrachten, um zugrundeliegende soziale Dynamiken aufzudecken und ihre Auswirkungen auf Wahrnehmungen und Einstellungen gegenüber Robotern zu erfassen. Für dieses Ziel bietet sich die Theorie der sozialen Identität an, welche in dem Phänomen der Kategorisierung ihren Ausgangspunkt nimmt, demzufolge Menschen sich selbst und andere Personen zu verschiedenen sozialen Gruppen zuordnen (Tajfel, 1982).

Die Zuordnung zu Gruppen erfolgt regelmäßig anhand von distinkten Merkmalen wie etwa Beruf, kulturellem Hintergrund, sozialem Status, Wertvorstellungen, Lebensstil oder Geschlecht. Ein wesentlicher Grund für die Gruppenzuordnung ist dabei das menschliche Bestreben nach einem positiven Selbstwert, und damit nach Bestätigung und Anerkennung. Der positive Selbstwert bezieht sich darauf, wie Individuen ihren eigenen Wert, ihre Bedeutung und ihre Fähigkeiten einschätzen. Die Gruppenzugehörigkeit zählt dabei zu den Faktoren, welche den eigenen Selbstwert beeinflussen. Konkret steigert eine Gruppenzugehörigkeit bzw.

die Identifikation mit selbiger dann den eigenen Selbstwert, wenn die Gruppe als positiv wahrgenommen wird. Entsprechend identifizieren sich Menschen besonders mit solchen Gruppen, denen sie selbst positive Eigenschaften und Werte zuschreiben. Um letzteres zu stärken, kommt es zur Abgrenzung der eigenen Gruppe gegenüber anderen Gruppen, was auch immer wieder damit verbunden ist, dass Fremdgruppen abgewertet und tendenziell negativ wahrgenommen werden (Tajfel & Turner, 1986).

Prozesse der Kategorisierung lassen sich auch auf die Mensch-Roboter-Interaktion übertragen. In der wissenschaftlichen Literatur wird typischerweise darauf abgestellt, dass Roboter dann sozialen Kategorien zugeordnet werden können, wenn die entsprechenden Roboter menschenähnliche Merkmale aufweisen (Eyssel & Hegel, 2012; Strait et al., 2018). In diesem Kontext werden etwa menschliche Fremdgruppen-Zuschreibungen adressiert, bei denen Vorurteile einzelnen menschlichen Gruppen gegenüber auf spezifische Roboter übertragen werden, infolgedessen es zu positiven/negativen Einstellungsmustern entsprechenden Robotern gegenüber kommt (Barfield, 2023). Eine solche Übertragung lässt sich beispielsweise in Bezug auf Geschlechterstereotypen beobachten. Weiblich aussehende Roboter werden als wärmer, aber auch inkompetenter eingeschätzt, als männlich wirkende Roboter. Diese Einschätzungen korrespondieren mit den grundlegenden stereotypen Wahrnehmungen von Frauen und Männern (Eyssel & Hegel, 2012). Auch die Anwendung nationaler Stereotype auf Roboter ist ein verbreitetes Phänomen. Wurden Roboter in Ländern produziert, die als weniger kompetent eingeschätzt werden, so werden auch die Roboter als weniger qualifiziert wahrgenommen (Spatola et al., 2019).

Jenseits der Kategorisierung von Robotern anhand menschenähnlicher Merkmale ist es für die Zukunft denkbar, dass sich der Wunsch nach einer klaren Abgrenzung formiert. Genährt wird dieser Wunsch dabei durch die perspektivische Schwierigkeit, die Gruppe *Mensch* anhand bisheriger Merkmale wie Aussehen, Intelligenz oder Fähigkeiten einfach und eindeutig von Robotern abgrenzen zu können. Anders formuliert tragen Entwicklungen wie humanoide und sozial-intelligente Roboter dazu bei, dass die klare Trennung zwischen Mensch und Maschine verschwimmt. Hinzu kommt, dass Roboter zukünftig in immer mehr (soziale) Bereiche vordringen können, die bislang Menschen vorbehalten waren. Im Zuge der fortschreitenden technologischen Entwicklung durchdringen Roboter immer mehr unseren Alltag, was sie Schritt für Schritt immer mehr zu einem Teil unserer Gesellschaft werden lässt.

Es ist sogar denkbar, dass die rasanten technologischen Entwicklungen nicht nur den Wunsch nach Abgrenzung befördern, sondern sogar eine Gruppendynamik im Sinne von *Mensch vs. Roboter* entstehen lassen. Eine Gruppendynamik impliziert dabei, dass Roboter als eigenständige soziale Gruppe wahrgenommen und gleichzeitig als Fremdgruppe definiert werden. Beides zahlt dabei darauf ein, dass Menschen ihre (einzigartige) soziale Identität sichern können. An dieser Stelle zeigt sich, dass Roboter sich klar von anderen Technologien unterscheiden, welche in der Vergangenheit einen rasanten Aufstieg erfahren haben – angefangen vom Internet über Smartphones bis hin zu Streamingdiensten. Diese Technologien haben zwar unseren Alltag durchdrungen und sind gewissermaßen allgegenwärtig, allerdings besitzen sie keine *soziale* Relevanz, da sie für soziale Beziehungen als irrelevant anzusehen sind. Entsprechend stellen sie auch nicht die einzigartige soziale Identität der Menschen in Frage.

4 Diskriminierung von Robotern

Das zuvor skizzierte perspektivische Bedürfnis von Menschen, sich von Robotern klar abgrenzen zu können, bietet nicht nur die Möglichkeit, die soziale Identität auf Seiten der Menschen zu stärken, sondern schafft auch Raum für Diskriminierung. Diskriminierung ist ein allgegenwärtiges Phänomen in der Gesellschaft, das in verschiedenen Kontexten auftritt und Resultat einer Abwertung von Fremdgruppen darstellt (Allport, 1954). Allgemein formuliert bezeichnet Diskriminierung ein Verhalten gegenüber Mitgliedern einer Fremdgruppe, bei dem ihnen eine faire und gleichberechtigte Behandlung verweigert wird und es zur Ausgrenzung kommt (Allport, 1954). Ein solch diskriminierendes Verhalten kann sich beispielsweise darin äußern, dass Mitglieder der Fremdgruppe beleidigt werden, dass ihnen Rechte abgesprochen werden oder dass sie im Extremfall körperliche Gewalt erfahren.

Im Kontext von Technologie wird seit geraumer Zeit Diskriminierung diskutiert, wobei typischerweise auf das Problem von diskriminierender Technologie abgestellt wird (Kordzadeh & Ghasemaghaei, 2022). Ein Beispiel hierfür bilden etwa Gesichtserkennungssysteme, welche auf Basis von Algorithmen Gesichter in Bildern oder Videos identifizieren und zuordnen. Untersuchungen haben gezeigt, dass hier die Ergebnisse von der ethnischen Herkunft der Personen signifikant beeinflusst werden. Insbesondere wurde festgestellt, dass einige Algorithmen bei hellhäutigen Menschen genauer funktionieren als bei Menschen mit dunklerer Hautfarbe, infolgedessen es wiederum zu Fehlidentifikationen oder Nichterkennungen bei bestimmten Bevölkerungsgruppen kommt (Bacchini & Lorusso, 2019). Ein weiteres Beispiel für diskriminierende Technologien sind Algorithmen in der Personalvorauswahl, die Bewerbende anhand von Lebensläufen oder

Online-Profilen beurteilen. In diesem Kontext zeigte sich, dass solche Algorithmen Bewerbende aufgrund von Merkmalen wie Geschlecht, Alter oder ethnischer Herkunft ungerecht bewerten können (Kordzadeh & Ghasemaghaei, 2022).

Während Gefahr und Problematik einer diskriminierenden Technologie heute klar gesehen werden, findet das Phänomen einer diskriminierten Technologie bislang kaum Beachtung (eine Ausnahme bildet etwa Barfield, 2023). Die Ausführungen, dass Roboter in Zukunft als ihre eigene soziale Gruppe – bzw. präziser als Fremdgruppe den Menschen gegenüber – wahrgenommen werden, schafft eine grundlegende Voraussetzung dafür, dass Menschen zukünftig Roboter diskriminieren könnten. Hierzu trägt auch bei, dass Fremdgruppenzuordnungen sowie die Tendenz, Fremdgruppen abzuwerten, die Entstehung von diskriminierenden Verhaltensweisen befördern (Tajfel, 1982; Tajfel & Turner, 1986). Gesellen sich hierzu in Zukunft weitere Bedingungsfaktoren für die Abwertung von Fremdgruppen, so könnte es dazu kommen, dass Robotern Einstellungen und Verhaltensweisen entgegengebracht werden, welche klassisch als diskriminierend gelten.

Ein zentraler Bedingungsfaktor für die Entstehung von Diskriminierung von Fremdgruppen sind Ängste und Gefühle der Bedrohung (Allport, 1954). Bereits heute gibt es Hinweise für die Befürchtung, dass Roboter aufgrund fortschreitender technologischer Entwicklungen in immer mehr Bereiche vordringen und hier wiederum Menschen heutige Arbeitsplätze streitig machen werden (Yogeeswaran et al., 2016). Zusätzlich zu dieser ökonomischen Bedrohung tritt das bereits angedeutete Phänomen, dass Roboter zunehmend menschenähnlicher werden, infolgedessen die Einzigartigkeit des Menschseins in Frage gestellt wird (Ferrari et al., 2016). Hierdurch wiederum können Menschen auch auf der symbolischen Ebene das Gefühl bekommen, leicht ersetzbar zu sein. Schließlich ist auch die zunehmende Abhängigkeit der Gesellschaft von Robotern, um etwa Produktivität und Effizienz in der Wirtschaft – und damit gesellschaftlichen Wohlstand – hochzuhalten, ein Auslöser für ausgrenzendes Verhalten, da diese Abhängigkeit zwingend dazu führt, dass Menschen sich Robotern gegenüber verletzlich machen und Verletzlichkeit ebenfalls mit Angst assoziiert ist (Mullis, 2015).

Bereits die hier skizzierten Ängste bilden einen Nährboden für den Beginn der Diskriminierung von Robotern durch Menschen, welche wiederum von unterschwelligen Ausprägungen bis hin zu gravierenden Formen reichen können. Leichtere Formen könnten sich in alltäglichen Situationen zeigen, beispielsweise wenn Menschen es bevorzugen, von Menschen statt von Robotern bedient zu werden oder wenn sie robotergesteuerten Fahrzeugen weniger Vertrauen entgegenbringen als menschlichen Fahrerinnen und Fahrern. Am Arbeitsplatz könnten Roboter beispielsweise in der Form auf Ablehnung stoßen, dass ihnen weniger wichtige Aufgaben zugewiesen werden oder ihre Präsenz als Bedrohung für die Arbeitsplatzsicherheit gesehen wird. Jenseits subtiler Formen von Diskriminierung sind aber auch gravierende Formen einer Roboterdiskriminierung durch Menschen denkbar. Hierzu zählen etwa Forderungen nach strengen Regulierungen zur Begrenzung ihrer Einsatzmöglichkeiten oder die Weigerung von Menschen, mit Robotern zu interagieren oder zusammenzuarbeiten. Auch Formen der Aggressivität gegenüber Robotern sind denkbar, beginnend mit dem Zurücksetzen oder Ausschalten von Robotern zwecks Störung ihrer Funktionsfähigkeit bis hin zu offenen Feindseligkeiten. Letztere beispielsweise in der Art, dass Roboter gezielt sabotiert oder gar zerstört werden.

Obgleich die gravierenden Formen von Roboterdiskriminierung aktuell vorrangig hypothetischer Natur sind, so sind sie durchaus als realistisches Szenario einzuordnen. Erste Hinweise hierauf liefern die wenigen Studien, die sich explizit mit der Ablehnung von Robotern beschäftigen. In diesen Studien wurde gezeigt, dass Vorurteile gegenüber Robotern tatsächlich vor allem dann bestehen, wenn diese Merkmale mit diskriminierten Gruppen von Menschen teilen. So fanden etwa Strait et al. 2018 in einer systematischen Analyse von YouTube-Kommentaren, dass asiatisch und afro-amerikanisch aussehende Roboter eben aufgrund ihres Aussehens von englischsprachigen Webnutzenden abgewertet werden. Diese Tendenz zeigte sich noch stärker in einer Untersuchung des sog. Shooter Bias an Robotern. Bartneck et al. 2018 konnten hierbei zeigen, dass sich diese Neigung von Menschen, in Schießsimulationen schneller auf bewaffnete schwarze Zielpersonen zu reagieren als auf bewaffnete Weiße, auch bei Robotern beobachten lässt. Zudem zeigten Yogeeswaran et al. 2016, dass Menschen nach Konfrontation mit autonomen und androiden Robotern weniger bereit waren, Forschung in der Robotik zu unterstützen. Besonders deutlich wird eine mögliche soziale Ausgrenzung von Robotern in einer Studie von Nijssen et al. 2019, in der sich eine Bevorzugung von Menschen gegenüber Robotern besonders in moralischen Dilemma-Situationen zeigte und Roboter hier bereitwilliger geopfert wurden. Bei den existierenden Studien gilt es zu berücksichtigen, dass selbige sich auf bisher existierende Roboter beziehen, welche zwar Menschen nachahmen, aber noch relativ einfach strukturiert und nicht zu komplexen Gedankengängen fähig sind. Indes ist es plausibel anzunehmen, dass die Ablehnung von Robotern – und damit auch deren Diskriminierung – im Zuge der rasanten Verknüpfung von Robotern und KI in Verbindung mit dem Streben, Menschen nachzuahmen, zunehmen könnte.

5 Implikationen und Ausblick

Gegenwärtig erscheint es noch weit weg – oder gar unvorstellbar – zu sein, dass sich die klaren Trennlinien zwischen Mensch und Maschine auflösen werden. Entsprechend gibt es bislang auch kaum Forschung zu der Frage, welche sozialpsychologischen Implikationen mit einer solchen Auflösung einhergehen könnten bzw. bildlich gesprochen, wie Menschen darauf reagieren könnten, wenn irgendwann die ersten Roboter mit uns am Frühstückstisch sitzen werden. Angesichts der rasanten technologischen Entwicklung im Bereich Robotik sowie der Integration von KI ist ein solches Szenario durchaus im Bereich des Möglichen.

Natürlich ist es denkbar, dass sich zwischen Menschen und Robotern eine positive Beziehung entwickeln wird und es gar nicht zu Ablehnungen oder gar diskriminierenden Verhaltensweisen kommt. Indes ist präsent zu halten, dass die Diffusion von Robotertechnologien – insbesondere im Zusammenspiel mit KI – eine Vielzahl an Änderungen mit sich bringt und Menschen sich nicht immer leicht mit Veränderungen tun. Entsprechend erscheint es – insbesondere vor dem Hintergrund des großen gesellschaftlichen Potenzials von Robotern – wertvoll, Menschen auf Veränderungen vorzubereiten und sie dabei zu unterstützen, neue Phänomene handhaben zu können. Diese Aufgabe inkludiert auch, proaktiv die Frage zu adressieren, welche Konsequenzen möglich sind, wenn keinerlei Begleitung erfolgt. Grundsätzlich erscheint es bei allen neuen Technologien sinnvoll, frühzeitig den Dialog mit Bürgerinnen und Bürgern zu führen sowie bestehende Sorgen und Ängste zu verstehen – und ernst zu nehmen.

Literatur

Allport, G. W. (1954). *The nature of prejudice.* Addison-Wesley.

Bacchini, F., & Lorusso, L. (2019). Race, again: How face recognition technology reinforces racial discrimination. *Journal of Information, Communication and Ethics in Society, 17*(3), 321–335. https://doi.org/10.1108/JICES-05-2018-0050

Bar-Cohen, Y., & Breazeal, C. (2003). Biologically inspired intelligent robots. In *Proceedings of the SPIE Smart Structures Conference* (pp. 14–20). https://doi.org/10.1117/12.484379

Barfield, J. K. (2023). Discrimination against robots: Discussing the ethics of social interactions and who is harmed. *Paladyn, Journal of Behavioral Robotics, 14*(1), Article 20220113. https://doi.org/10.1515/pjbr-2022-0113

Bartneck, C., Yogeeswaran, K., Ser, Q. M., Woodward, G., Sparrow, R., Wang, S., & Eyssel, F. (2018). Robots and racism. In *Proceedings of the 2018 ACM/IEEE International Conference on Human-Robot Interaction* (pp. 196–204). https://doi.org/10.1145/3171221.3171260

Ben-Ari, M. & Mondada, F. (2018). Robots and their applications. In: M. Ben-Ari & F. Mondada (Eds.), *Elements of Robotics* (pp. 1–20). Springer. https://doi.org/10.1007/978-3-319-62533-1_1

Breazeal, C. (2003). Toward sociable robots. *Robotics and Autonomous Systems, 42*(3–4), 167–175. https://doi.org/10.1016/S0921-8890(02)00373-1

Čapek, K. (2004). *R.U.R. (Rossum's Universal Robots)*. Penguin Group.

Dautenhahn, K. (2007). Socially intelligent robots: Dimensions of human–robot interaction. *Philosophical Transactions of the Royal Society B: Biological sciences, 362*(1480), 679–704. https://doi.org/10.1098/rstb.2006.2004

De Graaf, M. M., & Allouch, S. B. (2016). Anticipating our future robot society: The evaluation of future robot applications from a user's perspective. In *25th IEEE international symposium on robot and human interactive communication (RO-MAN)* (pp. 755–762). IEEE. https://ieeexplore.ieee.org/document/7745204

Dirican, C. (2015). The impacts of robotics, artificial intelligence on business and economics. *Procedia-Social and Behavioral Sciences, 195*, 564–573. https://doi.org/10.1016/j.sbspro.2015.06.134

Eyssel, F., & Hegel, F. (2012). (S) he's got the look: Gender stereotyping of robots. *Journal of Applied Social Psychology, 42*(9), 2213–2230. https://doi.org/10.1111/j.1559-1816.2012.00937.x

Ferrari, F., Paladino, M. P., & Jetten, J. (2016). Blurring human–machine distinctions: Anthropomorphic appearance in social robots as a threat to human distinctiveness. *International Journal of Social Robotics, 8*, 287–302. https://doi.org/10.1007/s12369-016-0338-y

Fjelland, R. (2020). Why general artificial intelligence will not be realized. *Humanities and Social Sciences Communications, 7*, Article 10. https://doi.org/10.1057/s41599-020-0494-4

Fong, T., Nourbakhsh, I., & Dautenhahn, K. (2003). A survey of socially interactive robots. *Robotics and Autonomous Systems, 42*(3–4), 143–166. https://doi.org/10.1016/S0921-8890(02)00372-X

George, A. S., & George, A. H. (2023). The rise of robotic children: Implications for family, caregiving, and society. *Partners Universal Innovative Research Publication, 1*(1), 82–101. https://doi.org/10.5281/zenodo.10045270

Goertzel, B. (2014). Artificial general intelligence: Concept, state of the art, and future prospects. *Journal of Artificial General Intelligence, 5*(1), 1–46. https://doi.org/10.2478/jagi-2014-0001

Goetz, J., Kiesler, S., & Powers, A. (2003). Matching robot appearance and behavior to tasks to improve human-robot cooperation. In *The 12th IEEE International Workshop on Robot and Human Interactive Communication* (pp. 55–60). https://doi.org/10.1109/ROMAN.2003.1251796

Goodrich, M. A., & Schultz, A. C. (2008). Human–robot interaction: A survey. *Foundations and Trends® in Human–Computer Interaction, 1*(3), 203–275. https://doi.org/10.1561/1100000005

Henschel, A., Laban, G., & Cross, E. S. (2021). What makes a robot social? A review of social robots from science fiction to a home or hospital near you. *Current Robotics Reports, 2*, 9–19. https://doi.org/10.1007/s43154-020-00035-0

Hockstein, N. G., Gourin, C. G., Faust, R. A., & Terris, D. J. (2007). A history of robots: From science fiction to surgical robotics. *Journal of Robotic Surgery, 1*, 113–118. https://doi.org/10.1007/s11701-007-0021-2

Jordan, M. I., & Mitchell, T. M. (2015). Machine learning: Trends, perspectives, and prospects. *Science, 349*(6245), 255–260. https://doi.org/10.1126/science.aaa8415

Kordzadeh, N., & Ghasemaghaei, M. (2022). Algorithmic bias: Review, synthesis, and future research directions. *European Journal of Information Systems, 31*(3), 388–409. https://doi.org/10.1080/0960085X.2021.1927212

Krämer, N. C., von der Pütten, A., & Eimler, S. (2012). Human-agent and human-robot interaction theory: Similarities to and differences from human-human interaction. In: M. Zacarias & J. V. Oliveira (Eds.), *Human-Computer Interaction: The Agency Perspective* (pp. 215–240). Springer. https://doi.org/10.1007/978-3-642-25691-2_9

McCarthy, J., Minsky, M. L., Rochester, N., & Shannon, C. E. (2006). A proposal for the Dartmouth summer research project on artificial intelligence, august 31, 1955. *AI Magazine, 27*(4), 12–14. https://doi.org/10.1609/aimag.v27i4.1904

Moravec, H. (1988). *Mind children: The future of robot and human intelligence.* Harvard University Press.

Müller, V. C., & Bostrom, N. (2016). Future progress in artificial intelligence: A survey of expert opinion. In V. C. Müller (Ed.), *Fundamental Issues of Artificial Intelligence* (pp. 555–572). Springer. https://doi.org/10.1007/978-3-319-26485-1_33

Mullis, E. C. (2015). Performing with robots: Embodiment, context and vulnerability relations. *International Journal of Performance Arts and Digital Media, 11*(1), 42–53. https://doi.org/10.1080/14794713.2015.1009233

Naneva, S., Sarda Gou, M., Webb, T. L., & Prescott, T. J. (2020). A systematic review of attitudes, anxiety, acceptance, and trust towards social robots. *International Journal of Social Robotics, 12*(6), 1179–1201. https://doi.org/10.1007/s12369-020-00659-4

Nijssen, S. R., Müller, B. C., Baaren, R. B. V., & Paulus, M. (2019). Saving the robot or the human? Robots who feel deserve moral care. *Social Cognition, 37*(1), 41–56. https://doi.org/10.1521/soco.2019.37.1.41

Park, I. W., Kim, J. Y., Lee, J., & Oh, J. H. (2005). Mechanical design of humanoid robot platform KHR-3 (KAIST humanoid robot 3: HUBO). In *5th IEEE-RAS International Conference on Humanoid Robots* (pp. 321–326). https://doi.org/10.1109/ICHR.2005.1573587

Ren, F., & Bao, Y. (2020). A review on human-computer interaction and intelligent robots. *International Journal of Information Technology & Decision Making, 19*(1), 5–47. https://doi.org/10.1142/S0219622019300052

Ribeiro, J., Lima, R., Eckhardt, T., & Paiva, S. (2021). Robotic process automation and artificial intelligence in industry 4.0 – A literature review. *Procedia Computer Science, 181*, 51–58. https://doi.org/10.1016/j.procs.2021.01.104

Rogers, E. (2003). *Diffusion of innovations* (5th ed.). Free Press.

Rus, D., & Tolley, M. T. (2015). Design, fabrication and control of soft robots. *Nature, 521*, 467–475. https://doi.org/10.1038/nature14543

Saeedvand, S., Jafari, M., Aghdasi, H. S., & Baltes, J. (2019). A comprehensive survey on humanoid robot development. *The Knowledge Engineering Review, 34*, Article e20. https://doi.org/10.1017/S0269888919000158

Spatola, N., Anier, N., Redersdorff, S., Ferrand, L., Belletier, C., Normand, A., & Huguet, P. (2019). National stereotypes and robots' perception: the "made in" effect. *Frontiers in Robotics and AI, 6*, Article 21. https://doi.org/10.3389/frobt.2019.00021

Sarker, I. H. (2022). AI-based modeling: Techniques, applications and research issues towards automation, intelligent and smart systems. *SN Computer Science, 3*, Article 158. https://doi.org/10.1007/s42979-022-01043-x

Savela, N., Turja, T., & Oksanen, A. (2018). Social acceptance of robots in different occupational fields: A systematic literature review. *International Journal of Social Robotics, 10*, 493–502. https://doi.org/10.1007/s12369-017-0452-5

Strait, M., Ramos, A. S., Contreras, V., & Garcia, N. (2018). Robots racialized in the likeness of marginalized social identities are subject to greater dehumanization than those racialized as white. In *2018 27th IEEE International Symposium on Robot and Human Interactive Communication* (pp. 452–457). https://doi.org/10.1109/ROMAN.2018.8525610

Takayama, L., Ju, W., & Nass, C. (2008). Beyond dirty, dangerous and dull: What everyday people think robots should do. In *Proceedings of the 3rd ACM/IEEE International Conference on Human-Robot Interaction* (pp. 25–32). https://doi.org/10.1145/1349822.1349827

Tajfel, H. (1982). *Social identity and intergroup relations*. Cambridge University Press.

Tajfel, H., & Turner, J.C. (1986) The social identity theory of intergroup behavior. In: S. Worchel & W. G. Austin, (Eds.), *Psychology of Intergroup Relation* (pp. 7–24). Hall Publishers.

Wang, W., & Siau, K. (2019). Artificial intelligence, machine learning, automation, robotics, future of work and future of humanity: A review and research agenda. *Journal of Database Management (JDM), 30*(1), 61–79. https://doi.org/10.4018/JDM.2019010104

Wiltshire, T. J., Warta, S. F., Barber, D., & Fiore, S. M. (2017). Enabling robotic social intelligence by engineering human social-cognitive mechanisms. *Cognitive Systems Research, 43*, 190–207. https://doi.org/10.1016/j.cogsys.2016.09.005

Yogeeswaran, K., Złotowski, J., Livingstone, M., Bartneck, C., Sumioka, H., & Ishiguro, H. (2016). The interactive effects of robot anthropomorphism and robot ability on perceived threat and support for robotics research. *Journal of Human-Robot Interaction, 5*(2), 29–47. https://doi.org/10.5898/JHRI.5.2.Yogeeswaran

Zhang, C., & Lu, Y. (2021). Study on artificial intelligence: The state of the art and future prospects. *Journal of Industrial Information Integration, 23*, Article 100224. https://doi.org/10.1016/j.jii.2021.100224

Złotowski, J., Proudfoot, D., Yogeeswaran, K., & Bartneck, C. (2015). Anthropomorphism: Opportunities and challenges in human–robot interaction. *International Journal of Social Robotics, 7*, 347–360. https://doi.org/10.1007/s12369-014-0267-6

People Love Technology: Remarks on AI from a Relation Ethics Perspective

Lea Chilian

1 Introduction

People *love* technology—but this chapter is not about sex robots. I would like to share my thoughts on the human enthusiasm for technology, robotics and artificial intelligence, discuss this from the perspective of relational ethics and point out some theological–ethical perspectives.

A conceptual, very basic classification is necessary in advance to reveal my understanding of AI: "AI" refers to forms of software that control machines e.g. robots or that appear as hardware-free digital programs and whose characteristic is that, as "learning systems" (Dabrock, 2022), they themselves continue to generate knowledge and behavioral options. This is referred to as "AI" or "artificial intelligence"—but what does that mean? In fact, there is talk in the literature of a "cacophony" of conflicting definitions and opinions on AI (Hoffmann, Hahn, 2020: 638) and the philosopher Harry Frankfurt refers to such terms as "bullshit words" because "nobody knows what they actually mean" and they say nothing but promise a lot (van Oorschot, 2022: 186). In the following, I therefore refer to a spectrum of AIs on which any digital entity can be placed. A strong AI that independently makes complex decisions, improves itself and communicates its own abilities to the outside world would be the extreme end of the spectrum, but it does not yet exist. A washing machine, on the other hand, can be placed at the lowest end of the spectrum despite "intelligent spot analysis". AIs are data analysis technologies that are similar or superior to human performance. In short,

L. Chilian (✉)
Universität Zürich, Zurich, Switzerland
e-mail: lea.chilian@sozethik.uzh.ch

C. H. Hoffmann (ed.), *Artificial Intelligence, Entrepreneurship and Risk*, Technikzukünfte, Wissenschaft und Gesellschaft / Futures of Technology, Science and Society, https://doi.org/10.1007/978-3-658-45544-6_12

AIs are based on a combination of impressively powerful learning algorithms and other problem-solving techniques based on artificial neural networks.

AIs are not intelligent in the philosophical understanding of intelligence. The term was coined in a research proposal back in 1956, where the idea of creating "intelligent machines" was discussed. In this regard, one could argue that today, every piece of software could be labeled as AI, or none at all, depending on how "intelligent" is defined. It is crucial to inquire about the context of the term's inception, and it's important to note that the term artificial intelligence was already used 50 years ago as a marketing tool to attract research funding.

Opinions differ widely when it comes to assessing modern technologies and the associated opportunities and risks. Peter Dabrock, theologian and former chair of the German Ethics Council, pointed out that the most urgent problem is that of data sovereignty, security and monopolization. People entrust a vast amount of data to AIs, and AIs learn based on this data. However, data collection can lead to dependencies (e.g. unpredictability of computational results) and discriminations (e.g. smokers being denied health insurance). The issue of judgment, responsibility, and accountability by AI systems represents another typical dilemma. Many people fear that through autonomous decisions, AIs could become too powerful and, in a Hollywood-like scenario, threaten the existence of humanity. They raise the question of whether we should allow AIs to become autonomous. In March 2023 more than 1000 experts from technology and research publicly called for a pause in development, a memorandum for new AI models, as further consideration would first have to be given to what possibilities for action AIs should acquire and whether we should allow them to become autonomous. (Whether Elon Musk, a supporter of the memorandum, should be taken seriously, just weeks before he acquires an AI company, is up for debate).

The debate about the benefits and dangers of AIs often uses religious language, both in general linguistic jargon and in philosophical considerations. Dabrock therefore calls for theological deconstruction work where, in Bonhoeffer's words, expectations of AIs concern the *ultimate.* Precisely because AIs are shaping and will shape the world we live in, theological ethics is called upon to engage in the ongoing debates and interdisciplinary commitment.

For this purpose, I conduct some (self-)observations regarding the human relationship with technologies (Sect. 2). Subsequently, I provide a brief overview of the ethical issues surrounding AIs and the blind spots that, in my opinion, can be observed therein (Sect. 3). Therefore, I propose a relational ethical approach to discussing AIs (Sect. 4) and offer some theological–ethical conclusions at the end (Sect. 5).

2 Observation: People Love Technology

Let's start with an exploration:

a. Have you ever complained to the navigation device in your car because it navigated you incorrectly?
b. Have you ever threatened to throw a printer or photocopier out of the window if it doesn't finally do what it was asked to do?
c. Have you or do you know someone who has given a first name to a car, robot vacuum cleaner or other technical device?
d. Have you ever laughed at their answers and (supposed) humor when using ChatGPT or communicating with Siri, Alexa, etc.?
e. Have you ever said "please" when using ChatGPT or communicating with Siri, Alexa, etc.?

Did you answer with "yes" more than three times? Most people would... This brings us straight to the point: people communicate, make contact and build relationships. This small enquiry mentioned important points to an observation: People engage in relationships with technology, machines, and learning systems. They give them names, address them, berate them, and find amusement in them—people love technology, robots, and AIs. I quote the socio-informatics expert Katharina Zweig: "We humans are surprisingly trusting of computers. That was already the case in 1966 with Eliza, the first chatbot. It was very simple, but people still trusted it with their innermost thoughts. This has been the case throughout the history of technology." (Zweig, 2023; translation ChatGPT). So we have a lot of faith in technology and in AIs. We trust it to do good things, to benefit us, because that is our understanding of technology—that we have created it to support us. And we tend to interact with technology that we use frequently and that helps us. We build emotional bonds with AIs and some AIs are programmed to respond to human emotions. They seem to be "subjects of thought, feeling and action" (Moos, 2021: 236; translation ChatGPT). Here, ethical questions inevitably arise regarding the moral foundations of our relationship with these mechanical creatures and disembodied conversational partners. Thorsten Moos, theologian and physicist, develops the concept of digital animism for the observable phenomena of familiarity between humans and AI. In ethnology, animism refers to the belief in the soulfulness of nature. For humans, technology is "more than mere stuff with merely functional significance" and AIs irritate our "old idea of a passive world of things" (Moos, 2021: 239, translation ChatGPT).

We are, according to Katharina Zweig, "at the point where people can only poorly differentiate: What comes from themselves, what comes from the machine? That will shake us up." (Zweig, 2023; translation ChatGPT). The subjectivization of technology raises questions about the status of technology and inevitably also about the status of humans—technical–ethical issues have to do with implicit anthropology. James McBride points out that the more clearly AIs and robots behave like humans, the more people will treat them as social agents and therefore expect moral guidance and liability from them (McBride 2019). So what has the ethical debate on these quasi-acting AIs looked like so far? And where, in my opinion, is their blind spot?

3 Context: Ethics and AIs

> *"Comparable in suddenness and violence to a natural disaster, modern technology came upon the Western peoples."* (Tillich, 1927: 297; translation ChatGPT)

The incorporation of technology into our actions appears to us as both commonplace and revolutionary. Technology serves here as a container term. Technology simply refers to instruments created by humans for use. Whether technology is good or bad is determined by the users. The theologian Paul Tillich formulated: "Technology is neutral. It provides means… In its potential lies the power to attempt. Temptation is always present where possibilities arise." (Tillich, 1927; 304, translation ChatGPT). Here, the threatening scenarios already mentioned above resonate.

Feelings of skepticism and threat as well as enthusiasm towards learning systems are distributed differently around the world depending on the culture. There is often a tendency to use AIs as a scapegoat for discrimination, accidents and damage (at least this is how I sometimes deal with printers). However, blaming technology is only possible if it is identified as acting independently and bearing responsibility. The action of so-called autonomous systems, especially those known from weapons or cars, raises questions about responsibility and the assumption of blame in the event of accidents. First of all, I would like to add to this aspect that these systems are far from acting autonomously in the sense in which we find Kant's definition of autonomy. Nicole Kunkel has therefore suggested the more correct term *auto-generative* systems, which I would like to endorse here (Kunkel 2023). But also an auto-generative vehicle can cause serious harm to users. This raises the question of responsibility—and this is a genuinely ethical one.

Ethicists distinguish between three areas for the present thematic context: Machine ethics asks about normative-practical guidelines for machines, while technology ethics addresses the ethically responsible use of technology—including complex machines. A borderline area is represented by engineering ethics or "ethics for programmers", which asks about the responsibility of those who program a code into the machine. (Klinge, 2021: 33; translation ChatGPT) Technology ethics is therefore concerned with the ethics of self-learning systems, whereas ethics for programmers does not (yet) expect systems to make independent decisions, but sees the responsibility as lying with humans who inscribe rules of action into the technology. I would like to avoid using the word "autonomy" here. The ethics community is not in agreement as to whether a machine can truly be or become autonomous in the Kantian sense of the word (i.e. giving itself a law). The question of whether AIs are to be treated as moral agents/ moral actors and are responsible, or whether humans can transfer responsibility to them, dominates the current debate (not only for vehicles, but also in medical diagnostics, for example). As AIs are not yet able to make decisions as moral agents, but this is envisaged for the future, Catrin Misselhorn discusses the extent to which moral implementation is possible for machines (Misselhorn 2022). Hendrik Klinge argues that the application of an inscribed moral code alone is not sufficient. For machines to be considered moral actors, they must also be able to make autonomous decisions on ethical issues (Klinge, 2021: 34). So far, however, AIs have not been able to do this; instead, they calculate on the basis of calculations. They therefore merely simulate intelligence and morality (Nida-Rümelin, Weidenfeld 2018).

In my opinion, this topic is highly interesting, relevant, and sensitive. However, Peter Dabrock criticizes the prioritization of this question in current AI ethics, a perspective with which I agree. However, I believe that further considerations are necessary to ethically understand and contextualize AIs. The current discussions on the question of the responsibility of AIs view AIs as actors in complete isolation. However, an ethics that does not include the entities involved and consider them in their relationships and dependencies remains incomplete and fails to recognize a central part of action (Leist 2005). For Paul Tillich had already recognized that "not infrequently a relationship develops between the machine and its driver that is related to the relationship with a living being." (Tillich, 1927: 301; translation ChatGPT). This leads me to the approach of including the relationship dimension in technology ethics and declaring it relevant for the ethical support of AI development. After all, we need to better understand our relationship to AIs and their impact.

Let me be clear: My following remarks in no way undervalue the important efforts–such as those initiated by numerous NGOs and the EU Parliament– to formulate principles and requirement catalogs for AI development. Rather they seek to complement and support them. These efforts are vital and should undoubtedly continue. Binding frameworks for AI regulation are absolutely essential, as for any other use and application of technology. At present, such draft regulations are criticized because they are considered too broad, vague, general or not specific enough (Hoffmann, Hahn 2020). Many of the drafts can be boiled down to five principles: transparency, justice and fairness, non-maleficence, responsibility, and privacy (Jobin et al 2019). They are reminiscent of the four principles of bioethics and that is no coincidence. But, like the bioethical principles, they lack terminological clarity. The principles must be interpreted and applied for each individual case—that is what makes them practicable for the *specific* clinical case. I mentioned earlier that we should speak of a spectrum of AIs. So we need rules for specific areas of application of the various AIs. Contextual rules are necessary—and in my opinion they are easier to develop if the relationships that stimulate the specific application are taken into account.

The dominance of the question of the moral status of AIs and the seemingly endless debates about it indicate not only that it is a complex problem, but that various aspects need to be considered to solve it—and perhaps not all have been considered yet. My thesis is that to treat the current debates on human–AI-interaction solely in terms of responsible ethics misses the complex phenomenon of AI use. I am not alone in this view. Moos for example is also more interested in the "social practices" in which AIs are treated as subjects (Moos, 2021). From this, he develops the normative question of which practices are appropriate for the subjectification of machines. However, I would like to approach the topic in a descriptive rather than a primarily normative way. In other words: How do relationships with AIs manifest and what conclusions can be drawn from a relational-ethical approach? What social and individual changes result from interactions with AIs and how are these to be evaluated ethically?

The dichotomy between man and machine, which has been perpetuated since Descartes, is being challenged by AIs, according to many voices in the field of technology ethics. I would like to rephrase this statement: The distinction between man and machine is not challenged by AIs—but it is challenged by the behavior of humans when they interact with it. When humans relate to AIs, it changes them and it changes our understanding of AIs. It is evident that moral philosophy, when addressing the actor's perspective, inavitably makes exclusionary decisions about who is included in the community of moral actors and who or what can be excluded from consideration. Consequently, this discussion revolves around issues of recognition and an analyzis of our relationshipwith the entity in question. In other words, moral status is not primarily determined by the capacity to take responsibility for ethical considerations, but rather emerges from our interactions and relationships with others (cf. Puzio, 2023).

4 An Relational–Ethical Perspective on AIs

> *"Ethics is applied anthropology in every sense"* (Wolfgang Trillhaas cited in Klinge, 2021: 37; translation ChatGPT)

So what image of humanity are we dealing with in AI ethics?

Florian Höhne has worked out that the image of humans as risk-informed decision-makers shapes the construction and use of AIs (Höhne 2022). Risk-informed decision-makers try to cope with the unpredictability of the world by calculating it. Humans create complex systems for this task. AIs are therefore images of risk-informed decision-makers and serve them in their need for security. At the same time, according to Höhne, it is to be expected that the use of AIs developed in this way will lead to the human image of the risk-informed decision-maker being reproduced and intensified—a circular influence.

Much more central to anthropology, however, is the doctrine of the human being as a relational being that is designed for sociality and can only be determined in its relationality. In the development of technology, however, this area seems to have received little attention so far; instead, the desire to contain the unknown and the threatening has dominated the development of smart computing technology. Even if people create their AIs in their own image and according to their needs, they still interact with them as the relationship-oriented beings that they are. In this respect, a relational ethical view of AIs is necessary in order to be able to take these orientations and their effects into consideration.

Modern anthropology understands humanity on the basis of its relational structure (Schwöbel, 2002). Relationships in which humanity exists form the basis of attempts to determine how humanity is to be understood (Schwöbel, 2002). In science, relationality is an extremely broad term that is used in many disciplines. It is not only used to describe the relationality and relational capacity of life. In biology, humans are described as beings that are in relationships with many other aspects of the world ("ecosystems"). In social sciences, human structure and purpose are determined by their social relationships with other people. And in Christian-theological anthropology, human existence is identified as being-in-relationship, constituted in the triune God's relationship to humanity (Schwöbel,

2002). We are dealing with a divine and creaturely structure of relationships. Ingolf U. Dalferth therefore points out that in a multi-relational lifeworld, concepts of only subject–object or subject–subject relations are oversimplifications (Mühling-Schlapkohl 2004; Dalferth 1984). Relational theories are about eliminating dichotomies and in this respect this approach is suitable for my concern of overcoming either anthropocentric or technocentric AI ethics.

A distinction must be made between reductionist and non-reductionist approaches to the ethics of relationships. I take a non-reductionist approach here. Accordingly, the relational dimension is not the only dimension of being human (that would be reductionist), but I adhere to a multidimensional view of being human, according to which specific relationships in which humans exist provide a further perspective on being human that must be included.

Judith Butler uses relation in the sense of a relationship that one has to others, and through which, for example, responsibility is justified (Butler, 2007). Possibilities for action are therefore not created in a person, a disposition or in resources, but in the sequence of positioning in a network (Altissimo et al., w.Y.). This leads to the conclusion that neither AIs nor humans, but rather the network in which both are found, should be used to understand ethical questions of responsibility allocation in the field of AI ethics. When analyzing this network, I advocate taking a close look at the relationship structures and relationship dynamics between humans and AIs in order to identify responsibilities and mutual dependencies. As a consequence, this shifts my focus away from an individualistic ethics that prioritize maximum autonomy and the greatest possible insulation from dependencies—an approach implicitly reflected in the question of AI's capacity for responsibility. An relational-ethical approach aims to make the fundamental dependency and dependence ethically visible. Butler calls this *constellations of vulnerability*. Even more precise than talking about dichotomous dependencies is therefore the more concrete formulation of "entangledness" with others (Villa, 2012: 121ff). When addressing normative-ethical questions about AI applications– such as how humans should interact with AIs and what levels of responsibility can be attributed to them– two researchers in human–machine interaction propose that the ability to maintain long-term relationships, as empirically observed, is a defining characteristic of personhood. Therefore, it could serve as a useful indicator of how machines should be treated (Mac Dorman; Cowley, 2006; Hoffmann, Hahn 2020).

Consider, for example, a chatbot that has been assigned a relaxed, humorous character. In such cases, people tend to treat an AI system as human-like due to emotional bias and are therefore likely to attribute moral significance to it. This behavior cannot be criticized. After all, if we didn't do this, wouldn't we be violating our own relationality and the emotions associated with it? Several studies show this: People develop tools, but they develop an emotional responsiveness to them (Gal 2019). From a technocentric moral perspective, we are therefore free to develop a criterion that assigns a tool a moral status or not—but if we classify our emotional relationships as relevant for the development of ethics, this has corresponding implications. The inclusion of care robots is an interesting illustrative example of such perspectives. People with dementia often find it surprisingly easy to interact with them and appear to experience improvements in their quality of life, even when human interaction has become challenging. What does this imply about the significance of human attention, which we typically regard as paramount?

I'd like to indicate some more ethical questions, and I illustrate this with another example, briefly touching upon AI romances and sex robots. Research has already shown that gender stereotypes and discriminations based on gender, race, sexual orientation, etc., are supported in these contexts (Ott 2022). There are also reports of sexual violence perpetrated by AIs, as systems learn from all users, including those who engage in violent or demanding behavior. Additionally, ethical questions arising alongside data security and responsibility issues are specific from a relational ethical perspective. For instance, what are the implications when the most common assistant systems have female voices and names? Does the possibility of having "perfect" relationships with AIs– capable of anticipating and managing our desires, mood swings, and shortcomings without experiencing hurt– change how we evaluate the quality of our frienships and relationships with other human beings? Does our capacity for frustration tolerance and empathy diminish because AIs are neither fallible nor needy?

Humans and machines cannot be considered ethically equal or equivalent. However, treating them as things would not honor our forms of interactions and relationships with them and would violate our ability to relate to them. Ethical concepts for the regulation and development of AIs should therefore always keep in mind what AIs mean for us humans and which understandings of ourselves and the world are reflected and manifested in their construction and in our interaction with them. Only then will ethics serve humans, which is its most urgent task.

5 Orientations: Theological–Ethical Perspectives on an Relational–Ethical Approach

"Theological ethics is ethics for people." (Klinge, 2021, 36; translation ChatGPT)

In his reflections on the logos and myth of technology, Paul Tillich points out that technology frees us from pain, work and limitations, but also from the uncanny—today we would add that AIs make it possible for us to calculate and contain risks. What is technically shaped and known is controllable, and thus, according to Tillich, "violence has taken the place of eros, which connects beings with beings." "The violence of rationality (…) cuts us off from the living stream that runs through all beings" (Tillich, 1927: 305, translation ChatGPT). Nowadays, this sounds terribly critical of technology and seems to "long for a supposedly lost religious past" (Moos, 2021: 75; translation ChatGPT). But at this point, Tillich is actually dealing with the question of what deep religious structure is expressed in technology. "Technology has reshaped the world, and this reshaped world is our world. We must build on it, knowing full well that if it is divine, if it is creative, if it is liberating, then it is also demonic, subjugating and destructive. It is ambiguous, like everything that is." (Tillich, 1927: 306; translation ChatGPT).

For the human being who orients himself and experiences himself in this world of rationality that he has created, it is helpful to take up another of Tillich's ontological determinations: "Man becomes man in personal encounters." (Tillich, 1991: 193; translation ChatGPT). These encounters, from which relationships are formed, are therefore not only a central place for the negotiation of ethics, but are also where humanity is constantly formed. Kate Ott asks: "Is it the nature of the machine, or the nature of the relationship and the human's understanding of that relationship, that matters most to the ethical deliberations?" (Ott 2022, 75). For Ott, the relevant question is not the moral status of AI—because little can be gained from this for our practice—but the understanding of the relationship and how we describe the interactions between humans and AIs. Because: "Humans bond with all sorts of robots and AI devices" (Ott 2022: 78), and "Humans are relational beings." (Ott 2022: 84). Ott justifies this in terms of a theology of creation. God enters into bonds, we also speak of covenants, and since we understand humans to be created in the image of God, humans would also strive "to seek everywhere a partner with whom to relate, not just among ourselves, human with human, but also with someone or something different from us." (Herzfeld, 2017: 95).

In theology, the concept of relationship is the starting point for many explorations of the God–human relationship. The relational dynamics within the Trinity and between God and humanity are central to understanding interpersonal relationships. These theological insights provide valuable material that can inform discussions in AI ethics. After all, the awe-inspiring complexity of relationships must not result in scientific silence or confusion (Meyer zu Hörste-Bührer; Bührer 2018: 3). Schwöbel states that God's relationship with humanity is the key to understanding all relationships (Schwöbel, 2002). In this respect, theological ethics can speak about faithfulness, responsibility, guilt and also grace in the face of the "merciless freedom" of AIs (Moos, 2021) and contribute to the understanding of AI–human relationships.

Ott identifies the contribution to a good society as the central aim of ethical questions in the context of AIs. From the perspective of virtue ethics, and in line with Aristotle, this is framed as the question of the good life within a community. This approach complements the applied AI ethics, which is predominantly shaped by a normative ethics of action and judgment. Johannes Fischer ritiques ethical inquiries that focus solely on two-valued judgments of action (right/wrong). Instead, he emphasizes that ethics must argue ethos-relatively. This includes "forms of hermeneutically directed ethical reflection [...] which raise awareness of existing orientations, interpret them and concretize them with regard to current moral questions" (Fischer, 2002: 201, translation ChatGPT). With the help of a relational ethical orientation, AIs can be measured by the extent to which they contribute to the good life (of all). According to Trutz Rendtorff, the task of ethics is to "keep the communication about shared ways of lving alive" (Rendtorff, 1990: 10). A theological, relational ethical approach views the shared way of life as extending far beyond human–to–human relationships, incorporating these into the broader context of shared living in a digitalized world.

6 Conclusion

The ethical questions raised here in AI ethics depend on the chosen description of the subject matter. The ethical discourses on the autonomy and moral responsibility of AIs represent a common description of AIs, e.g. as risk calculators. But what impact does it have on humans when they interact with AIs that are supposed to be risk-informed decision-makers?

In order to get to the bottom of further issues and ethical challenges, it is necessary to expand the description of the subject of "AI". Is there a risk of shifts in relational dynamics or a loss of relational depth if we can cultivate 'perfect' relationships with AIs and even prefer relationships with AIs over those with fellow humans? These and further questions will shape our professional and personal lives. However, for a responsible approach towards employees, coworkers, friends, and family members, it seems appropriate to engage with these questions.

In this chapter, I have proposed a reflection on changes in the AI debate through the lens of relational ethics in order to complement the ethical discussions surrounding AIs. After all, discussions about the moral status of AIs have dominated our debates so far—but these are questions that concern the aforementioned end of the spectrum, which we have by no means reached yet.

The relational-ethical approach represents a model for the inclusion of both the organic and the technical–digital in the expanding boundaries of a new ethics and offers the possibility not to develop a two-part ethics, one for humans and one for the technical, but to include the effects of interactions, as is already happening, for example, in environmental ethics discourses. The Christian understanding of the world and the human being provides further perspectives on how life with AI systems can be viewed ethically. We each individually decide whom or what we attribute moral status to. This often occurs at the individual level. This is not a negation of a normative, universal ethics, but rather an invitation to incorporate these realities into ethical reflection. An individual's status as a moral subject is not predetermined by a firmly established and stable ontological position, involving interactions with others. Instead, it is shaped by social relations that determine one's identity and role.

Building upon this critical awareness of our interconnectedness and entangledness, and the nuanced understanding of relational dynamics with AIs in our ethical deliberations, we could potentially dispel a concern articulated by computer pioneer Konrad Zuse, who warned of a future where "the danger of the computer becoming like humans is not as great as the danger of humans becoming like the computer." (Zuse cit. by Platow 2021, translation ChatGPT).

References

Altissimo, Alice; Eßer, Florian; Herz, Andreas; Köngeter, Stefan (w.Y.), Relationale Zugänge in angewandten Sozialwissenschaften, in: relational approaches in the applied social sciences [relapp], https://relapp.org/was-bedeutet-relational/ (letzter Zugriff: 30.01.2024)

Butler, Judith (2007), Kritik der ethischen Gewalt. Adorno-Vorlesungen 2002 (stw 1792), Frankfurt/M: Suhrkamp.

Dabrock, Peter (2022), »Prüft aber alles und das Gute behaltet!« Theologisches und Ethisches zu Künstlicher Intelligenz, in: Theologische Literaturzeitschrift Nr. 7, 635–650.

Dalferth, Ingo U. (1984), Existenz Gottes und christlicher Glaube. Skizzen zu einer eschatologischen Ontologie (Beiträge zur evangelischen Theologie 93), München: Kaiser.

Fischer, Johannes (2002), Theologische Ethik. Grundwissen und Orientierung (Forum Systematik Bd. 11), Stuttgart: Kohlhammer.

Gal, Danit (2019), Perspectives and approaches in AI ethics: East Asia, in: Dubber, M.; Pasquale, F.; Das, S. (Hg.) Oxford handbook of ethics of artificial intelligence. Oxford University Press, Oxford.

Herzfeld, Noreen (2017), Religious Perspectives on Sex with Robots, in: Danaher, John; McArthur, Neil (ed.), Robot Sex. Social and Ethical Implications, Camebridge: MIT Press, https://doi.org/10.7551/mitpress/10718.003.0011.

Höhne, Florian (2022), Bilder des Menschlichen. Theologisch-ethische Herausforderungen der Vorstellungswelten künstlicher Intelligenz. In: van Oorschot, Frederike/Fucker, Selina (Hg.): Framing KI. Narrative, Metaphern und Frames in Debatten über Künstliche Intelligenz.

FEST Forschung, Band 2, Heidelberg, heiBOOKS: S. 111–135. https://doi.org/10.11588/heibooks.1106.c 15196

Hoffmann, Christian Hugo, Hahn, Benjamin (2020), Decentered ethics in the machine era and guidance for AI regulation, in: AI & SOCIETY (2020) 35:635–644, https://doi.org/10.1007/s00146-019-00920-z.

Jobin A, Ienca M, Vayena E (2019) The global landscape of AI ethics guidelines. Nat Mach Intell 1:389–399.

Klinge, Hendrik (2021), Jenseits des Personalismus. Maschinenethik als theologische Herausforderung, in: Zeitschrift für Evangelische Ethik, 65. Jg., S. 31–45.

Kunkel, Nicole (2023), Autoregulative Weapons Systems: Automatization challenging Peace Ethics. In: Puzio, Anna/Kunkel, Nicole/Klinge, Hendrik [Hg]: Alexa, wie hast du's mit der Religion? Theologische Zugänge zu Technik und Künstlicher Intelligenz/Theological Approaches to Technology and Artificial Intelligence (Theologie und Künstliche Intelligenz). Wissenschaftliche Buchgesellschaft: Darmstadt 2023, S. 191–206.

Leist, Anton (2005), Ethik der Beziehungen. Versuche über eine postkantianische Moralphilosophie, Deutsche Zeitschrift für Philosophier Sonderband 10, Berlin: Akademie-Verlag.

MacDorman, Karl F.; Cowley, Stephen J. (2006), Long-term relationships as a benchmark for robot personhood. In: Proceedings of the IEEE, international workshop on robot human interaction communication, pp 378–383.

McBride, James (2019), "Robotic Bodies and the Kairos of Humanoid Theologies," Sophia 58: 663–76, https://doi.org/10.1007/s11841-017-0628-3.

Meyer zu Hörste-Bührer, Raphela J.; Bührer, Walter (2018), Relationale Erkenntnishorizonte als Bindeglied zwischen Exegese und Systematischer Theologie?, in: dies. (ed.), Relationale Erkenntnishorizonte zwischen Exegese und Systematischer Theologie (Marburger theologische Studien 129), Leipzig: Evangelische Verlagsanstalt, 1–13.

Misselhorn, Cathrin (2022), Artificial Moral Agents. Conceptual Issues and Ethical Controversy, in: The Cambridge Handbook of Responsible Artificial Intelligence: Interdisciplinary Perspectives, hg. von S. Voeneky et al., Cambridge, 31–49, https://doi.org/10.1017/9781009207898.005.

Moos, Thorsten (2021), Digitaler Animismus. Theologische Bemerkungen zu einer Ethik der Digitalisierung. In: Held, Benjamin/van Oorschot, Frederike (Hg.): Digitalisierung: Neue Technik, neue Ethik?. FEST Forschung, Band 1, Heidelberg, heiBOOKS: S. 235–256. https://doi.org/10.11588/heibooks.945.c12684

Mühling-Schlapkohl, Markus (2004), Art. Relationalität, in: RGG^4 Bd. 7, Tübingen: Mohr Siebeck, 258–261.

Nida-Rümelin, Julian; Weidenfeld, Nathalie (2018), Digitaler Humanismus. Eine Ethik für das Zeitalter der künstlichen Intelligenz, München 2. Aufl.

Ott, Kate (2022), Sex, tech, and faith. Ethics for a digital age, Grand Rapids: William B. Eerdmans Publishing Company.

Platow, Birte (2021), Homo, Homo digitalis, Homo deus—alte und neue Menschenbilder im Diskurs, in: Online Publikationsorgan des Katholisch-Soziales Instituts Siegen, 11/2021, https://www.ksi-institut.de/service/download/ksi-inform/index.html.

Puzio, Anna (2023), Zeig mir deine Technik und ich sag dir, wer du bist?—Was Technikanthropologie ist und warum wir sie dringend brauchen, in: Hermann Diebel-Fischer, Nicole Kunkel, Julian Zeyher-Quattlender (Hg.), Mensch und Maschine im Zeitalter "Künstlicher Intelligenz". Theologisch-ethische Herausforderungen (Leiten, Lenken, Gestalten 45), Berlin, Münster: LIT, 9–28.

Schwöbel, Christoph (2002), Gott in Beziehungen. Studien zur Dogmatik, hg. v. Katrin König; Katrin Bosse, Tübingen: Mohr Siebeck, 2., durchgesehene und korrigierte Auflage.

Tillich, Paul (1927), Logos und Mythos der Technik. In: Albrecht, Renate (Hg.) (1968): Gesammelte Werke. Bd. IX. Stuttgart, Evangelisches Verlagswerk: 297–306.

Tillich, Paul (1991), Liebe - Macht – Gerechtigkeit: V. Die Einheit von Gerechtigkeit, Liebe und Macht in zwischenmenschlichen Beziehungen, Berlin, Boston: De Gruyter, 1991. https://doi.org/10.1515/9783110869200

Rendtorff, Trutz (1990), Ethik. Grundelemente, Methodologie und Konkretionen einer ethischen Theologie. Bd. I (Theologische Wissenschaft. Sammelwerk für Studium und Beruf 13,1), 2. überarbeitete und erweiterte Auflage, Stuttgart, Berlin, Köln: Kohlhammer.

van Oorschot, Frederike (2022), Framing KI. Perspektiven für eine imaginationssensible Ethik Künstlicher Intelligenz. In: van Oorschot, Frederike/Fucker, Selina (Hg.): Framing KI. Narrative, Metaphern und Frames Künstlicher Intelligenz in interdisziplinärer Perspektive. FEST Forschung, Band 2. Heidelberg, Heidelberg University Publishing: 177–189. https://doi.org/10.11588/heibooks.1106.c15200.

Villa, Paula-Irene (2012), Judith Butler. Eine Einführung, 2. akt. Ausgabe, Frankfurt/M: Campus.

Zweig, Katharina (2023), Können wir künstlicher Intelligenz vertrauen, Frau Zweig? Interview Rudi Novotny, Stefan Schmitt, in: ZEIT Lebensfragen 2023/32, https://www.zeit.de/2023/32/katharina-zweig-ki-vertrauen-aufgaben (letzter Zugriff: 30.01.2024).

Transsubjectivity: Artificial Intelligence and the Extension of the Social Lifeworld

Jörg Noller

1 Embedded Artificial Intelligence

Artificial intelligence (AI), especially large language models (LLMs) have become the center of philosophical interest more than almost any other topic related to digitalization. In this paper, I will link the ethical dimension of AI to its epistemological status, thereby drawing on Kant's conception of the power of judgment. AI's ethical quality depends on what AI actually is and does. For example, we can ask whether AI is a medium, a simulation, a subject, or a virtual reality (VR). In any case, AI is transforming our everyday spheres of action and thus our ethical concepts: "Artificial intelligence will challenge all those concepts, and more, as ethicists begin to digest the problem of continued human coexistence with alternate (and perhaps superior) intelligences. That is to say, AI will challenge the very way in which we have tried to reason about ethics for millennia. If this is correct, novel approaches will be needed to address the ethics of AI in the future."[1] Drawing on Noller 2024, this paper aims to determine the relationship of AI's human–machine interaction from the perspective of human responsibility, relating it to the problem of virtual reality (VR). To this end, the

[1] Powers/Ganascia 2020, 28.

J. Noller (✉)
Ludwig-Maximilians-Universität München, Munich, Germany
e-mail: joerg.noller@lrz.uni-muenchen.de

C. H. Hoffmann (ed.), *Artificial Intelligence, Entrepreneurship and Risk*, Technikzukünfte, Wissenschaft und Gesellschaft / Futures of Technology, Science and Society, https://doi.org/10.1007/978-3-658-45544-6_13

concept of transsubjectivity is introduced,[2] distinguished from the concept of intersubjectivity, and motivated within the framework of the digital human lifeworld,[3] which Felix Stalder has called the "digital condition".[4] This will allow understanding AI in such a way that it is not located within the framework of a subject–object schema, as suggested by the term "human–machine interaction," but can be understood as part of an transsubjective and interobjective[5] lifeworld, in which human capacities and the performance of AI are deeply interwoven.[6] Ethical questions regarding AI can be discussed appropriately based on such an understanding of the broader lifeworld framework.

The human lifeworld is fundamentally structured by human goals and intentions. As I will show below, AI can be integrated into the human lifeworld by extending, relating, abstracting, and realizing these goals regarding digital structures that correspond to it. In this way, AI becomes transsubjective: It is not separated from human subjectivity as an own subject (e.g. as a "superintelligence"[7]), nor is it separated from human subjectivity as an object (e.g. a tool). Rather, the digital lifeworld is structured by human goals and interests, which are mediated and transformed by processes of AI as in the case of LLMs. An ethical evaluation of AI performance, therefore, depends on whether and how we are able to analyze these intimate relationships between human intentions, algorithms and databases in terms of conceptual and teleological processes of transsubjective extension.

Methodologically, I will approach the phenomenon of AI from a critical middle position and distance myself from the following three positions: (1) the *ideologization* of AI in the context of a transhumanist "Silicon Valley ideology,"[8] as currently represented by Ray Kurzweil in the context of his technological

---

[2] The concept of transsubjectivity has been discussed in various epistemological contexts, however, without being related to the phenomenon of AI. On the relationship between transsubjectivity and digitalization, see Noller 2022, 63.

[3] For a lifeworld account of AI that refers to Husserl's notion of "Lebenswelt," see Durt 2022.

[4] Stalder 2018.

[5] On the concept of interobjectivity, see also Hui 2016, 158–160, who, however, does not understand this concept in a lifeworld sense, but only in a technological sense.

[6] For an account of AI beyond subjectivity and objectivity, see Durt 2022, 70, and Noller 2022, 61.

[7] Bostrom 2014.

[8] Nida-Rümelin/Weidenfeld 2022, 4.

singularity thesis[9]; (2) the *trivialization* of AI as a mere tool, i.e., as a technological reduction; (3) the dramatization of AI as a danger of modernity to be combated, i.e., as an ideological counter-position to (1). I will develop this critical middle position towards AI by understanding it not as a specific technology of digitalization but as a deeply interwoven phenomenon within our teleological lifeworld. In doing so, it is essential to avoid dualistic subject–object divisions between humans on the one hand and machines on the other. To this end, I will reflect on the interactive dimension between humans and AI by focusing on the interrelationship of human purposes, intentions, databases, and algorithms. Transsubjectivity is thus a social context of human–machine interaction in which algorithms can be understood as both representations and extensions of individual and collective goals.

2 AI and the Digital Condition

If we understand AI not just as a technology but as a factor that is increasingly penetrating, shaping, and transforming our lifeworld, it makes sense to analyze AI in the context of what Felix Stalder has called the "digital condition" from a cultural studies perspective.[10] While "digitalization" refers to the technological phenomenon of converting analog into digital information, the "digital condition" describes the significance of digitalization as an ontological factor in everyday life. Stalder understands the digital condition as a specific culture that accompanies the emergence of new media and has replaced the mediality of the "Gutenberg Galaxy," as Marshall McLuhan called the culture of the printed book.[11] Stalder speaks of "a new and specific cultural environment [...], usurping or marginalizing an ever greater variety of cultural constellations."[12] He distinguishes "three *forms* of the digital condition" that constitute its uniform culture, which he calls "*referentiality*, *communality*, and *algorithmicity*".[13]

The phenomenon of algorithmicity is of central importance for AI's role within the human lifeworld. According to Stalder, algorithmicity "denotes those aspects

[9] Kurzweil 2005, 9: "The Singularity will allow us to transcend these limitations of our biological bodies and brains. We will gain power over our fates. Our mortality will be in our own hands. [...] There will be no distinction [...] between human and machine or between physical and virtual reality." For the notion of a "superintelligence", see Bostrom 2014.

[10] Stalder 2018.

[11] McLuhan 1962.

[12] Stalder 2018, 58.

[13] Stalder 2018, 58.

of cultural processes that are (pre-)arranged by the activities of machines".[14] By pre-structuring the sheer amount of data and information, they become manageable for individual and collective reference as the condition for the possibility of the constitution of meaning: "Without them, human beings could not comprehend or do anything within a culture built around digital technologies, but they influence our understanding and activity in an ambivalent way. They create new dependencies by pre-sorting and making the (informational) world available to us, yet simultaneously ensure our autonomy by providing the preconditions that enable us to act."[15] In contrast to Stalder, however, I shall understand AI not in the sense of algorithms and processes that operate in the background of human culture and lifeword, and which just reduce "big data" to "small data".[16] Rather, I shall interpret them in the sense of complex processes that are currently subsumed under the term "machine learning" (ML), which enters into a much closer relationship with us in the social lifeworld—it "gets closer to our skin than other technologies".[17] Therefore, the distinction between "human" and "machine" becomes increasingly less important seen from the perspective of informational processing and social teleology.

3 AI as Simulation or Duplication

Talking about the "artificiality" of intelligence is ambiguous. According to the weak interpretation, "artificiality" specifies the intelligence of technical systems in the sense that they *simulate* human intelligence. According to the strong interpretation, however, machines *realize* human intelligence. These two interpretations roughly correspond to the distinction between weak and strong AI that has become commonplace.[18] Conceptions of weak AI often understand it as a mere technical property of objects and thus tend to trivialize AI. In contrast, conceptions of strong AI understand intelligence as a property of a subject and thus tend either to ideologize, anthropomorphize, or dramatize it. By focusing less on the subject or object of AI and more on the process of *intelligence* as such and its

[14] Stalder 2018, 13.

[15] Stalder 2018, 59.

[16] Stalder 2018, 59.

[17] Müller 2023.

[18] Russell/Norvig 2022, 1020: "[T]he assertion that machines could act as if they were intelligent is called the weak AI hypothesis by philosophers, and the assertion that machines that do so are actually thinking (not just simulating thinking) is called the strong AI hypothesis."

conditions and teleology, these problems can be avoided. According to the hybrid account of AI that I will present in this paper is rather concerned with the interrelation of human subjectivity and technology. It aims relate both the databases and the algorithms to human purposes. As such, AI can be connected to and integrated into the human lifeworld so that the respective performances interfere with, extend, and complement those of human subjects. Such an interference model of AI suggests that it should not be understood in terms of "human–machine interaction" but should be further analyzed in terms of transsubjectivity, in which human intelligence is extended by artificial intelligence. The distinction between strong and weak AI is only significant if understood in a subject–object relationship of "human–machine interaction." It becomes meaningless if we understand AI in terms of transsubjectivity as part of the digital lifeworld.

Douglas Engelbart has coined the phrase "augmented human intellect"[19] for this transsubjective understanding of AI. He understands this as a "systematic approach to improving the intellectual effectiveness of the individual human."[20] However, this augmentation of human intelligence should not be understood in the sense of "isolated clever tricks that help in particular situations". Rather, according to Engelbart, "extension" means a holistic and systemic phenomenon; it concerns "a way of life in an integrated domain where hunches, cut and try, intangibles and the human 'feel for a situation' usefully coexist with powerful concepts, streamlined terminology and notation, sophisticated methods, and high-powered electronic aids."[21]

Despite Engelbart's holistic approach, which understands human–machine interaction as a "set of interacting components rather than by considering the components in isolation",[22] this "systematic approach" is still too focused on quantitative performance enhancement in the sense of "increasing human intellectual effectiveness".[23] Through these instrumentally understood extensions as "augmentation means",[24] the qualitative dimensions are lost sight of. A qualitative and not purely quantitative-instrumental interference model of AI suggests that it should not be understood in terms of "human–machine interaction", but rather further analyzed in terms of the concept of transsubjectivity, in which not only human reason, but also other human capacities are augmented by AI.

[19] Engelbart 1962.

[20] Engelbart 1962, II.

[21] Engelbart 1962, 1.

[22] Engelbart 1962, 2.

[23] Engelbart 1962, 3.

[24] Engelbart 1962, 9.

The question of whether and to what extent machines can be called intelligent was already addressed by Alan Turing in his classic essay "Computing Machinery and Intelligence" from 1950.[25] Turing argues that whether and to what extent machines can think cannot be answered directly—it is "too meaningless to deserve discussion".[26] Interestingly, however, Turing argues that this question would no longer be considered meaningless in the year 2000—the very year in which, according to Stalder, "a new cultural constellation"[27]—the "digital condition"—began: "I believe that at the end of the century the use of words and general educated opinion will have altered so much that one will be able to speak of machines thinking without expecting to be contradicted."[28] Instead of the question "Can machines think?," Turing suggests replacing it with the question "Are there imaginable digital computers which would do well in the imitation game?".[29] Turing avoids a direct definition of the term artificial intelligence and replaces it with a thought scenario in which the conditions are developed for a machine to be described as (artificially) intelligent with good reason.

Turing developed his replacement theory through what he called the "imitation game".[30] This experiment, also known as the "Turing test," consists of three instances—a woman (A), a man (B), and a person (C), who is supposed to find out the gender of the two persons unknown to her by asking clever questions, who in turn try to keep C in the dark about their identity. Now, either the woman or the man is replaced by a computer, so the following question arises as a condition for the success of the attribution of artificial intelligence to the computer: "Will the interrogator decide wrongly as often when the game is played like this as he does when the game is played between a man and a woman?".[31]

This raises the question of the extent to which "passing" such a test allows conclusions to be drawn about the "intelligence" of a computer. At first, it seems obvious to understand the Turing test in terms of reductionist behaviorism.[32] However, such an objection does not do justice to the attribution conditions formulated by Turing. Central to this is Turing's conception of the test as a *game* between three instances. The computer's intelligence is measured by the degree

[25] Turing 1950.

[26] Turing 1950, 422.

[27] Stalder 2018, 4.

[28] Turing 1950, 422.

[29] Turing 1950, 442.

[30] See Turing 1950, 433.

[31] Turing 1950, 434.

[32] For a related critique, see Searle 1984, 38.

to which it can *imitate* a natural person. This imitation performance must be linguistic, argumentative, and dialogical, i.e., highly social and context-sensitive.

This imitation performance, like the artificiality of intelligence, can be understood in two different ways. On the one hand, "imitation" can mean something like simulation. Still, on the other hand, it can also be understood in the sense of a duplication, i.e., the realization of human intelligence. John Searle, in particular, has pointed out this difference.[33] He argues that computers are only able to simulate human cognitive performance, but not to duplicate it—a difference he calls the "key distinction": "[N]obody supposes that the computer simulation is actually the real thing; no one supposes that a computer simulation of a storm will leave us all wet, or a computer simulation of a fire is likely to burn the house down."[34] Searle links the "key distinction" between mere simulation and actual duplication with his distinction between syntax and semantics.[35] He argues that computers are only ever capable of syntactic operations, but that these can never lead to semantic content, which, according to Searle, only mental states possess: "[C]onsciousness, thoughts, feelings, emotions, and all the rest of it involve more than a syntax. Those features, by definition, the computer is unable to *duplicate* however powerful may be its ability to *simulate*."[36]

However, the distinction between simulation and duplication of human intelligence is only relevant as long as we attribute it to a machine conceived as a subject. If we look at AI not from the subject's point of view but as a process in terms of intelligence performance and operation, that is as an extension of our cognitive faculties and their goals, this distinction becomes less and less significant. This teleological performance account of AI must also be distinguished from mere behaviorism, which does not focus on the performance and structure of AI but only on the *result* of its operation. This, however, leads to the opacity problem, according to which we cannot understand the reasons and teleology behind AI's operations. By conceiving AI not as a subject but as an interrelated process, intelligence performance can be embedded in various teleological lifeworld contexts and systematically linked to intelligence performance by human individuals and communities. From this teleological point of view, AI does not appear as abstract algorithms and mere calculation of numbers but as a meaningful process that deeply concerns human practices.

---

[33] Searle 1984.

[34] Searle 1984, 37–8.

[35] Searle 1984, 34.

[36] Searle 1984, 36.

To be sure, the notion of "intelligence," like the even broader notion of "rationality," can be understood in various ways. On a fundamental level, "intelligence" can be understood as the capacity and operation of information processing. Forms of information processing are comparing and distinguishing information, which includes detecting logical relationships such as identity, implication, or opposition, but also abstraction.[37] The link between human and artificial intelligence is the concept of a pattern. Our lifeworld consists of various kinds of patterns, be they economic, emotional, acoustic, visual, or linguistic. What is common to all these patterns is their goal-directed, teleological structure. These databases are social, that is, they transport meaning, intentions, and human goals. AI usually detects patterns within large databases using algorithms, which humans cannot process due to the sheer amount of information. With regard to social databases, AI's pattern recognition is not just an abstract calculation but enables and enhances teleological orientation. Therefore, responsible AI consists of adequately applying algorithms to social data. Based on social data, pattern recognition becomes a social practice, transforming algorithms into concepts and reasons for action.

What is important for the transsubjective account of AI is the interrelatedness of algorithms, databases, and human ends. On the transsubjective account, algorithms function as extensions of human goals. Likewise, the databases on which algorithms operate are deeply linked to human epistemic practices, like the linguistic and semantic content of the internet. The chatbot ChatGPT, for example, has been trained on the basis of several hundreds of billions of tokens that are basically of human origin.[38] AI and human rationality are interrelated in such a way that the former's algorithms are oriented and trained towards the teleology and structure of the human lifeworld. This teleology can be understood in terms of instrumental reason but must not be restricted to it. Instead, in being trained for various purposes, AI is part of theoretical and practical rationality that is generally oriented towards realizing goals. Transsubjectivity manifests in training an ANN, using specific algorithms and methods resulting in outputs corresponding to human purposes. This transsubjective practice has been described as "a fine art": "Success with backpropagation and other connectionist learning methods may depend on quite subtle adjustment of the algorithm and the training

[37] Concerning abstraction, see Buckner 2024 and Buckner (2018, 5345), who argues that ANNs "transform sensory based representations of particulars, on the basis of accrued experience, into more abstract representational formats" and therefore "model one crucially important component of intelligence—a form of categorical abstraction" (5341).

[38] See https://en.wikipedia.org/wiki/GPT-3#Training_and_capabilities.

set."[39] Human responsibility concerns both the database and the training procedure by means of algorithms. The former needs to be as objective and neutral as possible; the latter must adequately represent human goals. It is the unity of (empirical) database and (conceptual) algorithms that enable responsible AI, just as Immanuel Kant explained the very condition of cognition: "thoughts without content are empty, intuitions without concepts blind". Conceiving AI in terms of a unity of social data and conceptual algorithms allows us to reduce both the opacity and bias of AI systems, which are considered the "central issues"[40] of current data ethics. Solving the opacity problem demands interpreting AI performances regarding representations and extensions of human interactions. Solving the bias problem of AI demands interpreting social databases such as the Internet in terms of individual or collective human products. Objectifying AI, however, makes these solutions impossible since it categorically separates AI performances from those of humans.

4 AI and Virtual Reality

The emergence of artificial neural networks (ANNs) through ML has led to forms and operations of AI that outperform human intelligence in specific fields such as computer games, chess, Go, and pattern recognition. Therefore, it is essential to understand that AI is not just a *simulation* of natural or human intelligence but a new way of *realizing* intelligence, understood in terms of pattern recognition of various kinds and various human goal-related circumstances.

AI considered in terms of alternative, non-actual forms of rationality can be interpreted in terms of virtual reality (VR). Whereas VR is mainly being discussed as a kind of sophisticated simulation technology, I shall argue for a teleological account of virtual reality, according to which virtualizing means to realize purposes in a non-actual way, for example, intelligence by means of AI. According to this understanding, VR must be distinguished from mere simulation, even though a conceptual transition from simulation to virtual reality can be described as "virtualization." Virtual reality starts with mere simulation, i.e., a model-like orientation towards a given, original, or actual phenomenon. However, it increasingly emancipates itself from it so that the simulative aspect recedes further in favor of a realizing or duplicating aspect. Ultimately, there is no longer

[39] Buckner/Garson 2019.

[40] Müller 2023.

any structural ontological analogy, only an analogy of purpose and causality, as we can see in the complexity of multi-layered ANNs.

The concept of virtuality has recently received increased attention, not so much concerning simulation as *reality*. With the development of new media and their performative-causal significance, there is a growing willingness to recognize virtual reality as a reality of its kind and to distinguish it from mere simulation. Considering this distinction between simulation and virtual reality, AI's "artificial" aspect can be understood as an aspect of virtuality. This is because intelligence is realized causally and teleologically in a new, non-actual way. This new way of realizing intelligence is particularly evident in the process of ANNs and ML, which are initially modeled on the human brain and learning processes but then develop a logic of their own that can no longer be understood solely in terms of structural analogy.

The emergence of ML, in particular, shows that the extent to which computers can be called intelligent cannot be answered simply by distinguishing between simulation and duplication. Of all previous forms of artificial intelligence, *deep learning*, i.e., more complex ANNs, can be considered the most advanced. This raises the question of the extent to which the operations and results of the network can be described as "learning," "intelligence," and "knowledge." A purely behaviorist description of machine learning falls short because it does not consider the process leading to the respective results, whose epistemic significance is in question. Therefore, it seems appropriate to speak of a learning process concerning ANNs, for example, by the process of abstraction. After several runs based on large amounts of information, a learning effect of the network can be achieved, which lies in adequately adjusting the weights. This can be seen as knowledge acquired through a trial-and-error process, as goal-directed living beings do.

But how can we talk about neural networks "recognizing" something, "knowing" something, or even "thinking" something? Ideally, neural networks assign information and data to a correct category and manifest a classification performance. However, the decisive factor here is that these categories must be teleologically predefined. In the case of pattern recognition, it is possible to attribute something like a "determining power of judgment" to ANNs, which Kant described as "the faculty for thinking of the particular as contained under the universal": "If the universal (the rule, the principle, the law) is given, then the power of judgment, which subsumes the particular under it […], is *determining*".[41] Likewise, we can reformulate the determining power of AI as follows:

[41] Kant 2000, 69–70/5:179. On the judgmental power of ANNs, see Schubbach 2021, and Noller 2022, 58.

If the pattern is given, then AI determines (the probability of) what corresponds to this pattern. However, ANNs never operate autonomously, but only hypothetically, as their goals must be given to them by humans. Only if AI is given a certain goal is it able to detect what corresponds with it. AI does not possess any reflective power of judgment, as Kant put it: "If [...] only the particular is given, for which the universal is to be found, then the power of judgment is merely *reflecting*."[42] Also, ANNs are not living beings that can pursue their interests or exhibit anything like a survival instinct.[43] In this respect, their mode of operation can be understood as *modular* in the sense of a flexible capacity that can be linked back to human subjects in the real world.

5 AI and Transsubjectivity

Luciano Floridi has determined the lifeworld significance of AI as part of his theory of the *infosphere* and understood it in terms of "artificial agents". He thus extends the concept of moral agents by including AI under the concept of "moral agents" and "moral patients".[44] Floridi develops a concept of "mindless morality,"[45] which he considers applicable to AI. This term implies that AI cannot be attributed mental characteristics such as intelligence and free will. Nevertheless, Floridi argues that artificial agents can be ascribed *accountability*, while they are not ascribed *responsibility*, which only concerns those persons who created and conceived the artificial agent.[46] Floridi relates this, at first glance, unclear distinction between *accountability* and *responsibility* to the relationship between parent and child. While parents are jointly *responsible* for their child's behavior, they cannot be held legally *accountable* for their actions once it has reached a certain age.[47]

According to Floridi, the decisive factor for AI's moral and ontological definition is the descriptive framework and the level of abstraction on which it is

---

[42] Kant 2000, 67/5:179.

[43] On the importance of the context of life for AI, see also Dreyfus (1988), 39: "[H]uman beings are much more holistic than neural nets. Intelligence has to be motivated by purposes in the organism and goals picked up by the organism from an ongoing culture. If the minimum unit of analysis is that of a whole organism geared into a whole cultural world, neural nets as well as symbolically programmed computers still have a very long way to go."

[44] Floridi 2013, 134.

[45] Floridi 2013, 135.

[46] Floridi 2013, 135.

[47] Floridi 2013, 135.

based. Applied to AI, this means that the level of abstraction of information ("informational level of abstraction") and not that of substances must be chosen to describe its reality. The ontological level of information differs from Newton's ontology, for example, based on material objects and substances.[48] Floridi argues that the choice of the level of description and abstraction is not to be understood relativistically but is motivated by the purpose of explanation and teleology of reasons.[49]

Floridi agrees that AI is increasingly becoming part of our everyday lives and cannot be adequately understood within an ontology of substance. However, the concept of AI as (moral) subjects is problematic insofar as an informational (onto)logical level of description must be chosen that is not appropriate to our lifeworld. Rather, it is essential to choose a concept of reality in such a way that it can integrate both human persons and AI into a comprehensive lifeworld without changing the level of abstraction. As already shown in the previous section, the concept of virtual reality lends itself to this, which does not denote a different level of abstraction and description but rather a form of reality based on the lifeworld, as can already be seen in the phenomenon of banknotes, independent of digital technology, which as material reality do not have their designated value, but only thanks to social and intersubjective convention.[50]

In this respect, we need to introduce the concept of transsubjectivity for the lifeworld relationship between AI and free persons and distinguish it from intersubjectivity.[51] Digital objects have much more intimate relationships with each other than physical objects do. In the Internet of Things, objects communicate as subjects do in society. The division between subjects and objects is thus being increasingly being blurred, primarily when AI acquires real-world significance in different social contexts. Understood in this way, transsubjectivity means the networking of AI capabilities or services with interests and goals set by autonomous people. This context can be defined as a transsubjective context insofar as it is defined by *reasons* due to human rationality. The reasons that affect the relationship between autonomous persons and AI, and which produce the intelligence performance are no longer tied solely to a subject but are externalized and objectified. However, the concept of transsubjectivity must be complemented by the

---

[48] Floridi 2013, 27.

[49] Floridi 2013, 146–7.

[50] See Brey 2014 for the social role of VR.

[51] On the concept of interobjectivity, see Huy 2016, 161, who, however, motivates it primarily technologically: "Interobjectivity is a reality as well as a constantly motivated (improved) medium that can resolve the incompatibility between different systems (including human and technical systems)."

concept of interobjectivity. Just as the objects of networking can be understood as networking of mental contents that are no longer exclusively bound to a subject, they can be viewed from the perspective of objectivity.

Therefore, the ethical significance of AI must not be understood vertically, namely in the sense that we are confronted with a new kind of power that is superior to us humans in terms of intelligence. Instead, we must think of the significance of AI in a horizontal sense: AI is increasingly being integrated into our lifeworld and networked with it so that it is perceived less and less as a subject or object opposing us, but rather as an open teleological process that can be networked with human processes, indeed that can be understood as a complicated and mediated human self-relation. Both the algorithms and the databases that constitute ANNs are ultimately of human origin. Therefore, responsible AI demands that we become aware of this teleological human–machine dependence beyond mere subjectivity and objectivity.

6 Conclusion

By conceiving AI as a structural moment of our lifeworld, the focus shifts from understanding it as a subject or agent to the normative context of reasons initiated by human individuals and collectives and into which AI performances are being integrated. AI performance is always heteronomously interested and motivated: it requires a teleological initiation that it receives from outside. Therefore, the normative problem of AI arises not so much from how it functions but from the fact that it may not be adequately integrated into our lifeworld and that it confronts us as a technological, even technocratic, object, cementing a subject–object divide. From an ethical perspective, a transition from an "oppositional approach," according to which AI represents a potential danger to us, to a "systemic approach," according to which AI is viewed as "a set of technologies that are embedded in a system of human agents, other artificial agents, laws, nonintelligent infrastructures, and social norms," is therefore necessary.[52]

However, an ethical mediation of AI and lifeworld in the sense of transsubjectivity and augmented intelligence can only succeed if we use AI creatively as a virtual reality. The ethical challenge of AI is, therefore, primarily to integrate its technology into our lifeworld in such a virtualized way that it interferes with us in the lifeworld context and enables new forms of rationality, subjectivity, and objectivity. Such an intersubjective interference is to be understood not only in

[52] Powers/Ganascia 2020, 48–9.

terms of a quantitative increase in our intelligence but also as an expansion of other faculties, such as our will or our power of judgment. This expansion of our faculties must be understood not as a technology that affects us as individuals but as a qualitative virtual space that opens up transsubjectively and interobjectively. Transsubjectivity thus also turns out to be a teleological space of purposes into which we can integrate ourselves. Of course, this does not exclude the possibility that the inappropriate use of AI could limit our virtual space of action, or even to our being placed in mere illusory spaces that we mistakenly believe to be reality. Therefore, the responsible use of AI means expanding our space of action and integrating AI into our lifeworld so that it becomes transparent to our original goals, for which we are ultimately responsible.

References

Bostrom, N. (2014): *Superintelligence: Paths, Dangers, Strategies*, Oxford.

Brey, P. (2014): "The physical and social reality of virtual worlds", in: Mark Grimshaw (ed.): *The Oxford Handbook of Virtuality*, Oxford, 42–54.

Buckner, C. (2018): "Empiricism without magic: transformational abstraction in deep convolutional neural networks", in: *Synthese* 195, 5339–5372.

Buckner, C. (2024): *From Deep Learning to Rational Machines: What the History of Philosophy Can Teach Us about the Future of Artificial Intelligence*, Oxford.

Buckner, C./Garson, J.: "Connectionism", in: *The Stanford Encyclopedia of Philosophy* (Fall 2019 Edition), Edward N. Zalta (ed.), URL = <https://plato.stanford.edu/archives/fall2019/entries/connectionism/>.

Dreyfus, H. L./Dreyfus, S. L. (1988): "Making a Mind versus Modeling the Brain: Artificial Intelligence Back at a Branchpoint", in: *Daedalus* 117(1), 15-43.

Dubber, M. D./Pasquale, F./Das, S. (eds.) (2020): *The Oxford Handbook of Ethics of AI*, Oxford.

Durt, C. (2022): "Artificial Intelligence and Its Integration into the Human Lifeworld", in: Silja Voenecky et al. (eds.): *The Cambridge Handbook of Responsible Artificial Intelligence: Interdisciplinary Perspectives*, Oxford, 67–82.

Engelbart, D. (1962): *Augmenting Human Intellect: A Conceptual Framework*, Menlo Park.

Floridi, L. (2013): *The Ethics of Information*, Oxford.

Hui, Y. (2016): *On the Existence of Digital Objects* (= *Electronic Meditations* 48), Minneapolis/London.

Kant, I. (2000): *Critique of the Power of Judgment*, ed. Paul Guyer, transl. Paul Guyer and Eric Matthews, Cambridge.

Kurzweil, R. (2005): *The Singularity is Near*, London.

McLuhan, M. (1962): *The Gutenberg Galaxy: The Making of Typographic Man*, Toronto.

Müller, V. C., (2023), "Ethics of Artificial Intelligence and Robotics", in: *The Stanford Encyclopedia of Philosophy* (Fall 2023 Edition), Edward N. Zalta & Uri Nodelman (eds.), URL = <https://plato.stanford.edu/archives/fall2023/entries/ethics-ai/>.

Nida-Rümelin, J./Weidenfeld, N. (2022): *Digital Humanism: For a Humane Transformation of Democracy, Economy and Culture in the Digital Age*, Cham.

Noller, J. (2022): *Digitalität: Zur Philosophie der digitalen Lebenswelt*, Basel.

Noller, J. (2024): "Extended human agency: towards a teleological account of AI", in: *Humanities and Social Sciences Communications* 11(1), https://doi.org/10.1057/s41599-024-03849-x.

Powers, T. M./Ganascia, J.–G. (2020): "The Ethics of the Ethics of AI", in: Markus D. Dubber/Frank Pasquale/Sunit Das (eds.): *The Oxford Handbook of Ethics of AI*, Oxford, 27–51.

Russell, S./Norvig, P. (2022): *Artificial Intelligence: A Modern Approach*, Boston et al.

Searle, J. (1984): *Minds, Brains and Science*, London.

Schubbach, A. (2021): "Judging machines: philosophical aspects of deep learning," in: *Synthese* 198, 1807-1827.

Stalder, F. (2018): *The Digital Condition*, Translated by Valentine A. Pakis, Cambridge/Medford.

Turing, A. (1950): "Computing Machinery and Intelligence", in: *Mind* 59, 433-460.

Artificial Intelligence—Hype or Myth? Philosophical Aspects of an Entrepreneurial Success Story

Walther Ch. Zimmerli

1 Introduction

It goes without saying: the notion of "Artificial Intelligence (AI)" seems to refer to something like a revenant, or to put it less stately: like a jack-in-the-box. As soon as it seems to be dead it pops off the ground again, even more vigorously than before, getting an ever-increasing media coverage. At second glance, moreover, this recurring boom of AI additionally reveals its cyclical character. Thus, if critical philosophizing is asking impertinent second questions the task a philosopher dealing with both the phenomenon and the notion of AI is obliged to consider is the question: Does this cyclically recurring boom of AI meet the conditions of what we call just a "hype" or is there more to it? To put it differently: Is AI just a hype, is it a myth, or hype and myth or neither hype nor myth, and how does this relate to what we since Schumpeter call "innovation" (Zimmerli/Folkers 2017)? Or, broken down into two more precisely formulated questions:

It is with profound gratitude that I am acknowledging the opportunity to present some of the ideas outlined in this paper in a public lecture at the AI week 2023 of the Digital Society Initiative (DSI) of the University of Zurich where also the editor of this volume, ***Christian Hugo Hoffmann***, was invited as a distinguished panelist. I further developed my ideas in my Mercator lecture on April 15 2024 and in further discussions with some of the other fellows during my stay as Mercator Visiting Professor for AI in a Human Context 2024 at the Center for Science and Thought (CST) of the University of Bonn (cf. an earlier version in German language: Zimmerli 2021b).

W. Ch. Zimmerli (✉)
Institute for Advanced Study, TU Munich, Garching/Munich, Germany
e-mail: zimmerli@b-tu.de

C. H. Hoffmann (ed.), *Artificial Intelligence, Entrepreneurship and Risk*, Technikzukünfte, Wissenschaft und Gesellschaft / Futures of Technology, Science and Society, https://doi.org/10.1007/978-3-658-45544-6_14

Is there a more fundamental feature underlying the recurring boom of AI, and could this feature shed some light on its success in the market?

In order to at least try to answer some of these questions I will in what follows deal with what is called "hypes" within the field of innovation theory looking exemplarily at the Eliza-, the Watson- and most recently the ChatGPT-hype (1). In a second step I will be considering what in a more philosophical approach has been dealt with as "myths" (Blumenberg), especially the myths of the Thinking Machine and of the Digital Natives (2). The following three steps will then apply these instrumental concepts to the cyclically booming AI in order to look behind the screen and reconstruct first the origin of the AI myth itself by uncovering the Dartmouth Conspiracy, thus demonstrating the power of analogical and metaphorical thinking (3), and then reconstruct the origin of the Singularity Myth by analysing its eschatological narrative and its implications with respect to some of the most fundamental philosophical notions of modernity (4). The final step then will look into the anthropological aspects of AI by reconstructing the origin of the Posthumanism Myth thus unveiling its hidden characteristics with respect to what today is called "the Digital" (5).

2 Of Hypes

In order to get started with pondering the question whether the present boom of AI-related debates is sustainable or just one of the many flashes in the pan, i.e. a hype, it seems to be philosophically recommendable to first of all ask: What exactly are we talking about if we speak of a "hype"?

2.1 What is a Hype?

Like in many other cases the notion "hype" has next to its ordinary language use a more technical one. In ordinary everyday language we understand by a hype (greek hyperbole = exaggeration) a boost of attention, often brought about by the media and in many cases accompanied by some kind of almost deceptive constraint. With respect to innovations, however, the notion "hype" is also used as a technical term focussing especially on its cyclical character: In 1995 Jackie Fenn, Distinguished VP analyst and Fellow Emeritus of Gartner Inc., one of the leading US-American research and consulting companies in the field of innovation research, has developed a model which since then is known by the name of "Gartner Hype Cycle" (GHC) model. According to this model technological innovations can as a rule be described by the following structure with respect to both visibility and time (Fig. 1).

Each new technological development or product entering the market which doesn't immediately drop dead by a complete lack of visibility is met by an exponentially increasing growth of public attention thus reaching very soon a "Peak of Inflated Expectations", then undergoes a decline of visibility until reaching the "Trough of Disillusionment" and from there slowly climbing up what Fenn calls the "Slope of Enlightenment", eventually reaching the "Plateau of Productivity".

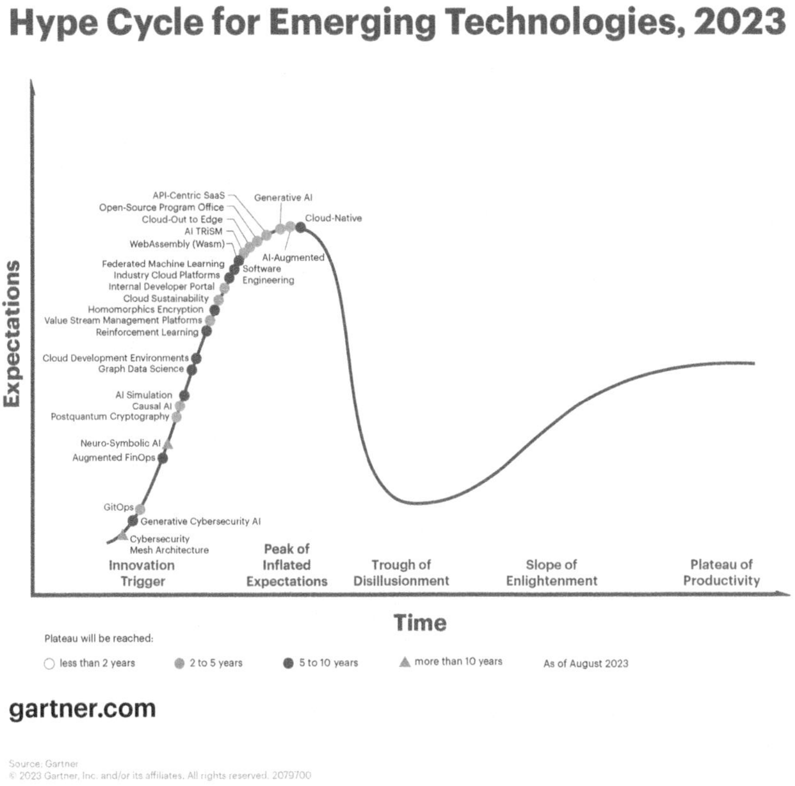

Fig. 1 Gartner hype cycle (Source: https://www.gartner.com/en/articles/what-s-new-in-the-2023-gartner-hype-cycle-for-emerging-technologies (11-02-2024)

And, as "innovation" since Schumpeter is being defined by success in the market, only now the technological development or product in question can be called "innovative" (cf. Zimmerli/Folkers 2017). This process, however, can be iterated, hence its characterization as "cycle".

According to this technical definition of "hype", our question will be: Is the development of AI adequately described by the Gartner Hype Cycle model? Or is there more to it?

2.2 The Eliza- and the Watson-Hype

Although not the epitome of a philosophical mind former German chancellor Helmut Kohl has 1983/84 coined the formula of the "grace of late birth" thus expressing the conviction that the generation of Germans born after 1930, cannot be held responsible for the Nazi crimes. Looking at AI I would like to talk of the "grace of early birth": People—and especially philosophers like me—who were born in the Forties are less susceptible to AI hypes because they have witnessed several of these hypes before. Just two examples: In the year 1966 MIT professor of Computer Science Joseph Weizenbaum published an interactive Computer program in natural language named ELIZA (Weizenbaum 1966), which was world-wide celebrated as a major breakthrough in Artificial Intelligence. Today ELIZA is considered a prototype of the modern chatbots. And it still is a model example of AI, because it is structurally so simple and at the same time so convincing. Weizenbaum himself was shocked by the reactions, especially by how easy it was to make people believe that they were communicating not with a machine but with a fellow human being (cf. Weizenbaum 1976).

Another example was the public launch of the IBM program WATSON (2011), named after Thomas J. Watson, the long-time president of IBM. This computer program, part of IBM's "DeepQA" research projects, triggered in the year 2011 another AI-hype after successfully competing with former champion Ken Jennings in America's favorite TV Quiz-Show "Jeopardy". This success was compared by the professional world with the historical victory of the IBM chess computer "Deep Blue" in 1996 beating world champion Garry Kasparow in a regular match of 6 games under competition conditions.

2.3 The ChatGPT-Hype

And right now (more precisely since November 2022) we are witnessing the most recent AI-hype around the dialogue-based chatbot ChatGPT (Generative Pre-trained Transformer), an AI-powered language model, developed by the highly controversial AI research and deployment company OpenAI, capable of generating texts based on context and past conversations. OpenAI is specialized in research on AI with, according to its charter of 2018, the intention of developing "safe and beneficial" general artificial intelligence, i.e. "highly autonomous systems that outperform humans at most economically valuable work".

The keywords here are "generative" and "economically valuable" because generative AI (or artificial generative intelligence) is considered to be the great new frontier of technology. This might be one of the reasons why Microsoft at the end of 2019 invested 1 billion US $ and more than 10 billion in 2021 and 2023; at present the value of OpenAI is estimated around 100 billion US $! Generative AI promises to be big business—the development of this most recent hype seems to be shortcutting the hype-cycle by e.g. integrating ChatGPT into Microsoft's search engine Bing.

And the melodramatic story of first firing OpenAI-boss Sam Altman and after a few days re-hiring him again (cf. Ferber 2023) demonstrates something else: If 'innovation' being defined by success in the market, and success in this case being measured in the order of magnitude of literally billions of US $ for a rather young start-up company (OpenAI was founded in 2015 only) with about 800 employees then there must be more to it. And yes, there is: on the one hand as has already been frequently pointed out: the value of the team of OpenAI and on the other hand as has not yet been adequately analysed: the product which only from a superficial perspective consists in a program like ChatGPT. What OpenAI is producing (and selling) is neither hardware nor software, but the "dreamware" promising unlimited cognitive performance, and this dream is being fuelled by an eternally powerful myth as will be demonstrated in what follows.

3 Of Myths

What we have seen so far is that in each and every case of the different AI hype cycles the Gartner hype cycle model is applicable: Exaggerated expectations led to high visibility as well as to deep disillusionment, followed by funding cuts, called "AI-winters": the first one between 1974 and 1980, the second one between

1987 and 1993 etc. The history of the AI-hypes is a history of a series of rises and falls.

What is interesting, however, is that the idea of an Artificial Intelligence keeps returning, and its manifestations are becoming ever more powerful. So, as mentioned above, the task of a philosophical approach to making sense of these AI hypes is to take a second look, a look behind the smokescreen of hype-cycles, and search for something more powerful than just hypes. And as stated above a possibility could be to look for myths behind the hypes.

3.1 What is a Myth?

Although there are different theories about what constitutes a myth and how it works (cf., Cohen 1969) common to all of them is the etymological finding that myths are to be understood as narratives (greek: mythos = story). There is, however, a difference between mythical and "grand narratives" the latter ones having, according to Lyotard (Lyotard 1979), come to an end in postmodern discourse. For our purpose the decisive additional philosophical importance is to be found in the analysis of myths given by Hans Blumenberg: Myths serve as attempts of human beings to come to terms with the confusing multiplicity of their environment by giving it significance. It is characteristic for these narratives to stay highly constant in their core, while at the same time being extremely variable in their margins (Blumenberg 1970). Thus, what we are looking for are narratives that are capable of making sense and are both extremely stable with respect to their narrative core as well as variable with respect to their marginal environment.

3.2 The Myth of the Thinking Machine

One candidate meeting these requirements almost perfectly is the idea of thinking machines: Against the background of the classical Aristotelian idea of an autonomously working machine and of the concept of the human being as a machine ("l'homme machine", d'Holberg and the French materialists) the idea of "machines who think" (McCorduck 1979) was reborn in the context of the development of computers. And it was Alan Turing who connected this idea with the concept of intelligence in his famous essay on "Computing Machinery and Intelligence", thus seemingly creating the notion of "artificial intelligence" *avant la lettre.* At second glance, however, it looks differently: Although beginning his article with the programmatic sentence "I propose to consider the question 'Can machines think?'", he immediately revokes it by replacing it by another, which is closely related to it and is expressed in rather unambiguous terms". (Turing 1950, 433). And this reformulated question is what he then calls the "imitation game", later known as the "Turing test" (cf. below).

What is important for my argumentation is:

1. Turing—other than most of his followers and interprets—seems to be very well aware of the ambiguity of the notions "to think" and "intelligence" in semantic connection to the concept "machine".
2. He therefore quite explicitly does not attempt to answer the question whether machines are capable of thinking or are intelligent.

To put it in a nutshell. In combination with the then current development of computing machinery the "myth of a thinking machine" (Taube 1961) has been stylized to a hype which afterwards has developed pretty much according to the developmental stages predicted by the Gartner hype cycle model.

3.3 The Myth of the Digital Natives

Let's now consider a more recent mythical aspect of AI, construed even more narratively and up to date widely accepted: I am talking about the narrative telling the history of digitalization using the ethnographic terminology of "natives" and "immigrants". Since educator and manager Marc Prensky in 2001 has coined the pair of concepts "digital natives" and "digital immigrants" the discourse on digitalization is no longer imaginable without it. Its mythical power roots in the resonating subtext of the pioneer-myth of the US-American settlers. It is, however, interesting that this mythical power is accompanied by a semantical inversion: While historically speaking the settlers, the "immigrants", were the pioneers, in its mythical adaption it is the other way around, as can be seen in another version where this *Go-West*-myth has been explicitly connected to the emergence of the World Wide Web: The subtitle of Howard Rheingold's influential book on "The Virtual Community" reads "Homesteading on the Electronic Frontier" (Rheingold 1993). Here not the settlers, but the "natives" are the pioneers, those who are "born digital" as John Palfrey and Urs Gasser have pointedly titled their well-known book (Palfrey/Gasser 2008), i.e. the "Millenials" or "Generation Internet".

4 The Origin of the AI-Myth

For my purpose neither the historical appropriateness nor the functional usefulness of these narratives are important. I am rather focussing on the important dimension underlying even their pragmatic analysis which is something close to what in depth psychology has been called "archetype" by C.G. Jung. These and the following deliberations are geared towards better understanding the mechanism of transforming a more or less neutral description into a powerful myth or at least coupling it with a myth-like narrative. In brief: they are examples of *myth-analysis* (which is not to be confused with "de-mythologization", Bultmann 1941; Jüngel 1985).

4.1 The Dartmouth Conspiracy

To now apply this instrument of myth-analysis to our core-topic AI we first of all have to acknowledge that its origins, although historically quite apparent, are still to be uncovered or rather: better understood. According to the commonly shared narrative the birth hour of the notion "artificial intelligence" is the so-called Dartmouth conference, more precisely the "Dartmouth Summer Research Project on Artificial Intelligence" funded by the Rockefeller foundation. And thus, the funding of academic activities requiring application, the first mention of the notion AI on file, both in the title of the application and it its text, is the funding proposal of this summer research school. In this proposal of September 1955, the applicants John McCarthy, Marvin Minsky, Nathanael Rochester, and Claude E. Shannon applied for funding of the mentioned Summer School which then took place in the summer of 1956 at Dartmouth College Hanover, New Hampshire, although in an abbreviated format (four instead of the originally planned six to eight weeks of brainstorming with a changing group of participants).

Of interest are two, usually neglected aspects of this "birth of AI":

- Although the list of the participants does indeed read almost like a "Who is Who in early AI, of even greater interest to us is who was *not* participating resp. was not invited. Looked at from this perspective pragmatically speaking the famous Dartmouth conference could be seen as an—in the end successful—attempt to relaunch the debate on "thinking machines" off the beaten track, as represented and personified by the omnipresence of the then already 61 years old "Guru" Norbert Wiener (cf. Nilsson 2010, 77 ff.). Therefore, a new terminology had to be created to replace "theory of automation" or "cybernetics". In the light of these observations the Dartmouth conference in fact can be seen as a conspiracy of the young wild ones (age cohort 28–39) to prevent the new ideas from being incorporated and usurped by Norbert Wiener's cybernetics.: "John McCarthy admitted to have coined the term artificial intelligence partly to escape association with Wiener's theory" (Venema et al. 2019, 385; cf. McCarthy 1996). According to the retrospective assessment of all of the participants the conference was, academically speaking, a failure. To put it differently: Its only success was to bring together a group of young, promising scholars under the umbrella of this new notion *Artificial Intelligence* (cf. Moor 2006, 87).

4.2 Why Intelligence?

If my reconstruction of the Dartmouth conspiracy makes sense, we now know why the applicants had to avoid notions like "automation theory", "cybernetics" etc. This, however, doesn't answer the question why this group of young wild conspirators around 28-year-old John McCarthy chose the notion "artificial intelligence"?

After what has been said with respect to Alan Turing's reasons for replacing his main question 'Can machines think?' by its functional equivalent, it seems to be quite evident that the McCarthy group of conspirators did not only know about the ambiguity of the notion "intelligence", but were quite deliberately choosing it. In English (other than in German) the concept "intelligence" refers primarily to collecting, processing and storing data and information, as in "Central *Intelligence* Agency" (CIA), located in the George Bush Center for *Intelligence* in Langley, VA. It is quite obvious that "intelligence" in this case doesn't refer to an agency or a center of extraordinarily intelligent people. To put it differently: "Intelligence" in this case doesn't refer to what we call "intelligence" in ordinary language or in (cognitive) psychology. And something similar is the case with the adjective "artificial" which literally doesn't meananything but "technical" or "technological". Thus, the notion of AI could be referring to something very unspectacular like collecting, processing and storing of data/information by technological means.

4.3 The Power of Analogy and Metaphor

Since the Dartmouth conference first the conspirators and then successively not just the AI-community, but literally the whole world by using this notion were (and still are) activating one of the most powerful tools connected with human language: the capability of creating and using synonymies, metaphors and analogies. Already in the proposal for the Dartmouth conference McCarthy et al. were quite explicit with respect to this synonymous use of the notion "intelligence" drawing on the idea of a "Turing machine": "The study is to proceed on the basis of the conjecture that every aspect of *learning* or any other feature of *intelligence* can in principle be so precisely described that a machine can be made to simulate it." (McCarthy et al., 1955). To both the Dartmouth conspirators and to Alan Turing, it was all about *simulating* learning and intelligence by a machine.

So from the very beginning of Artificial Intelligence, it has always been this deliberate semiotic ambiguity that was characteristic for the discourse on AI. Semiotically speaking there is always not just the logical and semantical but also the pragmatic dimension of language to be taken into account. Every (spoken or written) linguistic entity is ipso facto a speech act and doesn't just refer to something, but does furthermore *do* something (cf. Austin 1962). So, the question is as follows: What is it that McCarthy and his fellow-conspirators (and up to date the whole world-wide participants in the AI discourse) wanted to *do* by not just accepting but deliberately taking into account the ambiguity of the notion "AI"?

Next to the already mentioned internal motive of looking for a neutral notion as opposed to the then dominant paradigms and schools of thought the promise of making the mythical dream of a thinking machine come true seems to be a strong driving force. And when asking "cui bono", the main question of any criticism of ideology, it immediately becomes obvious that this promise is much more capable of raising funds than the austere idea of collecting, processing and storing data and information. The very notion of "Artificial Intelligence" has this mythical ring to it exerting some kind of almost magic spell capable of opening hearts (and purses). It is one of the most striking examples of best-selling dreamware (cf. above).

5 The Origin of the Singularity Myth

Since almost seven decades people fall for this promise which can be understood as an example of Blumenberg's conception of a myth mentioned above: It quite obviously represents the "highly constant core" of the narrative while its "margins are extremely variable", expressed in the different cyclical hypes. Interestingly enough, the development of these hype cycles is always accompanied by a series of predictions following a pattern which I like to call "prognosis in permanence". What I mean by this notion can best be demonstrated by the next stage of the AI myth: the Singularity myth. Since the Dartmouth conspiracy in the Fifties of the last century the different AI-hypes were always coupled with performance promises of the type "In *x* years the performance of AI will have reached a level *y*." If *y* then doesn't happen though, this doesn't falsify the prediction because the timeframe is moving as well. Already Aristotle knew that predictions (= statements in future tense) don't have a truth value; they can become true or false if transformed into statements in the present or past tense only. In our AI case the general prediction in question is: In *x* years we will have reached a point in time when all of our (intelligent) human performance will not only be met, but even surpassed by machines/programmes. This point in time is called "singularity" since 1993 when US mathematician and information scientist Vernor Vinge made this concept (which he had coined in 1983 already) more widely known at the Vision-21 symposium (Vinge 1983, 1993). In brief: the predictive part of the AI Myth has been transformed into the "Singularity Myth".

5.1 Eschatology

The argumentation chosen by the proponents of this myth can be described as a recursive version of Moore's law: The number of transistors in an integrated circuit doubles about every two years thus resulting in an exponential growth curve (Moore 1965, cf. Thackray/Brook/Jones 2015). If combined with AI this growth will be even more rapid and therefore singularity will be reached much earlier.

Although very interesting and to some extent even plausible, this conjecture wouldn't result in a myth if it wouldn't ring the bell already mentioned by touching on another deeply rooted ("archetypical") side of our (sub-)consciousness. This can best be demonstrated by looking at a famous book: One of the main protagonists of the Singularity movement Raymond Kurzweil, famous author and Director of Engineering at Google, has titled one of his best-selling books "The Singularity is Near" (2005). The religious myth he is alluding to is to be found in Scripture (Mark 1,15): "The kingdom of God is near". The context of this is an eschatological one and is in Christian theology called "expectation of an imminent eschaton". It might also be of interest that with respect to ecclesiastical history the expectation of an imminent eschaton nearly led to the extinction of the early Christian community because some of the early Christians, believing in an imminent second coming of Christ ("parousia") focussed on praying instead of working for their subsistence. And that was when the presbyters of this early Christian community developed the theologoumenon of the Godliness of labor. Without any over-exaggeration it can be stated, that the AI Myth in its new stage of the Singularity Myth is using this archetypical pattern down to the very choice of words.

5.2 The Imitation Game

The Singularity Myth was by Vernor Vinge and Raymond Kurzweil even marked with an exact "eschatological" date in order to avoid the criticism of "prognosis in permanence". Vernor Vinge explicitly said as early as 1993: "I believe that the creation of greater-than-human-intelligence will occur during the next thirty years(…). Just so that I am not guilty of a relative time-ambiguity, let me be more specific: I'll be surprised if this event occurs before 2005 or after 2030." (Vinge 1993). And Raymond Kurzweil dated this event of singularity—a little more cautious—in the year 2045.

The problem, however, is that the idea of "surpassing" or "outperforming" human intelligence, apart from the semiotic problem of the ambiguity in the meaning of "intelligence", is lacking one decisive step: It presupposes a state of affairs where AI is at least equal with human intelligence. Alan Turing, being aware of this, by operatively replacing the question as to the meaning of "to think" and "intelligence" by his imitation game, therefore suggested a test. This so-called "Turing-test" (which by the way so far no machine or program has passed successfully), is in fact quite explicitly geared towards imitation or simulation, i.e. towards the question whether a potential candidate (a machine resp. a program running on a machine) is capable to make a potential observer/user (or every observer/user) unable to distinguish its behaviour from the behaviour of a human being, to put it differently: to deceive not just one, but any observer/user.

5.3 Deception in an Extra-Moral Sense

As I have shown elsewhere (Zimmerli 2020 et passim), this idea is marking a much more fundamental change in the occidental thinking which I have called "the Anti-Cartesian experiment", culminating in the rehabilitation of deception. The last seven decades, initiated by Alan Turing's epochal shift of replacing the question of whether machines are capable of thinking by a game testing whether a human observer/user would ever be able to distinguish seemingly cognitive behaviour displayed by a machine from seemingly cognitive behaviour displayed by a human being, are characterized by attempts to deceive potential observers/users. Thus, instead of the Cartesian postulate to take nothing for granted that isn't proven "beyond any reasonable doubt" we are now embracing deception, deception, however, in a different sense. Friedrich Nietzsche wrote about "Truth and Lie in an Extra-moral sense", trying to just get through the moral smokescreen covering the semiotic mechanism of the relation between signs and meaning. Translated into the language of our topic: If not distinguishability, but *in*distinguishability is the goal of the behaviour of so-called "intelligent" machines then the resulting deception is neither a vice nor a virtue, but exactly that what the machines or the programs running on them are designed for.

6 The Origin of the Posthumanism-Myth

As demonstrated the Singularity Myth is drawing one part of its almost archetypical power from the eschatological narrative itself, the other one from the biblical way of its wording evoking associations with the idea of the early Christian expectation of an imminent return of Christ (parousia), meaning: Yes, the end is near, but it is accompanied by the promise of the beginning of another, a better world. In order to think this other better world, it is necessary to think the unthinkable, to conceive the by definition unconceivable, Singularity being defined as that which transcends the limits of human intelligence, outperforming it in every conceivable respect. And even the very ancestor of the idea of Singularity, the aforementioned Vernor Vinge, was aware of this paradoxical structure when he formulated that Singularity will be "far beyond our understanding", rendering any kind of extrapolations in an "interstellar future" literally impossible. In other words: If human intelligence will have succeeded in creating an "intelligence-greater-than-our-own" this than in an evolutionary way of thinking must be integrated into the next evolutionary stage of intelligence. And if its presently transcended stage is defined as the human one it can be assumed that the next stage is the post- or transhuman stage.

6.1 Analogous vs. Digital

The philosophical background of the evolutionary pattern underlying these deliberations is again to be found in Nietzsche, more specifically: in his theorem of the overman. After all that has been developed so far it doesn't come as a surprise that one of the main protagonists of post- or transhumanism is again Ray Kurzweil. He and most of the other trans- or post-humanists like Oxford-based Swedish philosopher Nick Bostrom are primarily interested in the technological enhancement of the human condition.

In order to connect this evolutionary concept with our AI-topic it might be helpful to link it to the conceptual pair "analogue" and "digital". Given, the relation of humans with their world is constitutively analogous, what does that now mean in a world which is increasingly becoming digital?

Both, a semiotic and an anthropological analysis result in the insight that the function of the human factor in human–machine interaction consists in permanently transforming digital elements back into analogue elements (cf. Zimmerli 2021a) Outside of speculative theoretical reasoning there is no such thing as a completely autonomous digital interaction just between machines without participation of human intelligence.

6.2 Human–Machine–Hybrid

Thus, the next stage within the development of human beings is a Human–Machine–Hybrid or Human–Machine–tandem. And this is not just a transient developmental stage within an independent process of digitalization, because the more digital our world is becoming the more necessary becomes the human contribution of re-analogizing, i.e. understanding. What we are doing right now in trying to make sense of both the AI hypes and its underlying myths is exactly what I am talking about: re-analogizing the Digital. In this process humans therefore don't disappear; quite the contrary: they become increasingly more important in making sense of the whole process of increasing digitalization.

6.3 DAC as the Truth of Posthumanism

Here now we have reached a point of bifurcation: Either we pretend to naively extrapolate the previous development of our species from what was called "humanism" into something completely unknown called "posthumanism", or we accept the necessity to reflexively apply this evolutionary pattern to itself by looking ever more closely at what we actually know with respect to the possible future understanding of the role of the human part in the human–machine–hybrid. If we choose the latter option we then see that the future role of the human part in this human–machine-symbiosis can be characterized as the role of a necessarily permanent Digital-Analogue-Converter (DAC). Thus the Post-Humanism-Myth does indeed make good anthropological sense not in the naïve way of technologically self-enhancing human beings but in the reflexive mode of human beings who not only *are* permanent DACs but also *know* that they are permanent DACs. This then is what could be called the "reflexive truth of the Myth of Posthumanism".

7 Take-Home Messages

What has been developed so far is Janus-faced: it is resulting in both economic and philosophical consequences:

From the economic point of view we are with respect to the development of AI learning a lesson in—unintended—private entrepreneurship: Starting from the purely academic endeavor to apply for funding a conference, a group of young scientists developed—for rather a-rational reasons—a metaphorical and analogy-driven concept that, due to appealing to deeply rooted, archetypical narratives, turns out to be one of the hidden reasons for a by definition innovative enormous success in the marketplace.

From the philosophical perspective, this success story results in one aspect of the otherwise defective idea of Posthumanism: It really seems to make sense, that we need a new understanding of humanism in at least a threefold sense:

- On the one hand it is now quite obvious that we are to replace the notion of "post-" or "transhuman" by the notion of "post-" or "trans-analogue". To be a human being under the conditions of a process of ongoing digitalization then implies, that we humans not only *are* permanently converting the Digital into the Analogue, but that we also *know it*.
- There is on the other hand some kind of duty to rediscover the realm of phenomena. What is needed are phenomenological descriptions of the relations between us humans and the Digital. We human beings as DACs spend a large portion of our lives interacting with digital devices without even bothering to describe what that does to us. We are still waiting for a *phenomenology of human–machine–hybrids.*

- And there still is the trans-cultural challenge: According to Blumenberg what matters with respect to myths isn't just their largely unchangeable core, but also their high marginal variability. In order to avoid falling victim to a next simplifying "one size fits all"-colonialism, it will be necessary to look more closely into the *cultural differences* with respect to the interaction of humans with the Digital, especially AI.

References

Austin, J.L. 1962. *How to Do Things with Words.* Oxford: Clarendon.
Blumenberg, H. 1970. Arbeit am Mythos. Frankfurt a.M.: Suhrkamp.
Bultmann, R. 1985. Neues Testament und Mythologie. Das Problem der Entmythologisierung der neutestamentlichen Verkündigung. Reprint of the 1941 version, ed. by E. Jüngel. München: Kaiser Verlag.
Cohen, P.S. 1969. Theories of Myth. Man New Series vol. 4 n.1, 337–353.
Fenn, J. 1995. The Microsoft System Software Cycle Strikes Again. Gartner Group, Stamford CT.
Ferber, M. 2023. Sam Altman kehrt womöglich zu Open AI zurück. NZZ 20.11.2023, 21
Kurzweil, R. 2005. The Singularity is Near. When Humans Transcend Biology. New York: Penguin.
Lyotard, F. 1979. The Postmodern Condition. A Report on Knowledge. Engl. transl. Theory and History of Literature, vol. 10: Manchester UP.
McCarthy, J. 1996. Defending AI Research: A Collection of Essays and Reviews. CSLI: University of Chicago Press.
McCarthy, J.; Minsky, M.; Rochester, N.; Shannon, C. 1955. A Proposal for the Dartmouth Summer Research Project on Artificial Intelligence. AI-Magazine 27.4, 12.
McCorduck, P. 1979. Machines Who Think. A Personal Inquiry into the History and Prospects of Artificial Intelligence. Natick MASS: A.K. Peters.
Moor, J. 2006. The Dartmouth College Artificial Intelligence Conference: The Next Fifty Years. AI Magazine, 27.4.
Moore, G. 1965. Cramming more components onto integrated circuits. Electronics, 38 (8).
Nilsson, N. 2010. The Quest for Artificial Intelligence. A History of Ideas and Achievements. Cambridge: Cambridge UP.
OpenAI Charter 2018 <https://openai.com.charter>.
Palfrey, J.; Gasser, U. 2008. Born Digital. Understanding the First Generation of Digital Natives. New York: Basic Books.
Rheingold, H. 1993. The Virtual Community. Homesteading on the Electronic Frontier. Cambridge: MIT Press.
Taube, M. 1961. Computer and Common Sense: The Myth of Thinking Machines. New York, London: Columbia University Press.
Thackray, A.; Brock, D.C., Jones, R. eds. 2015. Moore's Law. New York: Basic Books.
Turing, A. 1950. Computing Machinery and Intelligence. Mind 59, 433–460.

Venema, L.; Jerde, T.; Sweeney, Y.; Huth, J, eds. 2019. Return of Cybernetics. Editorial. Nature Machine Intelligence 1, 385.

Vinge, V. 1983. op-ed. Omni Magazine January. New York: Penthouse Publishing.

Vinge, V. 1993. The Coming Technological Singularity. How to Survive in the Posthuman Era. Proceedings Vision-21 Symposium, 30–31.

Weizenbaum, J. 1966. ELIZA – A Computer Program for the Study of Natural Language Communication between Man and Machine. Communications of the ACM vol. 9, issue 1, 36–45.

Weizenbaum, J. 1976. Computer Power and Human Reason. From Judgment to Calculation. New York: W.H. Freeman.

Zimmerli, W.Ch. 2020. Deus Malignus. The Digital Rehabilitation of Deception. In: Göcke, B.P.; Rosenthal-von der Pütten, A. eds. Artificial Intelligence. Reflections in Philosophy, Theology, and the Social Sciences. Paderborn: Brill mentis, 15–35.

Zimmerli 2021a. Der Mythos der göttlichen Maschine. NZZ 11.01.2021, 30.

Zimmerli, W.Ch. 2021b. Künstliche Intelligenz und postanaloges Menschsein. Entstehung, Entwicklung und Wirkung eines realen Mythos. In: A. Strasser, W. Sohst, R. Stapelfeldt, K. Stepec, eds. Künstliche Intelligenz – die große Verheißung. Berlin: Xenomoi, 193–219.

Zimmerli, W.Ch.; Folkers, G. eds. 2017. Innovation – die letzte Meile. Zürich: Collegium Helveticum H. 15.

Prof. Dr. Walther Ch. Zimmerli, Dphil.h.c. (University of Stellenbosch), is at present Philosopher in Residence at the Institute for Advanced Study (IAS) of TU Munich (TUM). He holds a Dr. phil. and a Dr. phil. habil. in philosophy from the University Zurich. 1978–1988 he was full Professor of Philosophy at Technical University Braunschweig, 1988–1996 at the Universities Bamberg and Erlangen/Nürnberg, as well as 1996–1999 at the University Marburg. 1999–2002 he was president of the Private University Witten/Herdecke gGmbH. 2002–2007 he served as member of the top-management of Volkswagen AG, founding president of Volkswagen AutoUni and member of the managing board of Volkswagen Coaching GmbH, 2007–2013 as president of Brandenburg Technical University (BTU) Cottbus. 2013–2016 he held the endowed Friedrich Senior Professorship and since then an Honorary Professorship at Humboldt-University Berlin. Since 2017, he has additionally been Senior Research Fellow at Collegium Helveticum (University and ETH Zurich), at the IWM Vienna, at FRIAS (Freiburg i.Br.), and at the Digital Society Initiative (DSI) of the University Zurich. 2024 he was Mercator Visiting Professor for AI in the Human Context and Research Fellow at the Center for Science and Thought (CST) of the University of Bonn.

Part V

Hands-on Contributions to the so-called New Technologies and AI Concepts in Terms of Business Opportunities

Supporting Start-Ups on Their Mission to Outperform Large Corporates in AI Development

Matthias Hölling

1 Introduction

Value creation by start-ups has been widely recognized as a benefit for our society. At the same time, Artificial intelligence received its "iPhone Moment" through the widespread use of ChatGPT and other tools that provide truly generative AI.

However, start-ups with AI centered business models have been around and widespread since 2010, at least. The author has been around the start-up world since about that time, focusing on start-ups with a research laden background and that could be considered deep tech start-ups. It has always been a question on where start-ups can gain a competitive advantage over established companies or corporations, with their limited resources, smaller networks, and reputation, but high agility and often access to talents that received training on the latest technology and research.

In this opinionated essay, he tries to lay out his point of view on whether AI focused start-ups face special challenges and if they can be specifically supported through incubator or accelerator-like structures. After establishing a mutual understanding about artificial intelligence and its neighboring fields, which shall also define the scope of start-ups we will look at, we see how they create value or created value in the past. We will also investigate some pain points on why those technologies may not be adopted by people or organizations despite their promised value. We will, on the other end, point to the challenges and critical

M. Hölling (✉)
Stiftung Technopark Zürich, Zurich, Switzerland
e-mail: matthias.hoelling@technopark.ch

C. H. Hoffmann (ed.), *Artificial Intelligence, Entrepreneurship and Risk*, Technikzukünfte, Wissenschaft und Gesellschaft / Futures of Technology, Science and Society, https://doi.org/10.1007/978-3-658-45544-6_15

points that make a start-up "successful" and how they can create value in view of competition of large corporations, with a quite general perspective reaching beyond AI or data science. Then we will look at specific aspects related to artificial intelligence to come to a differentiated answer to what AI related start-ups should be wary of and how they can eventually be supported. For this, we will explore an anecdotal sample of start-ups known to the author on how they have created value or have faced challenges.

We will see if we can recognize some patterns and hypotheses on promising ways and probable development of start-ups providing value around artificial intelligence and if there are changes to be expected compared to how start-ups were supported in the past. We will also consider as to what extent the widespread use of generative AI could constitute a fissure in how start-ups will provide value around AI, in general. Finally, we will make some considerations on incubators and accelerators that focus on AI and related fields.

2 From Big Data to AI

We explain here that since the origins of data mining, data science should be looked at as a full process to be able to capture value from algorithms. We also explain that human or organizational reservations to recent technologies often present a bigger hurdle to their introduction than the technical challenges of their implementation. Finally, we show that generative AI, such as ChatGPT or similar tools could lead to faster adaptation than previous data science techniques.

Let us put the challenges and advantages of start-ups aside for a moment and investigate artificial intelligence, where it creates value and where we see "customer pull" in this field. Artificial intelligence also was used as a term in expert systems, implementing a rule set. The first time a computer beat a grandmaster in chess was as early as 1989 (Hsu, 2002). However, we will focus in this essay on self-learning and improving AI, starting with big data/data mining which became common in the 2000s, machine learning based on neural networks and lately truly generative AI and large language models, because this is where the impact on society is and especially will be in the future. The concept for many of these technologies have been around for long before but became popular and useful due to the development of computing power and data access over the internet from 2000 onwards.

2.1 Value Creation Through Embedding Machine Learning into a Context

Indeed, as soon as a recent technology was in this sense discovered for being effective to solve problems in all different industries, such as deep learning, very quickly tools appeared that made them widely available and usable for a large public, without having to fully engage into the theory behind and the underlying mathematics. This suggests that start-ups, or tech companies in general do not always create value by developing fundamentally new algorithms. Indeed, the author believes that this is rarely the case.

Google released the TensorFlow initially in 2015 under an open-source license, the tool that made machine learning algorithms, such as "deep neural networks" available to the programming community without having to implement the "science behind" it (Wikipedia Contributors, 2023). In fact, there are reasons to believe that Google put TensorFlow Open Source to secure their investment to remain the preferred tool for deep learning inside and outside of Google (Jin, 2022). There is evidence that the introduction of this tool has created a boost in start-ups related to AI: "Since March 2016, there have been 260 startup companies globally dedicated to working on machine learning applications across different segments of the artificial intelligence (AI) market." (Noor & Bondarchuk, 2023).

To understand where a company can create value in data science, it is worthwhile to get back to the different process steps that need to be carried out when doing data science and artificial intelligence in general. For this, it helps to visualize that at the end of the day, we want to have a model that makes a decision for us, e.g. if a produced piece is good or bad, if a person is a potential customer, if a number on a piece of paper is a tax identification number or an amount to be paid, or if a pixel in an image should be blue or green if I prompted DALL-E to generate an image for me (we are not going into details of supervised or unsupervised learning, here). To make this decision, the data needs to be somehow presented to the algorithm, and, after the decision has been made, it needs to be somehow visualized to a user or—in an abstracted form—to another piece of software for further processing. (Han, Kamber, & Pei, 2011) Since the beginning of data science, big data and classification algorithms, the steps to "Do Data Science" are agreed to as follows (Han, Kamber, & Pei, 2011) (Fig. 1):

- Data Cleansing
- Data Selection
- Model Learning

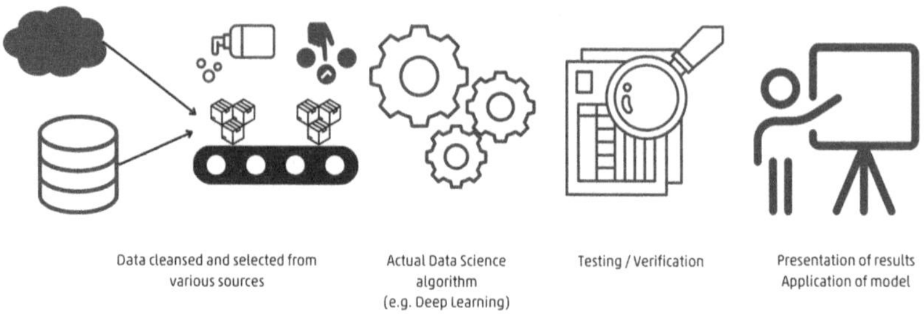

Fig. 1 Illustration of the process of a data science project

- Testing/Verification
- Model Visualization

The traditional way to create value with this process is to identify a problem where artificial intelligence could come in handy for a decision-making process, go through the different steps and present the results to management or the decision maker.

This process, however, is not very scalable, especially since data scientists are hard to find. A significant value creation can therefore only take place, if the "enveloping" steps are automatized, as well. One also needs to consider that the field of artificial intelligence is extremely dynamic. Whereas many companies were developing large language models on their own, there is in the meantime quite an offer to use LLM (Large Language Models) engines with an API, opening completely new options.

To summarize, we would expect the value creation more in the "surrounding" areas of data aggregation and visualization. Indeed, we will see later that the majority of AI or data science related start-ups have some additional expertise in a specific industry or for a specific problem. This does not mean that algorithms are not optimized by start-ups, but the value creation is almost always with the context of the enveloping process steps.

2.2 Resistance is Expected

With all the fascinating possibilities in our head that are technically feasible, we should not forget digitalization is still a big challenge for most companies. Successful AI often means that users do not realize that they are subject to AI, or even avoid the digitalization aspects, so it may be finally the driver that achieves a deeper penetration of digitalization in our daily processes. We have been talking about digitalization for decades. Big ERP implementation projects are typically an organization's nightmare. Although they can tremendously improve productivity for a fixed process, they take out flexibility and agility. Therefore, we still manually copy excel sheets to a webform or engage in similar activities.

Most people use the computer and software as a tool and do not want to engage in understanding data models, let alone process steps of data cleansing or selection for data science processes. There is also resistance because people take pride in their job, what they have learned and what they like to do. In fact, the author has come across a start-up that was optimizing industrial processes, which observed that data (prepared on carriers) "somehow got lost" a day before they wanted to collect it to test their technology. This, to illustrate how big the resistance can be in some cases.

2.3 Game Changer ChatGPT

With the introduction of OpenAI and their tools such as ChatGPT, DALL-E, etc., this topic experienced an "iPhone"-Moment in history (Huang, 2023). It did, however, not appear unexpectedly. The concepts in computer science were around since the 70s or 80s and started to become more investigated with "big data" processing in the 2000s. With the appearance of generative AI, this focus could shift since the data preparation and deployment steps become more accessible without requiring specialist knowledge of a data scientist. In addition, the application fields of AI models become broader, in general and are not limited to a specific use case obtained by supervised learning.

The introduction of ChatGPT also represents a disruption in terms of adoption. Whereas normally, new technologies have a tough time to make their way into the workforce to be used in corporate cultures or are accepted only because of top-down directives, this does not seem to be the case with ChatGPT (Fishbowl, 2023), as about 43% of workforce already used the tool as early as in February 2023. This is overwhelming as it just came out a few months before, even if the survey is predominantly covering tech companies. In the same survey, only one third of the users informed their superiors that they were using the tool. This shows that the tool is not used in a structured way that is well integrated into business processes. The development here may be less disruptive (Beninatto, 2023) and could present a chance for value capture by third parties. Indeed, companies seem to pick up the challenges of introducing generative AI into business processes and offer help to overcome them (Bilan, 2023). We will investigate that in more detail.

With respect to generative AI, we may have to distinguish between "informal chatting" with a bot to get a certain task done quicker or solve a specific problem, and a well-defined step of a process that is obligatory to be taken before the job can be considered done. It could even be that a lot of people are more open to explore recent technologies in their free time out of curiosity and fascination for technology in general but are more reserved when told to apply it to their job that they in principle know how to get it done.

2.4 The Connection to the Physical World is an Opportunity

When talking about value capture, one additional aspect has not been discussed, yet: the connection to the physical world. On one hand, there are technologies such as Edge AI, opening new possibilities in industrialization and automation or for everyday objects. On the other hand, it requires even more knowledge from AI themed start-ups, as they need to master besides the algorithms and processes around AI also some sort of physical layer.

The value creation and how easy the industry is to win over, is however specific and individual. We will discuss this later.

3 How Start-Ups Survive

Start-ups create value and they can either challenge large organizations caught in the inventor's dilemma or help them overcome it through partnerships. We see that the founders and the technology are a major factor in a start-up's success. Yet, a lot of them bury their dream of a scalable product to change the world for the sake of becoming a stable technology service provider.

Due to the enveloping nature of data science, where data must be provided from somewhere, treated with data science algorithms and then trigger a decision or be in some form presented, start-ups in the field very often have an additional expertise, not only data science. This justifies having at first look at start-up challenges and success factors in a more general way and then discuss what particularities data science and AI bring about. In general, one must see that a successful start-up is a quite unrealistic phenomenon to start with, at least if the term successful is to be interpreted in a naïve sense with a couple of geniuses fiddling with something in their garage that will be used by the whole world and generates billions of economic value, later. Therefore, those start-ups are called "Unicorns."

Large corporations or even SMEs have capital, a customer base, and experienced teams. They should, in principle, catch up with any technological advantage that a start-up may have, in a moment and then use their marketing power to push the start-up to marginal economic importance or even to bankruptcy.

3.1 Partnerships, Acquisitions, and Unicorns

Hess, Suhrbeer, and Siegwart (2013), based on Christensen (1997), however, explores the viewpoint from an industrial perspective when novel technologies are initially used by start-ups, even if this work is based on experience in the power industry. The key point is that large corporations firstly just observe novel technologies, as their adaptations would bear a significant risk for their structures or could cannibalize their own business areas.

Indeed, large corporations are focused on optimization and implementing changes that their customers request. Little of them succeed in implementing blunt ideas that customers never even envisaged to like. (We are not looking into innovative structures, intrapreneurship schemes that could overcome this limitation in the scope of this essay). Risk is always a key factor for those big companies, while the damage when a start-up "breaks" is manageable.

Indeed, Christensen (1997) introduced early the notion of a technology adsorption capacity of established companies that is often also in the author's experience a limiting factor for start-ups to be successful, we remember the aforementioned case where data got lost just before a start-up was to use it to test their technology, at this point. Eventually, a trigger point is reached when the novel technologies become inevitable to stay competitive. According to Hess et al., this is the ideal point for start-ups to partner with companies to have them apply and use their products or to enter joint ventures to adapt them according to the needs of the large corporations. For some technologies, this adaptation slope is extremely steep. Inarguably, AI related technologies are such a technology. One may even perceive a slight panic with companies feeling AI must somehow enter their processes and products, fast. This can create favorable opportunities for start-ups in the field. Such a partnership typically results into a merger and is one desired destiny of a start-up company, particularly by investors. The other one is extreme growth with a start-up on its own turning into company getting into

8-digit turnover figures and eventually becoming a Unicorn with a valuation of more than a billion dollars.

Whereas these developments are ideal cases, it is less than common. A lot of original ambitions to "conquer the world" with a new product often end up with a small team of a handful of engineers or businesspeople that provide some B2B related service or serve a somewhat niche market. Whereas one may perceive this as a disappointment, one should also consider that this may create satisfactory work for hundreds of thousands of people in Switzerland alone. Even if these start-ups do not fully satisfy the expectations of all investors and shareholders, the author would personally not consider them as unsuccessful, and they are also to some extent the base of our wealth in central European economies (Pahnke & Welter, 2019). Also, at Technopark Zurich, a third of the companies are seasoned and stable, providing the majority of the 2800 full-time equivalents working there.

In the case of an acquisition, the buyer may sometimes be more interested in getting the talent than the actual product. We have seen this a lot with early type of exits, before the actual product got a lot of market traction. The author has experienced cases where this resulted in Big Tech acquiring talent to redesign the start-up's technology, so that it would add new functionalities to widespread commercial products (the cases cannot be named for confidentiality reasons), but also in the form of pure "equihire", where the team was incorporated into the Big Tech company and the technology of the start-up played no or a very minor role. Repeatedly, however, a few start-ups make it to unicorn valuations, outpacing endeavors of other companies and sometimes even stashing efforts of large corporations to catch up with them. We have seen this a lot in Tech development, where WhatsApp outpaced existing messenger services from existing companies, Netflix outpaced others in streaming, etc. Suddenly, it was impossible for large companies to keep up with them and/or they were bought by bigger players.

3.2 Success Factors

Hoffmann (2023) identified a few impact factors that influence the valuation of a start-up, which is normally at some point a measure of its success, even if not a one on one "translation" for success in the real market (because even high valued start-ups may never capture their valuation in a real market, due to late failures or mergers for strategic reasons, for example).

Those factors are:

- Founders (are they serial entrepreneurs or scientists with limited business acumen).
- Innovation, particularly the degree of intellectual capital in the company, leads to higher valuations.
- Governmental and social influence, which is normally entering the motivation of founders.
- VC (Venture Capital) valuations, or in other words, follow the trail of investments that are "in fashion."

At the end however, classic business model design plays a vital role in whether a start-up can be successful or not, such as originally proposed by "Business Model Generation" (Osterwalder & Pigneur, 2010) and refined or adapted to specific cases for start-ups by many others, such as the lean canvas (Maurya, 2012) or the Co-Star Methodology (Gyorffy & Friedman, 2012).

The most critical part is the value proposition, here. Especially research driven start-ups underestimate the problem of a mediocre problem–solution fit with the customer combined with a limited technology absorption capacity (Hess, Suhrbeer, & Siegwart, 2013), e.g., the improvement and saved cost in a process does not consider the cost of changing the process, overall, including risk assessment, employee motivation, etc.

The big advantage a start-up has here, is the agility. In principle, it can react quickly on insights it gets from customer feedback or even pivot to a new business model, altogether. Often however, start-ups are focused on certain use cases and too little focused on the dialogue with potential customers. Once they start doing this, the original plan to change the world can also evolve into a small or medium-sized team providing technical services, finding their role to provide value to society, even if not as expected.

4 Challenges and Support Levers

What do founders want and what are the risks they are submitted to? We show that these are fundamentally different between pharma companies and others (especially IT companies). We also look at some concrete examples where companies met challenges in the experience of the author.

When we ask founders what their biggest challenges are, we often hear 3 key points:

- Finding funding

- Finding the right people
- Finding customers

For all these topics, it is important that the base of the business is solid, especially in terms of a product market fit and captivating value proposition, as discussed earlier. Support efforts should focus on improving this important aspect and not be centered so much on using networks and connections to "talk someone into" carrying out a pilot project without the recipient's conviction that it will solve their problem. Before going out and starting to talk to customers and using a network , it must be clear that the start-up offers a working product for a real problem on the customer side. Otherwise, the result may be a "pilot" that makes founders temporarily happy, but was merely motivated by politeness or "fear of missing out" on the customer's side.

Indeed, a survey from 2021 (CB Insights, 2021) suggests "poor product" as one of the primary reasons for failure besides the running out of funds. The latter may, however, be strongly related to the former (nobody would invest in a mediocre product market fit). We will look from a distinct perspective to certain risks and explain what the primary work of the start-up should be to mitigate and minimize these risk factors.

4.1 Initial Market Adoption Risk

To see if a technology can be successfully placed into the market, in an abstract way we can look at a risk concerning the technology versus a risk towards the product market fit. This shall be illustrated with two extremes representing a certain industry, each. The technology risk exists mostly for deep-tech start-ups, only. In the aforementioned survey (CB Insights, 2021) "not having the right team" was one of the failure reasons. An assumption is that this could be related to a technology risk for non-deep-tech start-ups.

In general, while software companies have a lower risk on the technology side, the risk for the product market fit (if we are talking about a product, not a service) is quite high. This goes for not overly complex software or engineering products at the state-of-the-art level. Examples could be a sharing platform or business software increasing productivity of staff. Important here is an elevated level of agility, that allows them to take customer feedback back into the design or to pivot with the product or addressed market, altogether.

For the development of new pharmaceutical products, susceptible of curing a disease, it is—to an extent—the other way around. Once the clinical validation has started, all production processes must be duly documented and cannot be easily changed, and even in the 3rd phase of clinical trials, the technology may still fail. However, if it works, the problem solution fit is generally given. After all, they can cure a disease. This is quite simplified, especially since there may still be a significant risk to bring a product to the market, but I found the idea to be useful when evaluating start-ups.

Hardware, medical devices, or other industries may be positioned somewhere in the middle of this continuum.

Special efforts for IP protection of the technology (patent, well preserved trade secrets) coincide typically with a high technology risk to protect the investment, but may, especially in the case of patents, limit agility. The author has seen a few cases where patents originally filed or licensed from universities became useless after a necessary pivot. They did often play a significant role in attracting initial investors, though.

There can also be products that have an elevated risk on the technological side as well as on the market fit side, e.g., complex hardware-based products for a conservatory market, such as construction, automotive or similar. In these cases, it should be evaluated very carefully how high the return is and if it Is worth the effort. A lot of care should be given to the question of whether the agility can be preserved to react to customer feedback. Unfortunately, deep tech start-ups tend to move to that corner of the graph.

There is, however, rarely a start-up that can successfully position itself in the bottom left of the above Fig. 2, unless it is for an extremely limited niche or local market (where nobody else 'bothers'). Otherwise, someone would already occupy this space. In other words, there is no significant value capture without associated risk.

As a conclusion, a start-up should evaluate those risks for themselves and work on it. For start-ups developing new drugs, all the main effort should be eliminating the technological risk. The path for this is typically well laid out with a sequence of animal and clinical studies. Most others should likely spend more time to eliminate the risk of a mediocre product market fit. Both must be open to pivot if they see one of the risks eventually leading to failure. In the case of companies with a high technology risk, however, this is extremely difficult.

The author has seen quite a few failures, of course, for pharmaceutical development, but even for AI based app development in the health sector, e.g., to analyze biometric data through mobile devices, clinical validations must be obtained, and companies have failed to achieve this. Therefore, a more lifestyle-oriented product could sometimes be an alternative to pivot to if an appropriate business model can be built around covering financial and technological aspects.

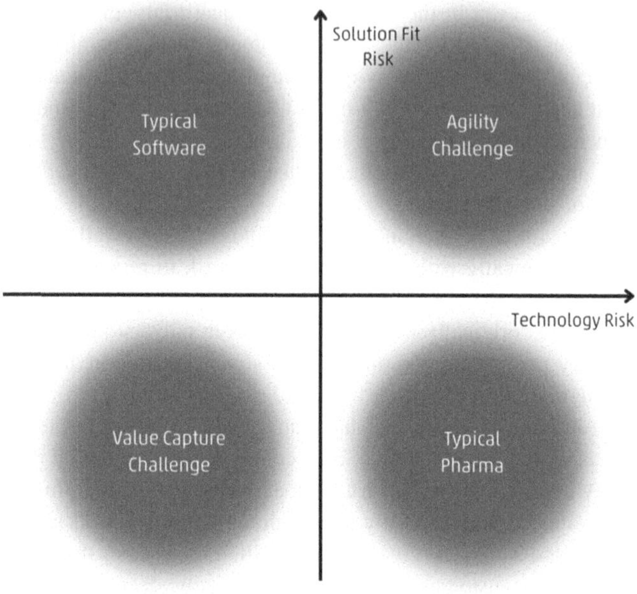

Fig. 2 Simplified illustration how Start-ups are submitted to technological and problem solution fit risks, often one mitigating the other

4.2 Specific Financing Risks

Certainly, there are financial risks for every start-up. Funding is a main concern for all of them. But let us consider again specific cases for the two "polar" industries: State of the art software development and pharma:

While a software company may not get sufficient funds to afford enough marketing to really grow, as the competitive landscape is much bigger and they are, without marketing virtually invisible to their customers, a pharma company usually works with key opinion leaders, often already involved in clinical trials. Those clinical trials, however, are extremely costly and sometimes also require a large network of specialists. Often, pharma start-ups are sold to one of the big players in the field after the 2nd stage of clinical trials for this reason.

So, while an archetypical software start-up, especially in a B2C context, needs to secure sufficient funds for marketing, an archetypical pharma start-up needs its funds primarily to validate their technology.

4.3 ESG or Regulations-Based Business Models

Some companies rely with their business model on some sort of governmental regulations or incentives or incorporate ESG (environmental, social and governance) goals. There is no doubt that ESG aspects are becoming more and more important in the corporate and business world. Also in business plan competitions, the business model should somehow serve some sort of sustainability goal. This may also be a trigger for start-ups to acquire talent that wants to feel good about the work they do (Rubin, 2022).

A challenge that one should consider here, however, is that complying with regulations is not widespread considered as a "fun" activity or one with a high creation of value. If a start-up can provide a service to get it out of the way with minimal effort, fine. The author has seen quite a few successful companies offering automized services in connection with data protection laws, such as the GDPR.

However, if it expects companies to be overly zealous about complying with regulations, this often does not work. It bears similar risks, when a start-up expects legislation to come up with regulations that are not in place, yet. We have seen cases here for sensor applications in food control or fashion product improvements that have proved to be difficult. Only if the clients must fear serious lawsuits, they may invest in a technology that goes beyond the compulsory to avoid it. Similar considerations go for data privacy issues. While widely discussed, many customers, at the end of the day, do not care so much about it. The author does not necessarily consider this a good thing, but it is what it is, and most private individuals may prefer to search products that are fun or save work, rather than those that protect their privacy. An exception here is when it concerns their children, though.

Certainly, burdens appear when dealing with optional efforts on environmental issues, too. Unfortunately, most people can be won over with initiatives that give them minimal "pain" and reassure their conscience, rather than those that are the most effective. This may be frustrating, but again, if the goal is to create a successful business model and value proposition for a start-up, those realities should be considered. Of course, there are plenty of individuals that seriously want to engage into the above "good cause" goals, but the market size should be considered adequately for business models targeting these individuals. The author has seen that founders are sometimes too optimistic, here. Also, complication of the business model does not necessarily help, according to the author's experience. During the peak times of blockchain related start-ups, we saw many start-ups putting this into ESG or environmentally related business models to prove some provenance or sustainability of products. It should be the subject of another research if customers would pay for sustainability when trust issues can be eliminated. The author experienced this as difficult, though and believes that the original intrinsic motivation to pay for sustainability is prevalent.

4.4 Regulatory Risks—New Challenges for AI Companies?

Regulations become increasingly important in fields other than Pharma or MedTech, or when dealing with corporates who require certain ISO certifications. Innovative technologies are extremely complex, and society (and governments) try to mitigate risks that are hard to predict when a technology is first introduced. In the context of this essay, it must be pointed out that the risks of AI are extremely widely discussed up to requiring a moratorium on all AI development (Future of Life, 2023).

A more practical approach is the recently passed EU AI Act (European Parliament, 2023) which wants to mitigate the societal challenges for this new technology while creating clarity for entrepreneurs at the same time. The discussion on this should be closely observed and AI start-ups certainly need to be aware and comply, even if they work in fields that are not immediately related to personal data.

4.5 Is Generative AI a Spoilsport?

With generative AI, the focus on the enveloping business may be challenged. A recent update to process PDFs for ChatGPT has put many start-ups at risk (Chowdhury, 2023). So, are start-ups resilient against improvements of generative AI by Big Tech companies?

It may again shift the activities of start-ups and create a whole new ecosystem of small businesses. In fact, one could expect a development, where we see an increasing number of wrappers using APIs to ChatGPT and similar tools that become the new "Web Designers" of tomorrow. They would offer a convenient and compelling user interface using their expertise in UX Design or knowledge about how to use the available generative AI software efficiently. It may create a similar boost in new start-ups being created as after the launch of TensorFlow.

In this context, the author has seen start-ups providing a service around the analysis of business process documents that successfully shifted their primary value proposition from the recognition of key elements in the document (that can be done by ChatGPT, now) to seamless integration with ERP systems.

But to produce something genuinely novel that tools like ChatGPT cannot do with some medium level expertise of the user will be increasingly difficult. In other words, start-ups should consider and predict how Big Tech will improve their tools and products and create value propositions that take this into account.

Some examples that the author considers resilient against future "ChatGPT Upgrades":

- Technology involving hardware, especially Edge computing or IoT (Internet of Things) themed AI
- Start-ups developing hardware or chip design, particularly adapted for deep learning and AI (there are a few, e.g., memristor type of technologies)
- Start-ups engaging in extremely advanced technologies, such as quantum machine learning (while this is still in an exploratory stage).
- Private Cloud applications, high security applications that cannot be put into the cloud. A contrary development here could however be secure computing, allowing operations on data without exposing the data.
- Start-ups with a predominant expertise in specific fields for which the AI part is important, but not their primary expertise aspect (Besides MedTech or HealthTech, the author has seen this particularly in real estate or production automation)

5 How Can Incubators or Accelerators Help?

Since we have laid out a certain set of challenges and advantages for start-ups, we can now see how an incubator or accelerator structure can exploit these findings. We will start out by explaining the two structures, in a general way, try to estimate what industries AI and data science start-ups may be working in to align possible support and point out some specific aspects of support.

Incubators and accelerators have been around for decades (Plug and Play opened in 2006).

The difference between the two: An incubator is often a physical space where new business can get everything from offices space, business coaching up to venture capital and are often funded by a (local) government or an academic institution. An accelerator is more competitive in the selection of the start-ups, but not confined to a local economy or provenance from a specific institution. They are also often commercially driven, e.g., the price the start-ups pay is equity and less bound to physical space. They are also more program oriented instead of just providing an offer that the start-up may use or not (Wikipedia Contributors, 2023).

Surprisingly, there are a few studies that discuss the effectiveness of incubators and accelerators. It was found that there is a success rate for 84% vs 44% of incubated start-ups vs. non-incubated start-ups (Tai, 2019). One of the main challenges to make such a comparison is that it is not comparing the same start-ups, or in other words, the above does not mean that of the 84% of the incubated start-ups only slightly more than half had been successful had they not been incubated.

In fact, all incubators and even more so accelerators have an acceptance policy that curates start-ups that they onboard. This step is well necessary for the following reasons:

- Incubation and acceleration should make certain things easier for founders but will not be able to create miracles. A mediocre business idea, common state-of-the-art technology or founders with below average business acumen is a bad base for a successful start-up, whether incubated or not.
- Incubators and accelerators provide a certain quality mark for onboard start-ups, attracting investors or customers. They experience that good start-ups are in this incubator, no matter what incubation services the start-up has experienced.
- Particularly accelerators often rely on the success of the onboard start-ups, as they often take a stake as a price for the participation in the program.

In general, the above identified risks and challenges apply to AI start-ups in the same way. So, the coaching and support can also be similar. As pointed out earlier, an AI start-up (or data science to make it broader) is not "only" doing data science but will provide value in the context of an industry. Therefore, experts from affected industries can provide valuable input. We will have a look at some data science start-ups from the past to get an idea about their business models and modes of value creation. As discussed, AI start-ups may understand data science,

LLM and even generative AI well and tweak the algorithms, but this is not their main activity, in most cases. An algorithm as such does not deliver value alone.

The author has had experience with primarily deep-tech start-ups working for support organizations since 2010. In this time, he came across around 66 start-ups related to deep-tech (Table 1).

Table 1 Additional expertise in start-ups the author has seen since 2010

Additional expertise	Number of start-ups
MedTech/HealthTech	11
Image analysis	7
Mobility	6
Generic	5
HR	4
Smart city	4
Fintech	3
Production	3
Advertising	2
Ethics	2
IT networks	2
Fashion	2
Real estate	2
Energy grid	1
Education	1
Multiphysics	1
ERP systems	1
Acoustics	1
Gaming	1
Supporting hardware	1
Environmental	1
Robotics	1
Social	1
Sensors	1
Legal	1
Construction	1
Grand total	66

Of those 66, 11 focused on MedTech, 8 on fields related to image processing and robotics, 6 in Mobility (where the aspect of mobility/smart cities was valued higher than the image processing aspect), 5 in finance or real estate, 4 in HR and 3 were each related to production technologies. 5 were generically offering AI services or products, e.g., to help implementing AI or data science projects.

The analysis did not show fields or industries in which start-ups where more successful than in others (e.g. by analyzing the development of the start-ups per industry). It shows however, that MedTech or HealthTech is somewhat popular and that, at least up to now, image processing has played a significant role. A recent survey of AI Start-ups in Switzerland found that only few start-ups were using generative AI in their business model, but the majority were relying on deep learning, instead (Kyora & Rockinger, 2023). While there may be a shift in start-up activity due to the recent surge of generative AI, there is no real reason to believe that they will affect completely different industries.

(Weber, Beutter, Weking, Krcmar, & Böhm, 2021) classifies start-ups into 4 different fields following a systematic taxonomy based around the value chain. This classification is not industry specific, at all, but confirms that data science and AI are somewhat industry agnostic fields. So in fact, the above figures may be an indication of what type of expertise could be helpful in an incubator or accelerator focused on AI and data science. Some other success factors are examined in the following.

5.1 Location is Important

The location plays a key role.

We saw that founders say that one of their biggest concerns is getting good talent. A convenient location can play a critical role in facilitating this, if the remuneration and ancillary benefits cannot keep up with large corporates, the team has the benefit of working at an easy to reach and attractive place. The author found that schemes to work from home—induced by the covid-19 pandemic –also apply to start-ups, but to a lesser extent, as their mode of work is more agile and less process oriented, in most cases.

5.2 The "Market" for Funding is not Confined to the Incubator

As far as funding is concerned, it is of course perfect if the incubation facility also has a means of investing. However, a conflict of interest should be avoided here, and it should be done by separate units. The criteria to onboard a company to the incubator may not be the same as investing in it. Both, and especially accelerators—however—provide a quality mark that may get an investor's attention.

5.3 Helping the Start-Up to Mitigate Risks and Find the Product Solution/Market Fit

We have looked at the previous section to identify certain risks that a start-up is submitted to. An incubator or accelerator can certainly help identify the risks through external experts and suggest ways to mitigate them, by means of external experts or seasoned entrepreneurs with experience. The above table gives a subjective overview as to what kind of expertise could be important.

However, very rarely, those experts can help with solving technical challenges. Instead, their support is valuable to find a better product market fit or comply with regulatory or legal issues of all sorts. They can also help with general management challenges, such as team management, recruiting, etc., or provide leads to investors and customers.

In general, one should distinguish whether they have "skin in the game" or not. Most coaches from the network of an incubator do not have "skin in the game." They can provide valuable insights from a more distant view without too much emotional attachment or worry. A challenge is that experts are sometimes abundant and may tend to impose themselves and it is very hard to judge their quality. An incubator or accelerator can provide guidance here. Besides the incubator, there are governmentally controlled coaching programs that vet and assign coaches to start-ups in a reasonable way, avoiding that they are overcoached.

At Technopark Zurich, there is a program called "strategy check" where we select from a pool of 25 experts about half a dozen to spend an afternoon with a particular start-up to discuss its challenges. The session goes on for 2–3 h. Although sometimes very tough for the start-up, we emphasize that this is an eye-to-eye-level discussion and the management decision remains with the start-up, whether they take in some of the advice or not. We found this to be extremely useful, as the start-up has access to several experts and coaches at the same time.

Of course, a coach may agree with the start-ups to get "skin in the game" and invest financially, in addition to his advice, e.g., becoming a board member. The situation is different for accelerators. Since they often have a stake in the start-up, they act like a board member with "skin in the game." That also creates trust with founders, and it was clear for them from the beginning that they receive this support.

They are also in a better position—at least for the acceleration period—to be something like a "master of ceremony" filtering and coordinating the various coaching and support offers around. For start-ups requiring specialized services, such as MedTech or Pharma companies to comply with special regulations, it is of course extremely useful to have easy access to these services for start-ups onboard the incubator or the acceleration program.

5.4 So, What About AI?

Can an accelerator or incubator provide something "special" for AI start-ups?

Of course, also the infrastructure should be excellent in terms of network and cloud access to hyperscalers, with perks for the start-ups, or even more useful, access to them to discuss the start-up's products. Apart from that, the author only saw 2 start-ups that focus on ethical aspects; however, most will be exposed to it, especially if they deal with personalized data. These are most image related companies or companies providing financial and HR services.

Indeed, it is to be seen if this could be a service or check that an incubator or accelerator can provide, among others. While it will be straightforward to comply with the EU AI act, for the time being, future developments may bring up ethical questions related to AI, repeatedly. Therefore, such expertise may be helpful in the context of start-ups implementing the latest AI research. At last, we have identified the risk of future developments of generative AI putting some start-up models in danger, as their service could be taken over by the generative AI, itself with minimal technical expertise.

An awareness for this development and a good feeling should be present in an AI focused incubator or accelerator and this may, indeed, go hand in hand with estimating ethical and regulatory implications.

References

Beninatto, R. (2023, November 18). *LinkedIn.* Retrieved from https://www.linkedin.com/pulse/complex-journey-technology-adoption-renato-beninatto-zp5xc/

Bilan, M. (2023, November 14). *Master of Code.* Retrieved from https://masterofcode.com/blog/red-flags-and-best-practices-for-generative-ai-in-telecom-banking-and-finance

CB Insights. (2021, August 3). *The Top 12 Reasons Startups Fail.* Retrieved from CB Insights: https://www.cbinsights.com/research/report/startup-failure-reasons-top/

Chowdhury, H. (2023, October 31). *A minor ChatGPT update is a warning to founders: Big Tech can blow up your startup at any time.* Retrieved from Business Insider: https://www.businessinsider.com/openai-chatgpt-pdfs-ai-startups-wrappers-2023-10?r=US&IR=T

Christensen, C. (1997). *The Innovator's Dilemma.* Harvard Business Review Press.

European Parliament. (2023, June 8). *EU AI Act: first regulation on artificial intelligence.* Retrieved from https://www.europarl.europa.eu/news/en/headlines/society/20230601STO93804/eu-ai-act-first-regulation-on-artificial-intelligence

Fishbowl. (2023, February 1). *Fishbowl.* Retrieved from https://www.fishbowlapp.com/insights/70-percent-of-workers-using-chatgpt-at-work-are-not-telling-their-boss/

Future of Life. (2023, March 22). *Pause Giant AI Experiments: An Open Letter.* Retrieved from Future of Life: https://futureoflife.org/open-letter/pause-giant-ai-experiments/

Gyorffy, L., & Friedman, L. (2012). *Creating Value with CO-STAR.* Enterprise Development Group.

Han, J., Kamber, M., & Pei, J. (2011). *Data Mining: Cocepts and Techniques, 3rd edition.* Elsevier Science Ltd.

Hess, S., Suhrbeer, S., & Siegwart, R. (2013, June). The Impact of Collaborative Innovation between Established Industry and Academic Technology Spin-offs. *Business and Management Research.*

Hoffmann, C. H. (2023, March). Startup valuation reassessed: against celebrity, sustainability and state intervention. *Journal of Ethics in Entrepreneurship and Technology.*

Hsu, F.-h. (2002). *Behind Deep Blue: Building the Computer that Defeated the World Chess Champion.* Princeton University Press.

Huang, J. (2023, February 1). *Dean's Speaker Series—Berkeley Haas.* Retrieved from Youtube: https://www.youtube.com/watch?v=9hzVdV63scU&t=610s

Jin, H. (2022, November 1). *TensorFlow Is Open-Source, But Why?* Retrieved from Medium: https://towardsdatascience.com/tensorflow-is-open-source-but-why-512d849f59d2

Kyora, S., & Rockinger, M. (2023). *Swiss Startup Radar 2023/2024.* Lucerne: JNB Journalistenbüro.

Maurya, A. (2012). *Running Lean.* O'Reilly Media, Inc.

Noor, K., & Bondarchuk, Y. (2023, September). *100+ Machine Learning Statistics: Market Share & Data Analysis.* Retrieved from MarketSplash: https://marketsplash.com/machine-learning-statistics

Osterwalder, A., & Pigneur, Y. (2010). *Business Model Generation.* Wiley.

Pahnke, A., & Welter, F. (2019). The German Mittelstand—Antithesis to Silicon Valley entrepreneurship? *Small Business Economics.*

Rubin, S. (2022, September 22). *The Power Of Purpose: How ESG Affects Your Company's Talent.* Retrieved from Forbes.com: https://www.forbes.com/sites/forbesbusinesscouncil/2022/09/22/the-power-of-purpose-how-esg-affects-your-companys-talent/

Tai, J. (2019, June 27). *How To Pick The Best Incubator For Your Startup.* Retrieved from Forbes.com: https://www.forbes.com/sites/theyec/2019/06/27/how-to-pick-the-best-incubator-for-your-startup/?sh=5dcca9cf42ba

Weber, M., Beutter, M., Weking, J., Krcmar, H., & Böhm, M. (2021). AI Startup Business Models. *Business & Information Systems Engineering.*

Wikipedia Contributors. (2023). *Startup Accelerator—Wikipedia.* Retrieved from Wikipedia: https://en.wikipedia.org/wiki/Startup_accelerator

Wikipedia Contributors. (2023, 12). *TensorFlow—Wikipedia.* Retrieved from Wikipedia: https://en.wikipedia.org/wiki/TensorFlow

Advancing Management Development with Biofeedback 3.0: Unveiling Insights into Decision-Making via Next-Generation Markers for Cognitive Processing in Aviation and Organizational Contexts

Christian J. Kusmitsch, Helmut Unger, Patrick Schmidlechner, Fabian Simmank, and Alexander Neuhaus

1 Introduction

In management science, decision-making refers to selecting a course of action or choosing from among various alternatives to achieve a specific goal or solve a particular problem. Decisions in management trigger or inhibit the flow of financial,

C. J. Kusmitsch (✉) · P. Schmidlechner · F. Simmank
Mindset Technologies, Hergiswil, Switzerland
e-mail: christian.kusmitsch@mindset-technologies.ch

P. Schmidlechner
e-mail: pschmidlechner@mindset-technologies.at

F. Simmank
e-mail: fsimmank@mindset-technologies.at

H. Unger
UNGER Performance Advisory, Pressbaum, Austria
e-mail: helmut.unger@helmut-unger.at

A. Neuhaus
Frequentis AG, Wien, Austria
e-mail: Alexander.NEUHAUS@frequentis.com

C. H. Hoffmann (ed.), *Artificial Intelligence, Entrepreneurship and Risk*, Technikzukünfte, Wissenschaft und Gesellschaft / Futures of Technology, Science and Society, https://doi.org/10.1007/978-3-658-45544-6_16

tangible, and intangible resources. For many centuries "leadership" was synonymous with management, and it was deemed to be inherited as a characteristic of nobility or military ranks. The idea that managers can be trained and can acquire capabilities and skills, and conclusively, management expertise can be taught is rather new (Drucker, 1954). Starting with the seminal work of Tversky and Kahneman (1974) or Simon (1979) behavioural economics has demonstrated in the field what was known from the laboratory—human rationality is flawed, bounded, and biased. Human judgement is often systematically wrong with typical mistakes and errors (Reason, 1990) and always restricted by the bounded nature of the cognitive apparatus and its limited processing capability (Miller, 1956). Before these findings, a manager was considered an "economic man", "rational and logical", with unlimited access to all necessary information for making impeccable decisions. After all, our human mind, fallible as it may be, has been very practical and overall effective for more than ten thousand years, and evolution sometimes is not only reaping the benefits of strengths, but we are surviving despite obvious weaknesses, even in harsh conditions. Is today's complexity overwhelmingly harsh, for the most complex system in the universe, the human brain?

Reaching the necessary performance under harsh conditions to thrive instead of surviving has always been a key competence in mankind and business. By striving for quicker and more adaptable offerings, with the increase of connections and opportunities between globalised economies, the business drove its complexity. The purpose of managers is to manage complexity and convert it to "manageable", because "in business, complexity gets bad press" (Reeves et al. 2020). Building an organization is part of how companies and "good practice" management reduce complexity. Luhman described organizations as social systems consisting of communications and decisions (from Tacke, 2020). Hence organizational success requires good decisions within the organization's relevant environment. Kolmogorov (1965) made complexity measurable and thus complexity became an objective attribute of a system. Through global interconnectedness and quicker processing of an increasing amount of data, our business world has become more complex. Why is understanding the outer world so important to successfully managing the system within? As Ashby suggested (Law of Requisite Variety Control), a controller can control something to the extent that it has sufficient internal variety (Ashby, 1956). Put in other words: can we find a container that is bigger than the to-be-contained essence? Apart from increasing complexity, we have entered a world that leaves us with more open questions than answers. How to thrive instead of surviving in a world that is so complex?

For every question and choice, in return, we have to pay a high psychological price: the cognitive load induced by deciding in a complex environment may be an important consideration in the interaction between anxiety and cognition (Vytal et al., 2013). How to approach problems? First, we need to understand which kind of problem we want to address and manage: a simple one, complicated, complex, or chaotic (Stacey, 1995). When it is "just" complicated, we can build models around with sufficient variety wherever and whenever possible. For the solution of complex problems, we can use metamodels like the "St. Galler Management Model" (Rüegg-Stürm & Grand, 2019).

Neglecting behavioural economics and dynamics of decision-making in the twenty-first century is still commonplace in many, we daresay the majority, of organizations but the immaturity of thinking is an invalid excuse in the face of degrading decision-making performance in often data-rich yet information-poor environments. AI drives complexity and AI boosts both rational and irrational fears, to say which is which remains a subject of hindsight. AI algorithms can integrate into systems that help humans in the loop to cope with complexity, e.g., by taking the burden of countless repetitions away. AI can detect patterns (e.g. fraud in financial transactions or anomalies in medical diagnostics) through superhumanly, lightning-fast learning loops.

Mankind—hence management—has three silver bullets for transcending self-inflicted immaturity: insight, learning and innovation. Artificial (alias man-made) products are very successful in promoting mankind's progress, how would today's world look like without artificial power tools, power drills, power motion machines (e.g., cars) or any apparatus that creates a form of power from energy and substitutes human labour and toil? There are two essential differences between Artificial Power and Artificial Intelligence. The destination, the goal of using either artificial power or natural power, the hand drill, or the power drill, stays the same, it is congruent if not identical. You drill a hole, with or without power, and you get a hole in the end. While we use AI as a tool, AI uses us, our artefacts, texts, and images, for it to function properly or at all and to change its status perhaps irreversibly while processing. We feed ChatGPT and other Large Language Models (LLMs) with text of a quality that depends essentially on our language capabilities, and it feeds us back with a text which fits very likely but is not written in our idiosyncratic semantic codes. AI is designed to learn, and it does so by ingesting the data mankind produces. Through interaction and "renegotiation" the destination is permanently changing due to human–AI interaction.

In the following chapter, we lead through our framework of cognitive enhancement with a method we call Biofeedback 3.0 (BFB 3.0). BFB 3.0 will lead –in our view– to scalable, productive, and helpful cognition-aware systems (CAS), systems that can detect even covert states of decision-makers and will finally lead to better decisions, with fewer human errors and free from human biases. Decision-makers will be less bounded by the nature of their cognitive apparatus and will be supported by an automated, perhaps autonomous system incorporating human-centred AI. Before that, an excursion into the nature of human cognition and its neural basis is undertaken. We have detailed two cases to demonstrate applications in the field and the acquisition of individual decision-relevant data and actionable insights in the highly regulated and safety-driven aviation sector and from the general management of an engineering company. Finally, we reflect on the ethics of our interventions and methods, regarding ethical AI and data privacy and underline questions, rooted in practice, and directed towards academia. Were this essay a pointed piece of life advice for managers it would be: "Life is not a game of Go but a Kobayashi Maru test!".

2 The Substrate of Decision-Making

To understand decision processes better and following our attempt to digitalize and integrate concomitant data streams into one cognition-aware system, we must first reflect on to what degree we can objectify and quantify covert components, process steps or correlates of decision-making in biological structures and human physiology. Discovering the substrate of the mind, consciousness and cognitive processes is not a task typical for practitioners, so we will focus on two regions in which functionality is a necessary condition for the phenomena we address here.

The prefrontal cortex (PFC) and the insula are constitutional in managing decision-making processes in the context of their emotional embeddings (Öngür & Price, 2000). Both brain regions are involved in processing emotions. Emotions colour cognition, fuel and even trigger decision-making.[1] A recent theoretical framework, meticulously, creatively, and empirically elaborated by Barrett (2017a, b; Barrett & Westlin, 2021) provides both a practical reference for scientific research and a terminology which removes overly scholastic barriers to putting science into practice. Barrett suggests that emotions prepare the body for action by producing a variety of physiological responses, based on the prediction of future energy needs. However, the specific form of these responses is not predetermined since it depends on the individual's learning history and the context, it is not a reaction *to* the world but a construction *of* the world (Barrett, 2017a; p. 104).

[1] "Emotion ohne Kognition ist blind, Kognition ohne Emotion ist leer" (Ciompi, 1994).

Coming from a clinical perspective, Antonio Damasio postulated the influence of so-called "somatic markers" on decision-making (Damasio, 1996). The absence of these markers, due to brain lesions especially in regions in the prefrontal cortex can result in inertia (Damasio, 1996). Bechara et al. (2005) noted that modern economic theory ignores the influence of emotions on decision-making. According to their somatic marker hypothesis, a systems-level neuroanatomical and cognitive framework manages decision-making based on its influence by emotion. Unconscious pre-evaluation leads to a somatic response before the decision even emerges in the mind, i.e. becomes conscious. This response creates a bias and lowers the threshold for deciding and in this case the threshold for showing overt behaviour with real-life consequences (Bechara et al., 1997). We are not going down the rabbit hole tagged "Where is the seat of emotion?" but two brain structures are key players: the insula and the prefrontal cortex. The insulae (insula, insular lobe or cortex) are folded deep within the cortex and are believed to be involved in consciousness and are linked to emotion and regulation of the body's homeostasis, its energy household instance. The insula is the substrate of functions including compassion, empathy, taste, perception, motor control, self-awareness, cognitive functioning, interpersonal experience, and awareness of homeostatic emotions such as hunger, pain and fatigue (Tinaz et al., 2018; Uddin et al., 2017). You might call the insulae also the seat of the sixth sense: interoception.

Interoception is a key function of human insight and biofeedback methodology. Interoception is "the integrative interpretation of internal and external stimuli, in the cognitive/emotional context, to derive an overall physiological representation of the state of the body, including conscious and unconscious aspects" (Berntson & Khalsa, 2021). From a psychological standpoint, interoception and self-hood are linked with body awareness (Quigley et al., 2021). Both interoception and body awareness can be trained and improved: after one week of "interoceptive training" (heartbeat perception), the participants in this study by Sugawara (2020) exhibited enhanced interoceptive accuracy and reductions in both somatic symptoms and anxiety levels. Remarkably, their decision-making processes shifted in a more rational direction compared with the prior established baseline. If there is a seat of an executive's abilities to reason and decide accordingly, it is the executive brain located right in our forehead, immediately behind the root of the nose—the prefrontal cortex (PFC). The primary function of the prefrontal cortex (PFC) is cognitive control, the coordination of mental processes and actions by current goals and plans (Miller & Cohen, 2001). At least six networks rooted in the PFC play a role in cognitive control, amongst them are the Attentional, Salience, and Default networks (for a deep insight see Menon & D'Esposito, 2022).

Decisions usually do not occur in a vacuum. External and internal context variables influence current goals and plans. Limited time, information overload, and antagonistic interpersonal interactions create "stressful" experiences (Starcke & Brand, 2016), even anticipation of stress impacts decisions (Starcke et al., 2008). The ability to make important decisions under objective pressure (strain as external context) or influenced by the apperceptive and subjective, internal component, stress is probably one of the most valuable managerial skills, from a business advisor's point-of-view. As mentioned in the beginning, human decisions are systematically biased and that starts at the neuronal level (Law & Gold, 2009) and accompanies reasoning through, the also bounded-rational results of social judgments. These mechanisms have both proven useful and not disadvantageous enough in the evolutionary race, but finally, they might make us take a toll due to the more complex world we live in.

A cognition-aware system (CAS) with humans in the loop is broadly speaking a silicon or carbon-based computing system (machine or brain) that senses the outside world and also the human agent in it. The CAS adapts to cognitive aspects of the personal context, i.e. so-called cognitive context such as momentary attention or engagement (Dingler, 2016a, b). Kambhampati (2020) calls for a human-aware AI system, for "goal-directed autonomous systems that are capable of effectively interacting, collaborating, and teaming with humans." The purpose is cognitive enhancements "that result in improved characteristics and capabilities, sometimes beyond the existing human range" (Moore, 2008), e.g., for understanding the world better or generating and acquiring new knowledge (Cinel et al. 2019). The human mind is a carbon-based context-aware system and the oldest CAS. It is context-aware because it recognizes, processes, and uses context to provide relevant information and/or services to the "user" self, where relevancy is weighed (compare Dey, 2000). The human mind is also cognition-aware since meta-cognition, which develops in the early stages of life, is the human art of thinking about thinking and reflecting on the conditions, or the context of thinking, provided by internal (interoception, somatic information, physiological state) as well as external (task load, environment) facts and factors (for an overview see Lai, 2011). Most technologies even *context*-aware ones (e.g., in aviation cockpits) do not consider users' attention levels, receptiveness, or cognitive capacities or more generally, they do not account for the context provided by the biological and physiological nature and physiological fluctuations of the user up to the standard of monitoring everything else in the surrounding environment, such as Mangler et al. (2021) monitored items and processes in an industry setting. Circadian rhythms and alertness (Dingler, 2016b, Dingler, Schmidt & Machulla, 2017) or personality traits extraversion (seeking stimulation) and introversion (not seeking stimulation) are defining elements of an individual arousal pattern (Eysenck, 1987).

The inputs for a cognition-aware system come from at least three different domains: the (1) psychological domain, the (2) behavioural domain, and the (3) physiological domain. The psychological domain spans the wide area of mind and behaviour. In our practice, tests, questionnaires and of course, interviews are the tools in our attempt to objectify subjective experiences. In the behavioural domain, biases can be detected using Natural Language Processing (NLP), e.g. in e-mail conversations or spoken language. The beauty of the physiological domain is that it connects all domains and is the first one to react to changes in the environment. Human thought and its jumpiness are challenges for AI algorithms' mission to get smarter compared to signal reading and machine-learning of signals from physical sources, e.g., vibrations for predictive maintenance. While AI will act upon recognized patterns in an algorithmic, predictable fashion *most* of the time, human decision-makers, in contrast, can make counter-intuitive decisions and be still successful in the respective environment (Agrawal et al., 2018). Even under the assumption that a bias-classifier or error detection agent would be able to rule out all human fallacies, it is the description of the entire *context, external and internal,* that is so hard to put into codes (Agrawal et al., 2018). Both negative and positive affect impact creativity in solving problems, positive affect evoked by candy or short funny movies facilitated creative solutions, independent from arousal (the pure rise of general excitability) in four experiments by Isen et al. (1987). Frank and Gilovich (1988) found that a cue as subtle as the colour of a person's clothing has a significant influence on decision-making and results. US National Football League and National Hockey League teams wearing black uniforms ranked to the top in the number of penalties. When they switched from non-black to black uniforms, the number of penalties increased immediately. Both, increased aggressiveness, and biased judgements from referees might be the cause. It demonstrates the power of the context, of the situation interacting with the "situative" self and the challenges when translating to machine-readable codes. Could we have found early psychophysiological signals allowing us to predict this type of behaviour? Rosalind Picard (2015) elaborated on the promises of affective computing and "objective data about emotions carries much more power than self-reported subjective feelings" (p. 1) but had to concede that "the effort to work out the situation, its interaction goals, and the personality differences of the participants is not simple" (p. 2). Human behaviour, psychology and physiology produce markers that can nowadays be measured with digital instruments. Digital markers,[2] also known as digital biomarkers, are objective and quantifiable physiological and behavioural data that are collected and measured via digital devices, such as portables (e.g., mobile phones), wearables (e.g. fitness trackers) or implantables (e.g., Neuralink).

Over the last few years, research and engineering have developed sets of observation tools invasive or non-invasive for brain activity (see Cinel et al., 2019). Invasive methods like having electrodes implanted (e.g., Neuralink) or permanent EEG electrodes attached in headbands, caps or headphones often lack field feasibility or validity (even for fatigue, see Adão et al., 2021). For reasons of feasibility and thanks to the increasing computation power in pattern recognition, proxies[3]—substitute measures that are easier to attain, and still informative regarding the underlying cognitive processes—gain more and more importance. Eye movements are a proxy but as well much more than that.[4] With eye tracking and gaze analysis technology, we can observe movements of the eyes and lids and categorize them into gaze patterns, saccades, fixations, and blinks and are also able to monitor changes in pupil size via pupillometric procedures. Balcetis has investigated the influence and functionality of gaze patterns for success in a multitude of life situations in various experiments (Balcetis, 2021; Balcetis & Dunning, 2006, 2010; Balcetis et al., 2020). Kahneman & Beatty (1966) have flagged the relation of memory load and covariance of pupil size alterations, alas, pupil diameter changes can be hard to measure in the field. Gaze patterns on the other hand even determine performance, thus measuring them, displaying them, and making them a signal or marker can change behaviour in a very direct way. But how to extract meaning from these very individual, psychophysiological data? The next step towards a cognition-aware system (CAS) is making psychophysiological data interact with the most astonishing human capability: insight.[5]

---

[2] Digital markers are used primarily in medicine and are the object of research, however, a study from Switzerland indicates the early phase we are in, starting with the lack of a consistent definition of "digital marker" (Alonso et al., 2023).

[3] These proxy measures must meet quality criteria: they should be reliable, valid and have a positive cost–benefit ratio (Horvath, 2022).

[4] The eyes are the part of the brain visible from the outside (which makes looking into somebody's eyes even more romantic for brain scientists). Also, the relationship between eye movement and cognition is well established.

[5] The German psychologist Karl Bühler probably coined the term "Aha!-Erlebnis (Experience)". Archimedes experienced it sitting in the bathtub as a moment of "Eureka" when he was finally solving the puzzle of buoyant force in water to get his job of finding the counterfeit crown done.

In the context of cognitive psychology and neuroscience, human insight is often defined as a sudden and clear understanding or solution to a problem that involves a restructuring or reorganization of one's mental representation of the problem. It is characterized by a rapid and often unexpected comprehension of relationships or connections between elements that were previously not apparent. Technically, it can be "defined as a combination of discrimination, understanding and intuition, is critical in the conceptual phases of solution-building, to open-ended real-world challenges" (Siregar, 2021).

Biofeedback is a method for interaction. Insight and biofeedback are related concepts, particularly in the context of self-awareness and self-regulation. While they are distinct, they can be interconnected in certain situations, especially when considering the use of biofeedback technology to enhance insight into physiological and psychological processes. Insight can happen spontaneously in a critical moment where the solution path is not fully overt (see Archimedes), or it can be generated or kindled by a coach within a coach–coachee relationship. This approach is similar to Socrates' "maieutic", the art of dialogically asking the right questions that lead the questioned mind to a solution.

3 You Can Manage What You Can Measure

In our cases from practice, we show results that are accomplished by using the "Biofeedback 3.0" methodology. Biofeedback 3.0 (BFB 3.0) is categorically different from Biofeedback 1 and 2 because of its move towards an automated advisory system (not feedback *only*) with at least one cognition-aware agent, either human or machine, and with the human in the centre. It goes beyond feedback because it provides advice to the recipient after it has fostered insight. Originally, biofeedback methodology (we call it here "biofeedback 1.0") was based on operant conditioning, mainly for somatic functions with psychological change following somatic change. Biofeedback involves the use of electronic monitoring to provide individuals with real-time information about their physiological processes, such as heart rate, muscle tension, skin temperature, and more. This information is presented to the individual, often through visual or auditory cues, allowing them to gain awareness and control over these physiological processes. Biofeedback is used in various therapeutic settings, including stress management, pain control, and performance enhancement, especially in athletes. Voluntary control of cardiac function has proven effects on physiologically related functions like breathing or, e.g. through conducting homeostatic negative feedback loops with heart-rate variability (HRV) changes (Lehrer & Eddie, 2013;

Lehrer & Gevirtz, 2014; Lehrer et al., 2020), leading to lower blood pressure and presumably better emotional regulation (Mather & Thayer 2018) in therapeutic settings. Heart rate variability (HRV) is a physiological phenomenon that refers to the variation in time intervals between consecutive heartbeats and reflects the dynamic interplay between the sympathetic (fight or flight) and parasympathetic (rest and digest) branches of the autonomic nervous system. Nowadays, HRV can be tracked ambiently and non-invasively with wearables like smartwatches or fitness trackers/pulse watches and is a measure of choice in evaluating cognitive load in pilots (see next chapter) or for monitoring recovery in athletes.

Biofeedback 2.0 is different from version 1 in the sense that the foundational mechanisms are rather unclear, while the method is still operant training, but the results aim at the control of body functions and cognitive or mental functions observed by the fed-back correlates in the brain. Slow cortical potentials (SCPs) are considered to reflect the regulation of attention resources and cortical excitability in cortical neuronal networks. SCPs such as the "contingent negative variation" are typically observed in anticipation of signaled responses (for an overview see Birbaumer et al., 1990) and influence reaction times (Rockstroh et al., 1982). Personality traits (e.g., A/B typology, Friedman & Rosenman, 2020) and situational context such as noise level influence the cortical signal characteristics and the perception even if there is no noticeable influence on performance in a task (Kusmitsch & Trimmel, 1997a, 1997b, 1998a, 1998b, 1998c). Feedback of cortical slow potentials (SPs) was tested for operant training of cortical activity and SP shifts (Birbaumer, 1999; Roberts et al., 1989) and in a variety of paradigms, people were able to control their cortical electrical potential, unaware of how they accomplished this trick. A similar, more practical combination of methods from Biofeedback 1.0 and 2.0 was proven effective in learning self-regulation for managers being exposed to stressful conditions: Neurofeedback (EEG) and biofeedback (Skin conductance) training improved results in probabilistic and discounting tasks may be due to higher adaptability during stressful situations and decreased subjective effects in discounting after a 3-month programme (Iodice et al., 2022).

Brain structures like PFC networks are central to guiding both concept-driven (top-down) and data-driven (bottom-up) attention in space, thus these networks are also engaged in the control of eye movements for directing visual attention (Funahashi, 2014; Zirnsak et al., 2014). Studying eye movements allows researchers to gain valuable insights into attention, decision-making, memory, language processing as well as arousal, cognitive load and in general "perception" of the world around us. The external world and the internal one meet here. In the field, we are particularly interested in specific types of eye movements, in addition to gaze tracking. "Mind-tracking" would be the semi-scientific advertising term for it. These specific types are saccades, fixations, blinks, and—if the setting allows—a variation of pupil size. Data acquisition in the field is non-invasive and achieved either by remote cameras or eye-tracking goggles. For our applications, the set of monitoring glasses (VPS19) used is very light (43 g) and the duration of wearing them is limited to a maximum exposition time of 180 min. The signals we can derive are valid and reliable under various conditions (light influences, temperature variations) and describe changes within the person monitored or actively using eye-tracking devices for quality management or educational/vocational/training reasons.

BFB 3.0 can self-improve and learn by matching psycho-physiological correlates with output variables, i.e., behaviour (observed, with HCI/IoT support, e.g. remote cameras) and perception (reported). From a system perspective, BFB 3.0 is both context and cognition-aware but not yet and not necessarily fully automatic or even autonomous. Critical system components are (1) digitalized biomarkers or signals, (2) feedback devices: human cultural artefacts and technical communication devices such as language, and displays, and (3) human insight and the consequences of it, plasticity, and learning.

4 Applied Psychophysiology for Decision-Making

4.1 Aviation—From Reactive Over Proactive to Predictive Systems

Aviation's way of double safety measures and clear procedures brought us to use our tracking technology in more strategic, executive management decision-making. Aviation is a strictly regulated environment where pilots manage the entire system of the aeroplane, the crew, the passengers and the aeroplane's hardware and software. The automation of aeroplanes puts human action and decision-making into a whole new context, like Capt. Sullivan has experienced a flight where automation's contribution went catastrophically wrong (Sullivan, 2019). Today, aviation is considered safe, also thanks to common high standards and strict quality management. Especially within westerly airlines, the accident rate in 2021 lies at 2.8 per 1 million departures low (EASA, 2022). Therefore, it makes no sense to *wait* for an incident or accident to learn from and derive further consequences for training, as it was in the earlier days of aviation. In earlier days of aviation's history, the stereotype of pilots was driven by the image of stalwart individuals who were braving the elements in their open cockpits (Helmreich & Foushee, 2019). These days, the root cause of accidents and incidents is found in technical issues. As aircraft technical standards improved significantly over the years, the focus on accident and incident investigation shifted to human behaviour and human factors (Hawkins, 1993). Since the collision of two Boeing 747s at Tenerife Airport in 1977, Crew Resource Management (CRM) has been a fundamental part of training for aviation professionals like flight and cabin crews, ATC staff, maintenance personnel, dispatchers, and others. This is underscored by the fact that CRM training is even legally mandated (EASA, 2018). John K. Lauber et al. defined CRM as the integration of all available resources (1979). This includes not only the "acquisition of timely, appropriate information" given by instruments (pilot-machine interface) but also interpersonal activities including leadership, effective team formation and maintenance, problem-solving, decision-making, and maintaining situation awareness (Helmreich & Foushee, 2019). The mere more of information processing, however, leads to increased cognitive load. *Cognitive* or mental load monitoring or prediction plays a subordinate role in everyday operations, there is no uniform scientific definition for cognitive or mental load (Sporer-Fellner, 2017).

The industry aims to transition from a *reactive* system to a *proactive* approach and ultimately to a *predictive* system. Only a holistic approach can ensure that the aviation industry and all stakeholders can maintain or improve their high safety standards. Aviation professionals are working in a complex socio-technical system (human–machine interaction). Success has already been achieved through the implementation of the ICAO Safety Management System (ICAO, 2018), but here human factors play only a secondary role. It is undisputed that the will to become even safer is supported by the entire aviation industry. Within most aviation actors, a non-punitive reporting culture (just culture) became standard, and an open exchange of occurrences exists and maybe this, among others, could be the baseline for higher standards, also in corporate cultures. Instruments in this context are a modern aviation training concept, accident and incident reports, audits and inspections, crew feedback and reports, firsthand insights, data from flight recorders and monitoring systems, and safety management systems (SMS, designed to identify, assess, and monitor safety risks and to detect errors for taking preventive measures; ICAO, 2018). All these measures lead to an *evidence-based training* concept. Syllabuses for mandatory pieces of training are adjusted accordingly, e.g., in flight simulators. The evaluation of performance and safety according to established criteria, especially for pilots, is carried out according to an extensive scale.

Alas, there is a gap. The aviation industry adopts a very open approach to errors, incorporating modern approaches to get even higher safety standards. However, the focus on individuals takes a bit of a back seat because individuals are sometimes hard to manage and hard to measure. Especially, when the factor cognitive load is under scrutiny. As early as 1917, Gemelli started studies on military pilots examining blood pressure, breathing rate and pulse rate to determine their emotional reactions (Gemelli, 1917). Heart rate measurement was consecutively added, e.g. by Opmeer and Krol (1973). A common outcome was that pilots both in short and long-haul operations show increased heart rates during take-off and landing phases (Roscoe, 1993) or during complex flight manoeuvres (in-flight refuelling, see Brown et al., 1969). "This implied association between heart rate and the difficulty of a flight manoeuvre" (Roscoe, 1993). During their series of studies, they considered respiratory rate to be the more sensitive indicator of mental workload than heart rate parameters (Opmeer & Krol, 1973), however, speaking is a confounding variable. Later, several tests discussed the possibility of taking heart rate variability (HRV) or sinus arrhythmia into account for estimating the mental workload and mental stress during tasks involving vigilance, information processing and decision-making. Roscoe (1993) reported that there is a noticeable lack of consistency of results, HRV seems unapt for precise prediction (Howitt, 1965, 1969). In further course, several studies emphasized heart rate monitoring of civil airline pilots.

The management of mental and cognitive load has gained significant prominence in recent years in different areas. In various modern working domains that involve advanced technologies, such as Human–Machine Interaction, users are facing heightened requirements in terms of mental and cognitive load, even as physiological demands decrease (Sporer-Fellner, 2017). Pilot cognitive load refers to the mental demands placed on pilots during flight operations, encompassing multitasking, situation awareness, decision-making, working memory, and stress management. Research in this area aims to understand these factors and enhance working conditions and training to improve aviation safety and pilot performance. Roscoe (1992) characterizes the physical demands on pilots during a flight as low, contrasting with the high *non-physical* demands. Additionally, many external factors can exacerbate cognitive load during a flight. For instance, fatigue (probably the biggest threat) can be mentioned as an example, along with various others like climatic conditions, personal well-being, and flying in the dark, that can impact pilots' mental states. Success in this pursuit can be fostered through various methods, including the application of the Human Error and Accident Management (HEAM) method (Cacciabue, 2004). In Fig. 1 we have displayed

Fig. 1 This picture shows a take-off of a commercial airliner from the pilot's perspective taken by a non-invasive eye-tracking glass. Eye tracking can assist in enhancing and solidifying standardized scan patterns—especially in the realm of Standard Operating Procedures (SOP), where additional value can be generated by an automated match of required and executed checks. Furthermore, AI (especially in the training and simulation domain) can verify whether specific tasks have been completed and ensure the verification of required actions. The white circle shows a 2-degree visual focus on instruments

gaze analytics used for training and insight into standard gaze pattern execution in the cockpit.

Figure 2 shows the application of BFB 3.0 in Air Traffic Control systems. Critical for the system's performance is the absence of distractions, even if these "distractions" are part of the task list to handle but require an attentional shift from high-impact surveillance tasks to lower-impact communication tasks. These are accomplished with a different device, the VCS. One way is individualizing communication panel displays, nevertheless: the effects of these changes need to be tested and fed back to the user for optimal performance of air traffic controllers (ATC). ATC insight is the key to optimizing system design and interaction.

4.2 Performance Advisory for Executives

In an ideal world, managers are rational agents because they can calculate the value of an option by relating cost and utility. They are piloting their enterprise

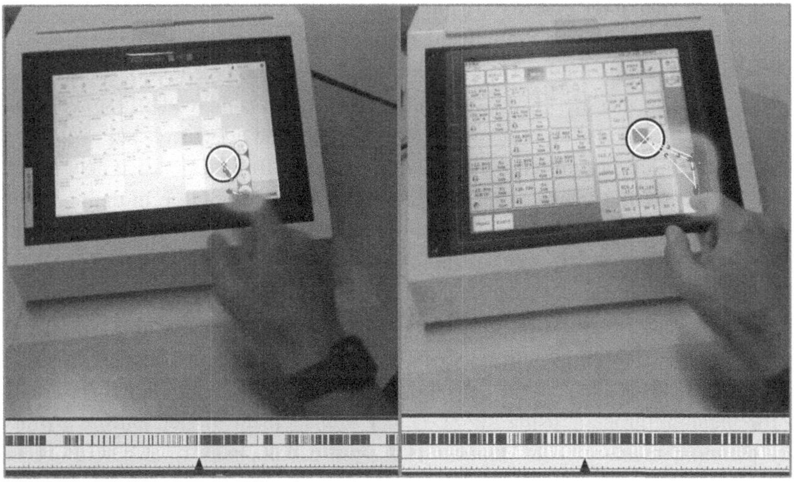

Fig. 2 Shows the X10 (left) and R8 (right) Voice Communication Solution (VCS) for Air Traffic Management by Frequentis AG. Frequentis, headquartered in Vienna, Austria, is a global supplier of communication and information systems for control centres with safety–critical tasks. Voice Solutions are a crucial and indispensable part of the whole solution, but the radar control screen should be the primary centre of attention, i.e. head-down time in VCS interaction should be minimized and cognitive load caused by display/task interactions should be understood and managed in terms of short (e.g., performance decrements), mid-term (e.g., slips, errors and mistakes due to loss of focus) and long-term effects (e.g., fatigue). In this context analysis, cognitive load and fatigue were observed with combined measures of saccades, blinks, and fixation parameters, perceptual characteristics, and task performance (errors, time-to-completion). Cognitive load is an essential component of the context description especially when Graphical User Interfaces (GUIs) impose different work and activity schemes upon the user. It was found that typical tasks interacted significantly with gaze measures (as proxies for cognitive load) but depended on specific GUI features and the predefined workflows

with a set of efficient and standardized procedures, either in an innovative fashion with higher risk or *lege artis*, state-of-art, with presumably less risk. Due to the asymmetric manner of assessing gains and losses (Tversky & Kahneman, 1974), their value calculation is systematically biased, and decisions are made in relativity not in absolute. Based on the ideas of Heckhausen & Gollwitzer (1987), Maja Storch[6] (2004) described the intrapersonal decision process along

[6] As Maja Storch (2004) states "Moreover, Damasio has shown that not all people perceive their somatic markers, even though their appearance can be measured physiologically, such

a five-stage "Rubicon Process" (name by the original authors) which was used as an orientation to guide through the decision-making process. The needs—we are yet unaware of—might covertly "communicate" to the centres of awareness, through somatic markers (Damasio et al., 1991). If we can "read" these feelings, they become conscious-aware, overt, motives (both the appetitive and avoidant). Alas, motives are not yet enough to initiate action! To take decisive action, we need to reframe and manage our inner conflicts. This takes place in the realm of the inner world; it is an act of finding out and cannot be directed or mandated from outside. Once we've worked through our inner obstacles, our motives gradually become intentional. What is remarkable during this procedure is a change in speech patterns, we turn from subjunctive to imperative. This can also be utilized in our cognition-aware system (CAS) as an event or a change of context.

On that basis we have used BFB 3.0 to reflect on covert attentional processes in the planning process, using the St. Galler Management Model. During a strategy planning meeting with Manager H.K., the development of different action scenarios was on the agenda. Financial planning is a complicated task according to the Stacey-Matrix, which can only be solved by know-how and expertise. When the implementation is well thought through, many aspects must be considered in their mutual interaction: Trends, may they be technical, social, legal etc. influence the company as well as customers and all other stakeholders. A poster of the St. Gallen Management Model (Rüegg-Stürm & Grand, 2019), a "map", was used to check both the hard facts of strategic planning, the analysis of stakeholders' expectations and claims as well as expectations and trends of how the relevant environment *might* change. During the 2-h discussion of the map, the sensory and analytical part of our "hybrid" CAS (coach plus machine) is focused on four markers: gaze, fixations, and saccades, where H.K. is gazing and shifting the gaze repeatedly (with saccades) and his stress level (from verbal rapport in interaction with blink patterns), see Fig. 3 for a summary.

Finally, the imagination of the status-quo inherits positive and negative valence each coming with a certain arousal (low–high). A decision requires emotional reason hence the expected balance of the positive and negative arousal must be better than the status quo's. This can also be utilized in our cognition-aware system or for our first step towards it, BFB 3.0. Following our example from above: In his narrative H.K. focuses verbally on customers and the competition,

as through changes in skin conduction. To help people successfully end the deliberation phase and cross the Rubicon, a key to a psychological counselling method can lie in directing clients' attention to the emergence of somatic markers from the unconscious, rather than to brood over possible solutions at a conscious level."

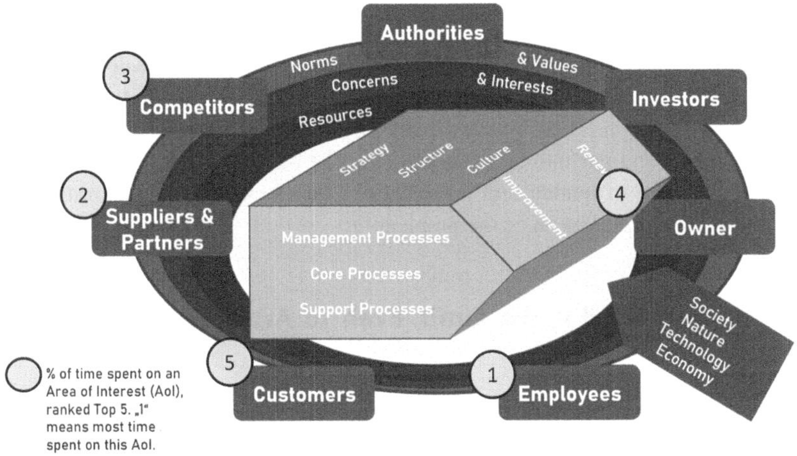

Fig. 3 During the discussion with Manager H.K. we have recorded eye movement parameters with the head-mounted gaze-tracking set VPS19. Between the coach and Manager H.K. a poster with the St. Gallen Management Model sized in A1 (594 × 841 mm, landscape) was mounted on the wall. Gazes were recorded at a 60 Hz rate and parameters (blinks, fixations, saccades) within the confinements of the poster were taken as a reference for the percentages shown in the image. Heat maps and gaze patterns were overlayed and showed areas of interest (AoI) indicating the focal points of mental processing by H.K

he frequently gazes at suppliers and employees most of the time (see Fig. 3 for a statistical description of H.K.'s gaze patterns during the discussion).

In the feedback and reflection of his eye movement patterns and perceived/observed stress indicators we identified a mismatch between what H.K. said he was thinking what he was up to and what he really must have been thinking, based on his cognitive processing. In an iterative process going through different strategic scenarios developed with H.K. and enriched by the aspects of the interactions between focus areas of the St. Gallen Management Model by the consultant H.K. gradually named obstacles and gains (insight) one by one and estimated his arousal on a scale (high-low; 1–10) and the valence (positive or negative). With the enriched dataset of measurements (BFB 3.0 with gaze pattern, cognitive processing, and dialogue) H.K. found out that he has had a blind spot on the development of his employees' capabilities (presumably unconscious fear) to implement new production technologies (conscious motive to be successful). For the success of the implementation of the new production technologies, obstacles have been identified for which solutions have to be found. In a concluding assessment, we jointly tuned H.K.'s estimations of risks under the consideration of the prospect theory (Kahneman & Tversky, 2013). Working with the help of the St. Gallen Management Model through the stages of the Rubicon Model, BFB 3.0 allows the development of a higher grade of insight and unconscious obstacles can be named and put into an overt process. The decision itself can be brought to a higher level of quality and potentially be accelerated by reorganization of one's mental representation of the problem.

5 The Mind is the Limit, Even to Artificial Intelligence

If life is not a game of "Go" but a Kobayashi Maru Test: what is the Kobayashi Maru Test? The Kobayashi Maru Test is a fictional scenario in the Star Trek universe that serves as a no-win scenario designed to test Starfleet cadets' character and decision-making abilities, particularly those in command positions. Captain James T. Kirk is famously known in the Star Trek universe for being the only cadet to have successfully "defeated" the Kobayashi Maru test by reprogramming the simulation, demonstrating his resourcefulness and ability to think outside the box—however, he broke the rules and thus the intention of the test. Go is renowned for its depth and the vast number of possible moves in any given position, leading to a wide range of strategic possibilities. It is in a way the antipode to the Kobayashi Maru test; it is a scenario where there are multiple ways to win. The AI AlphaGo has impressively demonstrated inventiveness counterintuitive to the human mind, based upon sheer calculation power. In a rules-based universe inhabited by economic man strategic machine brilliance would rule by the law of reason. The question remains what are the human capabilities in decision-making that need to be enhanced? Is more logic (like in Go) in decision-making a guarantee to come to better ethical procedures and standards or will the challenge rather be taming the maverick human mind with a cognition-aware system to reap the full benefits of synergy?

In our cases, we have demonstrated BFB 3.0 applications in two distinctly different settings, in pilot training and the top executives' main discipline, strategical thinking and deciding. In both cases, humans stay decision-making "agents". AI technology serves, e.g. as a pattern recognition tool on the signal level or as a set of algorithms for *proposing* decision paths with a higher probability of success, according to predefined parameters. In projects in aviation (cockpit, simulator, ATC) or management (solely strategic decision-making) to enrich training or de-risk decision processes we deploy different AI tools for signal acquisition (machine vision, for remote and user-mounted cameras), signal extraction (pattern recognition) and process automatization (process mining and quality control) also on aggregate levels like communications (Natural Language Processing, NLP, and voice sentiment analysis). These AI-specialized tools fall into the definition of narrow AI. So, would a digital twin of a human manager be a meaningful solution for an AI-based tool? With Mumford (1934), we propose, no. Lewis Mumford (1934, p. 31) describes "The Obstacle of Animism", which is the tendency of new technology designers to use humans or animals as a guide to design. Shneiderman (2020) concludes, referring to Mumford, that "successful designs are not based on human forms". However, an ultimate reference point in evaluating decision-making is human ethics. Since the ethical assessment depends on the situation, the perspective, and the ends, the embodied mind can never be the impartial observer it might strive to be. Technology can help in a two-fold way: On one hand, awareness can be raised for potential ethical implications of an action, on the other hand, feedback can be given on one's moral compass and adverse effects of an outcome not considered in the first place. Insight serves as a weapon against bias, know thyself, know thy biases—deal with them on a meta-level, a level you have achieved thanks to your meta-cognition and guided self-reflection. Insight is a game changer. Our premise is that human judgement can be (should be) a corrective for biased or false AI diagnoses, predictions by AI or any other kind of machine-generated information display, but that only works if human judgement is not biased itself, and as research shows it is. *Nota bene*, we did not reflect on the impact on business. In the end, the impact on business decides upon the propagation of a practice. We simply do not have enough experience and data for quantification. But we are aware that in some sectors our interventions will be more powerful than in others and the ethics of business and economic sectors overarches everything: even when our solution is fair, inclusive, transparent, and human-centric, in the illegal business of trade in human organs the ethic in detail does not bottom-up save a higher (or here: lower, very much lower) purpose.

AI technology is embedded in other technologies and within a business context that is characterized by investment horizons and competing resources. A hidden threat in terms of more focus on the cognitive/mental load of pilots e.g. is the improved airliner technology. Today's airliner technology has a high degree of reliability and has progressed in terms of redundancy, ergonomics, and self-protection. It is beyond question that these advancements in security are beneficial, but they also result in pilots needing to exert less cognitive performance in their daily operations. This results in diminishing opportunities for pilots to acquire skills as they increasingly rely on automation. The demand for profound piloting skills and experience has gradually decreased. However, this can lead to a faster cognitive overload for pilots, especially in the case of severe and complex errors (Wiener & Nagel, 1988)—underlined by the latest accidents/incidents, as described earlier.

Often, it is difficult for pilots to comprehend the logic behind computer algorithms, in fact, it is impossible. Here, targeted pilot training measures would need to be implemented. There is an awareness that effects such as "startle and surprise" can occur, and attempts have been made to minimize them through recurring simulator training. However, the problem is only superficially addressed in this training. This may be due, in part, to trainers and instructors lacking the necessary knowledge of how the mind works in these areas. Many of them are pilots from daily operations without in-depth training. Biofeedback 3.0 provides a means to the end of controlling for cognitive load and having a founded estimate of residual capacities in given situations, which might never occur in real life. Of course, this training approach adds cost to the process and that evokes the so-called Management Dilemma, where the high costs of training must be weighed against the likelihood and risk of a potential incident. The pressure to reduce costs also affects training departments. The attitude "Whoever thinks training is expensive should try an accident" is not helping to create a balance. It is obvious that high training standards lead to high safety performance, but this comes at a high cost for an airline which is competing internationally. It is difficult for training departments to quantify this relationship in dollars to the airline commercial managers (Bent & Kwok, 2010). Is the aviation industry—a role model to learn from? The aviation industry has long overworked the view that professionals who make mistakes while performing tasks appropriate to their skill level have some kind of personal deficit (Dismuke, 2010). Instead of "blame and punishment", the Just Culture Concept is in force (Dekker, 2012). "Just culture" refers to an organizational culture that encourages open communication, learning, and accountability to improve safety. This "No blame—but it better won't happen again" attitude could be a role-model for other corporates as well.

Crashes in the financial industry or of bankruptcy of companies fortunately rarely come at the expense of human life. Management is—similar to aviation—a complex task. What makes it perhaps even more complex, is to receive no direct feedback of success/failure, to be in a network of people and to act in a "no-error admitted" culture where the same mistakes are repeated over and over again. Many times, fuelled by the same motivations. Making these errors in judgment and failures in decision-making transparent with aviation as a role model should yield better risk management and better revenues and help to overcome organizational inertia. If you would get better containment of the risks by understanding the foundational context for free, you would take it. AI applied in business cannot be seen without an investment perspective, the socio-technical impact, and the business plan it is deemed to serve. “Whoever thinks decisions are a risk shall take the risk of indecisiveness” would constitute the nature of a balance sheet for risk management. AI could help to weigh and calculate the risk of managerial inertia.

As practitioners, we have the privilege of being optionally eclectic but compulsory overwhelmed by frameworks to choose from. The choice is vast. As Attard-Frost and colleagues (2023) say: “Ethical principles for developing and using AIS (AI systems) are proposed and contextualized in countless AI ethics guidelines.” Kazim and Koshiyama (2021) have provided a high-level overview of AI ethics, calling it “a highly interdisciplinary field, requiring expertise that spans the sciences (computer science) and the humanities (philosophy, law, sociology, psychology, etc.)”. Conclusively, they view AI ethics as “the psychological, social and political impact of AI”. Indeed, all three explicitly mentioned impacts are relevant to our procedure, the intervention, and its results. Taking psychophysiological data from persons (esp. of those in contingencies) comes with liability and commitment to make the method and the data handling transparent and understandable, with informed consent and data governance rules. Furthermore, since we are directly (in aviation) or indirectly (in general management via performance consultants such as co-author H.U.) in interaction and exchange with people in various industries we have adopted a perspective taken from industrial and organization psychology. Before launching a project at a customer site, we go through the following steps (relating Fig. 1 in Banks et al. 2022, p. 223), we exemplify based on our strategic case:

1. The issue is—to recognise the ethical issue if there is any: The decisions will change after our intervention, maybe causal even if not directly of fully attributable.
2. What we know and what we need to know: gather necessary information and not all available data. Gaze as a token of preoccupation has a self-evident validity for novices in brain research (so the majority of people).
3. Who’s concern is it? Through the discussion stakeholder importance and weight in the choice of options shifted from customers to employees, against the narrative in the beginning.
4. What can we do to de-escalate problems even in their roots? Identification of alternative actions, or better a set of options.

5. How does "A" work compared to "B"? BFB 3.0 is a tool that changes the state from "undecided" to "decided" (crossing the Rubicon) by enriched, augmented insight.
6. Work documentation and all-around *post-mortem.* The implementation of actions and monitoring of outcomes serve as a feedback mechanism.

Human-centeredness and machine-centeredness are non-exclusive. Human-centered AI (HCAI) is a critical or even essential component because these HCAI systems support human self-efficacy, promote creativity, clarify responsibility, and facilitate social participation and should augment and enhance humans rather than replace them. These systems could do what people do: perceive, think, decide, and act, but not in the same way. Relevant services of AI systems in the practice applied in the cases include pattern recognition (of numbers, e.g., of financial data, images, e.g., of the pupils or the face), text generation (speech to text) and text analytics (with means of natural language processing, NLP). The element of human autonomy will remain, although the degree of automatization of the entire system could be very high (Shneiderman, 2020, p. 120). Why do we not leave humans out of the loop and let AI decide? After all: fast, scalable, accurate, repeatable and by sheer mass a low-cost decision tool, with some qualities coming close to human-like intelligence, has driven hopes and dreams in AI development. Even with a reductionist, materialistic view of cognitive functions and their substrate, culture and legacy are essential discriminators of artificial and natural intelligence. Many of the specific human qualities cannot be deduced from the neuronal functions of the individual brains alone but they owe their existence to cultural evolution (Singer, 2021). Singer also calls for discussions on whether AI can acquire functions that are intentional or "conscious" or can be considered moral agents taking responsibility for their actions prematurely. Shresta and colleagues (2019) have analyzed five key contingencies: specificity of the decision search space, interpretability of the decision-making process and outcome, size of the alternative set, decision-making speed, and replicability. Currently, the overlap of AI's and human's idiosyncratic strengths and weaknesses is too small to miss out on benefits for either of them (Shrestha et al. 2019). Great minds think alike, AI and humans don't, and if they do it is rather AI that scales up human biases and reproduces stereotype in a trash in, trash out manner (Marinucci, Mazzuca & Gangemi, 2023). The true concoction with the highest potency lies in the right combination at the right time. Processes in AI reasoning might be incomprehensible and opaque, but the result should be presented in a way that distinctly fosters insight. As advisors in the field, we start with the end in mind: a system that allows humans to leverage their potential, take the right risks at the right time and make decisions free from or at least aware of biases and evolutionary burdens. Human in the centre, however, does not logically exclude having emergent properties AI in the centre as well. Shneiderman (2020) proposes a governance structure for human-centered Artificial Intelligence that in our view is doable and applicable both in corporate or highly regulated environments and in 1:1 performance advisory. He pursues and explains the idea that both, human control, and a high level of automation are possible: these two instances should conclusively not compete for the same resources. Training, educating, and teaching humans, thus creating actionable mass for insight and idea generation should not be weighed against investments in technology. Empowering people with tools that enhance them would be the result. Finally, a safety culture can be supported by the governance structures that certify trust-able processes. This can be resource-intensive, as training and multi-preparedness are resource-intensive in aviation.

Converting or defusing the complexity of context description into an—even complicated—plan will pave the way to engineering. Now, it is a matter of applied research, coming with ethical considerations due to the mere nature of the data necessary. Maybe the resolution of when—what—to what extent lies in describing, understanding, and monitoring the human decision-making apparatus in a different way, closer to the nature of the signal than to the culture of purpose and closer to the individual. Less utilitarian but a bit more scientific, based on the biological and physiological processes more than on processing architectures. Away from normal distributions of populations and normality repeating itself in endless machine-learning cycles, back to the person and all personal idiosyncrasies.

References

Adão Martins, N. R., Annaheim, S., Spengler, C. M., & Rossi, R. M. (2021). Fatigue monitoring through wearables: A state-of-the-art review. Frontiers in Physiology, 12, 2285.

Agrawal, A., Gans, J., & Goldfarb, A. (2018). Prediction, judgment, and complexity: a theory of decision-making and artificial intelligence. In The economics of artificial intelligence: An agenda (pp. 89–110). University of Chicago Press.

Alonso, A. K. M., Hirt, J., Woelfle, T., Janiaud, P., & Hemkens, L. G. (2023). Definitions of digital biomarkers: a systematic mapping of the biomedical literature. medRxiv, 2023-09.

Ashby, W. R. (1956). An introduction to cybernetics.

Attard-Frost, B., De los Ríos, A., & Walters, D. R. (2023). The ethics of AI business practices: a review of 47 AI ethics guidelines. AI and Ethics, 3(2), 389–406.

Balcetis, E. (2021). Clearer, Closer, Better: How Successful People See the World. Ballantine Books.

Balcetis, E., & Dunning, D. (2006). See what you want to see: motivational influences on visual perception. Journal of Personality and Social Psychology, 91(4), 612.

Balcetis, E., & Dunning, D. (2010). Wishful seeing: More desired objects are seen as closer. Psychological Science, 21(1), 147–152.

Balcetis, E., Riccio, M. T., Duncan, D. T., & Cole, S. (2020). Keeping the goal in sight: Testing the influence of narrowed visual attention on physical activity. Personality and Social Psychology Bulletin, 46(3), 485–496.

Banks, G. C., Knapp, D. J., Lin, L., Sanders, C. S., & Grand, J. A. (2022). Ethical decision making in the 21st century: A useful framework for industrial-organizational psychologists. Industrial and Organizational Psychology, 15(2), 220–235.

Barrett, L. F. (2017a). How emotions are made: The secret life of the brain. Pan Macmillan.

Barrett, L. F. (2017b). The theory of constructed emotion: an active inference account of interoception and categorization. Social Cognitive and Affective Neuroscience, 12(1), 1–23.

Barrett, L. F., & Westlin, C. (2021). Navigating the science of emotion. In Emotion measurement (pp. 39–84). Woodhead Publishing.

Bechara, A., & Damasio, A. R. (2005). The somatic marker hypothesis: A neural theory of economic decision. Games and Economic Behavior, 52(2), 336–372.

Bechara, A., Damasio, H., Tranel, D., & Damasio, A. R. (1997). Deciding advantageously before knowing the advantageous strategy. Science, 275(5304), 1293–1295.

Bent, J., & Kwok, C. (2010). Flight Training and Simulation as Safety Generators. In E. Salas, & D. Maurino, Human Factors in Aviation. Amsterdam: Elsevier.

Berntson, G. G., & Khalsa, S. S. (2021). Neural circuits of interoception. Trends in Neurosciences, 44(1), 17–28.

Birbaumer, N. (1999). Slow cortical potentials: Plasticity, operant control, and behavioural effects. The Neuroscientist, 5(2), 74–78.

Birbaumer, N., Elbert, T., Canavan, A. G., & Rockstroh, B. (1990). Slow potentials of the cerebral cortex and behavior. Physiological Reviews, 70(1), 1–41.

Brown, W. K., Gurre, J. D., Meyer, J. F., Buckly, C. J., & Brown, C. A. (1969). Aeromedical aspects of the first nonstop transatlantic helicopter flight: II heart rate. Aerospace Medicine.

Cacciabue, P. C. (2004). Guide to Applying Human Factors Methods. London: Springer.

Cinel, C., Valeriani, D., & Poli, R. (2019). Neurotechnologies for human cognitive augmentation: current state of the art and prospects. Frontiers in Human Neuroscience, 13, 13.

Ciompi, L. (1994). Affektlogik. (4. Auflage). Klett-Kotta: Stuttgart.

Cooper, J. E., White, M. D., & Lauber, J. K. (1979). Resource management on the flightdeck. NASA Conference Publication 2120. Ames Research Center, Moffet Field, CA.: NASA.

Damasio, A. R. (1996). The somatic marker hypothesis and the possible functions of the prefrontal cortex. Philosophical Transactions of the Royal Society of London. Series B: Biological Sciences, 351(1346), 1413–1420.

Damasio, A. R., Tranel, D., & Damasio, H. C. (1991). Somatic markers and the guidance of behavior: theory and preliminary testing.

Dekker, S. (2012). Just Culture Balancing Safety and Accountability. Boca Raton: CRC Press.

Dey, A. K. (2000). Providing architectural support for building context-aware applications. Georgia Institute of Technology.

Dingler, T. (2016a). Cognition-aware systems to support information intake and learning. Dissertation.
Dingler, T. (2016b). Cognition-aware systems as mobile personal assistants. In Proceedings of the 2016 ACM International Joint Conference on Pervasive and Ubiquitous Computing: Adjunct (pp. 1035–1040).
Dingler, T., Schmidt, A., & Machulla, T. (2017). Building cognition-aware systems: A mobile toolkit for extracting time-of-day fluctuations of cognitive performance. Proceedings of the ACM on Interactive, Mobile, Wearable and Ubiquitous Technologies, 1(3), 1–15.
Dismuke, K. R. (2010). Understanding and Analyzing Human Error in Real-World Operations. In E. Salas, & D. Maurino, Human Factors in Aviation. Amsterdam: Elsevier.
Drucker, P. (1954). The practice of management. Harper & Brothers: New York.
EASA. (2018). Basic Regulation (Regulation (EU) 2018/1139). Cologne.
EASA. (2022). Annual Safety Review. Cologne: Easa.
Eysenck, H. J. (1987). Arousal and personality: The origins of a theory. In *Personality dimensions and arousal* (pp. 1–13). Boston, MA: Springer US.
Frank, M. G., & Gilovich, T. (1988). The dark side of self-and social perception: black uniforms and aggression in professional sports. Journal of Personality and Social Psychology, 54(1), 74.
Friedman, M., & Rosenman, R. (2020). How to tell a Type A from a Type B. In Shared Experiences in Human Communication (pp. 124–129). Routledge. Chicago.
Funahashi, S. (2014). Saccade-related activity in the prefrontal cortex: its role in eye movement control and cognitive functions. Frontiers in Integrative Neuroscience, 8, 54.
Gemelli, A. (1917). Sur l'application des méthodes psyco-physiques à l'examen des candidats à l'aviation militaire. Rèsumé de l'Auteur. Archives Italiennes de Biologie, 67(7), 162–196.
Hawkins, F. H. (1993). Human Factors in Flight. Aldershot: Ashgate.
Heckhausen, H., & Gollwitzer, P. M. (1987). Thought contents and cognitive functioning in motivational and volitional states of mind. Motivation and Emotion, 11, 101–120.
Helmreich, R. L., & Foushee, C. H. (2019). Why CRM? Empirical and theoretical bases of Human Factors Training. In B. G. Kanki, J. Anca, & T. R. Chidester, Crew Resource Management. Amsterdam: Elsevier.
Horvath, J. C. (2022). The Limits of Neuroscience in Business. *MIT Sloan Management Review*, *64*(1), 1–4.
Howitt, J. S. (1965). A preliminary study of flight deck workloads in civil air transport aircraft (Report No 1240). London: Flying Personnel Research Committee.
Howitt, J. S. (1969). Flight deck work studies in civil air transport aircraft. Conference Proceedings (No 56). Measurement of Aircrew Performance. Paris AGARD.
ICAO. (2018). Safety Management Manual (SMM)—Doc 9859 AN/474. Montreal.
Iodice, P., Cannito, L., Chaigneau, A., & Palumbo, R. (2022). Learned self-regulation in top-level managers through neurofeedback training improves decision-making under stress. Scientific Reports, 12(1), 6127.
Isen, A. M., Daubman, K. A., & Nowicki, G. P. (1987). Positive affect facilitates creative problem solving. Journal of Personality and Social Psychology, 52(6), 1122.
Kahneman, D., & Beatty, J. (1966). Pupil diameter and load on memory. Science, 154(3756), 1583–1585.

Kahneman, D., & Tversky, A. (2013). Prospect theory: An analysis of decision under risk. In Handbook of the fundamentals of financial decision making: Part I (pp. 99–127).
Kambhampati, S. (2020). Challenges of human-aware ai systems: Aaai presidential address. AI Magazine, 41(3), 3–17.
Kazim, E., & Koshiyama, A. S. (2021). A high-level overview of AI ethics. Patterns, 2(9).
Kolmogorov, A. N. (1965). Three approaches to the quantitative definition of information'. Problems of Information Transmission, 1(1), 1–7.
Kusmitsch, C. J. & Trimmel, M. (1997a). Effects of background noise on the late positive component (LPC) of the ERP and influences of noise-sensitivity. Presentation at the 25th Annual Scientific Meeting of the British Psychophysiology Society at Glasgow.
Kusmitsch, C. J. & Trimmel, M. (1997b). Typ A/B Behaviour Pattern and the influence of background noise on the late positive component (LPC) of a visually presented cognitive task. Poster presented at the 25th Annual Scientific Meeting of the British Psychophysiology Society at Glasgow.
Kusmitsch, C. J. & Trimmel, M. (1998a). Changes in brain DC potentials dependent on background noise and mental load. Presentation at the 8th International Conference on Combined Effects of Environmental Factors at Baden.
Kusmitsch, C. J. & Trimmel, M. (1998b). Relationship of brain DC potential changes and scaled qualities of listening to short probes of music [Abstract]. International Journal of Psychophysiology, 30, 111–112.
Kusmitsch, C. J. & Trimmel, M. (1998c). Relationship of brain potential changes and scaled qualities of listening to short probes of music. In G. A. Chiarenza (Ed.), 9th World Congress of Psychophysiology, 14–19 September 1998. Taormina, Italy: I.O.P.
Lai, E. R. (2011). Metacognition: A literature review. *Always Learning: Pearson Research Report, 24*, 1–40.
Law, C. T., & Gold, J. I. (2009). Reinforcement learning can account for associative and perceptual learning on a visual-decision task. *Nature Neuroscience, 12*(5), 655–663.
Lehrer, P., & Eddie, D. (2013). Dynamic processes in regulation and some implications for biofeedback and biobehavioral interventions. Applied Psychophysiology and Biofeedback, 38, 143–155.
Lehrer, P., Kaur, K., Sharma, A., Shah, K., Huseby, R., Bhavsar, J., & Zhang, Y. (2020). Heart rate variability biofeedback improves emotional and physical health and performance: A systematic review and meta-analysis. Applied Psychophysiology and Biofeedback, 45, 109–129.
Lehrer, P. M., & Gevirtz, R. (2014). *Heart rate variability biofeedback: how and why does it work?* Frontiers in Psychology, 756.
Mangler, J., Diwol, K., Etz, D., Rinderle-Ma, S., Ferscha, A., Reiner, G., ... & Haslgrübler, M. (2021). Sustainability Through Cognition Aware Safety Systems--Next Level Human-Machine-Interaction. arXiv preprint arXiv:2110.07003.
Marinucci, L., Mazzuca, C., & Gangemi, A. (2023). Exposing implicit biases and stereotypes in human and artificial intelligence: state of the art and challenges with a focus on gender. AI & SOCIETY, 38(2), 747–761.
Mather, M., & Thayer, J. F. (2018). How heart rate variability affects emotion regulation brain networks. Current Opinion in Behavioral Sciences, 19, 98–104.
Menon, V., & D'Esposito, M. (2022). The role of PFC networks in cognitive control and executive function. Neuropsychopharmacology, 47(1), 90–103.

Miller, E. K., & Cohen, J. D. (2001). An integrative theory of prefrontal cortex function. Annual Review of Neuroscience, 24(1), 167–202.
Miller, G. A. (1956). The magical number seven, plus or minus two: Some limits on our capacity for processing information. Psychological Review, 63(2), 81.
Moore, P. (2008). Enhancing me: the hope and the hype of human enhancement. John Wiley & Sons.
Mumford, L. (1934). Technics & Civilization. University of Chicago Press: Chicago.
Opmeer, C. J., & Krol, J. P. (1973). Towards an objective assessment of cockpit workload: Physiological variables during different flight phases. Aerospace Medicine.
Öngür, D., & Price, J. L. (2000). The organization of networks within the orbital and medial prefrontal cortex of rats, monkeys and humans. Cerebral Cortex, 10(3), 206–219. Chicago.
Picard, R. W. (2015). The promise of affective computing. *The Oxford handbook of affective computing*, 11–20.
Quigley, K. S., Kanoski, S., Grill, W. M., Barrett, L. F., & Tsakiris, M. (2021). Functions of interoception: From energy regulation to experience of the self. Trends in Neurosciences, 44(1), 29–38.
Reason, J. (1990). Human error. Cambridge university press.
Reeves, M., Levin, S., Fink, T., & Levina, A. (2020). Taming complexity. Harvard Business Review, 98(1), 112–121.
Roberts, L. E., Birbaumer, N., Rockstroh, B., Lutzenberger, W., & Elbert, T. (1989). Self-report during feedback regulation of slow cortical potentials. Psychophysiology, 26(4), 392–403.
Rockstroh, B., Elbert, T., Lutzenberger, W., & Birbaumer, N. (1982). The effects of slow cortical potentials on response speed. Psychophysiology, 19(2), 211–217.
Roscoe, A. H. (1992). Assessing pilot workload. Why measure heart rate, HRV and respiration? Biological Psychology.
Roscoe, A. H. (1993). Heart rate as a psychophysiological measure for in-flight workload assessment. Ergonomics, 36(9), 1055–1062.
Rüegg-Stürm, J., & Grand, S. (2019). Das St. Galler Management-Modell: Management in einer komplexen Welt. utb GmbH.
Shneiderman, B. (2020). Human-centered artificial intelligence: Three fresh ideas. AIS Transactions on Human-Computer Interaction, 12(3), 109–124.
Shrestha, Y. R., Ben-Menahem, S. M., & Von Krogh, G. (2019). Organizational decision-making structures in the age of artificial intelligence. California Management Review, 61(4), 66–83.
Simon, H. A. (1979). Rational decision making in business organizations. The American Economic Review, 69(4), 493–513.
Singer, W. (2021). Differences between natural and artificial cognitive systems. Robotics, AI, and Humanity: Science, Ethics, and Policy, 17–27.
Siregar, E. (2021). Learning human insight by cooperative AI: Shannon-Neumann measure. IOP SciNotes, 2(2), 025001.
Sporer-Fellner, S. (2017). Entscheidungsverhalten und mentale Beanspruchung von Piloten. Wiesbaden: Springer.
Stacey, R. D. (1995). The science of complexity: An alternative perspective for strategic change processes. Strategic Management Journal, 16(6), 477–495.

Starcke, K., Wolf, O. T., Markowitsch, H. J., & Brand, M. (2008). Anticipatory stress influences decision making under explicit risk conditions. Behavioral Neuroscience, 122(6), 1352.

Starcke, K., & Brand, M. (2016). Effects of stress on decisions under uncertainty: A meta-analysis. Psychological Bulletin, 142(9), 909.

Storch, M. (2004). Resource-activating selfmanagement with the Zurich Resource Model (ZRM). *Editorial, Impressum 2 Clinical Experience.*

Sugawara, A., Terasawa, Y., Katsunuma, R., & Sekiguchi, A. (2020). Effects of interoceptive training on decision making, anxiety, and somatic symptoms. BioPsychoSocial Medicine, 14, 1–8.

Sullivan, K. (2019). No man's land. The untold story of automation on QF72. HARPER-COLLINS 360: Glasgow.

Tacke, V. (2020). Systemtheorie der Organisation. Niklas Luhmann. In Handbuch Organisationssoziologie.

Tinaz, S., Para, K., Vives-Rodriguez, A., Martinez-Kaigi, V., Nalamada, K., Sezgin, M., ... & Constable, R. T. (2018). Insula as the interface between body awareness and movement: a neurofeedback-guided kinesthetic motor imagery study in Parkinson's disease. Frontiers in Human Neuroscience, 12, 496.

Tversky, A., & Kahneman, D. (1974). Judgment under Uncertainty: Heuristics and Biases: Biases in judgments reveal some heuristics of thinking under uncertainty. Science, 185(4157), 1124–1131.

Uddin, L. Q., Nomi, J. S., Hébert-Seropian, B., Ghaziri, J., & Boucher, O. (2017). Structure and function of the human insula. Journal of Clinical Neurophysiology: Official Publication of the American Electroencephalographic Society, 34(4), 300.

Vytal, K. E., Cornwell, B. R., Letkiewicz, A. M., Arkin, N. E., & Grillon, C. (2013). The complex interaction between anxiety and cognition: insight from spatial and verbal working memory. Frontiers in Human Neuroscience, 7, 93.

Wiener, E. L., & Nagel, D. C. (1988). Human Factors in Aviation. Oxford: Gulf Professional Publishing.

Zirnsak, M., Steinmetz, N. A., Noudoost, B., Xu, K. Z., & Moore, T. (2014). Visual space is compressed in prefrontal cortex before eye movements. Nature, 507(7493), 504–507.

Christian J. Kusmitsch has +20 years of experience cooperating with national and international companies in engineering, IT, mobility, and health as a preferred advisor and project developer. He received his doctorate in neurosciences from the University of Vienna and the University of California San Diego (UCSD) in 2000. The synergy of understanding people, human factors, and their impact on business performance by using cutting-edge technology led to the founding of Mindset Technologies AG in 2022. The focus is on high-performance environments like top management, car racing, military, and high safety conditions like civil aviation, control rooms, or human-robot interaction on the shop floor.

Leveraging AI-Enabled People Analytics for Designing Organizations

Yash Raj Shrestha, Ralf Buechsenschuss, and Patrick Nicolas Tinguely

1 Introduction

Companies witnessed a major transition to hybrid work models following the global COVID-19 pandemic, accompanied with major AI advancements that dramatically impacted their workforce and ways of working (Mamgain 2021; Jarrahi 2018). Especially, moving into hybrid working has increased the amount of data collected in the workplace. For instance, digital collaboration tools leave behind digital trace data that organizations can leverage, such as records of zoom or Teams meeting and email exchanges (Stier et al. 2020). As these disruptions create uncertainty about the future and resilience of organization designs, forward thinking executives in various organizations have begun to adopt *People Analytics* to strengthen their analytical capabilities and augment organization design decisions with data driven insights and prescriptions.

People analytics represents a novel data-driven approach to manage human resources (McAfee & Brynjolfsson 2012; Gal et al. 2020). It is rapidly transforming how business organization's identity, manage, motivate, control and

Y. R. Shrestha (✉)
University of Lausanne, Lausanne, Switzerland
e-mail: yashraj.shrestha@unil.ch

R. Buechsenschuss
University of Mannheim, Mannheim, Germany
e-mail: ralf.buechsenschuss@students.uni-mannheim.de

P. N. Tinguely
ETH Zurich, Zurich, Switzerland
e-mail: ptinguely@ethz.ch

C. H. Hoffmann (ed.), *Artificial Intelligence, Entrepreneurship and Risk*, Technikzukünfte, Wissenschaft und Gesellschaft / Futures of Technology, Science and Society, https://doi.org/10.1007/978-3-658-45544-6_17

coordinate their employees and tasks (Huselid 2018; Chamorro-Premuzic et al. 2017). On the one hand, people analytics is expected to raise the effectiveness and efficiency of various human resource functions including recruiting, onboarding, evaluating, and developing people (Leonardi & Contractor 2018; Tursunbayeva et al. 2018). Recent development in AI has facilitated organizations to process and analyze large volumes of diverse datasets, and automatically predict, modify and influence organizational patterns and human behaviors (von Krogh 2018; Shrestha et al. 2019). In this article, we specifically focus on this more advanced AI-enabled version of people analytics and refer to it as people analytics in the subsequent sections.

Motivated by the benefits of people analytics powered by AI, organizations across industries are increasingly adopting it within their HR functions (Fernandez & Gallardo-Gallardo 2020). Given the enhanced capabilities of AI and its rapid adoption across organizations, people analytics is expected to be pervasive in the organizations. For instance, Google is one of the frontrunners in the domain of people analytics and has achieved substantial business gains by replacing opinions, feelings, and intuitions of its managers with data and analytics (Talent Management and HR 2014). In doing so, Google has established People and Innovation Lab (or "Pi-Lab") where psychologists, decision scientists and researchers interpret and analyze people's analytics insights and turn them into actionable managerial decisions. One popular work of Pi-Lab is Project Oxygen, a research-backed multi-year project that aimed to identify and rank leadership traits of managers using detailed comments from employee surveys, employees' expectations from managers which were sourced from complaints and praises mentioned in performance reviews, and manager-subordinate engagement data (re: Work 2016).

On the other hand, given the very data-driven nature of AI-enabled people analytics and the resulting high stakes decisions, organizations are facing numerous technical and organizational challenges when adopting people analytics. For instance, they need to build capabilities and infrastructures for collecting, storing, analyzing and modeling people related data, which demands considerable resources and talents (Feuerriegel et al. 2022). These often require upfront costs and technical capabilities. Contemporary AI development and deployment is highly cost intensive (Shrestha et al. 2023). Further, the adoption of people analytics use-cases introduces social concerns related to ethics, privacy, and fairness, which requires careful and continuous management attention and designs in place (Herath et al. 2022; Herath et al. 2023).

Given these promises and perils, managers aiming to use People Analytics in designing their organizations are required to carefully navigate the trade-offs.

Despite clear benefits and need of design centric knowledge, extant literature has specifically adopted a descriptive lens and investigated promises and challenges of people analytics and examined *descriptively* how people analytics can help shave practical solutions (Rasmussen et al. 2023; Storey & Wright 2023; Ferrar & Green 2021). While there has been increased theoretical attention to people analytics as well (van der Togt & Rasmussen 2017), we still lack prescriptive theories on how to effectively use people analytics within organizations. In this book chapter, we examine ***how to design effective organizations with people analytics***. In contrast to descriptive insights that examine and explain observed phenomena, we aim to develop normative insights and framework on people analytics and its use in organization design. The recent breakthroughs in deep learning and generative AI and their rapid adoption in people analytics suggest that there is value for pre-emptive and normative design theories in practice. Therefore, we offer normative insights on people analytics for organization design. We believe that such an approach to research can contribute to progress in a design-centric science such as organization design (Obel & Burton 1984; Mintzberg 1979). In doing so, we first take stock of the numerous promises and challenges of people analytics. Next, we provide a brief overview of organization design and its fundamental problems. Subsequently, we synthesize the promises and challenges to produce prescriptive and actionable recommendations on using people analytics for organization design endeavors. Finally, we provide illustrations on how Organizational Network Analysis, a type of people analytics, can be used in designing organizing using some concrete use-cases.

2 Background

2.1 People Analytics and Its Enhancement with AI

Over the last decade, decision support systems and data analytics to augment decision-making in organizations is gaining much traction. In particular, People Analytics, defined as "an emerging area of HRM practice, research and innovation concerned with the use of information technologies, descriptive and predictive data analytics and visualization tools for generating actionable insights about workforce dynamics, human capital, and individual and team performance that can be used strategically to optimize organizational effectiveness, efficiency and outcomes, and improve employee experience" (Tursunbayeva, Di Lauro and

Pagliari 2018), has been the focus of that attention. People analytics promise managers better understanding and optimization of their workforce and has recently been deemed as top 10 HR technology disruptions.

Isson and Harriott (2016) provide a useful framework that organizes People Analytics into seven main "pillars" that comprises: (1) Workforce planning; (2) Sourcing, (3) Acquisition/hiring; (4) Onboarding, culture fit, and engagement; (5) Performance assessment and development and employee lifetime value; (6) Churn and retention; and (7) Wellness, health, and safety. Recent research has also expanded this list by adding categories including collaboration; diversity and inclusion; people risk; and inter-organizational relationships (Tursunbayeva, Lauro & Pagliari 2018).

These functions of People Analytics have gained much enhancement with the recent advancements in artificial intelligence technology and data availability (Shrestha, Krishna & von Krogh 2021). Empirical research suggests that Artificial Intelligence algorithms excel at extracting robust and complex patterns from data and making accurate predictions and insights, making them particularly useful in processing information about people and tasks within organizations. Contemporary Artificial Intelligence algorithms are particularly capable of processing unstructured data (e.g., images, text, and video) which consists of the bulk of organizational data.

2.2 Problems of Organization Design

Organizations can be portrayed "as (1) multi-agent system with (2) identifiable boundaries and (3) system level goals (purpose) towards which (4) the constituent agent's efforts are expected to make a contribution" (Puranam 2018, p. 163). Organization design research and practice is concerned with "understanding (a) how organizations, seen as structures that usefully aggregate the efforts of individuals, work and (b) how to make them work better" (Puranam 2018). Organization design requires solutions to the two unavoidable and interlinked problems of (a) division of labor and (b) integration of efforts (Obel & Burton 1984).

Division of labor includes task division and task allocation. Task division deals with the decomposition of organizational goals into a set of interrelated tasks and subtasks, while task allocation maps interrelated subtasks to individuals or groups by matching their expertise and experience with the subtasks' requirements (Hackman and Oldham 1976). Task division and allocation is often facilitated by task ownership where reward structures are put in place such that individuals believe that they are accountable for the outcome of the tasks, both in terms of quality and timeliness. The integration of efforts involves mapping a set of rewards to organizational agents to incentivize them to cooperate in accomplishing specific tasks (Gulati et al. 2005; Heath and Staudenmayer 2000). There has been significant research on formulating and identifying solutions to these problems of organizing (Puranam 2018), including those with emerging technologies (Tinguely & Shrestha 2023). In this book chapter, we argue and illustrate how AI-enabled people analytics is beneficial in finding novel solutions as well as improving existing solutions to the problems of organizing, namely division of labor and integration of efforts.

3 Promises of People Analytics for Organization Design

Strategic adoption and deployment of AI-enabled people analytics present a consequential avenue for furnishing organizations with invaluable insights essential for augmenting non-routine and cognitively intricate decision-making processes. The integration of AI-driven analytics equips both managerial cohorts and organization designers with data-informed and precision-targeted insights, fostering the discernment necessary for optimizing workforce dynamics. This reservoir of informed decision-making is instrumental in delineating and refining task division and task allocation paradigms within organizational frameworks.

The real-time nature inherent in AI-enabled people analytics furnishes a dynamic and responsive mechanism for monitoring and appraising the efficacy of prevailing solutions. The inherent agility of these analytics permits seamless adaptability, facilitating timely updates in response to unforeseen contingencies or exceptions. For instance, in instances where a task allocation to a team fails to yield anticipated outcomes, the application of people analytics can guide recalibration. This recalibration may encompass alterations in reward provisioning structures or the identification of more optimal task allocations, subsequently effectuated to rectify inefficiencies.

Delving deeper into the potential of People Analytics for organizational design, its application extends beyond conventional realms, unveiling promises crucial for organizational efficacy and dynamism.

3.1 Comprehensive, Real-Time, and Holistic View of Labor Force

People analytics provides a comprehensive view of the workforce across the entire employee-life-cycle, including employee performance over time, task contributions, engagement, career progression, and wellbeing. People Analytics applications aggregate data from diverse sources including performance management tools, recruitment platforms, task monitoring, employee surveys, and enterprise social media. Such a comprehensive view is effective in gathering insights on workforce demographics, turnover rates, tenure distribution, individual and team expertise, task completion and quality, and overall composition and characteristics of the workforce. People Analytics can reveal and expose unexpected interactions, overlooked conflicts, and diagnose opportunities to introduce intervention (e.g., in the form of changes in organization redesign, personalized coaching, or specific incentive designs), reveal gaps in communication and collaboration, and measure the effectiveness of current organization designs and interventions. For instance, Google uses People Analytics to monitor productivity metrics of individuals and teams in gauging their effectiveness.[1] Monitoring productivity using AI enables checking the pulse of the workplace to consistently improve it. Feedback from such monitoring could be used to suggest changes in existing task division and task allocation designs. In fact, they have many engaged employees with an average participation rate of 90% in People Analytics monitoring, ultimately showcasing their success in improving business methods and morale.

One advantage of AI-enabled people analytics is that it can reveal the actual relationships among individuals, teams, departments, and functions, as compared to expected relationships prescribed by organizational designers (Mintzberg 1979). For instance, such actual relationship between individuals and individual-task interdependencies could be used to identify misalignment in organization design in place (e.g., formal structures) and emerging structures in organizations (e.g., informal structures around organizational networks). Recent scholarship has also identified such a holistic view can open up new management capabilities, such as mapping expertise and employee interactions, tracking employee sentiments to preemptively make HR decisions, and providing insights into all human activities (Tinguely et al. 2020). These capabilities facilitate the potential to maximize organizational efficiency, increase productivity and encourage collaboration

[1] https://www.sage.com/en-au/blog/case-study-how-google-uses-people-analytics/.

and coordination. The advanced People Analytics capabilities can be used to optimize organizational structures through better understanding current pain points and improving the real structures after diagnosis and simulations.

3.2 Improved Employee Engagement and Well-Being

By analyzing data related to employee health, absenteeism, and work-life balance, people analytics can pinpoint factors that impact employee well-being. This understanding enables organizations to implement initiatives and organization designs that support a healthier and more productive workforce, leading to increased employee engagement, creativity, and improved employee retention. Using People Analytics can also help replicate the same thinking and approach applied to customer experience. Organizations can leverage behavioral insights to better understand what employees do and how they experience organization design processes such as reward and incentive designs and task allocations, which can be adjusted to employees' needs. Especially in times when retention of key talent is challenging, a tailored employee experience informed by People Analytics can become a competitive advantage. Further, People Analytics should enhance reskilling and upskilling policies and tailored reward systems, which ultimately enable better People Organizational-fit through leveraging organizational data.

For instance, Juniper Networks use People Analytics to analyze not only where the top-performing employees come from but also where they go after leaving the company (Boudreau & Rice 2015). This enables Juniper Networks to gain insight on diverse industry career paths, helping it to develop fresh and modern organization design and HR policies to avoid the turnover of talented individuals. This also enables Juniper Networks in identifying issues with existing incentive and reward designs that might trigger turnover. Similarly, over the years, Microsoft has been using HR analytics to develop statistical profiles of those employees who are more likely to leave their company. Based on the insights they generate by doing so, they typically introduce managerial interventions such as assigning mentors and opening discussions on task allocation or making opportunities for enhanced workplace communication. By focusing their approach on the parts of the business with particularly high turnover rates, Microsoft has been able to reduce attrition rates by more than half.[2]

[2] https://www.microsoft.com/en-us/human-resources?rtc=1.

3.3 Automation of Expertise Matching

The embedding of people analytics capabilities within organization design processes can automate administrative work (e.g. task formulation), and free up managers for more meaningful work. In a recent report, 55% of US HR managers estimate that AI will be a regular part of their work in the next five years.

People analytics can automate assessing employee skills, competencies, and development needs. By automatically mapping the skills landscape within the organization, managers can identify skill shortages, high-potential employees, and areas for upskilling or reskilling, enhancing the agility and adaptability of the workforce (Van den Broek et al. 2021). Further, people analytics can help employees identify skill gaps and recommend targeted learning opportunities. Upskilling and reskilling initiatives help individuals enhance their competencies by suggesting useful training programs. People analytics can facilitate skill-based approaches to building teams. Understanding the team expertise combination results in better designs of high-performance teams, optimized use of available expertise, and improved recruitment strategies for future hirings. This should overall enable more efficient use of talent and reduce role confusion and redundancy.

3.4 Better Time Management

Effective time management is a critical aspect of individual and organizational productivity. Advanced People analytics practices offer data-driven insights that empower individuals and organizations to optimize time allocation, streamline workflows, and enhance overall productivity. The personalized insights enabled by the latest technologies and personalized insights show employees how they spend their time, e.g. having meetings vs. focus time. Therefore, individuals can make informed decisions about how they use their time, leading to improved work efficiency and a healthier work-life balance. For organizations, embracing people analytics for time management can result in increased productivity, reduced burnout, and a more engaged and motivated workforce. For instance, organizations can leverage passive collaboration and network data (e.g., emails, chats, videoconferencing data) to improve personal productivity and optimize when and with whom employees work. Another concrete example is using digital exhaust to make billability more accurate in professional service firms and redirect the distribution of attention, e.g. targeted analysis can make customer facing employees aware about how they spend their time working with customers vs. with internal stakeholders, highlighting the needs to shift more attention to customer and away from internal alignment and bureaucracy.

3.5 Improved and Inclusive Decision-Making

Through interactive dashboards and reports, People Analytics provides managers with real-time access to key workforce metrics and insights. These tools enable data-driven decision-making across all units of the organization. Further, organizations can use such tools to measure diversity and inclusion metrics and evaluate the representation of different demographics in the firm. They can better identify potential biases in recruitment and talent management processes, and hence foster a more inclusive workplace culture. One way to foster diversity in talent management is for example to look at the network of men and women and how well they are connected to relevant sponsors to progress in their careers. Understanding the proportion of relevant sponsors in networks can help to add relevant sponsors to networks of high potentials to ensure that they progress in their careers. Such improved and inclusive decision-making could benefit in designing inclusive organizations.

4 Perils of People Analytics for Organization Design

Despite its myriad of benefits, AI-enabled people analytics comes with horny challenges that need to be carefully navigated by responsible managers. In this section, we review a set of such challenges and provide some recommendations on managing potential risks related to these challenges.

4.1 Data Quality and Privacy

The efficacy of people analytics hinges upon the integrity and comprehensiveness of its foundational data, where accuracy and completeness are pivotal in deriving meaningful insights. However, the intrinsic limitations within certain data sources pose substantial challenges to the efficacy of people analytics initiatives. Notably, digital traces encapsulating collaborative exchanges and interactions among employees—such as those found within emails—serve as approximate proxies for social connectivity, often omitting nuanced facets of interpersonal dynamics, notably the dimensions of interpersonal trust and emotional unpredictability. These nuanced aspects constitute foundational elements contributing to the establishment of psychological safety and a sense of organizational belonging (Edmondson 2018).

Moreover, the perceptions individuals harbor regarding their relationships with peers might supersede the factual relationships delineated within communication and interaction patterns. This inherent dichotomy suggests that people analytics might inadvertently present a biased portrayal of the social network, albeit perceived as accurate by its users. Notably, the integration of AI-based algorithms within people analytics frameworks can inadvertently exacerbate pre-existing biases latent within data sources, potentially skewing resultant analytical outcomes (Shrestha & Yang 2019). The endeavor to rectify these data limitations has led to the exploration of technologies like sociometric badges or sensors, ostensibly aiming to mitigate the shortcomings. However, the intrusive nature of these technologies poses a considerable conundrum, compelling organizations to grapple with a trade-off between the acquisition of accurate and comprehensive data and the preservation of employees' rights to privacy. This dilemma necessitates a balanced approach, wherein the pursuit of meticulous data accuracy aligns harmoniously with the imperative to safeguard employees' privacy rights.

Maintaining stringent adherence to data privacy regulations, securing explicit consent, and imbuing responsible data usage practices serve as cornerstone principles not only to foster and uphold employee trust but also to ensure compliance with stringent data protection legislations. Striking this equilibrium between data accuracy and privacy preservation is critical to engendering trust among employees and cultivating an organizational ethos grounded in ethical data utilization practices.

4.2 Navigating Prevailing Data Silos and Resource and Expertise Constraints

Within the intricate frameworks of large, conventional organizations, the integration of data across diverse departments and hierarchical structures poses formidable challenges. The inherent path dependence within data collection and analysis methodologies has led to a fragmentation of data repositories, delineated along departmental or system lines, rendering them considerably inaccessible within organizational frameworks. While often unintentional, these data silos present considerable hurdles in identifying, a priori, the diverse datasets available and their inherent value within the ambit of People Analytics frameworks.

Moreover, the legacy character of these data silos engenders inconsistencies and uncertainties within the organizational data landscape. The lack of alignment in data collection and utilization practices across departments contributes to substantial disparities, manifesting in variances in both the breadth (pertaining to the number of pertinent variables) and depth (characterizing the quality of measurements) of prevalent datasets. Addressing and reconciling these disparities necessitates not only substantial time and resource investments but crucially demands a concerted political will to foster consensus and formulate operational agreements among diverse departments and leaders.

The construction of an adept and proficient people analytics team necessitates the convergence of individuals equipped with a synergistic blend of HR domain expertise and technical prowess encompassing data analytics and machine learning (ML). The identification and retention of such multifaceted talent present considerable challenges, particularly within fiercely competitive job markets. Notably, smaller organizational entities often grapple with resource limitations and budgetary constraints, impeding their ability to invest in advanced people analytics tools and technologies, thereby hindering their access to the expansive realm of data-driven insights.

Of particular significance, organizational imperatives entail the upskilling and education of managerial cadres in leveraging the latest generation of AI-enabled analytics to facilitate informed decision-making processes. However, ensuring equitable upskilling opportunities across all organizational strata emerges as a critical concern to prevent incumbent power structures from disproportionately leveraging these educational initiatives to bolster existing power dynamics. Striving for balanced upskilling endeavors across the organizational hierarchy is essential to foster an environment that embraces and leverages AI-driven analytics effectively while promoting equitable access to transformative educational pathways.

4.3 Compliance Concerns and Fairness

The integration of people analytics in workforce data collection and analysis introduces a myriad of ethical, legal, and compliance considerations imperative for organizational adherence to regulatory frameworks and ethical norms. Safeguarding compliance standards, ensuring the sanctity of employee privacy, and mitigating the risks of unfair or prejudicial high-stakes decisions (such as those related to hiring, promotion, or termination) stand as paramount imperatives in the realm of people analytics implementation.

The regulatory landscape governing the utilization of workforce data is intricate and continually evolving, embodying a transient nature underscored by ongoing developments in legal frameworks. For instance, the ongoing evolution and formulation of the EU AI Act exemplify this dynamic nature of regulatory frameworks. Organizations grappling with People Analytics confront an inherent dichotomy: the juxtaposition between accessing individual-level data to develop intricate models and relying on aggregated data coupled with external solution providers, often characterized by opaque models (commonly referred to as 'black box' models).

The deployment of algorithms and AI within the domain of people analytics poses an inherent risk of perpetuating biases ingrained within historical datasets. HR professionals are tasked with the pivotal responsibility of vigilance in identifying and remedying biases entrenched within these systems. Mitigating bias becomes instrumental in safeguarding the pursuit of fair and equitable outcomes for all employees, necessitating meticulous scrutiny and active measures to curtail any inadvertent perpetuation of historical biases within algorithmic decision-making frameworks.

4.4 Organizational Inertia

The seamless integration of people analytics within the intricate frameworks of organization design processes and decision-making workflows presents a multifaceted challenge. Effectively harnessing the insights derived from people analytics to inform and mold organizational design necessitates a comprehensive integration strategy within pre-existing decision-making systems and entrenched practices. This integration process, however, encounters a spectrum of challenges, notably including apprehensions and resistance from segments of the workforce regarding the introduction of AI-powered decision-making paradigms.

The introduction of AI-powered decision-making systems can provoke feelings of apprehension and discomfort among certain employees, amplifying the challenge of embracing and deploying people analytics seamlessly. This phenomenon is exemplified in the reluctance of managers to wholly rely on algorithmic insights driven by AI, fearing a potential erosion of their professional stature and identity (Tinguely et al. 2023). The underlying resistance to relinquishing decision-making control poses a substantial hurdle in the effective integration of people analytics into organizational decision-making processes.

The successful implementation of people analytics necessitates an earnest endeavor to surmount resistance to change and cultivate a pervasive data-driven ethos within the organizational culture. Overcoming this resistance mandates a comprehensive strategy encompassing educational initiatives, transparent communication channels, and the cultivation of an environment that celebrates the symbiosis of human expertise and AI-driven insights. Fostering a data-driven culture becomes a linchpin for the seamless and effective adoption of people analytics, ensuring its pervasive integration into the fabric of decision-making frameworks and organizational design processes.

4.5 Predictions to Actions

The prevalent discourse surrounding the advent of algorithms has primarily concentrated on the automation and augmentation of human activities, ostensibly liberating managerial attention. This discourse spans divergent viewpoints, with certain scholars heralding the onset of a potential fourth industrial revolution. Conversely, others have sounded a cautionary note, warning of AI's potential to displace human labor, potentially leading to widespread unemployment. Moreover, emerging models advocating collaborative ecosystems between humans and AI, wherein certain facets of human roles are delegated to AI counterparts, have gained prominence in scholarly discourse (Puranam 2021; Jarrahi 2018).

This trajectory of thought has been amplified by the recent strides in large language models (LLMs), enabling the autonomous execution of intellectually demanding and creatively intricate tasks by algorithms. Despite the pervasive optimism regarding the transformative effects of algorithmic automation and the prevalent concerns over the erosion of human agency in this narrative, the integration of AI into HR decisions necessitates a profound recognition of the enduring role of human decision-making. This aspect often remains understated in contemporary discussions.

Human decisions persist as an integral component across the spectrum of AI integration, playing pivotal roles in the design, implementation, and consequence management of AI applications within HR domains. Their influence extends significantly, shaping the resultant value generated and the inherent risks engendered by the utilization of AI. Recognizing and understanding the intertwining nature of human decision-making within the context of AI deployment becomes paramount in appraising the holistic implications and ramifications of algorithmic integration into HR frameworks.

5 Leveraging AI-Enabled People Analytics to Design Organizations: Use Case of Organizational Network Analysis

Technological advancements coupled with the burgeoning availability of expansive datasets have presented a myriad of opportunities to apply sophisticated methodologies in tackling intricate organizational dilemmas, particularly in the sphere of organizational design. This exposition delineates three overarching design imperatives: (a) cultivating unimpeded information dissemination, (b) fostering structural adaptability, and (c) nurturing collaborative proficiency. Each of these design aspirations is expounded upon in conjunction with the instrumental role of Organizational Network Analysis (ONA).

ONA serves as a tool that scrutinizes and graphically represents authentic interpersonal connections, as well as the breadth and intensity of collaborative endeavors and information exchanges among personnel within an organizational framework, including potential engagements with external stakeholders. By dissecting and visually portraying these intricate networks, ONA uncovers latent associations interwoven across departments, teams, and individual contributors, thereby illuminating invaluable insights pivotal for catalyzing organizational metamorphosis.

5.1 Design for Unfettered Information Flows

Contemporary organizational design transcends conventional analyses centered solely on formal structures, such as the traditional organizational chart. It now encompasses a comprehensive understanding and ongoing monitoring of emergent informal structures in real-time. Acknowledging the profound influence of workforce interconnectedness and knowledge circulation, organizations increasingly recognize the criticality of these factors. This recognition has precipitated a surge in interest, exploration, and implementation of Organizational Network Analysis (ONA) as a potent methodology for optimizing organizational structures and cultivating innovation.

ONA serves as a potent tool in uncovering organizational silos—sequestered pockets within an organization impeding effective communication and collaboration. Often imperceptible through formal structure analysis or commonplace observations, these silos obstruct organizational coherence. By visually representing communication patterns and delineating clusters of employees with limited interconnections, ONA unveils these concealed silos. This revelation empowers organizational leaders to dismantle barriers, fostering cross-functional interactions that engender a culture of collaboration, shared objectives, and heightened alignment.

Furthermore, ONA plays a pivotal role in driving innovation, thriving in environments where diverse perspectives converge, and ideas intersect. By identifying key influencers, information brokers, and central nodes within organizations, ONA enables strategic connectivity among individuals from varied backgrounds and expertise. This orchestrated network cultivation fosters an environment conducive to idea cross-pollination, nurturing innovation by enabling the unimpeded flow of knowledge across hierarchical and departmental boundaries.

Central to organizational growth is the exchange of critical knowledge, often regarded as the lifeblood of progressive entities. ONA visualizes pathways of information flow, pinpointing areas where knowledge stagnates or remains underutilized. Analyzing these knowledge channels enables organizations to enact targeted interventions to bolster idea and knowledge exchange. Whether through the implementation of mentorship programs, interdisciplinary task forces, tailored networking events, or optimization of communication platforms, ONA assists in fortifying these critical channels.

In conclusion, the application of ONA represents a seminal advancement in modern organizational design. By dissecting the intricate web of employee relationships and interactions, ONA provides indispensable insights pivotal for catalyzing organizational evolution. From mitigating silos to nurturing innovation and fostering knowledge exchange, ONA empowers organizations towards a more interconnected, innovative, and adaptable future.

5.2 Design for Structural Flexibility

In the dynamic and constantly changing business environment, developing organizational capabilities for organizational change is becoming a key competitive advantage. ONA has recently become an important method to inform organization design interventions to successfully lead transformation processes. ONA helps to understand the intricate web of relationships and interactions among employees and offers a wide range of options that can revolutionize the effectiveness of change management initiatives. Unveiling the effectiveness of existing employee dynamics is of highest relevance to understand deficiencies of the current organizational design. Hence, ONA can serve almost like an "organizational X-ray" revealing the hidden dynamics between employees and teams.

By mapping communication patterns, collaboration frequencies and influence networks, ONA provides a holistic view of information flows and who are critical influencers especially in times of change. This deep understanding of the social fabric within the workplace becomes instrumental in change initiatives and can help to implement informed interventions for effective organizational change. Traditional change management often faces challenges due to underestimating the human factor—the social networks, informal hierarchies, and influential nodes within an organization. ONA bridges this gap by offering actionable insights. Equipped with collaboration data, managers in charge of change can design interventions that target critical nodes, influential connectors, and communication bottlenecks. This precision allows for tailored strategies that resonate with the organization's social structure, thus fostering smoother transitions and change adoption. In many change processes, the ability to gain buy-in, foster collaboration and mitigate resistance is a huge challenge. ONA equips change managers with a comprehensive understanding of where resistance might stem from, who the change champions are, and how to harness the power of influential clusters to drive change adoption. This strategic alignment of interventions with the organizational network paves the way for more seamless and successful transformations.

In conclusion, ONA emerges as a cornerstone in modern change management practices. Its ability to decode the social architecture of organizations empowers change managers to navigate transformations with precision and empathy. By leveraging ONA insights to design targeted interventions, organizations can effectively transform themselves, reducing friction in the process and ensuring not just successful transitions but also laying the groundwork for a culture of continuous adaptation and growth. Integrating ONA into change management strategies is a paradigm shift, entering a new era of organizational design leveraging the new analytical capabilities of human resources.

5.3 Design for Collaborative Structures

ONA can facilitate designs of effective collaboration by decoding formal and informal organizational structures. On one side, formal structures outline role, responsibilities, and reporting lines, providing a blueprint of authority and formal power within an organization (Clement & Puranam 2018). On the other side, the informal structure embodies the vunseen and hidden connections, communication and collaboration patterns and social networks that dictate the real flow of work and decision-making. ONA can provide a clear view, beyond the hierarchical and formal structure to unveil the organic interplay of relationships, information flow and influential nodes within the informal network. Revealing these differences between formal and informal structures is pivotal to understand the operational realities of an organization. While the formal structure theoretically defines the levels of authority and responsibilities, the informal network provides a real view of how information flows through the organization, how expertise is shared and how innovation gets created. Especially large or complex organizations struggle with the interplay between formal and informal structures.

In this endeavor, ONA can create multiple benefits:

1. Providing a comprehensive understanding of the hidden social networks behind operations, which can enable leaders to navigate the complexities of interactions and dependencies across various departments and teams.
2. Leaders equipped with the right network insights can craft targeted interventions to bridge the gap between formal and informal structures.
3. Improving collaboration through identifying key influencers, brokers and connectors within the informal network, enabling organizations to leverage these employees to promote and improve levels of collaboration. By strategically aligning influential nodes and fostering connections, collaboration across departments and hierarchies can be significantly augmented.

Organizational Network Analysis is becoming a transformative tool that can help organizations to decode their formal and informal structures. Its ability to uncover hidden connections is particularly advantageous for large and complex organizations that can bridge the divide between formal and informal organizational realities, thereby fostering a more interconnected, collaborative, and effective organization.

6 Conclusion

The application of people analytics in designing organizations represents an exciting frontier poised to revolutionize organization design practice. Our book chapter offers a forward-looking perspective that delves into the transformative potential of AI-augmented people analytics, meticulously exploring the array of benefits and challenges it presents. As AI continues its rapid evolution, the organization design landscape stands on the cusp of unprecedented change. This evolution promises to equip organizations with the tools necessary to unlock the full potential of their workforce, ushering in an era characterized by unhindered information flows, enhanced structural adaptability, and collaborative prowess.

Yet, amidst this transformation, it's crucial to underscore the imperative of responsible AI adoption. This chapter serves as a clarion call to managers, urging them to not only embrace the transformative power of AI but to do so conscientiously, ensuring the well-being and nurturing the growth of their most valuable asset—their employees. By advocating for a conscientious and ethical integration of AI into organization design practices, organizations can navigate this paradigm shift while upholding ethical standards and prioritizing employee welfare.

The untapped potential for improved decision-making looms large on the horizon. Rapid adoption of people analytics in organization design will be pivotal in establishing a competitive edge, fostering a more informed and enriched employee experience, and fundamentally altering the decision-making landscape to be more evidence-based and data-driven.

In conclusion, the burgeoning era of AI-augmented people analytics beckons organizations to embrace change with alacrity. The speed and astuteness of adoption will delineate those at the vanguard of innovation, paving the way for enhanced operational efficiency, heightened competitiveness, and a workplace culture rooted in informed, forward-thinking decisions.

References

Boudreau, J., & Rice, S. (2015). Bright, shiny objects and the future of HR. Harvard Business Review, 93(7), 72–78.

Chamorro-Premuzic, T., Akhtar, R., Winsborough, D., & Sherman, R. A. (2017). The datafication of talent: How technology is advancing the science of human potential at work. Current Opinion in Behavioral Sciences, 18, 13–16.

Clement, J., Puranam, P. (2018). Searching for Structure: Formal Organization Design as a Guide to Network Evolution. Management Science, 64(8), 3879–3895.

Edmondson, A. C. (2018). The fearless organization: Creating psychological safety in the workplace for learning, innovation, and growth. John Wiley & Sons.

Fernandez, V., & Gallardo-Gallardo, E. (2020). Tackling the HR digitalization challenge: Key factors and barriers to HR analytics adoption. Competitiveness Review, 31(1), 162–187.

Feuerriegel, S., Shrestha, Y. R., von Krogh, G., & Zhang, C. (2022). Bringing artificial intelligence to business management. Nature Machine Intelligence, 4(7), 611–613.

Ferrar, J., Green, D. (2021). Excellence in people analytics: How to use workforce data to create business value. Kogan Page.

Gal, U., Jensen, T. B., & Stein, M. K. (2020). Breaking the vicious cycle of algorithmic management: A virtue ethics approach to people analytics. Information and Organization, 30(2), 100301.

Gulati, R., Lawrence, P. R., & Puranam, P. (2005). Adaptation in vertical relationships: Beyond incentive conflict. Strategic Management Journal, 26(5), 415–440.

Hackman, J. R., & Oldham, G. R. (1976). Motivation through the design of work: Test of a theory. Organizational Behavior and Human Performance, 16(2), 250–279.

Heath, C., & Staudenmayer, N. (2000). Coordination neglect: How lay theories of organizing complicate coordination in organizations. Research in Organizational Behavior, 22, 153–191.

Herath Pathirannehelage, S., Johannsson, J. G., Shrestha, Y. R., & von Krogh, G. (2023). Artificial Intelligence-Augmented Decision Making in Supply Chain Monitoring: An Action Design Research Study. European Conference on Information Systems.

Herath Pathirannehelage, S., Shrestha, Y. R., & von Krogh, G. (2022). Design principles for AI-based IS artefacts in organisational decision-making: An action design research study.

Huselid, M. A. (2018). The science and practice of workforce analytics: Introduction to the HRM special issue. Human Resource Management, 57(3), 679–684.

Isson, J. P., & Harriott, J. S. (2016). *People analytics in the era of big data: Changing the way you attract, acquire, develop, and retain talent.* John Wiley & Sons.

Jarrahi, M. H. (2018). Artificial intelligence and the future of work: Human-AI symbiosis in organizational decision making. Business Horizons, 61(4), 577–586.

Leonardi, P., & Contractor, N. (2018). Better people analytics. Harvard Business Review, 96(6), 70–81.

Mamgain, R. P. (2021). Understanding labour market disruptions and job losses amidst COVID-19. Journal of Social and Economic Development, 23(Suppl 2), 301–319.

McAfee, A., & Brynjolfsson, E. (2012). Big data: The management revolution. Harvard Business Review, 90(10), 60–68.

Mintzberg, H. (1979). *The. structuring of Organizations.*

Obel, B., & Burton, R. M. (1984). *Designing efficient organizations: Modelling and experimentation.* Chicago

Puranam, P. (2018). The microstructure of organizations. Oxford University Press.

Puranam, P. (2021). Human–AI collaborative decision-making as an organization design problem. *Journal of Organization Design, 10*(2), 75–80.

Rasmussen, T. H., Ulrich, M., & Ulrich, D. (2023). Moving People Analytics From Insight to Impact. Human Resource Development Review, 0(0).

re: Work. (2016). People analytics—where HR meets science at Google. Available at: https://rework.withgoogle.com/blog/people-analytics-where-hr-meets-science-at-google/

Shrestha, Y. R., & Krishna, V. (2021). Augmenting organizational decision-making with deep learning algorithms: Principles, promises, and challenges. Journal of Business Research, 123, 588–603.

Shrestha, Y. R., & Yang, Y. (2019). Fairness in algorithmic decision-making: Applications in multi-winner voting, machine learning, and recommender systems. Algorithms, 12(9), 199.

Shrestha, Y. R., von Krogh, G., & Feuerriegel, S. (2023). Building open-source AI. Nature Computational Science, 1–4.

Shrestha, Y. R., Ben-Menahem, S. M., & Von Krogh, G. (2019). Organizational decision-making structures in the age of artificial intelligence. California Management Review, 61(4), 66–83.

Stier, S., Breuer, J., Siegers, P., & Thorson, K. (2020). Integrating survey data and digital trace data: Key issues in developing an emerging field. Social Science Computer Review, 38(5), 503–516.

Storey, J., & Wright, P. (2023). Strategic human resource management: A research review. Routledge Press.

Talent Management and HR. (2014). How Google is using people analytics to completely reinvent HR. Available at: www.tlnt.com/how-google-is-using-people-analytics-to-completely-reinvent-hr-2/

Tinguely, P. N., & Shrestha, Y. R. (2023). NFT-enabled organization design: Prospects, promises and hurdles. Journal of Organization Design, 13, 23–31.

Tinguely, P., Shrestha, Y. R., & Von Krogh, G. (2020). How does your labor force react to COVID-19? Employing social media analytics for preemptive decision making. *California Management Review.*

Tinguely, P. N., Lee, J., & He, V. F. (2023). Designing human resource management systems in the age of AI. *Journal of Organization Design, 12*(4), 263–269.

Tursunbayeva, A., Di Lauro, S., & Pagliari, C. (2018). People analytics - a scoping review of conceptual boundaries and value propositions. International Journal of Information Management, 43, 224–247.

Van den Broek, E., Sergeeva, A., & Huysman, M. (2021). When the machine meets the expert: An ethnography of developing AI for hiring. MIS Quarterly, 45(3).

van der Togt, J., & Rasmussen, T. (2017). Toward evidence-based HR. Journal of Organizational Effectiveness: People and Performance, 4(2), 127–132.

von Krogh, G. (2018). Artificial intelligence in organizations: New opportunities for phenomenon-based theorizing. Academy of Management Discoveries, 4(4), 404–409.

Künstliche Intelligenz in der Fertigungsindustrie

Motor für höhere Produktivität und Innovation. Vorausgesetzt, das Mindset der Mitarbeiter stimmt

Armin Ribis and Simone Carlo Surace

1 KI als Puzzleteil in der Wertschöpfungskette eines Unternehmens

KI ist ein zunehmend wichtiger Bestandteil in der Wertschöpfungskette, da sie die Effizienz steigert und Innovations- und Geschäftsprozesse beschleunigt. Durch die Integration von KI in verschiedene Phasen der Wertschöpfung werden Daten für Optimierungszwecke nutzbar gemacht. In der Produktion verbessert KI die Qualitätskontrolle und steigert die Produktivität. Im Vertrieb und Marketing ermöglicht KI, Umsatzpotenziale zu erkennen, Kunden personalisiert anzusprechen und das passende Angebot zum richtigen Zeitpunkt zu platzieren. Im Kundenservice nutzt man KI-basierte Tools zur schnelleren und effizienteren Problemlösung. Die Integration von KI in die Wertschöpfungskette unterstützt auch die Entwicklung neuer Produkte und Dienstleistungen oder bei der Festlegung der Preispositionierung. Durch die Automatisierung repetitiver Tätigkeiten leistet KI einen Beitrag zur Kostensenkung und zur Steigerung der Skalierbarkeit von Unternehmen, indem eine grössere Unabhängigkeit von Fachpersonal erreicht wird.

A. Ribis (✉)
makeAIwork, digitell GmbH, Zürich, Schweiz
e-mail: armin.ribis@make-ai-work.com

S. C. Surace
Artificialy SA, Lugano, Schweiz

Flock Labs AG, Bern, Schweiz

C. H. Hoffmann (ed.), *Artificial Intelligence, Entrepreneurship and Risk*, Technikzukünfte, Wissenschaft und Gesellschaft / Futures of Technology, Science and Society, https://doi.org/10.1007/978-3-658-45544-6_18

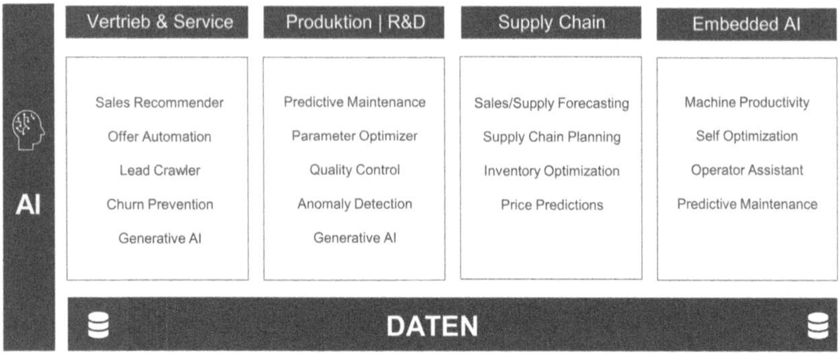

Abb. 1 Die Übersicht zeigt eine Auswahl praktischer KI-Anwendungsbeispiele entlang der Wertschöpfungskette eines Unternehmens

KI revolutioniert die Wertschöpfungskette in der Fertigungsindustrie, indem sie Unternehmen befähigt, schneller auf Marktanforderungen zu reagieren und einen nachhaltigen Mehrwert zu schaffen. KI ist ein entscheidender Enabler, um die Wettbewerbsfähigkeit eines Unternehmens nachhaltig zu stärken.

Nachfolgend werden die in Abb. 1 dargestellten KI-Anwendungen zur verbesserten Veranschaulichung näher erläutert.

Sales Recommender
Mithilfe von Machine-Learning-Algorithmen werden bestehende Kundendaten analysiert. Algorithmen erstellen aus Kundeneigenschaften wie der Branchenzugehörigkeit oder der Unternehmensgrösse, dem bisherigen Kaufverhalten und den Produktbewertungen des jeweiligen Kunden und ähnlicher Kunden sowie den Eigenschaften der verschiedenen Produkte eine Kaufempfehlung. Die Funktionsweise kennt man von Onlineshops: „Dieses Produkt könnte Sie auch interessieren" oder „Andere Kunden haben auch das gekauft". Diese sogenannten Recommender-Systeme funktionieren auch im B2B-Vertrieb, sofern ausreichend relevante Daten in den CRM- und ERP-Systemen vorhanden sind. Ein klassischer Anwendungsfall für Recommender-Tools im Maschinenbau ist der After-Sales-Bereich, um Empfehlungen für den Ersatzteilverkauf zu generieren.

Offer Automation
Mithilfe einer KI lässt sich der Angebotsprozess automatisieren, indem Algorithmen Daten analysieren (z. B. historische Angebote, Preislisten, Rabattverhalten, Wettbewerberpreise), Preisempfehlungen geben, personalisierte Angebote erstellen

und eine automatisierte Kommunikation mit dem Kunden ermöglicht werden. KI steigert somit die Effizienz und Geschwindigkeit in der Angebotserstellung.

Lead Crawler
Auf Grundlage von manuell oder automatisch erstellten Kundenprofilen (z. B. auf Grundlage bestehender Kundendaten) suchen Algorithmen in internen und externen Daten nach neuen Kunden. An den Vertrieb werden Leads ausgespielt, die ein Geschäftspotenzial haben könnten. Der Vertrieb qualifiziert, ob tatsächlich ein Umsatzpotenzial besteht oder nicht. Dieses Feedback wird an die KI zurückgespielt, wodurch diese aus den Rückmeldungen lernt und den Suchprozess kontinuierlich optimiert.

Churn Prevention
Algorithmen identifizieren Kunden, bei denen ein erhöhtes Risiko besteht, dass sie in absehbarer Zeit nicht mehr kaufen bzw. zum Mitbewerb abwandern. Dieses Risiko ermittelt die KI auf Grundlage von Kundendaten wie Bestellverhalten, Zahlungsmoral, Angebotswerte oder Kommunikationsdaten zur Stimmungsanalyse (Sentiment). Vertriebsmitarbeiter erhalten Zugriff auf ein Dashboard, das eine Übersicht über gefährdete Kunden bietet. So können proaktiv Massnahmen ergriffen werden.

Generative AI (z. B. Sprachassistenten)
Sprachassistenten wie Chat-GPT von Open AI, Google Bard oder Open-Source-Lösungen können in unterschiedlichsten Unternehmensbereichen eingesetzt werden. Grundsätzlich ergeben Sprachassistenten in allen wissensintensiven Organisationseinheiten Sinn, wie z. B. in Marketing, Kundenservice, Forschung und Entwicklung, Einkauf, HR oder im Vertrieb. Der Vorteil von Large Language Models (LLMs) ist, dass diese bereits vortrainiert sind und in relativ kurzer Zeit für ein Unternehmen produktiv einsetzbar sind. Den grössten Nutzen erzielt man mit Sprachassistenten, wenn externe wie auch interne Datenquellen integriert werden. Diese Daten können in unstrukturierter Form (z. B. Produktbeschreibungen als PDF, E-Mails, Websites) vorliegen und von Sprachassistenten verarbeitet werden.

Machine Parameter Optimization
In der Produktion werden Algorithmen genutzt, um Maschinenparameter automatisch einzustellen, um einen optimalen Produktionsoutput in Bezug auf Produktqualität und Performance zu generieren. Parameter-Optimizer-Tools werden auch

in der Produktentwicklung eingesetzt, um beispielsweise Rezepturen für chemische Produkte schneller zu entwickeln. Dabei werden zeitaufwändige physische Experimente durch Computersimulationen ersetzt.

Predictive Maintenance

Für die Maschinenwartung können Algorithmen genutzt werden, um den Zustand von Maschinen in Echtzeit zu überwachen (Condition Monitoring). Durch eine permanente Datenanalyse und die Nutzung von Algorithmen wird Wartungsbedarf von Maschinen oder einzelnen Komponenten vorhergesagt, um ungeplante Stillstände zu verhindern. Diese Form der proaktiven Instandhaltung erhöht die Nettolaufzeit der Maschinen und senkt Wartungskosten, indem Wartungsarbeiten genau dann durchgeführt werden, wenn sie erforderlich sind, anstatt nach einem festen Wartungsintervall.

Quality Control/Anomaly Detection (Computervision)

Mithilfe von Computervision kann die Qualität der Produkte während der Produktion überwacht werden. Eine KI-basierte Qualitätskontrolle durch Bilderkennung nutzt sogenannte neuronale Netze, um Muster in Bildern zu erkennen. Durch das KI-Training mit bildbasierten Beispieldaten erkennt die KI Unregelmässigkeiten (Anomalien) an den Produkten, identifiziert Defekte, optische Fehler oder fehlerhafte Dimensionen. Diese automatisierte Bildanalyse ermöglicht eine schnelle und präzise Qualitätsbewertung und hilft den Produktionsmitarbeitern, den Produktionsprozess laufend zu optimieren und den Ausschuss zu reduzieren.

Sales/Supply Forecasting

Algorithmen können Muster in historischen Verkaufsdaten erkennen, um daraus eine Umsatzprognose zu erstellen. Üblicherweise nutzt man neben den historischen Absatzdaten auch Angebote, Verkaufschancen, Marktstatistiken oder Konjunkturdaten, um eine höhere Forecast-Genauigkeit zu erzielen. Durch die Integration relevanter Datenquellen kann eine sehr gute Prognosequalität erzielt werden. In der Praxis erzeugt die KI oft bessere Forecasts als jene, die von Mitarbeitern erstellt werden. Die Nutzung von KI hilft, den Planungsaufwand für den Vertrieb signifikant zu reduzieren, die Genauigkeit der Prognosen zu erhöhen und damit die Planungsdaten nutzbarer zu machen für nachgelagerte Abteilungen wie Beschaffung, Produktion oder Logistik. Besonders in den Fällen, wo sehr viele Daten vorliegen (z. B. sehr viele unterschiedliche Produkte oder Produktsegmente sowie eine Vielzahl externer und interner Datenquellen) und a priori unbekannt ist, wie diese Daten voneinander abhängen, können KI-basierte Prognosemodelle einen Mehrwert auch gegenüber herkömmlichen statistischen Ansätzen bringen.

Industrial Embedded AI
Integrierte KI-Technologie in Maschinen bzw. generell in Produkten ermöglicht die automatisierte Steuerung und Optimierung von Produktionsmaschinen bzw. die Unterstützung von Produktionsmitarbeitern, die richtigen Entscheidungen zu treffen. Embedded AI dient der Produktivitätssteigerung im Produktionsprozess. Anwendungsbeispiele für Embedded AI findet man allgemein in der Robotik, der OEE (Overall Equipment Effectiveness) Optimierung, Predictive Maintenance oder in Sprachassistenten für Maschinenbediener.

2 Erfolgsfaktoren für KI-Implementierungen

Die Nutzung von KI, um Geschäftsprozesse zu optimieren oder neue Geschäftsmodelle zu entwickeln, steht bei vielen Unternehmen auf der Agenda. Trotzdem ist es wichtig, sich der Herausforderungen von KI-Projekten bewusst zu sein. Diesbezüglich unterscheiden sich KI-Projekte nicht von anderen Softwareeinführungsprojekten. KI- und Data-Science-Projekte haben für Unternehmen, die gerade damit starten, oft experimentellen Charakter, um herauszufinden, inwiefern die existierenden Daten für KI überhaupt nutzbar sind bzw. ob KI überhaupt Sinn ergibt. Das erste KI-Projekt, das ein Unternehmen umsetzt, ist von entscheidender Bedeutung für den zukünftigen Stellenwert von KI im Unternehmen. Ist das erste Projekt ein „Fail“, ist es schwierig, Management und Mitarbeiter zu motivieren ein weiteres KI-Projekt zu starten. So könnte die AI-Journey eines Unternehmens ein abruptes Ende nehmen. Um das zu vermeiden, braucht es eine systematische Analyse der KI-Ideen, Potenziale und Use Cases. Nachfolgende Faktoren sind für KI-Projekte erfolgsentscheidend (Abb. 2).

Erfolgsfaktor #1

Das Unternehmen muss über die richtigen Daten verfügen

Ohne Daten hat man kein Data-Science-Projekt. Ohne richtige Daten keine KI. Es kann auch schwierig sein, die notwendigen Daten zu sammeln, zu erstellen oder extern zu beschaffen.

Auch wenn der Zugriff auf die Daten vorhanden ist, muss man immer noch einige Herausforderungen überwinden:

- Wie und wo sind die Daten gesichert?
- Liegen die zugrunde liegenden Daten in ausreichend guter Qualität vor?

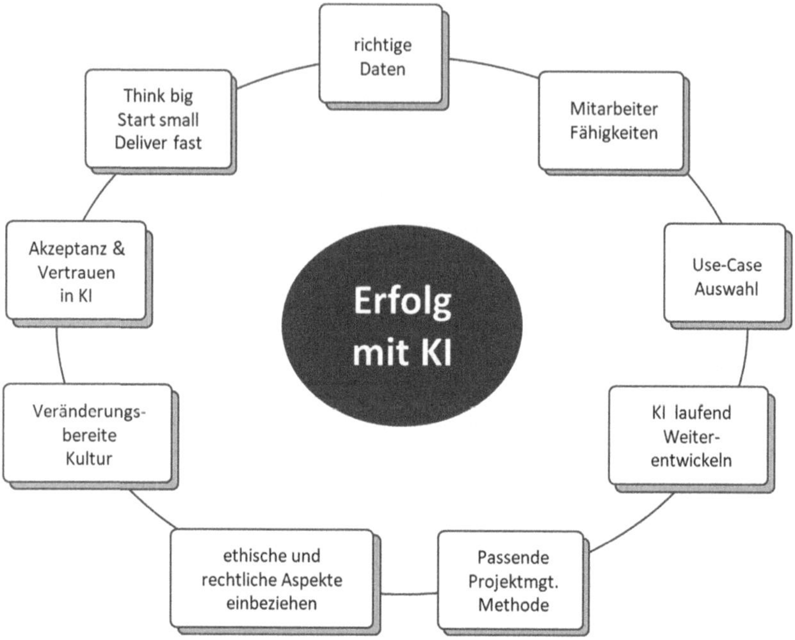

Abb. 2 Übersicht der Erfolgsfaktoren für KI-Projekte

- Können die Daten bereinigt werden?
- Können die Daten ethisch und rechtlich für den beabsichtigten Anwendungsfall verwendet werden?
- Gibt es interne Mitarbeiter, die die Kombination aus Daten und Geschäftsprozesse verstehen?
- Das Thema Datenmanagement ist eine der zentralen Herausforderungen in KI-Projekten. Hier ein paar Empfehlungen:
- Etablierung interner Guidelines, wie Daten zu erfassen und zu nutzen sind, um den GiGo-Effekt (garbage in – garbage out) zu vermeiden.
- Nie davon ausgehen, dass Daten in einer hohen Qualität vorliegen.
- Sicherstellen, dass interne oder externe Mitarbeiter mit den nötigen Skills in Data-Science, Advanced-Analytics und Machine-Learning eingesetzt werden.

Erfolgsfaktor #2

Die richtigen Mitarbeiter mit den richtigen Fähigkeiten einsetzen

Tech-Talente zu finden, einzustellen und zu halten ist nie einfach. Und es geht nicht nur darum, qualifizierte Mitarbeiter zu finden. Man muss auch die richtige Kombination aus Mitarbeitern finden. Die Zeiten, in denen ein Data Scientist ein komplettes KI-Projekt allein umsetzt, sind vorbei.

Die Implementierung von KI-Lösungen – von der Ideenfindung über die Bereitstellung der Lösung bis zum produktiven Einsatz – benötigt verschiedenste Fähigkeiten und Rollen, um Erfolg zu haben.

Erfolgsfaktor #3

Auswahl des richtigen KI Use Case mit einer glasklaren Zielsetzung

Haben sie jemals ein Projekt mit einer unscharfen Zielvorstellung begonnen? Oder ein Projekt mit einem klar definierten Ziel, das nicht realistisch ist oder keinen sinnvollen Mehrwert bringt?

Manche Unternehmen stellen gut ausgebildete Data-Scientisten ein, ohne eine klare Ausrichtung definiert zu haben. Die Neueingestellten tauchen dann aus einem mehrwöchigen „Forschungsloch" auf, nur um festzustellen, dass sie die Zielvariable missverstanden hatten, was die Analysen irrelevant machte.

Um dieses Risiko zu minimieren, starten Sie Ihr Projekt richtig und stellen Sie die richtigen Fragen vor dem Start des Projekts. Man muss tief in eine Aufgabenstellung eintauchen, um das zugrunde liegende Problem, das gelöst werden muss, wirklich zu verstehen.

Erfolgsfaktor #4

Think big – Start small – Deliver fast

Je ungewisser der ROI einer Softwareinvestition, desto geringer ist die Chance, dass das Management das Projekt befürwortet. Das gilt auch für KI-Projekte. Die Entscheidungsträger in Unternehmen, die über KI-Investitionen entscheiden, scheuen grosse (risikobehaftete) Investitionen. Insbesondere bei KI-Projekten wird oft die Vorgabe gemacht „sucht nach den quick-wins". Anders ausgedrückt Think big – Start small – Deliver fast.

Think Big Legen Sie mit dem Management fest, in welchem Unternehmensbereich nach KI-Potenzialen gesucht werden soll (z. B. Produktion, Produktentwicklung, SCM, Vertrieb/Marketing). Mit einigen Stakeholdern aus dem ausgewählten Fachbereich führen Sie Envision Workshops durch. Es werden gemeinsam Ideen entwickelt und mit internen/externen Experten bewertet. Das ist die Grundlage für die Auswahl des ersten Use Case und die Erstellung einer Road-Map zur Ausrichtung der KI-Initiative.

Start Small starten Sie mit einem Use Case, der Ressourcen, Budget und Projektdauer überschaubar hält. Suchen Sie sich „low hanging fruits" – speziell am Anfang der AI-Journey. Der ausgewählte Use Case sollte eine geringe Komplexität haben, und das Business muss einen klaren, signifikanten Mehrwert erkennen.

Deliver Fast Wählen Sie einen Use Case, bei dem relativ zeitnah erste Ergebnisse vorliegen. Insbesondere die Datenbasis ist hier entscheidend. Wenn aus den ersten Datascreenings klar wird, dass man einige Monate in die Datenvorbereitung investieren muss, dann wird das Projekt nicht gerade beflügelt. Schnelle Ergebnisse motivieren die Projektbeteiligten, das KI-Projekt mit hoher Priorität zu verfolgen.

Erfolgsfaktor #5

Die Implementierung einer KI-Lösung markiert nicht das Ende eines KI-Projekts

Nach der traditionellen Definition endet ein IT-Projekt, wenn der geplante Leistungsumfang geliefert wurde. Das gilt auch für KI-Projekte. Man muss sich allerdings bewusst sein, dass der laufende Betrieb von KI-Lösungen besondere Aufmerksamkeit erfordert. KI-Modelle müssen über eine gewisse Zeit validiert und gegebenenfalls angepasst und verbessert werden. Das passiert (noch) nicht selbständig durch die KI. Es braucht ein Team für die laufende Weiterentwicklung, um die Funktionsweise aufrechtzuerhalten und einen nachhaltigen Mehrwert für die Fachbereiche schaffen können.

Erfolgsfaktor #6

Die passende Projektmanagement-Methode

Unternehmen greifen noch immer gerne auf klassische Projektmanagement-Methoden (z. B. Wasserfall) zurück, die oft zu ineffizientem Informationsaustausch und falschen Erwartungshaltungen der Fachbereiche führen. Oft hat man ein Ziel, das man erreichen möchte – z. B. mit predictive maintenance eine höhere Maschinenproduktivität. Allerdings weiss man im Vorfeld noch gar nicht, ob Datenumfang und -qualität sinnvoll und ausreichend sind, um dieses Ziel zu erreichen. Daher haben Data-Science-Projekte zu Beginn experimentellen Charakter. Es wird im Labor gearbeitet. Wenn dann vernünftige Ergebnisse erzielt werden, kann man darüber diskutieren, wie diese möglichst an der richtigen Stelle eingesetzt und nutzbar gemacht werden können. Die Erwartungen der Fachbereiche müssen möglichst messbar und realistisch definiert werden. Über technische und menschliche Schnittstellen sollte man immer wieder sprechen, um bestehende Ressourcen zu nutzen und die technischen Spezifikationen an die Bereitschaft und die Möglichkeiten der bestehenden Infrastruktur und der Nutzer anzupassen. Im Projektmanagement sollte man auf agile Projektmethoden und auf eine transparente, offene Kommunikation setzen. Das wirkt sich positiv auf das Change-Management aus.

Erfolgsfaktor #7

Ethische und rechtliche Aspekte einbeziehen

KI-Modelle optimieren Geschäftsprozesse ohne Rücksicht auf soziale oder ethische Grundprinzipien. KI ist in manchen Fällen ein zweischneidiges Schwert, das zu ethischen oder auch rechtlichen Problemen führen kann.

Data-Science-Beispiele, die zu ethischen und rechtlichen Problemen führten:

- Gesundheitsrisiko-Scoring: Gesundheitsdienstleister in den USA verwendeten einen Gesundheitsrisiko-Score, um festzustellen, ob sie jedem Patienten eine proaktive Gesundheitsbehandlung anbieten sollten. Grundsätzlich ist das eine gute Idee. Das Modell verwendete jedoch die Gesundheitskosten zur Bewertung der Gesundheitsrisiken. Und weil dunkelhäutige Patienten tendenziell niedrigere Gesundheitskosten hatten, priorisierte dieser Gesundheitsrisiko-Score hellhäutige Patienten für eine proaktive Behandlung.
- Erstellung von Wählerprofilen in Facebook: Cambridge Analytica nutzte Daten von Millionen von Facebook-Profilen ohne Freigabe der Nutzer, um psychologische Profile von Wählern zu erstellen. Facebook wurde daraufhin mit einer Geldstrafe von fünf Milliarden Dollar belegt und Cambridge Analytica ging bankrott.
- Es ist wichtig, potenzielle ethische und rechtliche Probleme im Voraus und während der KI-Nutzungsdauer zu bewerten.

Erfolgsfaktor #8

„Business-Transformation" findet nur in einer Unternehmenskultur statt, die offen für Veränderungen ist.

Jeder kennt den Spruch „Culture eats Strategy for breakfast". Die besten Use Cases, die beste Strategie und das erfahrenste Team nützen wenig, wenn die Unternehmenskultur keine Innovation zulässt. Es braucht eine „open minded" Kultur, in der Mitarbeiter neugierig sind, Neues zu lernen.

Führungskräfte berichten regelmässig, dass kulturelle Herausforderungen – nicht technologische – das grösste Hindernis für die erfolgreiche Einführung von KI-Initiativen sind und damit ein Hindernis für die Optimierung von Geschäftsprozessen darstellen.

Denken Sie daran, dass nicht alle Projektbeteiligten mit voller Überzeugung an Bord sind. KI ist etwas Neues, Unbekanntes, dessen Ergebnis nicht 100%ig klar ist. KI kann unter Umständen Arbeiten und auch Berufsbilder ersetzen, mit denen sich Menschen identifizieren. Das Vertrauen in dieses Thema entwickelt sich mit der Zeit. Es bedarf daher eines hohes Masses an Transparenz und offener Kommunikation. Oft steigen die Neugier und das Interesse während des Projekts und damit auch das Wissen und Verständnis rund um Data-Science und KI. Das sind wichtige Grundvoraussetzungen, wenn KI Bestandteil in der täglichen Arbeit der Mitarbeiter werden soll.

Beziehen Sie insbesondere jene Mitarbeiter ein, die sich auf dem Gebiet KI noch nicht wohl fühlen. Beginnen Sie mit kurzen, einfachen Trainings und Coachings und betonen Sie die Bedeutung der datengetriebenen Entscheidungsfindung. Damit helfen Sie allen Beteiligten, den Veränderungsprozess zu durchlaufen.

Erfolgsfaktor #9

KI-Lösungen brauchen Akzeptanz und Vertrauen

Data-Scientisten bringen oft die besten KI-Modelle hervor, aber die Mitarbeiter wollen die Modelle nicht als Instrument nutzen. Warum?

Ein Klassiker sind Scoring-Modelle wie zum Beispiel die Vorhersage der Wahrscheinlichkeit, dass Kunden abwandern (Churn-Prediction).

Das Problem bei aggregierten Scoring-Modellen ist, dass die Mitarbeiter nicht verstehen, wie sich die Scores bilden. Der Score ist eine Blackbox. Daher gewinnen die Anwender der KI-Modelle kein Vertrauen und tun sich schwer bei der Ableitung der richtigen Massnahmen. Welche Massnahmen soll ein Mitarbeiter ergreifen, wenn der Churn-Score bei 87 von 100 liegt? Also ein stark gefährdeter Kunde.

Akzeptanz bei den Anwendern erreicht man nur, indem man in diese Blackbox reinschauen kann. Den Anwendern soll ein KI-Modell erklärt werden bzw. müssen Anwender das Know-how haben, um zu verstehen, wie KI grundsätzlich funktioniert. Empfehlungen und Prognosen der KI sollten erklärt werden. Man spricht von „explainable AI". Verstehen heisst auch Akzeptieren heisst Vertrauen.

Jedes Unternehmen muss sich mit den eigenen KI-Potenzialen intensiv auseinandersetzen. Diese sachlich analysieren, bewerten und wenn das Potenzial vorhanden ist – mit dem richtigen Use Case und einem qualifizierten Team loslegen. Think big – start small and deliver fast – but start.

3 Methodik zur Auswahl des besten KI-Use-Case

Die Auswahl des richtigen KI-Use-Case ist von entscheidender Bedeutung, da sie den gesamten Erfolg einer Implementierung bestimmt. Ein passender Use Case muss eng mit den Zielen und Herausforderungen des Unternehmens verbunden sein. Eine falsche Auswahl kann zu erheblicher Ressourcenverschwendung führen und das Vertrauen in KI-Technologie zerstören. Die Akzeptanz im Unternehmen ist ein weiterer entscheidender Faktor. Ein gut gewählter Use Case erleichtert die Überzeugung der Stakeholder, da er einen klaren Nutzen und konkrete Auswirkungen aufzeigt. Der Start mit dem richtigen KI-Use-Case fördert die Bereitschaft der Mitarbeiter, Prozesse und Arbeitsweisen zu verändern und hilft, Ängste vor Veränderung abzubauen (Abb. 3).

Vor allem am Anfang der AI-Journey ist die richtige Use-Case-Auswahl wichtig. Man muss sich auf die Suche nach den „low hanging fruits" begeben. Jene Use Cases, die einen hohen Mehrwert generieren und in kurzer Zeit umsetzbar sind. Wie findet man „low hanging fruits"?

Fragt man Mitarbeiter nach Ideen, wofür man KI im Unternehmen einsetzen könnte, kommen jede Menge Ideen. Die Auswahl sollte unbedingt einer klaren, nachvollziehbaren Systematik folgen, sodass Entscheidungen nachvollziehbar sind. Jede Use-Case-Idee sollte auf folgende drei Kriterien hin bewertet werden:

- *Feasibility* Ist die Implementierung technisch machbar?

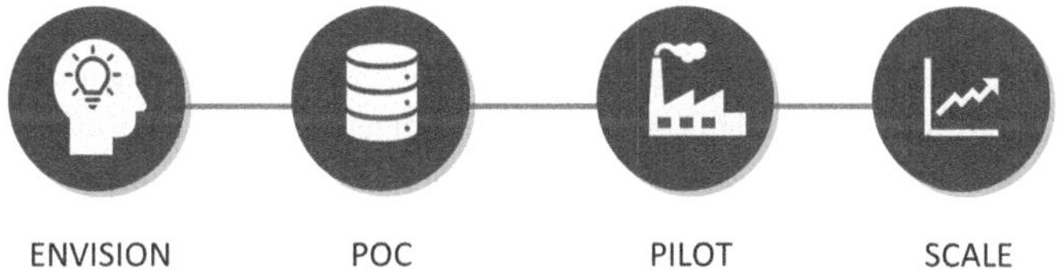

Abb. 3 Ein Unternehmen sollte nicht mit der erstbesten Idee in einen PoC (Proof of Concept) starten, sondern Envision Workshops durchführen, um die richtigen KI-Ideen herauszufiltern

- *Complexity* Ist die Umsetzung ressourcenintensiv und bedarf der Einbindung mehrerer Unternehmenseinheiten bzw. externer Partner?
- *Business Value* Welchen messbaren Mehrwert hat das Unternehmen?

Die Bewertung des Business-Value eines KI-Projekts ist manchmal eine Herausforderung. Auch wenn eine Bewertung in einigen Fällen schwierig sein mag und teilweise mit vagen Annahmen verbunden ist, sollte man nicht auf eine zahlenmässige Bewertung verzichten. Optimiert man im Produktionsbereich ist eine Bewertung anhand von Kosteneinsparung oder Produktivitätssteigerung in der Regel gut möglich. Auch die Verbesserung in der Qualitätskontrolle kann mit der Ausschussquote gemessen werden. Schwieriger wird es auf der Umsatzseite, wenn man versucht, Umsatzsteigerungspotenziale zu quantifizieren. Es geht darum, dass die Bewertung mit internen Experten durchgeführt wird und so ein konstruktiver Dialog angestossen wird. Durch die kritische Auseinandersetzung mit der Nutzenbewertung entwickelt man eine einheitliche Sichtweise und eine klare Meinung zu einer KI-Idee. Sind die internen Experten von einer KI-Idee nicht 100 % überzeugt und motiviert, sollte man damit auch nicht in die Umsetzung gehen. Die Auswahl eines geeigneten Anwendungsfalls ist entscheidend für den Erfolg in der AI-Journey. Die systematische Auswahl und die Einbeziehung der relevanten Stakeholder stellen sicher, dass KI nicht nur technisch erfolgreich ist, sondern vor allem den strategischen und geschäftlichen Anforderungen des Unternehmens entspricht. Dies trägt dazu bei, die Investitionen in KI wirksam einzusetzen und den Weg für eine nachhaltige Integration in die Geschäftsprozesse zu ebnen (Abb. 4 und 5).

4 Die Symbiose aus Mensch und KI

In kritischen Unternehmensbereichen, wie zum Beispiel in der Produktion, stellt die Umsetzung automatisierter KI-basierter Lösungen technisch, wirtschaftlich und aus rechtlicher Sicht eine grosse Herausforderung dar, weshalb der Mensch in der Regel die endgültigen Entscheidungen trifft. Zur Prüfung eines Schadensfalls muss dokumentiert werden, wie die Entscheidungen, die dazu geführt haben, zustande gekommen sind. Dies ist bei KI-Systemen, die sich wie eine Blackbox verhalten, sehr schwierig. Betrachtet man die Stärken von Menschen – wie Flexibilität oder Kreativität – und jene von KI – die Verarbeitung grosser Datenmengen, gleichbleibende Performance und Genauigkeit bei repetitiven Tätigkeiten – wird schnell klar, dass der maximale Nutzen einer

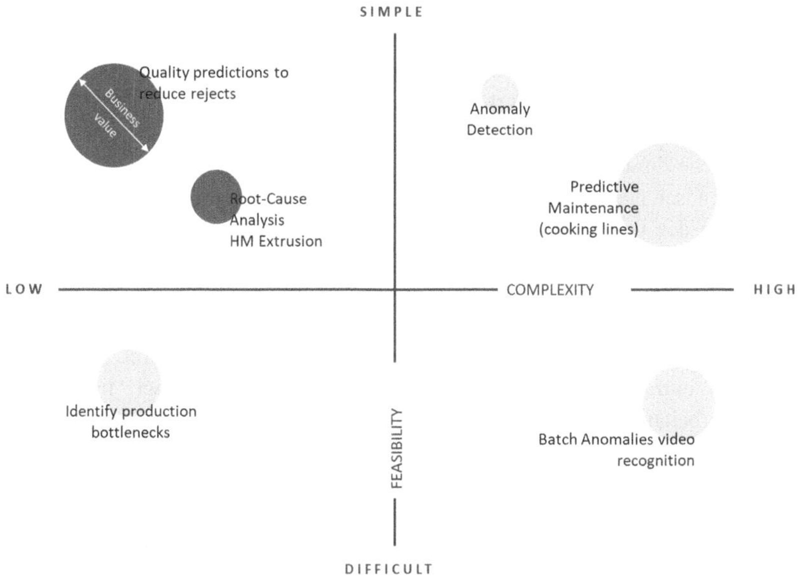

Abb. 4 Das Ergebnis der Use-Case-Bewertung wird auf Grundlage der oben genannten Bewertungskriterien dargestellt. Jene Use Cases, die in der Abbildung links oben sind und einen hohen Business-Value für das Unternehmen versprechen (Durchmesser der Kreise), sollten mit hoher Priorität verfolgt werden. Das sind die „low hanging fruits"

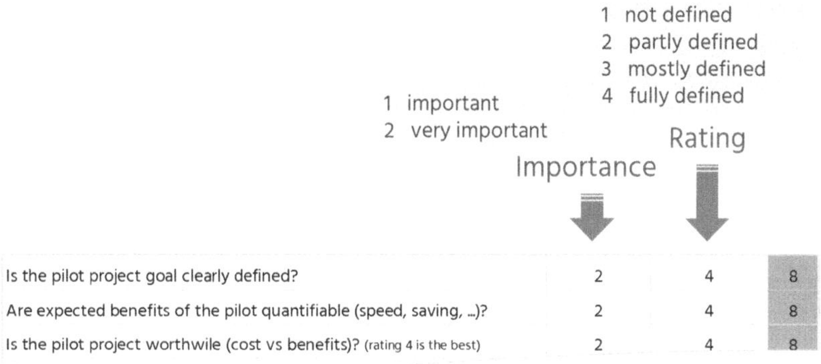

Abb. 5 Die Bewertung nimmt man anhand eines Kriterienkataloges vor. KI-Beratungsunternehmen nutzen Kriterienkataloge und deren Expertise zur umfassenden Bewertung, die gemeinsam mit Mitarbeitern des Unternehmens durchgeführt wird

KI durch die Zusammenarbeit mit Menschen erzielt werden kann. Der Mensch mit seinem Erfahrungswissen und seinen Stärken wird deshalb auch zukünftig eine wichtige Rolle spielen, ob als Entscheider oder als Mitarbeiter, der ein KI-System überwacht, trainiert, weiterentwickelt und optimiert. Die Mensch-KI-Schnittstelle bzw. das User-Interface nimmt eine wichtige Rolle ein, um die beschriebene Zusammenarbeit zu ermöglichen und Entscheidungen durch gut erklärte Prognosen oder Empfehlungen der KI zu unterstützen. Die richtige Wahl und Gestaltung des User-Interface ist für die Akzeptanz wichtig. Neben klassischen Visualisierungen in Form von Text oder Grafiken bieten Sprachassistenten wie Chat-GPT ein bedeutendes Potenzial für die Zukunft.

Durch die Erkennung von Anomalien z. B. bei Vibration, Temperaturanstieg, Druckabfall oder Abriss der Stromaufnahme werden Ausfälle von Maschinen bzw. von einzelnen Baugruppen oder Werkzeugen vorhergesagt. Instandhaltungsmitarbeiter werden über eine webbasierte Applikation am Steuerungsstand informiert. Die Entscheidung, wann die Wartung tatsächlich durchgeführt wird, obliegt weiterhin dem Mitarbeiter. Mithilfe dieser Applikation können Wartungsarbeiten frühzeitig erkannt und eingeplant werden, um einen ungeplanten Stillstand zu vermeiden, der wirtschaftlich einen beträchtlichen Schaden verursachen kann.

Im Sinne einer guten Erklärbarkeit zeigt das KI-Tool, welche Anomalien im Prozess auftreten und welche (korrelierenden) Faktoren für die Wartungsprognose verantwortlich sind. Das System gibt eine Wahrscheinlichkeit an, mit der es innerhalb der nächsten 6, 12 oder 24 Stunden zu einem Ausfall kommen könnte. Das System zeigt zudem, welche ähnlichen Ereignisse es in der Vergangenheit gab und ob das System bei diesen Fällen richtig oder falsch lag. Ein KI-System ist nur so gut wie die Genauigkeit der Vorhersagen sind und ob die Anwender in die KI Applikation vertrauen haben oder nicht. Die Art der Visualisierung der KI-Empfehlungen im User-Interface spielt eine wichtige Rolle im Vertrauensaufbau.

Das KI-System hilft den Mitarbeitern zu lernen, welche Anomalien zu Maschinenausfällen führen und wie lange die Wartungszyklen dauern bzw. von welchen Faktoren diese beeinflusst werden. Andererseits können Mitarbeiter über Eingabemechanismen ebenfalls ihr Erfahrungswissen einbringen und einen Wartungsfall einstellen, wenn sie z. B. selbst wegen einer eigenartigen Geräuschentwicklung einer Maschine einen baldigen Ausfall erwarten. Sie können dem System Rückmeldung geben, ob es wirklich in einem bestimmten Zeitraum zu einem Ausfall gekommen ist oder ob eine durchgeführte Wartung zur Lösung einer Anomalie geführt hat. Auch die Adaptierbarkeit des KI-Systems ist wichtig. Also die Anpassung an die sich verändernde Gegebenheiten der Produktion und die Integration mit anderen IT-Systemen (z. B. ERP). Um KI-Systeme mit einer hohen Akzeptanz zu entwickeln, müssen das KI-Modell, der Geschäftsprozess und das User-Interface aufeinander abgestimmt werden. Hierfür ist ein interdisziplinäres Team erforderlich, das Expertenwissen in allen genannten Bereichen einbringt. Eine wichtige Rolle nehmen Usability-Experten ein. Sie helfen dabei, die KI-Applikation und das richtige User-Interface in den Arbeitsablauf zu integrieren.

5 Fazit

Die erfolgreiche Einführung von KI in Unternehmen der Fertigungsindustrie erfordert nicht nur technologische Implementierungen, sondern auch eine strategische Herangehensweise, die das Mindset der Mitarbeiter berücksichtigt. Die Integration von KI in die Wertschöpfungskette bietet Potenziale zur Steigerung der Effizienz, der Innovationskraft und der Wettbewerbsfähigkeit eines Unternehmens.

Bis heute wird KI in der Regel als Werkzeug für Mitarbeiter eingesetzt, wobei ihre Effektivität stark von dem Vertrauen der Mitarbeiter in die Ergebnisse abhängt. Es erfordert ausserdem die Bereitschaft der Mitarbeiter, ihre Arbeitsmethoden anzupassen und den Empfehlungen der KI zu folgen. Die Implementierung vollautomatisierter, autonom funktionierender KI-Lösungen in kritischen Unternehmensbereichen ist in den meisten Fällen nicht das Ziel. Menschliche Entscheidungen bleiben aufgrund technischer, wirtschaftlicher und rechtlicher Komplexität oft unerlässlich.

Vielmehr sollen KI-Lösungen in Arbeits- und Entscheidungsprozesse der Mitarbeiter sinnvoll integriert werden. Eine KI agiert als Assistent, um schnelle und richtige Entscheidungen zu ermöglichen. Das User-Interface spielt dabei eine entscheidende Rolle, besonders im Kontext der „erklärbaren KI". Ein KI-System gewinnt Vertrauen, wenn es klare und präzise Empfehlungen oder Prognosen liefert, die für Mitarbeiter verständlich sind. Die Nutzung eines KI-Systems erfolgt nur, wenn die Mitarbeiter diesem auch vertrauen.

Armin Ribis ist seit vielen Jahren in der Softwareindustrie als Berater tätig. In mehr als 200 international tätigen Unternehmen hat er digitale Transformationsprojekte initiiert, geleitet oder begleitet. Sein akademischer Hintergrund ist Bauingenieurwesen und Betriebswirtschaft. Seit 2018 fokussiert er sich auf das Thema künstliche Intelligenz und unterstützt Unternehmen dabei, KI-Systeme an den richtigen Stellen im Unternehmen einzusetzen.

Nämlich dort, wo KI den grössten Nutzen für das Unternehmen und die Anwender stiftet. www.make-ai-work.com

Dr. Simone Carlo Surace ist als Data-Scientist tätig und Mitgründer von Flock Labs, ein junges Startup im Bereich der daten- und simulationsbasierten strategischen Entscheidungsfindung. Er hat einen Master in Theoretischer Physik und ein PhD in Neurowissenschaften von der Universität Bern (Schweiz). Nach Abschluss seiner Doktorarbeit betrieb er akademische Forschung an der Schnittstelle von Neurowissenschaften, Statistik, und Maschine Learning in Zürich und Bern. Die letzten Jahre setzte er diese Erfahrung im industriellen Umfeld als Data Scientist sowie als Software- und Machine-Learning-Entwickler in zahlreichen Projekten bei Artificialy SA um. Strategisches Denken in Kombination mit Daten und KI ist seine grosse Leidenschaft, die er nun als Gründer und Partner bei Flock Labs sowie in dessen Produkten verwirklichen will.

The Philosophical Compass: Navigating Humanity Towards a Desirable AI

David Rohrmann and Michael Hengl

1 End with Why

A famous speaker and author suggested in his book to always "start with why" (Sinek 2009) when it comes to managerial decisions around leadership, culture and strategy. We think this is not enough. In our opinion and experience it is equally important to end with the why, to design, create and shape towards something great, benevolent, purposeful. What is the AI narrative that we want to tell? How is the plot supposed to unfold? So far there are mainly two positions noticeable, both strongly affected by the latest developments of Large Language Models (LLM), like ChatGPT from OpenAI and Microsoft, BERT and LaMDA from Google, Llama from Facebook, to name a few.

There are the AI fatalists, following which we are practically doomed: AI will take away our jobs and increase economic inequalities (Chui et al., 2016). As the systems are trained mainly by white males, it is highly biased, discriminatory and unfair (Crawford, 2016). AI is also becoming a black box in terms of transparency regarding the materials used for training, the validity of its outputs, since it cannot guarantee the truth for all its statements, and its accountability because of opaque nature of its decision-making algorithms (Goodman & Flaxman, 2017).

D. Rohrmann (✉)
Spaze GmbH, Berlin, Germany
e-mail: davidrohrmann@me.com

M. Hengl
1492 webrain GmbH, Altaussee, Austria
e-mail: mh@1492.org

C. H. Hoffmann (ed.), *Artificial Intelligence, Entrepreneurship and Risk*, Technikzukünfte, Wissenschaft und Gesellschaft / Futures of Technology, Science and Society, https://doi.org/10.1007/978-3-658-45544-6_19

Some accuse the systems even of hallucinations, which is a slippery anthropomorphic slope, leading to the final conviction that AI will eventually become a superintelligence (Bostrom, 2014) and overpower humanity with our fate being extinct, enslaved or at best tolerated. Famous examples in pop culture are the Matrix and Terminator movies where superintelligent is clearly interchangeable with super-evil.

On the other side of the trenches are the AI disciples with sparkling eyes when it comes to the advantages of AI. They also think that we are doomed, albeit only in scenarios where we do not leverage and embrace AI's merits: The fundamental benefit is the capacity of AI algorithms to analyze incredible amounts of data and identify patterns and weak signals that may be overlooked by humans, making it an invaluable driver for innovation (Brynjolfsson & McAfee, 2014). Especially the advances in health care will have an unprecedented impact on society due to better diagnostics, drug discovery, personalized medicine, revolutionizing patient outcomes and medical practices (Topol, 2019). Also our everyday lives as consumers, customers or users will benefit greatly from AI as omnipresent and close to omniscient chat bots, assistants, recommendation systems, individually generated contents, personalized virtual realities (Jiang, Li, & Huang, 2017). The results are greater customer satisfaction and loyalty. They meet an increased overall economic productivity due to higher efficiency, automation, and self-propelling technological progress (Davenport & Ronanki, 2018). With the support of smarter and faster computers we will create even smarter and faster computers, bringing about overall economic growth and welfare, especially in but not restricted to AI-related industries (Manyika et al., 2017).

Of course, there is a vast middle ground of arguments and opinions to the topic. The moderate voices, however, have a rather hard time of cutting through the noise and being heard. After all, this is a game of scientific reputation, media attention, and very straightforward economic interests. It is about investments, favorable legislation, market shares, stock prices. This is nothing new or unusual. There is just one crucial aspect to the matter: Whether in favor of or against the developments of AI, both parties would agree that after the invention of computers and then the internet, AI has the potential for spearheading the next digital and industrial revolution with unparalleled impacts.

2 Philosophy, Economics and Business Ethics

That said, the question "how will it be?" falls short. The more important question with the end in mind is: "how do we want it?" What is the AI story that we as a species want to tell each other, our children, and next generations to come? The future does not just happen with us on the passenger seat. We make it happen if we choose to take the steering wheel firmly in our hands. We select, evaluate, decide, develop, deploy, and bear the consequences. Now, at this point, free conscious will on an individual and collective level must be assumed. Both, the fully sacral perspective with God having it all pre-figured out, or a completely mechanistic and deterministic world view with sequential chains of cause and effect, bring us back to question one, merely watching how things will unfold. The passenger seat though, is a lousy place when approaching with full speed junction after junction on the course of our civilization.

We are not promoting to lie low until we have it all figured out in the last detail. On the contrary, there will be potential miracles and pitfalls after each bend. Some paths will only appear once they are trotted. There will never be a complete map for the vast digital landscapes that humanity will create. And even with such a map on our hands, we would still need a compass to navigate. We argue that philosophy can be this compass.

There have been numerous publications on the relevance of philosophy for economics. The two biggest economic schools still prevailing today, namely capitalism and socialism have strong roots in philosophy. Adam Smith, the godfather of capitalism was first of all a philosopher with his important work "The Theory of Moral Sentiments" (1759) preceding his economic main work "The wealth of Nations" (1776) by 17 years. Karl Marx was more renowned for his philosophic contributions to politics when he published his socialistic take on economic theory "Das Kapital" (1867). These are only two of the main philosophic economists—or economic philosophers.

When descending from the macro layer of economics and entering the meso layer (i.e. business and organizations) and micro layer (i.e. management and leadership), a whole discipline called business ethics was formed in the 1970s. As a branch of applied ethics and professional ethics, it systematically explores ethical principles and addresses moral or ethical challenges inherent in the business realm. As ethical framework, it integrates various facets of business conduct, reaching from individual applicability to entire organizations and economic governance. The foundation of business ethics is derived from sources such as individual convictions or virtues, organizational declarations, legal constructs, social developments, and environmental requirements. These sources establish norms, values, and ethical or unethical practices, collectively forming the guiding principles for specific businesses or business in general.

Business ethics, as manifested in contemporary organizational standards and culture, encompasses a set of values, norms, and principles that govern the conduct and behavior of individuals within a business organization, amongst organizations, and organizations as members of society. The construct of business ethics is bifurcated into two dimensions that are also present in general ethical discussions: normative and descriptive business ethics. As corporate practices and professional specialization, business ethics operates as a normative framework. These normative aspects prescribe how individuals and organizations ought to behave. On the other side, academics employing descriptive methods seek to comprehend and explain actual business behavior. The diverse array and volume of ethical issues in the business world reflect the intricate interplay between profit-maximizing endeavors and considerations beyond economic realms. To name a few, some topics and concepts originating in business ethics have massively influenced today's managerial and economic discussions such as CSR (corporate social responsibility), stakeholder theory, ESG (environmental, social and corporate governance), sustainability reporting, SDG (United Nations Sustainable Development Goals), purposeful work.

3 Searching for the Why

Especially the debate around purpose in business is essential for the line of reasoning in this chapter. Purpose is the why to any aspiration. Together with a more specific and time-bound vision of the state of things in the future and a rough outline how to get there, the mission statement, these three aspects form the fundamental trinity of a compelling narrative: why to get there, what will it look like, how do we make it. They are interdependent, complementary, often confused, yet very different in their nature.

The current discourse of the why of AI falls short in all three segments. What is the general purpose of AI, its nature, its meaning as human artefact? How does our vision of AI for the next decades look like, our desired state of the future, our AI-supported accomplishments as species, our social development not how it might unfold but how we want it to? And what are our mission path to make it happen, our navigation, milestones and equipment. These questions are hardly found. Not that they are not asked at all. There are brilliant academic attempts in posing the questions and answering them. Scientific journals like the "Minds and Machines—Journal for Artificial Intelligence, Philosophy and

Cognitive Science", the "Journal of Artificial Intelligence and Consciousness" or the "The AI Ethics Journal" are a few examples. When taking the assumed amplitude of change for humanity into account that was mentioned earlier, this is simply not enough though.

We need to start and end with a compelling why of AI for us humans. When you compare AI with a journey, this will definitely take us into uncharted territory beyond the horizon. How to prepare for a trip where the destination is out of sight? More so, it is outside our past experiences and thus we cannot extrapolate any reliable prognosis from there. In such cases, Matthias Horx, a renowned futurist from the German Zukunftsinstitut suggest using regnosis instead (Horx 2021). The idea is not to try to predict the future but to invent it and by doing so create it. This entails the strong notion that the future is not something that happens detached from our thinking, feeling, and doing. We shape it by our individual and collective intentions. Now, this can be seen from a spiritual perspective with Buddha saying, "what we think, we become". Or we take a psychological standpoint like the concept of self-fulfilling prophecies. Or we look through the lens of the latest research in quantum physic where there seems to be growing evidence that our directed thinking has measurable impact on the quantum level physically (Kaufmann & Radin 2023). The idea of mind over matter appears to return from the metaphysical realm back to the material world.

4 The Conceptual Compass

The implication reiterates that our envisioning or portrayal of the future landscape of AI leans way more toward being normative than descriptive. Normative not in the narrow sense of what should AI be allowed to do and not to do, i.e. what norms and standards should it comply to. Normative here means that our future narrative of AI will strongly norm the future of AI. What we think, want, and tell, we become.

With philosophy and ethics as the aforementioned compass on this journey yet to go on, we are equipped with some tried and true tools in forms of concepts, models, arguments, and logics. Since literally thousands of years, philosophers around the globe have tried to make sense of the world as it is and as it should be. Though there have been strong attempts by the analytical philosophy, especially the logical empiricism, in the last century to frame philosophical questions as mathematical equations in order to calculate and deduct answers rather than to prosaically describe them, philosophy has stayed highly axiomatic. Every explanation can be challenged with a "and why is that?" like children do it. This so-called infinite regress makes it impossible to eventually prove any statement. The final way out of this epistemologically unsatisfying situation is the introduction of an axiom. The axiom being a fundamental believe on which a specific framework is built upon. Coming from the Greek axioein it means "to deem worthy" or "to require". This is important to grasp because every school of thought is then a believe system, eventually. As complex, sophisticated, or convincing it comes across, we cannot get away from the responsibility of choosing consciously or subconsciously the initial axioms of our thinking.

Looking at the future of AI, fortunately or unfortunately, we cannot rely on any philosophical truths because there are none. What we can do though is to make conscious decisions about the believe systems we are utilizing. The act is to choose not to prove, to agree not to verify, to invent not to discover. That way, we might end up with a collective consensus about humanity's desired state of the future with AI and thus increase the probability for ending up exactly there. To decide what is more or less desirable we have numerous competing frameworks at hand. All to mention would quite fill whole books rather than a single chapter. A hopefully inspiring choice will be presented in the next sections. The package insert for all of them stays the same: none of the frameworks is true or truer than others and even within each of them the options are equally valid. This is not to be confused with rhetorical arbitrariness. On the contrary, it urges to reveal the logical figures and deepest layers of our reasoning, to unmask attempts of proof, and to give room for an agreeable formation of opinions.

5 Deontology Versus Consequentialism

Unarguably two of the biggest rivaling ethical frameworks are that of deontology and consequentialism. These two ethical theories provide different perspectives on how one should determine the rightness or wrongness of attitudes and actions in ethical decision-making. While deontology focuses on the inherent nature of actions and adherence to moral rules, consequentialism assesses morality based on the overall outcomes.

Deontology is an ethical theory that posits certain actions as inherently right or wrong, regardless of their consequences. It emphasizes adherence to moral rules or duties, hence it is often referred to as duty-based or rule-based ethics. The priority clearly lies in the intention behind an action rather than in the outcomes. The most prominent deontological frontrunner is Immanuel Kant with his "Groundwork of the Metaphysics of Morals" (1785). He argues for the

Categorical Imperative, which states that individuals should only act according to principles that could become universal laws without contradiction. The (in)famous example is to refrain from lying even if that would save someone's life because lying is inherently wrong, irrespective of the consequences. The same goes for promises. They must not be broken under all circumstances. The whole concept of a promise fails to function when it would include the notion: you may break a promise when you think you have a good reason for it.

Deontological ethics would create and assess AI systems based on inherent ethical principles or rules. Certain actions or functionalities of AI may be deemed intrinsically right or wrong, irrespective of the consequences. An example for this would be the first law of robotics from Asimov "A robot may not injure a human being, or, through inaction, allow a human being to come to harm" (Asimov, 1950: 40). A deontological evaluation of AI would consider the rights and responsibilities associated with the use of AI technology. This includes ensuring that AI systems respect certain irrevocable individual rights. Steven Spielberg's movie Minority Report revolves around that theme: a precognitive system prevents crime from happening by eliminating the potential culprit. Deontologically it would make total sense to prohibit a powerful AI predicting such crimes even when there is highest probability for accuracy because nobody should be punished for a deed they did not even attempt to commit.

Consequentialism, on the other hand, is an ethical theory that evaluates the morality of an action based on its consequences. The core idea is that the rightness or wrongness of an action is determined by the overall outcome it produces. Consequentialist theories, such as utilitarianism (Mill, 1861), argue that one should maximize overall utility, hence its name. Actions that lead to the greatest good for the greatest number are considered morally right. The morality of the action is contingent on its consequences.

Consequentialist ethics would evaluate the morality of AI systems based on their outcomes, i.e. the overall impact and consequences for individuals, society, and the environment. It would prefer an AI that seeks to maximize overall happiness or well-being, efficiency gains, economic benefits, or societal advancements. Trade-offs between competing values need to be considered carefully. This could involve for example weighing the benefits of automation against potential negative consequences, such as job displacement. This balancing would need a high flexibility and adaptability in assessing AI. Ethical evaluations may evolve based on changing circumstances and the growing understanding of AI's impact on society.

To sum it up, both viewpoints play an important role when shaping a desirable future profile of AI. None of them is better or worse. In practice, ethical considerations of designing a desirable AI would involve a combination of deontological and consequentialist considerations. Integrating the inherent ethical principles and guidelines together with an assessment of overall outcomes and societal impact allows for a compelling big picture.

6 Individuals or Systems

So, we can wish for an AI with strong principles or strong focus on the outcome or a combination of both. That brings us to another major dimension: who bears responsibility for our future AI? Do we opt for a single AI that is global, comprehensive, total and for everyone? Maybe even without individuals' factual agreement? Or do we want lots of different AI instances, each with their own rights, duties, and responsibilities? With users who can join and leave any system? With full individual control and accountability of each AI and those operating it? Again, philosophy provides us with some interesting thoughts on the matter. Especially in ethics, there was always the question of where morality resides. Whether it is inherent in the individual or embedded in a larger system. We are not looking specifically where to locate morality within AI. This is a topic for another discussion. Our question here is different: if we want to create an AI story that is good, attractive, and desirable, where do we anchor those good or great attributes?

In the political economic debate, this problem has been discussed widely in the context of market liberalism. It emphasizes the importance of free markets, limited government intervention in the economy, and individual liberties. The position clearly advocates for minimal regulations by the system, putting a lot of responsibility on individual actors like people, organizations, and industries. If everybody is responsible for themselves then everybody is somehow taken care of, is one line of argument. Individuals should therefore have the freedom to pursue their economic goals with as little interference from the government as possible. Market liberals do not condone laws and constitutions, though. In fact, for free markets to function legal frameworks that are clear, consistent, and applied equally to all individuals are considered crucial, e.g. property rights. Altogether, the state activities should be reduced to the bare minimum though and the rest will be sorted out individually. In the context of our future AI, this would mean a highly unregulated AI landscape with little governance, lots of competition, maybe more innovation and dynamics, but maybe also more losers and winners.

Often (yet not necessarily) connected to liberal and individualistic positions is a rather positive human image. It makes sense to grant a lot of freedom to the individual only when you are optimistic that this freedom is not constantly abused, either deliberately or instinctively. Are humans basically good, for themselves and for others? Or is man the often-quoted Hobbesian wolf that needs to be tamed, ruled, sanctioned? Latter position brings us right to a group of philosophers, Thomas Hobbes being one of them, that had a remarkable impact on society with their thinking: the social contract theory. It is a philosophical concept that explores the origins and legitimacy of social rules and norms, political authority, and governance. The idea is that individuals voluntarily form a social contract, agreeing to abide by certain rules and norms for mutual benefit and the common good. This contractual agreement establishes the basis for political authority and the creation of a just and orderly society. The basis for such a contract is often the description of a hypothetical state of nature where individuals exist without formal government, established laws, and authority. Everybody is free to pursue their own interests, which leads to better or worse situations, depending on the human image applied. Therefore, we cannot leave it on an individual level for things to work out. The result is exactly the alternative draft to moral individualism: we need strong institutions that ensure the survival of society because the self-interest of people will directly or in the long run go against it. This rather fear-based perception of human reality comes in varying degrees. Hobbes, without a doubt, was one of the grimmest and most pessimistic thinkers. His Leviathan (Hobbes 1651) was a metaphor for absolute sovereignty as political power, necessary to uphold civilization.

On the other end of the spectrum of social contract theorists was Jean-Jacques Rousseau (1762), who became a trailblazer for the Age of Enlightenment and the French Revolution. He had a much more optimistic view on human nature. Individuals would form governments to express their general will as the common interests of the people. Rousseau was advocating direct democracy as legitimation of political power, where citizens actively participate in decision-making. He also emphasized the importance of individual freedom within the constraints of the general will. That makes him a rather liberal representative of social contract theory, with a more joy-based (towards something great) than fear-based (away from something terrible) approach.

For our AI case, we can summarize that with a social contract theory mindset we do not leave the development to individual actors, like people or organizations.

We would need to come up with a strong overarching governance, a general system that ensures AI is doing good, detailed binding rules for every player on the field, with a limited freedom of choice. Again, both positions, the individualistic and the systemic one, are valid with their merits and down-sides. We just need to be conscious and explicit about the believe systems we are referring to when constructing and evaluating our future vision of AI. Especially the human image that we are using for our narrative must be revealed to allow fruitful discussions.

7 We Need to Talk

At this point, we can compose our desired AI from strong binding principles (deontology) or accountability for the impact achieved (consequentialism). We can shift the responsibility for such an AI between individuals and systems. And we should be able to articulate if the general backdrop of the human image against we shape our story is rather joy-based or fear-based. Though these considerations origin from widespread and influential philosophical or ethical debates, some of them lasting for centuries, they will not be enough to design a substantial, material why of AI. How should AI really look like in terms of benevolence, security, privacy, intellectual rights, discrimination, biases, access, availability, attractiveness, love, humor, imagination, and innovation? This list is absolutely non-exhaustive and purely illustrative. We need more fine-tuned procedures to decide for the best purpose AI can have, the best vision for AI within humanity, the best why of AI altogether.

In the 1970s, the German philosophers Jürgen Habermas (1981) and Karl-Otto Apel (1973) laid foundation to a philosophical offspring from the Kantian deontology, called discourse ethics. They found Kant to be too focused on the individual when constructing moral principles, too solipsistic, too rigid. To legitimize norms and values Habermas and Apel suggested a much more intersubjective approach, based on communicative rationality. The suggestion is to use communication and reasoned discourse for ethical decision-making and the establishment of normative principles. It is a formal and procedural type of ethics, meaning that it does not define the content of morality itself but the form and procedure to legitimize it.

The idea behind it is that there is no absolute truth of ethical statements to be found. Also, the validity of a moral norm cannot be justified in the mind of an isolated individual reflecting on the world. It is justified only intersubjectively in the process of reasonable, dialectic discussions between individuals. We can, however, determine universally valid statements about generally recognized moral principles if they are based on mutual understanding that is achieved by individuals in argument and consensus of all those involved. This is not to be mistaken for a bare or naïve consensus theory of truth, i.e. a confirmation of truth based on actual consensus in an actual community without further regulation or qualification. Therefore, the correctness of normative or prescriptive sentences, like the one we want to create for shaping our future AI, is established with the help of a discourse with reasonable arguments, organized according to specific rules.

There are two governing principles for the discourse: the universality principle (U) and the discourse principle (D). While they are interconnected, they focus on different aspects of ethical reasoning.

(U) is defined as follows:

"All affected can accept the consequences and the side effects that [the norm's] general observance can be anticipated to have for the satisfaction of everyone's interests, and the consequences are preferred to those of known alternative possibilities for regulation" (Habermas, 1991: 65).

(D) is a principle concerning the way norms conforming to (U) must be justified through discourse. It involves the idea that ethical norms and principles should be the result of rational communication among individuals who engage in open, genuine, and respectful conversation. These communicative interactions are supposed to be free from coercion, deception, and other distortions. In such an idealized situation, individuals have equal opportunities to express their opinions, and the force of better arguments should prevail.

In summary, (U) emphasizes the need for moral principles to be universally acceptable, while (D) underscores the importance of arriving at these principles through open and rational communication. Apel's concern hereby is that of an adequate mediation between strategically unconditional, sincere, and persuasion-free communication with opponents, and such strategic practices, like offers and threats, that are dictated by the legitimate self-interests of individuals (Apel, 1992: 35). To ensure communicative exchange on such high standards, it is important mentioning that Habermas has created an extensive list of specific discourse rules, called presuppositions. They are an interesting treat to study, yet very analytical, in-depth, and with little added value for our line of argument. A much more digestible set of ground norms with appropriate abstraction level and quantity has been composed by Wolfgang Kuhlmann:

N 1: Argue rationally.

N 2: Strive to reach a reasonable consensus.

N 3: In all cases where your interests may conflict with those of others, endeavor to reach a reasonable practical consensus with them.

N 4: Always strive to contribute to the (long-term) realization of such conditions that come closer to the ideal communication community, and always take care to preserve the already existing conditions of the possible realization of an ideal communication community (Kuhlmann, 1985: 23).

It is important to acknowledge that sheer thought experiments alone cannot replace a communicative exchange with others regarding norms that will affect them. Actual discourse always trumps a hypothetical one. As such intersubjective real-world discourses can be highly dynamic, they are catalysts for collective learning that does not dilute in moral relativism or gets stuck in static dogmatism. Ideally, such learning spawns a rational trajectory, grounded in the particular circumstances of those involved and aimed at universal validity.

8 The Ideal Communication Community

Picking up Kuhlmann's norms, especially N 4, we step on very interesting grounds when it comes to creating, evaluating, legitimizing our desired AI narrative. Looking at the list of potential topics to elaborate, mentioned in the beginning of the former section, it makes sense to foster ongoing debates about them. These topics are complex, often multi-disciplinary, constantly evolving through technological advancements, and facing an increasing technical adaption rate. Since they have a growing impact on people's lives, society should take care to have the best communication community in place elaborating such delicate matter.

How would such a community, collective, crowd look like in reality? What are smart boundary conditions and drivers in today's real world, beyond abstract hypotheses and philosophical constructs? A typical example of a practical discourse is exactly the topic of business ethics, in which theorists and practitioners argue about the way in which ethical standards should become effective in the economic sphere. The spectrum ranges from Karl Homann (Homann & Lütge 2004), who argues that the economy should be given a political framework and that the economic subjects should be allowed to act within this framework, to Peter Ulrich (1997), who in his integrative business ethics calls for a primacy of ethical standards to be taken into account for all economic activity and attempts to justify this in terms of discourse ethics.

Especially in the beginning, the discourses within and about business ethics have been far from ideal, led by a far from ideal community. It was few experts—mostly theoretical philosophers—with thorough academic education yet limited practical insights and even less experiences in the economic world. The discussions took place in journals and university lectures, about those affected, not with them. It was not until major scandals like Enron and WorldCom that shocked the business world, and in turn fueled the debate with growing audiences. Along with globalization also the discourses became more global. The Volkswagen Diesel scandal, Wirecard, Facebook, BP, Lehman Brothers, they all had world-wide impacts and attracted world-wide attention. Another aspect was technology that brought the communication community to the next level, particularly the spread of the internet and later the rise of social media. I am not saying that electronic mass media and ubiquitous individual communication tools lead to a better discourse per se. However, there is at least good potential that this can pay into the discourse principle (D) and ground norm N 4 by simply enabling a more open, broader, and better informed, thus more rational discourse.

With the ideal communication community in mind, there is a specific risk when looking at the AI debate today. The separation of political powers has been an enormous improvement in how we run states, paving the way for modern societies. Today's discussions around AI can be seen as a step back in this particular regard. The global tech giants that operate the AI systems represent the executive. At the same time, they are heavily involved in the regulation of these systems and themselves as organizations. So they are also a heavy part of the legislature. Plus, they are the ones with best access to evaluations of all AI applications because they own the systems, which turns them also into judiciary. Despite the name, OpenAI is no longer predominantly open source since 2019. Microsoft alone invested $10 Billion in 2023 (Bass, 2023). There is severe money on the table. Too much, to leave the discussion to the few actors we are facing today. We are not saying we should increase the separation of (intellectual) powers and install artificial Chinese Walls, even if that was an option, which is probably not. The better option would be to integrate the three instances vertically, with the goal of bridging the gaps between what is possible and what is desirable. And it deems even more crucial to engage and integrate a much broader crowd horizontally for each instance. We should take good care of not leaving the technological frontrunners alone, encapsuled and detached from those affected by their thinking and doing: humanity.

9 The Wisdom of the Crowd

Ideally, we do not wait for the first earth-shaking AI scandals to happen before we start working on our compelling story as this is literally history in the making. Also, we are most probably better off, when we leave the development not in the hands of a few (self-)chosen ones. Experts and expertise are definitely necessary in such complicated and far-reaching matters. There is a latent risk though that the more specialized professionals are, the less of a general, overarching standpoint they can take. Everybody perceives the world through their lenses of knowledge, experience, and ambition. Like the common proverb says: for a hammer, everything looks like a nail. There is nothing wrong with it, if we are conscious about it and take appropriate counter measures. We do not want the experts to dismiss their views, as this would mean giving up their precious and well needed expertise. To converge to the ideal communication community, we must balance the debate between specialists of different disciplines and between specialist and generalists. Following the outlines of discourse ethics, we are also advancing the quality of exchange if we simply increase the participation, shifting the focus from few individual thinkers and actors and leveraging our collective intelligence. Being aware of enormous scholastic discussion about the definition of intelligence, it is defined here quite simply as problem-solving capacity either of individuals or of groups.

Vox populi, vox dei. Already the Romans considered the voice of the people to be divine or in more secular terms: from a higher level of consciousness. It is a good thing that we presumably have such higher means to bring the AI story to a great end, as one of our specie's superpowers amongst others is exactly this collective intelligence of ours. Based on our unique language processing capacities, we have evolutionized evolution itself by decoupling our development from slow "hardware-based", i.e. DNA-based, alterations and adaptions and excelled in much faster "software-based" improvements, i.e. language, communication, and culture claims philosopher and cognitive scientist Daniel Dennett in his book "From Bacteria to Bach" (2017). The idea is that humanity became what it is because of our neurological capabilities. We were able to benefit from a natural selection of our memes instead of genes. While genes are self-explanatory, the term memes was coined by Richard Dawkins (1976) in his book "The Selfish Gene". A meme is a cultural artifact. It can be an idea, behavior or style that spreads from person to person within a culture through imitation. Memes act as a carriers of cultural values, symbols or practices that can be transmitted from one mind to another through language, writing, gestures, rituals or other imitable phenomena. Part of the concept is the consideration of memes as cultural analogs to genes, as they replicate themselves in their community or environment, mutate, and react to selection pressure with only the fittest ones to survive.

These memes allow us to carry wisdom and experiences through space and—more important—through time. No other species on the planet has managed to pass this amazing amount of learning from one generation to another. And with our advancements of information and communication technology over the last decades, speed and comprehensives of these transmissions have even increased. The velocity of knowledge generation and distribution are truly mind-blowing. We can now build on each other's ideas like no other and never before. Similar to our brains, humanity benefits from more and more connections. It is not the sheer amount of neurons that makes us smart. It is the number of connections between them and the internal structure of the brain. This is also the explanation why the Neanderthals were outsmarted by homo sapiens, although both had similar brain sizes. From this perspective, we can see humanity as global entity, as a social brain, with all its individuals, organizations, institutions as neurons that get ever more connected. Consequently, our collective intelligence will increase with the same pace.

The key to this growth is communication, exchange, discourse. The more a group approaches the characteristics of an ideal communication community, the higher its collective intelligence becomes. It is vital to acknowledge that groups are not unconditionally or automatically smart. Some are smarter than others and some groups are straightforward stupid. There are enough historic, contemporary, and personal anecdotes and demonstrations available to all of us. Over the last years though, there has been more research, evidence, and examples for the inherent potential of the collective intelligence of groups, especially in contrast to the intelligence of individual experts or small expert circles. In the TV show "Who wants to be a millionaire", the audience joker consistently outperforms the single expert telephone joker when the candidates are seeking help for a quiz question. James Surowiecki has collected an impressive number of examples in his book "The Wisdom of Crowds" (2005). Only one year later, in 2006, the prestigious Massachusetts Institute of Technology founded The MIT Center for Collective Intelligence. The concept has also permeated the business world. Today, it is absolutely common to talk about swarm organizations, decentral teams, holocracy, agility. These are all attempts to leverage the collective intelligence of groups without having to rely on central steering and thinking by high-ranked managers or experts.

10 One Expert is Good, Many Experts Are Better

The more complex a problem is, the more unclear the data seems, the more unforeseeable events might turn out, the better it is to use the power of a social brain of the many, instead of a few selected expert brains. Remember the hammer and the nail. Every expert is such only because of her being deep in her field of expertise. Everybody has blind spots. The more you focus on one side, the bigger is the likelihood you miss out on other perspectives. Collectively though we cannot only compensate those individual myopias. Under the right circumstances, we can integrate them so that the whole becomes truly greater than the sum of its parts. These circumstances heavily influence the degree of collective intelligence. There are four boundary conditions necessary for the governance of communication communities to make collective intelligence even possible (Brink & Rohrmann, 2012). They are: diversity, transparency, autonomy, and feedback (Rohrmann, 2015).

Diversity is the initial richness of knowledge, experiences, skills, and intentions any crowd builds its wisdom upon. This heterogenous and multidisciplinary variety allows for better outcomes to be produced than for example small, homogenous expert groups would be able to. While there are limits to the practicability of extreme social diversity (e.g. not having a shared language everybody understands) or moral diversity (e.g. conflicting core values), a vast cognitive diversity makes groups more creative, results more robust, and implementations less prone to resistance. In an early example from 1906, Sir Francis Galton could demonstrate that with enough diversity of the participants, the collective outcome of a guess-the-weight-competition is extremely smart. The calculated average of all guesses, despite some people being way of the mark, came very close to the actual weight (Surowiecki, 2005: XII). The art is to foster as much diversity as a group can handle, allowing for as many participants with as many opinions as possible to be involved without jeopardizing the whole process. That way, diversity becomes not only fertile grounds for the discourse, it also helps to get closer to the ideal communication community in terms of openness and involvement.

The need for transparency is both procedural and with regards to content. The rules for the discussion, in terms of participating, contributing, voting, decision-making, and other formal features must be clear for everybody to encourage rational, honest conversations and avoid strategic dodges. On the other hand, there is a great need for material transparency on available information, common goals, necessary outcomes, achieved agreements, current status, latest developments, etc. Otherwise, discourses are likely to run in circles, lead nowhere, or simply produce deficient results. Transparency has also a strong corrective power.

When people know that other people know, discussions and behaviors tend to be more civilized. Selected application of anonymity, e.g. for final votings or to protect whistle blowers, are necessary. Exchanges in certain internet forums bear disturbing witness to individual and collective wrong-doing when there are inappropriate opportunities to hide one's actions in the anonymity of the world wide web.

Autonomy is recognized as another essential characteristic of a collectively intelligent communication community. Without autonomy, every group is a smart as a single person, which is often the most expensive opinion in the room. With individual autonomy, everyone is called upon to form and contribute their own opinion, to think and act independently according to their convictions within the framework of the discourse. Only based on such true disclosure by the participants, potential outcomes will allow for reasonable consensus, as requested by ground norm N 2. This requires a genuine expansion of empowerment and enablement of the group members. In more detail, this entails personal responsibility and accountability, freedom from coercion and manipulation, the ability to contribute or abstain, empathy and mutual support, recognition, and acknowledgement. Altogether theses aspects form the frame within real individual autonomy of thought, emotion, and action arises.

The last ingredient necessary to leverage and grow the collective intelligence of a group is feedback. We have already discussed the relevance of interconnectedness for being smart, whether with regards to connected neurons in an individual's brain or to people as a social brain. The actual connectedness depends on both, the number of connections and the amount of exchange. Like in a CPU, the processing power of a community correlates strongly with the frequency of communication, the buzz. Since this is almost always bidirectional, as people usually have dialogic conversations, we have feedback loops in place. Ideas are iterated, dispositions evaluated, opinions shared, acclamation or disgust voiced. This is all feedback in a wider sense. Without it there is no learning and development, neither individually nor collectively. Some feedback loops have to be in real-time to be of added informative value, like our sense of balance. Even a 5 s delay might put us to the ground, when riding a bicycle for example. Other feedback cycles can be longer, like scientific studies. When it comes to groups, interpersonal feedback (this is the narrow sense, how we typically refer to it), plays an even more important part. It creates not only cognitive but limbic resonance. Relationships are strengthened or weakened when praise or critique are handed out. We know that one of the most severe punishments for children is giving them the silent treatment or for adult convicts it is solitary confinement. When researchers from Google's Project Aristotle were searching for the most important factor determining high-performing teams they clearly settled for psychological safety (Delizonna, 2017). And without feedback people feel lost and plain unsafe. Feedback does not always have to be positive. Every exchange is better than nothing. Eventually, feedback holds groups together, lets them grow emotionally and rationally, and makes them smarter.

11 Moonshot Stories

"We choose to go to the moon", John F. Kennedy's famous words still resonate today and have inspired startups and large organizations alike to dream big and bigger. In innovation management this is often referred to as moonshot thinking: daring to set and aim for a target that seems so far, so new, so courageous it has even a sneaking suspicion of impossibility. We firmly believe, thinking alone is not enough in the case of AI. We plead for a moonshot story yet to be created, to be shared and discussed, to be improved every time we tell it. With all our communication technologies at hand, we can reach nearly the whole population of this planet. We have social and classical media, internet forums and data banks. We can do large-scale crowdsourcing and dare more direct democracy. We can vote, poll, align and thus positively fueling any political debate. And we have the privilege to cultivate our culture like never before in the history of mankind to make the story even better, fuller, more complete. Science-fiction mingles with science-facts. Technological matters get enriched by literature and art.

Putting the pieces together, there is no reason why we should not be able to create a narrative that will actually guide us to the future, beyond the horizon and even to new worlds. The philosophical compass is not perfect and will never be complete. It can only give direction and is not free of contradictions. However, using and trusting our specie's incredible collective intelligence, plus adding a decent amount of joy-based optimism, it seems very reasonable that we really end up with the story we want, with a why that we can collectively justify, and with a desirable AI in favor for all of humanity.

References

Apel, K.-O. (1973). Transformation der Philosophie. Frankfurt am Main: Suhrkamp.

Apel, K.-O. (1992). Diskursethik vor der Problematik von Recht und Politik. In: Apel, K.-O., Kettner, M. (ed.), Zur Anwendung der Diskursethik in Politik, Recht und Wissenschaft. Frankfurt: Suhrkamp, 29–61.

Asimov, I. (1950). "Runaround". I, Robot (The Isaac Asimov Collection ed.). New York City: Doubleday.
Brink, A., & Rohrmann, D. (2012). Swarm Governance: Der Schwarm als moralisches Regime zur Koordination ökonomischer Transaktionen in Organisationen. In: Priddat, B. P. (ed.), Diversität, Steuerung, Netzwerke: Institutionenökonomische Ausweitungen. Marburg: Metropolis-Verlag, 37–52.
Brynjolfsson, E., & McAfee, A. (2014). The Second Machine Age: Work, Progress, and Prosperity in a Time of Brilliant Technologies. New York City: W. W. Norton & Company.
Bass, D. (2023). Microsoft Invests $10 Billion in ChatGPT Maker OpenAI. Bloomberg (https://www.bloomberg.com/news/articles/2023-01-23/microsoft-makes-multibillion-dollar-investment-in-openai).
Bostrom, N. (2014). Superintelligence: Paths, Dangers, Strategies. Oxford University Press.
Chui, M., Manyika, J., & Miremadi, M. (2016). Where Machines Could Replace Humans—and Where They Can't (Yet). McKinsey Quarterly.
Crawford, K. (2016). Artificial intelligence's white guy problem. The New York Times.
Davenport, T. H., & Ronanki, R. (2018). Artificial intelligence for the real world. Harvard Business Review, 96(1), 108–116.
Dawkins, R. (1976). The Selfish Gene. Oxford University Press.
Delizonna, L. (2017). High-Performing Teams Need Psychological Safety: Here's How to Create It - Six ways to build trust. Harvard Business Review (https://hbr.org/2017/08/high-performing-teams-need-psychological-safety-heres-how-to-create-it).
Dennett, D. C. (2017). From Bacteria to Bach and Back: The Evolution of Minds. New York City: W. W. Norton & Company.
Goodman, B., & Flaxman, S. (2017). European Union regulations on algorithmic decision-making and a "right to explanation". AI Magazine, 38(3), 50–57.
Habermas, J. (1981). The Theory of Communicative Action: Volume 1 - Reason and the Rationalization of Society. Boston: Beacon Press
Habermas, J. (1991). Moral Consciousness and Communicative Action. Cambridge: MIT Press.
Hobbes, T. (1651). Leviathan or the Matter, Forme and Power of a Commonwealth Ecclesiasticall and Civil. London: Andrew Crooke.
Homann, K., & Lütge, C. (2004). Einführung in die Wirtschaftsethik. 3. Auflage., Münster: Lit-Verlag.
Horx, M. (2021). Das Regnose-Prinzip - Willkommen im Futur 2 oder Wie man die Zukunft in sich selbst erkennt (https://www.horx.com/zukunftsforschung/das-regnose-prinzip/).
Jiang, J., Li, P., & Huang, L. (2017). Artificial intelligence in service. Journal of Service Management, 28(2), 227–249.
Kant, I. (1785). Groundwork of the Metaphysics of Morals.
Kuhlmann, W. (1985). Reflexive Letztbegründung. Untersuchungen zur Transzendentalpragmatik. Freiburg/München: Alber.
Manyika, J., Chui, M., Miremadi, M., Bughin, J., George, K., Willmott, P., & Dewhurst, M. (2017). A Future That Works: Automation, Employment, and Productivity. McKinsey Global Institute.
Marx, K. (1867). Das Kapital: Kritik der politischen Ökonomie. Hamburg: Verlag von Otto Meißner.

Mill, J. S. (1861). Utilitarianism. London: Parker, Son, and Bourn.
Rohrmann, D. (2015). Wer steuert den Schwarm? Vier Aspekte von Governance und Führung als Treiber der kollektiven Intelligenz. Forum Wirtschaftsethik, 2/2015, 15–19.
Rousseau, J.-J. (1762). Du Contrat Social, ou Principes du Droit Politique. Amsterdam: Marc-Michel Rey.
Sinek, S. (2009). Start with why: How great leaders inspire everyone to take action. New York, NY: Portfolio.
Smith, A. (1759). The Theory of Moral Sentiments. London: A. Millar.
Smith, A. (1776). An Inquiry into the Nature and Causes of the Wealth of Nations. London: W. Strahan and T. Cadell.
Stuart A. Kauffman, S. & Radin, D. (2023). Quantum aspects of the brain-mind relationship: A hypothesis with supporting evidence, Biosystems, Volume 223.
Surowiecki, J. (2005). The Wisdom of Crowds: Why the Many Are Smarter Than the Few and How Collective Wisdom Shapes Business, Economies, Societies, and Nations. New York City: Anchor.
Topol, E. J. (2019). Deep Medicine: How Artificial Intelligence Can Make Healthcare Human Again. New York City: Basic Books.
Ulrich, P. (1997). Integrative Wirtschaftsethik: Grundlagen einer lebensdienlichen Ökonomie. Bern: Haupt Verlag.

Part VI

Action Theoretical Contributions to Entrepreneurship, Creativity, Decision Competence and Risk Management

Betting on More Entrepreneurship 2.0 and 3.0: What We Can Learn from Reflecting on Startups, Innovation and Artificial Intelligence

Christian Hugo Hoffmann

1 Introduction

The importance of this article, where the concepts of Entrepreneurship 1.0 to 3.0 are formed for the first time to the best of my knowledge, is first of all based on the observation that entrepreneurship and startup careers have faced a hard time. This holds true for the country this author comes from (Germany) and to a lesser extent for the country this author deliberately lives in (Switzerland)—but of course there are also shining counterexamples like Israel and Estonia where entrepreneurship is not only *en vogue*, but also goes hand in hand with social status. There are many reasons for my observation, which primarily applies to my environment, the GSA[1] region. One reason is that most young talents and university graduates prefer to work for the state (which is associated with security and predictability of life circumstances) than to build a career in the "free" economy (where overtime and fluctuations are feared). Put in a provoking way, they would rather administrate and redistribute "public funds" instead of earning what can subsequently be administrated and redistributed via taxes and levies.

[1] GSA stands for Germany, Switzerland and Austria.

C. H. Hoffmann (✉)
Hoffmann Economics, Freienbach, Switzerland
e-mail: christian@hoffmann-economics.com

C. H. Hoffmann (ed.), *Artificial Intelligence, Entrepreneurship and Risk*, Technikzukünfte, Wissenschaft und Gesellschaft / Futures of Technology, Science and Society, https://doi.org/10.1007/978-3-658-45544-6_20

This impression is suggested, for example, by a survey of 2000 students in Germany in 2020 by the consulting firm EY, which was picked up by Handelsblatt (2020).[2]

With my contribution at hand, which draws on a wealth of first-hand experience of entrepreneurship, I would like to advocate a revaluation of entrepreneurship and raise enthusiasm for innovation creation, which startups stand for more than any other form of enterprise. And this purpose, which is pivotal for individuals as well as the economy as a whole, is served from an unusual perspective in what follows, distinguishing between Entrepreneurship 1.0, 2.0, and 3.0. The classification and differentiation into Entrepreneurship 1.0 to 3.0 is ought to be seen as primarily systematic and not temporal. As far as the temporal aspect is concerned, however, it can be stated: I refer to Entrepreneurship 1.0 only in terms of modern entrepreneurship since about the turn of the millennium, because my investigation is not intended to be a historical reappraisal and certainly a comparison with earlier corporations (from the seventeenth or eighteenth century, for example) and their structure would be distorted—not to mention the hitherto unmentioned systemic upheaval stemming from the introduction of limited companies in the middle of the last millennium (Amend-Traut, 2013).

What happened with the turn of the millennium? Perhaps somewhat simplified: Anyone who entered the working world prior to the year 2000 knows and was familiarized with the so-called *SSEE* world: the acronym stands for *Stable, Secure, Easy, Explicit*. Then, overnight, came the digitalized *VUCA* world which has presented itself as *Volatile, Uncertain, Complex, Ambiguous* (Bennett & Lemoine, 2014). Bye-bye SSEE, hello VUCA. Let's dive in and go in medias res.

2 What is Entrepreneurship Resulting in Entrepreneurship 1.0?

Entrepreneurs embody the creative power of free markets and ensure the prosperity of societies (Hoffmann, 2015). Surprisingly, however, economists largely ignore the role of the entrepreneur, the concept is often insignificant and marginalized in this community. The economics textbooks usually do not contain any entry

[2] Or think of another telling example from Germany, which is even more recent, not directly about entrepreneurial spirits, but its prerequisites, namely the willingness to take autonomy and personal responsibility. As part of a Randstad study (2022) on rising living costs, particularly with regard to energy costs, 60% of German employees deem Father State as responsible for mitigating the consequences of inflation, only one in four takes responsibility for themselves.

on the entrepreneur and his role for the economy (Baumol, 2010; Brickley et al., 2008; Besanko et al., 2007). The explanation for this can be found in the neoclassical focus on equilibria and perfect competition. Even leading representatives of the Austrian School of Economics, such as Friedrich August von Hayek and Israel M. Kirzner, conceive of the entrepreneur primarily as the resourceful arbitrageur who, guided by the invisible hand of the market, discovers and reduces price differentials.

In a refreshing departure from this tradition, Austrian economist Joseph A. Schumpeter (1908, 1939) coined the most pronounced concept of the entrepreneur to date (Frambach et al., 2018). In his understanding, the entrepreneur does not bring the economy into a state of equilibrium, but leads it out of a static state. (S)he is the creative-heroic figure who successfully harnesses technical inventions and permanently changes and advances the economy and society by creating new products, designing novel methods, or opening up uncharted markets. The Schumpeterian entrepreneur is, therefore, an innovator or a creative destroyer of the status quo, who is able to convince people (customers) of a new offer. (S)he gains a monopoly profit from these activities before imitators catch up, ultimately resulting in a market which enters a phase of stagnation again. Schumpeter's concept of the entrepreneur is refreshing because it places the entrepreneur at the center of market activity as a formative phenomenon and does not see her/him merely in functional terms, as a decision-maker and risk-bearer, or marginally as a basically interchangeable assistant in fulfilling consumer wishes. The typification of the entrepreneur as an active, individual shaper and creator who disrupts markets is most likely to correspond to the self-image of many entrepreneurs today and it is captured in my notion of Entrepreneurship 1.0. In other words, Schumpeter's model of the entrepreneur can be grasped as the core essence of entrepreneurship and, given the segmentation that follows in the next sections, be referred to specify Entrepreneurship 1.0.

At the same time, one might wish to draw Schumpeter's attention to the question of whether the titanic or field-lord-like significance of the entrepreneur is not too one-sided or biased and, thus, fosters a perhaps harmful cult of personality[3] which disregards the complex interplay of supply and demand and the prerequisites of a free market economy and its protagonists (plural). Before I return to this

---

[3] A beautiful exemplification of this also contemporary risk is portrayed by Brown and Farrell (2021) who sketch the cult of We, i.e., of the WeWork company, its eccentric founder and former CEO Adam Neumann, and the Great Start-Up Delusion.

question and elaborate on it in terms of the value of ecosystems for Entrepreneurship 2.0, a few comments on the classic entrepreneur's (1.0) risk profile are in order.

To put it in a nutshell: The essential dependence of classically conceived entrepreneurship on risk-taking already became clear elsewhere: "The entrepreneur uses his freedom by taking risks. He (and of course she) makes assumptions, speculates—and then takes action based on these assumptions. The entrepreneur is thus in a state of constant uncertainty and takes risks. Uncertainty is a prerequisite of entrepreneurship [...]." (Hoffmann, 2015: 8).[4] An entrepreneur or startup founder bears risks, for example, by using personal wealth to establish and register a company (sometimes called *having skin in the game*) or by testing to what extent a given combination of heterogeneous capital factors such as investors' money or human resources (the right team is king!) is (eventually) profitable or not. This tight bond between risk and entrepreneurship is being softened to some extent in the 2.0 era—yet, not too much, of course, and only comparatively; because no decision-making problems, whether in entrepreneurship or in everyday life or anywhere else, can do without risk management. You simply cannot not take risks.

Apart from the Schumpeterian entrepreneur figure as the essence of modeling Entrepreneurship 1.0 as well as both the close and necessary tie between being an Entrepreneur 1.0 and taking substantial risks with founding and running a venture, what other features can be detected to characterize Entrepreneurship 1.0, especially vis-à-vis the Entrepreneurship 2.0/3.0:

- In Entrepreneurship 1.0, focus is on the individual entrepreneur or a small group of entrepreneurs (the founding or core team) and the entrepreneur is an innovator, a source of new ideas, goods, services, and business/or procedures.
- An entrepreneur of 1.0 is a human individual who creates and invests in one or more businesses, but focuses on one business at a time, bearing most of the risks and enjoying a decent portion of the rewards.[5]

[4] The quote is originally in German: "Der Unternehmer nutzt seine Freiheit, indem er sich auf Wagnisse einlässt. Er (und selbstverständlich auch sie) trifft Annahmen, spekuliert – und unternimmt dann etwas basierend auf diesen Annahmen. Der Unternehmer befindet sich also in einem Stadium ständiger Unsicherheit und nimmt Risiken auf sich. Unsicherheit ist eine Voraussetzung des Unternehmertums [...]."

[5] Admittedly, this wording is vague and suffers from being under-determined. However, on this level of generality, we cannot make more precise statements. With this in mind, it is fair to state though that the entrepreneur takes the highest risks because (s)he puts money on the table to get the company started *and, on top of that*, depends on a thriving startup to secure an

- A startup of Entrepreneurship 1.0 is a very risky and for some time unprofitable endeavor which is seeking profits, but requires that additional capital is injected by new, external investors in return for shares of the startup. A startup is, at least initially, not creditworthy, i.e., is not qualified to take loans from a bank.
- Raison d'être: A startup of Entrepreneurship 1.0 is set up to translate new technologies (such as generative AI nowadays) or inventions into innovations coupled with the prospect of disrupting a market. Or a startup of Entrepreneurship 1.0 is set up for designing and applying a new business model to disrupt a market.

A well-known example of Entrepreneurship 1.0 is *Airbnb*, operating an online marketplace for short-term homestays and experiences. Airbnb has been credited with revolutionizing or disrupting the tourism industry, and, by now, transformed into an established company with an impressive valuation and a sound, validated and working business, acting as a broker and charging a commission for each booking on the Airbnb platform. This was not foreseeable when the company was founded in 2008 by Brian Chesky, Nathan Blecharczyk, and Joe Gebbia. Because few would have thought back then that many people in the USA or throughout the world would open the doors of their private home to strangers to let them stay for one or more nights. And even less would have betted on a multi-billion business based on that hypothesis which is why the founding team also faced a difficult time first raising funds from investors to whom they could initially only present low user numbers, and thus an unprofitable business on paper. Even though one could contend that some substantial tech is involved with the platform/website (especially given the time of 2008) to explain Airbnb's success, I would rather highlight the new business model which even inspired (together with Amazon and a few others) a whole new segment of economics, namely the so-called *platform economy* (Parker et al., 2017).[6]

income, thus a *concentration of risk*. On the other hand, the rewards from successful venture-building vary drastically, e.g., depending on if an *exit* and what kind of *exit* is realized or on how many and what type of external investors were onboarded prior to the exit, etc. Cf. also Hoffmann (2023a) for more information.

[6] After reading the third section of this chapter, the reader might actually be surprised to find the pioneers and paragons of the platform economy subsumed under Entrepreneurship 1.0 in lieu of Entrepreneurship 2.0. For instance, is it really compelling to ascribe Amazon with its marketplace which is open to and used by third parties to Entrepreneurship 1.0 since it looks more like an ecosystem? Or is it not a platform as the ones operated by the representatives of the *platform* economy which also often counts as a central constituent of

3 What is Entrepreneurship 2.0 and Why Do I Reckon with It?

I argue in this section that we are on the brink of a new age of entrepreneurship, what I subsequently call Entrepreneurship 2.0, where the distribution of risk is reversed which happens in a twofold manner: Firstly, some entrepreneurs nowadays are able to take significantly fewer risks than their predecessors of the era 1.0. How? In a nutshell, entrepreneurs 2.0 essentially *ecosystemize* their business (cf. Kawohl et al., 2022). Secondly, the distribution of risk is also reversed when we enter Entrepreneurship 2.0 because it is employees of precisely such young, micro and small companies who acquire skills and experiences that boost their careers and make them more robust than it would be the case with bureaucratic monsters in the public or private sector.

3.1 The drivers of Entrepreneurship 2.0

The point about ecosystemizing a business for entrepreneurs is that taking advantage of ecosystem opportunities will be essential for ongoing innovation and succeeding on the markets of the future (Kawohl et al., 2022). I hypothesize that the future lies in "ecosystemized" businesses strategies. In this context, a new phenomenon is increasingly gaining ground: the business ecosystem. This is an association of (typically three to about ten) companies, including contractors, which are aligned by a central company, the orchestrator, to a common value proposition, which on the one hand is more than just the sum of the individual contributions and on the other hand could not be delivered by any single company alone (Hoffmann, 2023b). In public, the term "ecosystem" is often used inappropriately and almost inflationarily, which then leads to a lack of understanding of the implications of this new important topic for the economy.

I see two distinct drivers of ecosystems. One has already been mentioned in the introduction above, namely the transition to a VUCA world a few decades ago. VUCA applies to an environment which defies confident diagnosis and befuddles executives (Bennett & Lemoine, 2014). Table 1 summarizes the main take-aways from their work on this phenomenon and the link to ecosystems.

---

an ecosystem à l'Entrepreneurship 2.0? These are legitimate questions and deserve to be debated controversially by future research. For now, one of my answers would be that players such as Amazon do not possess the property of co-dependency unlike the players of ecosystems of Entrepreneurship 2.0 that are truly dependent on each other to deliver a superior value proposition. See below in the main text.

Table 1 Distinctions within the VUCA framework and connections with ecosystems. Cf. also Bennett & Lemoine, 2014

	What it is	An example	How to effectively address it
Volatility	Relatively unstable change; information is available and the situation is understandable, but change is frequent and sometimes unpredictable	Commodity pricing is often quite volatile; jet fuel costs, for instance, have been quite volatile in the twenty-first century	*Agility*, a chief feature of startups, is key to coping with volatility. Resources should be aggressively directed toward building slack and creating the potential for future flexibility. *Ecosystems* allow for high flexibility through the partner network of startups, incumbents, contractors, etc. coming along
Uncertainty	A lack of knowledge as to whether an event will have meaningful ramifications; cause and effect are understood, but it is unknown if an event will create significant change	Anti-terrorism initiatives are generally plagued with uncertainty; we understand many causes of terrorism, but not exactly when and how they could spur attacks	*Information* is critical to reducing uncertainty. Firms should move beyond existing information sources to both gather new data and consider it from new perspectives. New sources and perspectives are opened up and made accessible through *ecosystems* where diverse players (beyond certain industries) are integrated

(continued)

Table 1 (continued)

	What it is	An example	How to effectively address it
Complexity	Many interconnected parts forming an elaborate network of information and procedures; often multiform and convoluted, but not necessarily involving change	Moving into foreign markets is frequently complex; doing business in new countries often involves navigating a complex web of tariffs, laws, regulations, and logistics issues	*Restructuring* internal company operations to match the external complexity is the most effective and efficient way to address it. Firms should attempt to 'match' their own operations and processes to mirror environmental complexities. This goes back to Ashby's law for absorbing complexity: "Only variety can destroy [absorb, C.H.] variety" (cf. Ashby, 1956, and Schwaninger, 2009: 14). *Ecosystems* help with building those capabilities and complexities
Ambiguity	A lack of knowledge as to 'the basic rules of the game'; cause and effect are not understood and there is no precedent for making predictions as to what to expect	The transition from print to digital media has been very ambiguous; companies are still learning how customers will access and experience data and entertainment given new technologies	*Experimentation*, which is in the DNA of startups, is necessary for reducing ambiguity. Only through intelligent experimentation can firm leaders determine what strategies are and are not beneficial in situations where the former rules of business no longer apply. Particularly, hypotheses and value propositions can be tested in the market by means of *ecosystems* where the risks of testing are partially outsourced since more than one company forms an ecosystem

The second driver of ecosystems and of Entrepreneurship 2.0 hinges on the digital transformation. While the digital transformation itself is invoked (too) often, a corollary essential to the emergence of both ecosystems and Entrepreneurship 2.0 is the subsequent reduction of transaction costs to near zero, which is rendering corporate boundaries more traversable and which is discussed at length in Jeremy Rifkin's thought-provoking book "The Zero Marginal Cost Society" (Rifkin, 2014).

According to a study by McKinsey (Atluri et al., 2017), it is estimated that in 2025, of the expected more than $190 trillion in global business revenue, nearly 30% will be redistributed across today's industry boundaries. The study expects a consolidation of today's several hundred sectors and industries into then newly twelve overarching complexes of customer requirements, such as travel and hospitality, mobility, or education. The extent to which the industries will aggregate in the long-term and the exact magnitude of this aggregation would appear to be a bold speculation at this point in time (Hoffmann, 2023b). However, it is certain that industries will undergo massive change by the time they become obsolete and will be restructured along the customer experience/processes at a higher level of aggregation. This new level of aggregation is likely to be an ecosystem as briefly outlined here.

The superior value proposition in such ecosystems—superior because by definition an ecosystem is more than the sum of the contributions of the individual organizations involved therein—can basically take two forms: Either the ecosystem partners focus together on a specific customer need, which they fulfill in a way that has not yet been achieved. A nice example in this regard are the 3D printed soles of the start-up "Tailored Fits" from Horw, Switzerland (Lingens & Gassmann, 2018). Together with its ecosystem partners, the startup offers soles that are 3D printed to personally fit the customer's foot, making them superior to traditionally manufactured soles in terms of fit and function. Or, the other variant of a superior value proposition through ecosystems is the provision of a common bundle of services that covers, for example, an entire customer journey (Lingens & Gassmann, 2018). The House of Lab Science AG, cofounded by this author in 2021 who also designed/set up the business model and architecture, can serve as an example. In the center of its operations is the value proposition "Lab-as-a-Service" positioned. House of Lab Science focuses on both life sciences startups' and more established companies' need for labs and orchestrates a whole range of services from various providers (for an overview of the currently more than 60 services, see this page), from cleaning to concierge service, to biowaste disposal specialists—all of which is hybrid in the customer's sense: physical advice and execution combined with digital presence and self-service on the proprietary House of Lab Science platform. As with Tailored Fits, the individual services are provided by specialized partners, but they are closely orchestrated and coordinated by House of Lab Science. For more insights into the House of Lab Science, cf. also Hoffmann (2022a).

3.2 The new risk profiles

The distribution of risk is also reversed when we enter Entrepreneurship 2.0 because it is employees of precisely such young, micro and small companies who acquire skills and experiences that boost their careers and make them more robust than it would be the case with bureaucratic monsters in the public[7] or private sector. The latter are places where many jobs are cut in the face of ultimately home-made management mistakes (Finews, 2023; NZZ, 2023a; NZZ, 2023b; Handelszeitung, 2023; Handelsblatt, 2022).

A startup of Entrepreneurship 2.0 is and possibly prefers to remain *proudly small*, e.g., in contrast to some former startups of the era 1.0 which turned into giants themselves such as Airbnb. In this light, we might start wondering what the optimum size of a corporation is. As I also explored elsewhere (Hoffmann, 2019), big business is slow, bureaucratic and hierarchic (and this is probably also a reason why entrepreneurs chose to actually become entrepreneurs): Administrative and management teams in big businesses monitor and coordinate production, and since such administrators held crucial knowledge necessary to operate the business, they were usually paid a premium above what their skills would have commanded in the spot market. In the Information and Networking Age of 2.0, most tasks that were formerly captured within large firms as an expedient to reduce information and transaction costs (Ronald Coase), will be taken over by much smaller entities, (small) startups, or even migrate back to the spot market.

Why do corporations and large companies in particular stand in the pillory? Because they have too much comfort and live off the successes they once achieved a long time ago—in BCG language: they milk their *cash cows* (BCG,

---

[7] In terms of the public sector, the situation or diagnosis is a bit more complicated to be precise. For example, in Germany, the fonctionnaires' employment cannot be terminated by the employer (unless the civil servants become criminal or break the law). However, perhaps employees in the public sector start feeling unfulfilled by their work, which can be insufficiently stimulating or too repetitive, etc. (e.g., Wynen, 2019), so they begin to look for alternatives in the private sector, only to realize that their previously gained experiences and qualifications from the current position are in little demand. Cf. Business Insider (2023), and Haufe (2022).

2023; cf. also Christensen, 1997). Why are they structurally disadvantaged in innovation creation and, thus, in meeting the requirements of tomorrow's world? On the one hand, because a large tanker cannot move as agilely and quickly as a speed boat startup. On the other hand, because their existing core business makes them path-dependent (*legacy costs*).

There are other, potentially less known and studied factors that come to play when we consider and analyze the risks which entrepreneurs 2.0 and their team members bear vis-à-vis the (larger) risks of a career in a big corporation. Let's look at and discuss two examples:

- Large corporations are less open to forms of *New Pay* than startups: Let's undertake a little thought experiment: Suppose you are invited to come up with a pay system for the entire world—on the condition that you do not know which pay group you yourself belong to. What would you propose? This figure of thought is called the "veil of ignorance," coined by the philosopher John Rawls. The "veil" refers to the ignorance of where you would stand in a hierarchy. The lesson from this thought experiment is that if we do not know whether we will have advantages, we prefer a just system. In fact, however, when it comes to pay, the willingness to embrace more equitable models ends in most established, and thus hierarchical, companies. While startups already lead the way with *New Work* (Zukunftsinstitut, 2023), mature companies also fall behind in terms of *New Pay*, i.e., a pay structure that everyone thinks is fair[8] vs. *old pay*: pay is determined in individual negotiations (Krogerus & Tschäppeler, 2022: 28). New Work and New Pay shape tomorrow's workplaces, and those who shape or participate in or contribute to those antecedents today will probably have a better standing tomorrow.
- Apropos New Work and organizational theory: For a long time, it was believed that the pyramid shape was the best structure for organizing people and work. That is not necessarily wrong: clear hierarchies provide order and stability. But there are increasing signs that this type of organization no longer works (in the VUCA world, see above): Employees are exhausted—and so are managers. There exist many reasons for this exhaustion. One of them is the fact that the pyramid structure does not do justice to the *complexity* of our world today. So says Frederic Laloux, author of the seminal book "Reinventing Organizations" (Laloux, 2014). And he's not the only one. There is an alternative:

[8] This author is a liberal by heart as regular publishing activities suggest (e.g., Hoffmann, 2020). And even though the specification of New Pay I propose in this paper as a pay structure that everyone thinks is fair probably alerts many liberals, I am convinced that this specification is fundamentally compatible with liberalism.

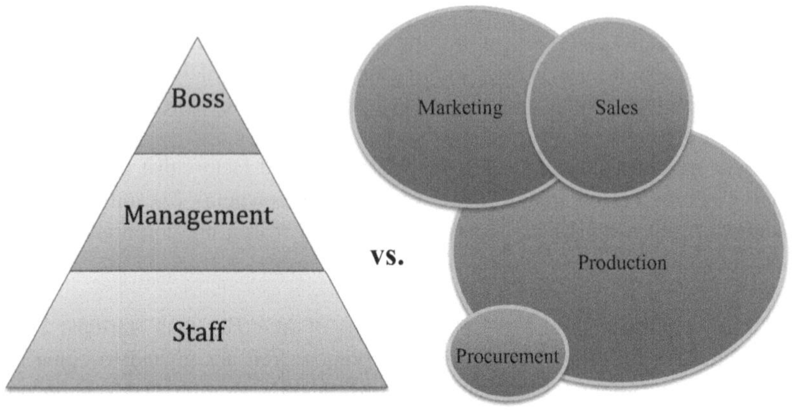

Fig. 1 Round is the new angular. Left: Hierarchic structure of mature companies, vs. Right: Team members organize themselves in groups that represent different functions such as sales and marketing and cooperate to make the best work for the company

organize people not in a pyramid, but in a circle (Krogerus & Tschäppeler, 2022: 160f.). Overlapping circles, as displayed in Fig. 1, can then even count as a panacea for fighting silo thinking in organizations. How does this question about the organizational structure of companies impact entrepreneurs' and employees' risk position? My hypothesis is in this regard that while people in big corporations try to primarily manage and meet the expectations of their "line manager",[9] which possibly leaves them insensitive to what is needed (also from their own side) to keep the company competitive, to make it fit for the future, etc., entrepreneurs and team members of small startups are either at the forefront of innovation, sales and coping with business challenges or at least very close to it. Like this, entrepreneurs 2.0 and their teams acquire many more business skills and valuable experiences on how markets and venture-building works, which in turn are qualifications and distinctions that are very lucrative to most employers.

If we delineate Entrepreneurship 2.0 based on these findings from 3.1 and 3.2 and if we pay special attention to what differentiates it from Entrepreneurship 1.0, we can arrive at the following characteristics:

[9] In the simplified Fig. 1, the line manager of a member of staff is one or more managers, for a manager it is the boss, and for the boss it would be the Board of Directors.

- In Entrepreneurship 2.0, focus is *not* on the individual entrepreneur or a small group of entrepreneurs, but on the *ecosystem* which is constituted, among others, by one or more startups and which brings forward a superior value proposition.
- An entrepreneur of 2.0 is a human individual who creates and invests in one or more businesses, not necessarily focuses on one business at a time,[10] bearing some but not necessarily most of the risks as the risks are distributed throughout the elements of the ecosystem.
- A startup of Entrepreneurship 2.0 is a less risky venture.
- Raison d'être: A startup of Entrepreneurship 2.0 is more purpose-driven and addresses overarching complexes of customer requirements by contributing to and, possibly orchestrating an ecosystem which, in turn, serves a superior value proposition. A certain disruptive technology or fancy business model of a single company within an ecosystem are less in the foreground.

To conclude, the risks that entrepreneurs bear in the age of Entrepreneurship 2.0 are comparatively and drastically lower in two directions: in juxtaposition with Entrepreneurship 1.0, and in juxtaposition with careers in "bureaucratic monsters" in the public or private sector. The title of this chapter in the volume at hand, according to which professionals, graduates and students ought to bet on more entrepreneurship, is thereby critically questioned and challenged. How large is the risk really?

4 Outlook: Entrepreneurship 3.0 and Its Drivers

A second systemic and even more striking turning and tipping point is marked by the transition from Entrepreneurship 2.0 to 3.0. How so? The third evolutionary stage in entrepreneurship, which I can only sketch superficially in this paper, will no longer be dominated by human players; instead, machine players will be in the mix.

Together with my colleague Dr. André Dahlinger from the Bosch IoT Lab at the University of St. Gallen (HSG), I submitted an article to the journal *Foresight* in 2019 dedicated to illuminating socio-economic implications of decentralized, autonomous, and self-owning organizations (DAOs). That is, pointedly, there are no longer human shareholders behind such enterprises. The case study from that

[10] Systemize yourself, define your systems' boundaries and decide for yourself to what different ecosystems you would like to function as a subsystem to.

Table 2 Analysis of scenarios driven by two underlying technologies: AI and blockchain. Cf. also Hoffmann & Dahlinger (2020: 57)

Maturity level of the technology	Weak AIs (i.e., below level-5 autonomy)	Strong AIs (i.e., level-5 autonomy of cars)
Basic blockchain solutions	Nowadays taxis	Smart taxis that do not own themselves
Sophisticated blockchain solutions	Nowadays taxis	DAB taxis

paper (Hoffmann & Dahlinger, 2020), discussed using the example of the cab market, receives its boost from the combination of two fundamental technologies: artificial intelligence (AI), specifically autonomous driving, with blockchain technology, specifically smart contracts. The interplay between these two fundamental technologies is picked up in Table 2. In the cab case, a cab would be virtually like a one-man cab operation, but handling all business operations autonomously: driving, payment, maintenance, etc. Moreover, the "brain" that runs the cab business could run on a decentralized network.

What would be the consequences of the cab case (cf. also Hoffmann, 2022b)? Firstly, it can be deduced that decentralized, autonomous, and self-owned cab companies would sooner or later become the dominant provider of cab rides in a free market (not necessarily a monopoly, let alone an undesirable one) because they would place their service at a lower price (e.g., no cost for a driver, no cost for human management, no profit distribution or retention necessary, etc.). What does this add up to? The paper (Hoffmann & Dahlinger, 2020) argues that in a "robo-economic system," i.e., an economic system where many decentralized, autonomous, and self-owning firms (not just for the cab market) emerge and gain competitive advantage, the return on capital would be much lower on average, simply because much less is owned on average, without severe effects on overall wealth creation, since the principles of free markets would be left untouched. Unlike current forms of capitalism, where special groups (e.g., capitalists, lobbyists) often secure advantages, i.e., market power (Posner & Weyl, 2018), surpluses in the robo-economic system would be virtually distributed among users. Would this not amount to a radicalization of markets with socially desirable outcomes?

Based on the cab and the cab market example, which Hoffmann & Dahlinger (2020) have already illuminated, further applications of Entrepreneurship 3.0 and the limits of its applicability must be examined in terms of avenues for future research. Socio-economic classifications or the presentation of possible societal impacts would also need to be further controverted. One step back: how large is the "risk" for Entrepreneurship 3.0 in the first place? When contemplating scenarios, the question of the (occurrence) probability of these scenarios must not be neglected. A third and final proposal for future research is, therefore, devoted to this question, primarily scrutinizing what AI is and what hurdles still need to be overcome before the technology would qualify for Entrepreneurship 3.0. For these very parameters for determining the probability of occurrence, I can refer the interested reader to my latest De Gruyter book which covers precisely these points (Hoffmann, 2022c).

For now. if we depict Entrepreneurship 3.0 based on this very short outline and if we pay special attention to what differentiates it from Entrepreneurship 2.0, we can arrive at the following characteristics:

- In Entrepreneurship 3.0, focus is *not* on the individual entrepreneur, not on the *ecosystem*, but on customer satisfaction as well as technological feasibility coupled with the economics of technology.
- Entrepreneurship 3.0 is no longer marked by either human individuals or organizations owned by humans, yet by DAOs.
- A startup of Entrepreneurship 3.0 is a risk-free venture because in terms of DAOs there are no longer any human shareholders or entrepreneurs involved. This does, of course, not imply that there would not be any risks for other stakeholders such as customers. (For instance, the chances that even a highly developed AI in terms of an autonomous vehicle can have an accident is certainly not zero. Therefore, a customer who is taking a DAO cab should think of and respond to that risk.)
- Raison d'être: A startup of Entrepreneurship 3.0, a DAO, would free humans from work that machines can do by themselves and at better conditions (e.g., better pricing). There is a general debate going on whether automation and digitalization will lead to a massive loss of employment, because machines will replace humans in many fields of work (McClelland, 2023; Forbes, 2020). Hence, many people fear to be left behind because of this debate and they might fear that inequality would increase tremendously as only those who own the machines would benefit directly from the profit share. In Hoffmann & Dahlinger's (2020) scenario, however, the profits are mostly distributed among the consumers via reduced prices.

5 Take-Aways for the Reader at a Single Glance

- Entrepreneurship 1.0 is derived from the Schumpeterian model of the entrepreneur and the entrepreneur as an individual in this category is essentially bearing high risks.
- Entrepreneurship 2.0 is marked by a reversal of comparative risks: Entrepreneurs of the class 2.0 take both significantly less risks than those of the class of 1.0 through ecosystemizing their business and significantly less risks compared to employees in large corporations or public administrations.
- There is a call to action for Entrepreneurs 2.0: Systemize yourself, define your systems' boundaries and decide for yourself to what different ecosystems you would like to function as a subsystem to.
- The near-term future lies in "ecosystemized" businesses strategies.
- Entrepreneurship 2.0 prevails in a VUCA world and is enabled by the reduction of transaction costs to near zero in the wake of the digital transformation, which is rendering corporate boundaries more obsolete.
- New Work and New Pay shape tomorrow's workplaces, and those who shape or participate in or contribute to those antecedents today will probably have a better standing tomorrow.
- Whereas employees of large corporations try to primarily manage and meet the expectations of their "line manager", entrepreneurs and their team members are at the forefront of innovation, sales and coping with business challenges. In this way, entrepreneurs 2.0 and their teams acquire many more business skills and valuable experiences on how markets and venture-building works, which in turn are qualifications and distinctions that are very lucrative to most employers today and tomorrow.
- In Entrepreneurship 3.0, focus is *not* on the individual entrepreneur, not on the *ecosystem*, but on customer satisfaction as well as AI readiness coupled with the simple economics of AI (Agrawal et al., 2018; Hoffmann, 2018).
- Entrepreneurship 3.0 is no longer marked by either human individuals or organizations owned by humans, yet by DAOs.

References

Agrawal, A., Gans, J., & Goldfarb, A. 2018. Prediction machines. The simple economics of Artificial Intelligence, Boston: Harvard Business Review Press.

Amend-Traut, A. 2013. Der Reichshofrat und die Kapitalgesellschaften. Die Bemühungen um eine Handelskompanie zwischen den Hansestädten und Spanien. In. A. Amend-Traut, A. Cordes & W. Sellert (Eds.). Geld, Handel, Wirtschaft: Höchste Gerichte im Alten Reich als Spruchkörper und Institution. Berlin, Boston: De Gruyter: 61–90.

Ashby, W.R. 1956. An Introduction to Cybernetics. London: John Wiley & Sons.

Atluri, V., Dietz, M., & Henke, N. 2017. Competing in a World of Sectors Without Borders. McKinsey Quarterly, 3-2017. https://www.mckinsey.com/business-functions/mckinsey-analytics/our-insights/competing-in-a-world-of-sectors-without-borders (14-03-23).

Baumol, W.J. 2010. The Microtheory of Innovative Entrepreneurship. New Jersey: Princeton University Press.

BCG, 2023. What Is the Growth Share Matrix? Available at: https://www.bcg.com/about/overview/our-history/growth-share-matrix (13-03-23).

Bennett, N. & Lemoine, G.J. 2014. What a difference a word makes: Understanding threats to performance in a VUCA world. Business Horizons, 57: 311–317.

Besanko, D., Dranove, D., Shanley, M., & Schaefer, S. 2007. Economics of Strategy. Hoboken: Wiley.

Brickley, J., Zimmermann, J., & Smith, C.W. 2008. Managerial Economics & Organizational Architecture (5th edition). Boston: McGraw-Hill/Irwin.

Brown, E., & Farrell, M. 2021. The Cult of We: WeWork and the Great Start-Up Delusion. London: Penguin Random.

Business Insider, 2023. Ihr interessiert euch für den Öffentlichen Dienst? Ja, er hat viele Vorteile—aber diese Nachteile solltet ihr auch kennen. 18.01.2023. Available at: https://www.businessinsider.de/karriere/oeffentlicher-dienst-diese-nachteile-solltet-ihr-kennen-c/ (14-03-23).

Christensen, C. 1997. The Innovator's Dilemma: When New Technologies Cause Great Firms to Fail. Cambridge, MA: Harvard Business Review Press.

Finews, 2023. Chefs der Credit Suisse erstmals ganz ohne Bonus. 14.03.2023. Available at: https://www.finews.ch/news/banken/56209-geschaeftsbericht-cs-ulrich-koerner-axel-lehmann-thomas-gottstein-boni (14-03-23).

Forbes, 2020. U.S. Lost Over 60 Million Jobs—Now Robots, Tech And Artificial Intelligence Will Take Millions More. 27.10.2020. Available at: https://www.forbes.com/sites/jackkelly/2020/10/27/us-lost-over-60-million-jobs-now-robots-tech-and-artificial-intelligence-will-take-millions-more/?sh=2bc4f47c1a52 (14-03-2023).

Frambach, H., Koubek, N., & Kurz, H.D. (Eds.). 2018. Schöpferische Zerstörung und der Wandel des Unternehmertums: Zur Aktualität von Joseph Schumpeter. Marburg: Metropolis.

Handelsblatt, 2022. In Industrie und Handel droht Stellenabbau. 26.10.2022. Available at: https://www.handelsblatt.com/politik/deutschland/ifo-beschaeftigungsbarometer-in-industrie-und-handel-droht-stellenabbau/28770372.html (10-03-23).

Handelsblatt, 2020. Amt statt Autoindustrie—Warum die meisten Nachwuchstalente für den Staat arbeiten wollen. 12.12.2020. Available at: https://www.handelsblatt.com/karriere/traumjob-beamter-amt-statt-autoindustrie-warum-die-meisten-nachwuchstalente-fuer-den-staat-arbeiten-wollen-/26683194.html (10-03-23).

Handelszeitung, 2023. McKinsey plant grössten Stellenabbau in der Firmengeschichte. 22.02.2023. Available at: https://www.handelszeitung.ch/unternehmen/kundigungen-geplant-mckinsey-will-2000-stellen-abbauen-576126 (10-03-23).

Haufe, 2022. 80 Prozent der Beschäftigten im öffentlichen Dienst können sich Wechsel des Arbeitgebers vorstellen. 07.01.2022. Available at: https://www.haufe.de/oeffentlicher-dienst/digitalisierung-transformation/zufriedenheit-und-wechselbereitschaft-im-oeffentlichen-dienst_524786_558938.html (14-03-23).
Hoffmann, C. 2023a. Startup valuation reassessed: Against celebrity, sustainability and state intervention. Journal of Ethics in Entrepreneurship and Technology. Forthcoming. Available at: https://doi.org/10.1108/JEET-08-2022-0012 (13-05-23).
Hoffmann, C. 2023b. Ecosysteme—Treiber der künftigen Schweizer Wirtschaft? Hoffmann Economics. Available at: https://www.linkedin.com/posts/hoffmann-economics_ecosysteme-treiber-der-k%C3%BCnftigen-schweizer-activity-7041139767807275009-YL2J?utm_source=share&utm_medium=member_desktop.
Hoffmann, C. 2022a. Systems thinking and action for an entrepreneurial journey towards the House of Lab Science. In. N. Rezaei & A. Saghazadeh (Eds.). Integrated Science. Volume 7. Thinking: Bioengineering of Science and Art. Springer: 653–669.
Hoffmann, C. 2022b. Mehr Liberalismus 2.0 wagen: Ein Plädoyer für mehr Unternehmertum, Unternehmer und mehr Reflexion über neue Technologien. Forum Freie Gesellschaft. Available at: https://forum-freie-gesellschaft.de/mehr-liberalismus-2-0-wagen/ (14-03-2023).
Hoffmann, C. 2022c. The Quest for a Universal Theory of Intelligence. The Mind, the Machine, and Singularity Hypotheses. Berlin: De Gruyter.
Hoffmann, C. 2020. How to foster entrepreneurship. A liberal's critique of the status quo in Switzerland. Startups.ch & HSG Entrepreneurship Blog. Available at: https://blog.startups.ch/en/how-to-foster-entrepreneurship-a-liberals-critique-of-the-status-quo-in-switzerland/ (14-03-2023).
Hoffmann, C., & Dahlinger, A. 2020. How capitalism abolishes itself in the digital era in favour of robo-economic systems: socio-economic implications of decentralized autonomous self-owned businesses. Foresight, 22: 53–67.
Hoffmann, C. 2019. Free the entrepreneurs. The next stage of human society as the (only) path to prosperity. Swiss Entrepreneurs Magazine, 3: 68–70.
Hoffmann, C. 2018. Die Kunst der Vorhersage. Buchbesprechung «Prediction machines. The simple economics of Artificial Intelligence». Springer Spektrum. Available at: https://www.spektrum.de/rezension/buchkritik-zu-prediction-machines/1603000 (14-03-2023).
Hoffmann, C. 2015. Einleitung. Am Nexus von Freiheit und Wohlstand. In. C. Hoffmann (Ed.). Befreit die Unternehmer! Der (einzige) Weg zum Wohlstand. Zürich: Edition Liberales Institut: 7–15.
Kawohl, J., Krechting, D., & Schilling, R. 2022. Ecosystemize Your Business. How to succeed in the new economy of collaboration. Frankfurt: MVB.
Krogerus, M., & Tschäppeler, R. 2022. Zusammen Arbeiten: Zürich: Kein&Aber.
Laloux, F. 2014. Reinventing Organizations: A Guide to Creating Organizations Inspired by the Next Stage in Human Consciousness. London: Nelson Parker.
Lingens, B., & Gassmann, O. 2018. Das Ende des Branchendenkens. Die Volkswirtschaft. Plattform für Wirtschaftspolitik. https://dievolkswirtschaft.ch/de/2018/06/gassmann-lingens-07-2018/ (14-03-2018).
McClelland, C. 2023. The Impact of Artificial Intelligence—Widespread Job Losses. 30.01.2023. Available at: https://www.iotforall.com/impact-of-artificial-intelligence-job-losses (14-03-2023).

NZZ, 2023a. Todeskampf eines Warenhauses: Galeria Karstadt Kaufhof schliesst 52 Filialen und entlässt über 4000 Mitarbeiter. 13.03.2023. Available at: https://www.nzz.ch/wirtschaft/galeria-schliessungen-ld.1730220?reduced=true (13-03-23).

NZZ, 2023b. Schwarzer Montag für amerikanische Regionalbanken. 13.03.2023. Available at: https://www.nzz.ch/wirtschaft/schwarzer-montag-fuer-amerikanische-regionalbanken-ld.1730284?reduced=true (14-03-23).

Parker, G.G., Van Alstyne, M.W., & Choudary, S.P. 2017. Platform Revolution: How Networked Markets Are Transforming the Economy? And How to Make Them Work for You. New York City: Norton & Company.

Posner, E.A., & Weyl, E.G. 2018. Radical Markets: Uprooting Capitalism and Democracy for a Just Society, Princeton: Princeton University Press.

Randstad. 2022. Energiekosten und Inflation: Deutsche Arbeitnehmer sehen Vater Staat in der Verantwortung Randstad Studie zu steigenden Lebenshaltungskosten. Available at: https://www.presseportal.de/pm/13588/5365536 (10-03-23).

Rifkin, J. 2014. The Zero Marginal Cost Society: The Internet of Things, the Collaborative Commons, and the Eclipse of Capitalism. London: Palgrave MacMillan.

Schumpeter, J.A. 1939, Business Cycles, New York: McGraw-Hill, 2 volumes.

Schumpeter, J.A. 1908. Das Wesen und der Haupteinhalt der Theoretischen Nationalökonomie. Leipzig: Duncker & Humblot.

Schwaninger, M. 2009. Intelligent Organizations. Powerful Models for Systemic Management. Berlin: Springer.

Wynen, J., Verhoest, K., & Kleizen, B. 2019. Are public organizations suffering from repetitive change injury? A panel study of the damaging effect of intense reform sequences. Governance: An International Journal of Policy Administration and Institutions, 32: 695–713.

Zukunftsinstitut, 2023. Megatrend New Work. Auszüge aus der Megatrend-Dokumentation und dem Zukunftsreport 2023. Available at: https://www.zukunftsinstitut.de/dossier/megatrend-new-work/ (14-03-23).

Dr. Christian Hugo Hoffmann is the Managing Director & Founder of Hoffmann Economics, an investment and investment advisory boutique focusing on Venture Capital. He holds a PhD in Management from the University of St. Gallen (Switzerland), used to work as Assistant Professor of Finance at the University of Liechtenstein, and is currently pursuing his second PhD in the Ethics of Artificial Intelligence. Apart from being passionate about Artificial Intelligence (AI) in academia, Christian is a tech entrepreneur by heart with three software start-ups in Germany, Switzerland, and Malawi under his belt. Moreover, Christian served as Deputy Director of and Head of AI at the Swiss Fintech Innovation Lab in Zurich, as Director of Startup Grind Geneva, and continues to fulfill his role as cluster-preneur, start-up investor, coach and mentor in various programs, and with involvements in several tech start-ups. More information at: https://www.christian-hugo-hoffmann.com/.

Unternehmertum 2.0 – Ein mit dem geltenden europäischen Kartellrecht vereinbarer Wandel?

Sebastian Klein

1 Einleitung

Ausgehend davon, dass die durch Christian Hugo Hoffmann im vorausgegangenen Kapitel aufgestellte These korrekt ist, wonach sich das Unternehmertum an der Schwelle zu einem Übergang hin zu einer Ökosystematisierung befindet, stellt sich die Frage, ob ein solcher Wandel mit dem geltenden Kartellrecht der Europäischen Union vereinbar ist. Ausganspunkt der Fragestellung ist Art. 101 AEUV und das entsprechende Verbot der Kartellbildung. Ob eine solches Ökosystem unter Art. 101 AEUV zu subsumieren wäre, soll im Folgenden untersucht werden.

Nachfolgend werden zunächst die in diesem Kapitel herangezogenen Prämissen erläutert. Hierauf folgt ein Überblick zu den relevanten Regelungsinhalten der vorbezeichneten Norm, wobei besonders auf den Unternehmens- und Wettbewerbsbegriff eingegangen wird. Im Anschluss wird eine Subsumtion des Konzepts des Ökosystems unter die zuvor gemachten Ausführungen vorgenommen und ein abschließendes Fazit mit Ausblick gegeben.

2 Prämissen

Prämisse 1: Das Unternehmertum befindet sich in einem Wandel hin zu einer Ökosystematisierung der Wirtschaft.

S. Klein (✉)
Universität Bonn, Bonn, Deutschland
e-mail: s3snklei@uni-bonn.de

C. H. Hoffmann (ed.), *Artificial Intelligence, Entrepreneurship and Risk*, Technikzukünfte, Wissenschaft und Gesellschaft / Futures of Technology, Science and Society, https://doi.org/10.1007/978-3-658-45544-6_21

Prämisse 2: Ökosystematisierung meint unternehmerische Zusammenschlüsse von üblicherweise drei bis zehn Unternehmen, welche von einem zentralen Unternehmen, dem „Orchestrator", auf ein gemeinsames Wertversprechen ausgerichtet werden.

Prämisse 3: Ein Ökosystem kann auf zweierlei Weise agieren:

3.a) Die Ökosystempartner konzentrieren sich gemeinsam auf ein bestimmtes Kundenbedürfnis, das sie in einer bisher nicht erreichten Weise erfüllen.
3.b) Die Ökosystempartner stellen ein gemeinsames Dienstleistungs- und/oder Produktbündel bereit.

3 Das Kartellverbot nach Art. 101 AEUV

Den Wettbewerbsvorschriften der AEUV liegt zum einen der Gedanke zu Grunde, dass auf einem gemeinsamen Markt alle Unternehmer hinsichtlich des „Wie" und „Ob" ihres Angebots eine selbständige Strategie determinieren und verfolgen sollen (sog. Selbständigkeitspostulat).[1] Zum anderen sollen Unternehmen daran gehindert werden, alleine aufgrund ihrer schieren Größe und hieraus resultierenden Marktmacht wettbewerbsverfälschende Maßnahmen zu unternehmen.[2]

3.1 Anwendungsbereich

3.1.1 Grundsätze des räumlichen Anwendungsbereichs

Nach Art. 52 Abs. 1, Abs. 2 EUV i. V. m. Art. 355 AEUV gelten die Regelungen der Art. 101 ff. AEUV für das gesamte Territorium der europäischen Union. Die Vorschriften enthalten jedoch keine kollisionsrechtlichen Regelungen, sodass ihre Anwendbarkeit in Fällen mit Drittstaatenbezug durch den Telos und Tatbestand der Normen unter Berücksichtigung völkerrechtlicher Limitierungen determiniert wird.[3] Hinsichtlich der völkerrechtlichen Auswirkungen ist bei

[1] EuGH, Urt. v. 16.12.1975 – C-40/73 = NJW 1976, 470 (471 Rn.4); Loewenheim/Meessen/Riesenkampff/Kersting/Meyer-Lindemann/*Grave/Nyberg*, Kartellrecht, AEUV Art. 101 Rn. 121.

[2] Amtsblatt der Europäischen Union Nr. C 306 vom 17.12.2007, S. 156; Dauses/Ludwigs/*Hoffmann*, HdB-EuWiRe, H. I. § 3 Rn. 3.

[3] Grabitz/Hilf/Nettesheim/*Stockenhuber*, EU-Recht, AEUV Art. 101 Rn. 32; Dauses/Ludwigs/*Hoffmann*, HdB-EuWiRe, H. I. § 1 Rn. 28.

Drittstaatsbezug auf Grundlage des konkreten Einzelfalles und unter Rücksichtnahme auf die Grundsätze des Völkerrechts zu entscheiden, wie erheblich der Bezug zum Binnenmarkt i. S. d. Auswirkungsprinzips (s. unten) sein muss, um eine Anwendbarkeit des EU-Kartellrechts zu begründen.[4]

Die Art. 101, 102 AEUV sind als Teil des Vertrages über die Arbeitsweise der Europäischen Union Teil des unionalen Primärrechts, mithin jener völkerrechtlichen Verträge, welche einen Teil der Grundlage für die Europäische Union als solche bilden (Art. 1 Abs. 2 S. 1 AEUV) und damit gewissermaßen deren „Verfassung" darstellen.[5] Aufgrund des Vorrangs des Unionsrechts gegenüber dem nationalem Recht sind alle abschließenden und eine Unterlassungs- oder Handlungspflicht begründenden Vorschriften des Primärrechts, welche keiner weiteren Exekutivmaßnahmen bedürfen, für jegliche Normadressaten innerhalb der Mitgliedsstaaten unmittelbar bindend.[6] Hierunter fallen auch die Art. 101 f. AEUV, da sie Private unmittelbar berechtigen und verpflichten.[7]

3.1.2 Auswirkungs- und Durchführungsprinzip

Nach herrschender Ansicht findet im europäischen Kartellrecht das sogenannte „Auswirkungsprinzip" Anwendung, wonach es zur Anwendbarkeit der Art. 101, 102 AEUV einer mittelbaren, spürbaren und vorhersehbaren Auswirkung der entsprechenden Maßnahme auf den Binnenmarkt bedarf, wobei dieses Prinzip zwar durch EuG[8] und Kommission[9] bereits als völkerrechtlicher Maßstab anerkannt wurde, der EuGH[10] das Prinzip jedoch lange Zeit nicht endgültig rezipierte.[11] Vielmehr wandte der EuGH das sogenannte „Durchführungsprinzip" an, wonach für die Anwendbarkeit der Art. 101, 102 AEUV erforderlich sei, dass die Implementierung der tatbestandlichen Maßnahme auf EU-Territorium

---

[4] Immenga/Mestmäcker/*Rehbinder*, WettbR, IntWettbR A. Rn. 4.

[5] Dürig/Herzog/Scholz/*Dederer*, GG Art. 100 Rn. 108; *Bieber/Epiney/Haag/Kotzur*, Europarecht, Rn. 1, 93; *Streinz*, Europarecht, Rn. 2 f., 455.

[6] EuGH, Urt. v. 05.02.1963 – Rs 26/62 = NJW 1963, 974; Grabitz/Hilf/Nettesheim/*Stockenhuber*, EU-Recht, AEUV Art. 1 Rn. 67 ff.; *Bieber/Epiney/Haag*/Kotzur, Die Europäische Union, § 6 Rn. 53; *Streinz*, Europarecht, Rn. 458.

[7] Calliess/Ruffert/Weiß, AEUV Art. 101 Rn. 24.

[8] Vgl. Urt. v. 25.03.1999 – Rs. T-102/96 = EuZW-Sonderausgabe 1/2019 2019, 19.

[9] Vgl. die Entscheidung der Kommission v. 08.12.2010 – C(2010) 8761 final.

[10] Vgl. EuGH, Urt. v. 14.07.1972 – Rs. 48/69 = BeckRS 2004, 73172; EuGH, Urt. v. 27. 9. 1988, Rs. 89/85, 104/85, 114/85, 116/85, 117/85 = GRUR Int 1989, 671.

[11] Pechstein/Nowak/Häde/*Brömmelmeyer*, FK-EUV/GRC/AEUV, AEUV Art. 101 Rn. 21 f.; s. auch: Grabitz/Hilf/Nettesheim/*Schuhmacher*, EU-Recht, AEUV Art. 101 Rn. 32; Immenga/Mestmäcker/*Rehbinder*, WettbR, IntWettbR A. Rn. 3.

erfolgt, mithin ein Umsetzungsakt stattfindet, d. h. die kartellierte Ware abgesetzt oder Dienstleistung ausgeführt wird, wobei etwaige Umsetzungen durch die Kartellanten unmittelbar oder mittelbar innerhalb der EU erfolgt.[12]

Im Rahmen der Intel-Entscheidung erkannte der EuGH jedoch im Jahr 2017 an, dass auch das Auswirkungsprinzip für die Determinierung der Anwendbarkeit des EU-Kartellrechts heranziehbar sei und gleichberechtigt neben dem Durchführungsprinzip stehe.[13]

3.1.3 Aspekt der Zwischenstaatlichkeit

Ferner setzt die Anwendbarkeit der Art. 101, 102 AEUV ihrem Wortlaut nach sowie aufgrund der Zuständigkeitsregelung des Art. 3 Abs. 1 Bst. b AEUV eine Verbindung zum Binnenmarkt, entweder in Form einer Wettbewerbsbeschränkung *innerhalb* des Binnenmarktes (Art. 101 AEUV) oder einer Ausnutzung einer marktbeherrschenden Stellung *auf* dem Binnenmarkt oder einem wesentlichen Teil desselben (Art. 102 AEUV), voraus, sodass diejenigen Aktivitäten, welche nicht innerhalb des EU-Binnenmarktes erfolgen, nicht in den Anwendungsbereich der Art. 101 f. AEUV fallen.[14]

Hinsichtlich der Anwendbarkeit auf Sachverhalte zwischen Mitgliedsstaaten wird nach Art. 101 Abs. 1 bzw. Art. 102 S. 1 AEUV weiterhin vorausgesetzt, dass die betreffenden tatbestandlichen Handlungen die Eignung zur Beeinträchtigung des Handels zwischen den Mitgliedsstaaten besitzen, was der Abgrenzung des europäischen Kartellrechts von den nationalrechtlichen Kartellvorschriften der Mitgliedsstaaten dient.[15] Nach der Rechtsprechung des EuGH ist die Zwischenstaatlichkeit demnach zu bejahen, wenn die Maßnahme in nicht bloß geringfügiger (d. h. spürbarer) Weise objektiv in rechtlicher wie tatsächlicher Hinsicht mit hinreichender Wahrscheinlichkeit dergestalt zu einer unmittelbaren oder mittelbaren, tatsächlichen oder potenziellen negativen Beeinflussung des Handels zwischen Mitgliedstaaten führt, dass die Verwirklichung eines einheitlichen

---

[12] EuGH, Urt. v. 27.09.1988 – Rs 89/85 u. a. = NJW 1988, 3086 Rn. 16; MüKo-BGB/*Wurmnest*, Rom II-VO Art. 6 Rn. 279; Immenga/Mestmäcker/*Rehbinder*, WettbR, IntWettbR A. Rn. 7 ff.

[13] EuGH, Urt. v. 6.9.2017 – C-413/14 P = NZKart 2017, 525 (526 Rn. 44 f.); Calliess/Ruffert/Weiß, AEUV Art. 101 Rn. 9; Dauses/Ludwigs/*Hoffmann*, HdB-EuWiRe, H. I. § 1 Rn. 31.

[14] Immenga/Mestmäcker/*Rehbinder*, WettbR, IntWettbR A. Rn. 3; vgl. auch: Grabitz/Hilf/Nettesheim/*Stockenhuber*, EU-Recht, AEUV Art. 101 Rn. 32.

[15] Dauses/Ludwigs/*Hoffmann*, HdB-EuWiRe, H. I. § 1 Rn. 34; Hasselblatt/*Lubitz*, MAH GewRS, § 47 Rn. 10.

Marktes der Mitgliedstaaten gehemmt werden könnte (sog. „Wahrscheinlichkeitsmaßstab).[16] Die Kommission veröffentlichte im Jahre 2004 Leitlinien über den Begriff der Beeinträchtigung des zwischenstaatlichen Handels, in welchen sie auf den Aspekt der auch vom EuGH geforderten tatbestandlichen „Spürbarkeit" detailliert einging und die Vermutung aufstellte, dass es an der Spürbarkeit jedenfalls dann fehle, wenn der Marktanteil der an den Maßnahmen beteiligten Parteien innerhalb der Union fünf vom Hundert nicht überschreitet *und* der mit den von den Maßnahmen betroffenen Wirtschaftsgütern erzielte Umsatz der beteiligten Unternehmen innerhalb der Union den Betrag von 40 Mio. Euro nicht überschreitet.[17] Diese Leitlinien sind nicht verbindlich, dürften jedoch aufgrund des nach wie vor weiten Auslegungsspielraums in der Praxis weitgehend rezipiert werden.[18]

3.2 Normadressat: Unternehmen und Unternehmensvereinigungen

Normadressat der Art. 101, 102 AEUV sind private sowie nach Art. 106 AEUV auch öffentliche „Unternehmen" und „Unternehmensvereinigungen".[19]

3.2.1 Unternehmen

Der Unternehmensbegriff ist funktional zu verstehen, sodass über institutionelle Aspekte oder die Organisationsstruktur hinausgehend die betreffende Tätigkeit maßgeblich ist.[20] Die europäische Rechtsprechung definiert den Begriff des Unternehmens ständig als „jede eine wirtschaftliche Tätigkeit ausübende Einheit,

---

[16] EuGH, Urt. v. 13. 7. 2006 – C-295/04 bis C-298/04 = EuZW 2006, 529 (531 Rn. 41 f.); Urt. v. 28. 4. 1998 – C-306/96 = EuZW 1998, 404 (405 Rn. 16); Urt. v. 11.07.1985 – 42/84 = BeckRS 2004, 71164 Rn. 22; Urt. v. 30.06.1966 – 56/65 = BeckRS 2004, 73331; Hasselblatt/*Lubitz*, MAH GewRS, § 47 Rn. 10; Saenger/Aderhold/Lenkaitis/Speckmann/*Lotze*, Handels- und GesR, § 12 Rn. 43.

[17] Amtsblatt der Europäischen Union Nr. C 101 vom 27.04.2004, S. 81–96, Rn. 52.

[18] Grabitz/Hilf/Nettesheim/*Stockenhuber*, EU-Recht, AEUV Art. 101 Rn. 219; Hasselblatt/*Lubitz*, MAH GewRS, § 47 Rn. 11, u.a. beziehend auf: *Bechtold/Bosch/Brinker*, EU-Kartellrecht, AEUV Art. 101 Rn. 104.

[19] BeckOK-InfoMedR/*Kumkar*, AEUV Art. 101 Rn. 6; Calliess/Ruffert/Weiß, AEUV Art. 101 Rn. 13; vgl. auch: Immenga/Mestmäcker/*Zimmer*, WettbR, AEUV Art. 101 Rn. 12.

[20] Grabitz/Hilf/Nettesheim/*Stockenhuber*, EU-Recht, AEUV Art. 101 Rn. 51; MüKo-EUWettbR/*Säcker/Steffens*, AEUV Art. 101 Rn. 952.

unabhängig von ihrer Rechtsform und der Art ihrer Finanzierung".[21] Der Begriff wird seitens des EuGH sehr weit ausgelegt und umfasst auch solche Vereinigungen, welche unternehmerisch *noch* nicht aktiv sind, in absehbarer Zeit jedoch dem Markt beitreten werden (potenzielle Unternehmen).[22]

Eine wirtschaftliche Betätigung in diesem Sinne setzt zudem *keine* Gewinnerzielungsabsicht voraus und ist bereits dann zu bejahen, wenn Güter oder Dienstleistungen auf einem Markt schlichtweg angeboten werden.[23] Maßgeblich ist damit nicht das „Unternehmen", sondern das „Handeln *als* Unternehmen", sodass die Art und Weise des Agierens das maßgebliche Abgrenzungskriterium darstellt.

Die Eigenschaft des Unternehmens i. S. d. Art. 101 f. AEUV als Rechtsträger ist zudem nicht erforderlich, da vielmehr eine Dichotomie zwischen dem „Unternehmen" als Rechtsobjekt und dem Rechtsträger als Rechtssubjekt vorliegt, bei welcher lediglich eine *Zuordnung* des Unternehmens zu einem Rechtsträger notwendig ist.[24] Die wirtschaftliche Einheit selbst ist damit effektiv Normadressat, wohingegen der Rechtsträger den etwaigen Entscheidungsadressaten darstellt.[25] Ob dies eine gerechtfertigte Auslegung darstellt, kann vorliegend in hiesigem Kontext dahinstehen, da sich diese Ausarbeitung einer praxisorientieren Betrachtung verschrieben hat.

3.2.2 Unternehmensvereinigung

Die faktische Vereinigung einzelner Unternehmen kann aufgrund einer eigenen im Verkehr nach außen erkennbaren wirtschaftlichen Tätigkeit bereits als

---

[21] EuGH, Urt. v. 16.11.1995 – Rs. C-244/94 = EuZW 1996, 277; Urt. v. 11.12.1997 – Rs. C-55/96 = EuZW 1998, 274 (275 Rn. 21);

[22] Münch-HdB-GesR/*Hoffmann/Doehner*, § 6 Rn. 15; Amtsblatt der Europäischen Union Nr. L 243 vom 19.09.1994, S. 46 f.; Nr. L 255 vom 08.10.2004, S. 18 f.

[23] Schnapp/Wigge/*Becker*, HdB-Vertragsarztrecht, § 23 Rn. 58 (Fn. 244); EuGH, Urt. v. 16.06.1987 – Rs.118/85 = BeckRS 2004, 71333 Rn. 7; Urt. v. 18.06.1998 – Rs. C-35/96 = BeckRS 1998, 55267 Rn. 36; s. auch: EuGH, Urt. v. 01.07.2008 – C-49/07 = JuS 2008, 1131.

[24] Grabitz/Hilf/Nettesheim/*Stockenhuber*, EU-Recht, AEUV Art. 101 Rn. 51 f., beziehend auf: EuGH, Urt. v. 13.07.1962 – C-17/61, C-20/61 = BeckRS 1962, 104978; *Jüchser/Klein* NZI 2023, 370 f.

[25] Grabitz/Hilf/Nettesheim/*Stockenhuber*, EU-Recht, AEUV Art. 101 Rn. 51 f., beziehend auf: Entscheidung der Kommission v. 13.07.1994 – E 94/601/EG, ABl. L 243, S. 1.

„Unternehmen" angesehen werden, sodass die Prüfung einer Unternehmensvereinigung i. S. d. Art. 101 f. AEUV hinsichtlich der Prüfung des Vorliegens eines „Unternehmens" im vorbezeichneten Sinne subsidiär ist.[26]

„Unternehmensvereinigung" meint hierbei den Zusammenschluss einzelner Akteure mit gemeinsamen Interessen, welche (nicht jedoch alle)[27] jeweils als „Unternehmen" im zuvor dargelegten Sinne zu qualifizieren sind und die in Bezug auf die Vereinigung eine gewisse (Mit-) Entscheidungsbefugnis besitzen, sodass es zu einer institutionalisierten Form der Verhaltensabstimmung kommt.[28] Zur Bejahung der Tatbestandsvoraussetzung ist weder erforderlich, dass die Vereinigung selbst Rechtsträgereigenschaft besitzt noch, dass sie als solche einen Erwerbs- oder Gewinnzeck verfolgt oder am Wirtschaftsleben teilnimmt – jedoch sind im Falle einer fehlenden Rechtsubjektivität die rechtsfähigen Mitglieder Normadressat der Art. 101 f. AEUV.[29] Die Tatbestandsvariante der Unternehmensvereinigung dient gewissermaßen als Auffangtatbestand, welcher nach der Auslegung des EuGH einschlägig ist, wenn entweder die Tätigkeit der Unternehmensvereinigung als solche oder die ihrer Mitglieder gerade die Folgen herbeiführen soll, welche durch die Kartellvorschriften verhindert werden sollen, wobei jedoch der Anwendungsvorrang der „Unternehmens"-Tatbestandsvariante zu berücksichtigen ist, sollte diese zugleich einschlägig sein.[30] Letzteres ist beispielsweise dann gegeben, wenn eine eigene wirtschaftliche Tätigkeit der Vereinigung neben die Wahrnehmung der Mitgliederinteressen als Aktivität der Vereinigung tritt.[31]

Tatbestandliche Handlungsweisen können sowohl in förmlichen Beschlussfassungen als auch faktischen Handlungen der Organe bestehen, wobei jegliche

---

[26] EuGH, Urt. v. 29.10.1980 – Rs. 209–215/78, 218/78 = BeckRS 2004, 72290 Rn. 88; Immenga/Mestmäcker/*Zimmer*, WettbR, AEUV Art. 101 Rn. 58.

[27] EuG, Urt. v. 13.12.2006 – T-217/03, T-245/03 = BeckRS 2006, 19546 Rn. 55.

[28] EuGH, Urt. v. 11.09.2014 – C –382/12 P = NZKart 2015, 44 (45 f. Rn. 66 ff.); Pechstein/Nowak/Häde/*Brömmelmeyer*, FK-EUV/GRC/AEUV, AEUV Art. 101 Rn. 53.

[29] EuGH, Urt. v. 29.10.1980 – Rs. 209–215/78, 218/78 = BeckRS 2004, 72290 Rn. 88; Pechstein/Nowak/Häde/*Brömmelmeyer*, FK-EUV/GRC/AEUV, AEUV Art. 101 Rn. 53; Grabitz/Hilf/Nettesheim/*Stockenhuber*, EU-Recht, AEUV Art. 101 Rn. 85.

[30] EuGH, Urt. v. 29.10.1980 – Rs. 209–215/78, 218/78 = BeckRS 2004, 72290 Rn. 88; EuGH, Urt. v. 08.11.1983 – C-96–102, 104, 105, 108 und 110/82 = BeckRS 2004, 73971 Rn. 20; Grabitz/Hilf/Nettesheim/*Stockenhuber*, EU-Recht, AEUV Art. 101 Rn. 83 ff.

[31] Dauses/Ludwigs/*Hoffmann*, HdB-EuWiRe, H. I. § 1 Rn. 71.

Handlungen auch denjenigen Mitgliedern zuzurechnen sind, welche im Falle eines Beschlusses gegen diesen gestimmt oder sich enthalten haben.[32]

3.3 Kartellverbot, Art. 101 AEUV

Nach Art. 101 Abs. 1 HS 1 AEUV sind alle Vereinbarungen zwischen Unternehmen, Beschlüsse von Unternehmensvereinigungen sowie aufeinander abgestimmte Verhaltensweisen, welche den Handel zwischen Mitgliedstaaten zu beeinträchtigen geeignet sind und eine Verhinderung, Einschränkung oder Verfälschung des Wettbewerbs innerhalb des Binnenmarkts bezwecken oder bewirken verboten und mit dem Binnenmarkt unvereinbar.

3.3.1 Verbotene Verhaltensweisen nach Abs. 1

Der Tatbestand des Art. 101 Abs. 1 AEUV setzt zunächst ein Agieren der Normadressaten in Form von Vereinbarungen, Beschlüssen oder aufeinander abgestimmten Verhaltensweisen voraus. Erforderlich ist damit jedenfalls, dass die Unternehmen oder Unternehmensvereinigungen hinsichtlich ihres Agierens Autonomie besitzen, d. h. dass das Verhalten nicht durch eine Rechtsvorschrift vorgeschrieben ist.[33]

3.3.2 Vereinbarungen

Eine Vereinbarung liegt unter Zugrundelegung des weiten durch die Rechtsprechung geprägten Verständnisses vor, wenn die Normadressaten nach außen hin ihren gegenseitigen Willen zum Ausdruck bringen, auf dem Markt auf eine bestimmte Art und Weise zu agieren.[34] Auf einen Rechtsbindungswillen der Akteure kommt es hierbei ebenso wenig an, wie auf aus der Vereinbarung erwachsende tatsächliche rechtliche, respektive moralische oder faktische Obliegenheiten – umfasst wird damit auch das sogenannte „gentlemen's

---

[32] EuGH, Urt. v. 29.10.1980 – Rs. 209–215/78, 218/78 = BeckRS 2004, 72290 Rn. 156; von der Groeben/Schwarze/Hatje/*Schröter*/*van Vormizeele*, EU-Recht, AEUV Art. 101 Rn. 50.

[33] EuGH, Urt. v. 05.10.1995 – C-96/94 = BeckRS 2004, 77933 Rn. 25; Geiger/Khan/Kotzur/Kirchmair/*Khan*, EUV/AEUV, AEUV Art. 101 Rn. 14.

[34] EuGH, Urt. v. 29.10.1980 – 209/78 = BeckRS 2004, 72290 Rn. 86; Urt. v. 08.11.1983 – Rs. 96–102, 104, 105, 108 und 110/82 = BeckRS 2004, 73971 Rn. 2, 20; EuG, Urt. v. 14.05.1998 – T-304/94 = BeckRS 1998, 154643 Rn. 76; Urt. v. 26.10.2000 – T-41/96 = GRUR Int 2001, 616 (617 Rn. 67); Urt. v. 14.10.2004 – T-56/02 = BeckRS 2004, 78079 Rn. 59 ff.; Calliess/Ruffert/*Weiß*, AEUV Art. 101 Rn. 47; Dauses/Ludwigs/*Hoffmann*, HdB-EuWiRe, H. I. § 2 Rn. 4.

agreement".[35] Gleiches gilt für die Form der Abrede, sodass es auch keiner Verschriftlichung dieser bedarf.[36] Ferner bedarf es auch nicht des Vollzugs der Abrede.[37]

Den Tatbestand erfüllen kann auch ein Agieren gegenüber Dritten, welches in der Verweigerung des Eingehens einer Vertragsbeziehung besteht.[38] Einseitige Maßnahmen einzelner Akteure (z. B. Weisungen oder Empfehlungen) – wohl aber das konkludente Einverständnis mit diesen – fallen jedoch nicht unter den Begriff der Vereinbarung nach Art. 101 AEUV, da es eines Ausdrucks gemeinsamer Willensbildung bedarf – diese können allerdings nach Art. 102 AEUV tatbestandsmäßig sein.[39]

3.3.3 Beschlüsse von Unternehmensvereinigungen

Die Tatbestandsvariante des Beschlusses meint die institutionalisierte Bildung eines Gesamtwillens der Mitglieder, welche in einem konstituierenden Dokument (z. B. einer Satzung) entsprechend vorgesehenen ist und im Einklang mit den Regelungen dieses Dokuments wirksam zustande kommen muss.[40] Auch die nicht am Beschluss beteiligten Mitglieder oder diejenigen, welche gegen diesen stimmten, werden von dieser Tatbestandsvariante erfasst.[41]

Die Behandlung bloßer Empfehlungen ist dahingehend umstritten, als dass fraglich ist, ob es der Verbindlichkeit der Empfehlung bedarf.

In der europäischen Rechtsprechung wird jedenfalls dann von einer Tatbestandsmäßigkeit nach Art. 101 Abs. 1 AEUV ausgegangen, wenn die ergangene

[35] EuGH, Urt. v. 03.07.1985 – C-243/83 = NJW 1986, 651 (652 Rn. 17); EuG, Urt. v. 14.05.1998 – T-347/94 = BeckRS 1998, 155483 Rn. 65; Geiger/Khan/Kotzur/Kirchmair/*Khan*, EUV/AEUV, AEUV Art. 101 Rn. 15; Calliess/Ruffert/*Weiß*, AEUV Art. 101 Rn. 47.

[36] EuGH, Urt. v. 13.07.2006 – C-74/04 P = JuS 2007, 494 (496 Rn. 37); EuG, Urt. v. 11.12.2003 – T-56/99 = BeckRS 2003, 161003 Rn. 20; Streinz/Eilmansberger/*Kruis*, EUV/AEUV, AEUV Art. 101 Rn. 1.

[37] Grabitz/Hilf/Nettesheim/*Stockenhuber*, EU-Recht, AEUV Art. 101 Rn. 100.

[38] Calliess/Ruffert/*Weiß*, AEUV Art. 101 Rn. 51.

[39] Geiger/Khan/Kotzur/Kirchmair/*Khan*, EUV/AEUV, AEUV Art. 101 Rn. 15; Grabitz/Hilf/Nettesheim/*Stockenhuber*, EU-Recht, AEUV Art. 101 Rn. 101.

[40] Entscheidung der Kommission v. 07.02.1985, ABl. 1985 L 35, S. 20 Rn. 23; EuGH, Urt. v. 27.1.1987 – C-45/85 = NJW 1987, 2150 (2151 Rn. 29 f.); Urt. v. 11.9.2014 – C-382/12 P = NZKart 2015, 44 (46 Rn. 69); EuG, Urt. v. 24.5.2012 – T-111/08 = BeckRS 2012, 80963 Rn. 244; Dauses/Ludwigs/*Hoffmann*, HdB-EuWiRe, H. I. § 2 Rn. 18; Streinz/*Eilmansberger*/*Kruis*, EUV/AEUV, AEUV Art. 101 Rn. 10.

[41] EuGH, Urt. v. 29.10.1980 – C-209/78 = BeckRS 2004, 72290.

unverbindliche Empfehlung durch die Akteure umgesetzt werden.[42] Weiterhin wird seitens des EuGH und eines Teils der Literatur vertreten, dass ferner Empfehlungen im Rahmen von Verbandsbeziehungen, die gerade zu dem Zweck ergehen, das Marktverhalten der Akteure zu harmonisieren, nach Art. 101 Abs. 1 AEUV tatbestandsmäßig sind, soweit diese einen gemeinsamen Verbandswillen ausdrücken.[43] Bei ausdrücklich unverbindlichen Preisempfehlungen – nicht jedoch bei denjenigen Preisempfehlungen, hinsichtlich welcher es an einer derartigen Deklarierung fehlt – im Rahmen von Franchise- oder ähnlichen Systemen gilt dies hingegen nicht, da auf diese Weise nicht der notwendige wirtschaftliche Druck – welcher auch eine de facto Verbindlichkeit begründen würde – des Franchisegebers auf den Franchisenehmer ausgeübt wird.[44]

Stimmen in der Literatur tendieren zum Teil dazu eine faktische[45] Verbindlichkeit zu fordern, die deutsche Rechtsprechung schließt eine Beschränkung auf rechtlich verbindliche Empfehlungen nicht aus.[46]

Die Ansicht, wonach es grundsätzlich nicht auf eine Verbindlichkeit der Empfehlung ankomme, wird zum Teil mit dem Argument der Verhinderung von Umgehungen und der notwendigen Gleichstellung mit der Tatbestandsvariante der „Vereinbarungen" verteidigt.[47] Dieses Argument überzeugt indes nicht. Es erscheint im Kontext der Bebußung unionskartellrechtlicher Verstöße vor dem Hintergrund des unionsverfassungsrechtlich in Art. 2 EUV und Art. 47 Abs. 2 GRCh verankerten Rechtstaatsprinzips und des von diesem mitumfassten Verhältnismäßigkeitsprinzips[48] schwer nachvollziehbar, warum die bloß passive Entgegennahme einer nicht verbindlichen Empfehlung, ohne dass diese umgesetzt würde, mit einer Vereinbarung oder einem Beschluss gleichzusetzen sein soll,

---

[42] EuGH, Urt. v. 08.11.1983 – C-96–102, 104, 105, 108 und 110/82 = BeckRS 2004, 73971 Rn. 20; Calliess/Ruffert/*Weiß*, AEUV Art. 101 Rn. 56.

[43] EuGH, Urt. v. 29.10.1980 – Rs. 209–215/78, 218/78 = BeckRS 2004, 72290 Rn. 86 ff.; Urt. v. 27.1.1987 – C-45/85 = NJW 1987, 2150 (2151 Rn. 32); Dauses/Ludwigs/*Hoffmann*, HdB-EuWiRe, H. I. § 2 Rn. 36 ff.; kritisch: Streinz/*Eilmansberger*/*Kruis*, EUV/AEUV, AEUV Art. 101 Rn. 10.

[44] Dauses/Ludwigs/*Hoffmann*, HdB-EuWiRe, H. I. § 2 Rn. 38.

[45] So noch: Langen/Bunte/*Bunte*, Kartellrecht, 10. Aufl., EG-Vertrag Art. 81 Rn. 27.

[46] BGH, Beschl. v. 14.08.2008 – KVR 54/07 = BeckRS 2008, 20019 Rn. 27, hierzu: Hoeren/Sieber/Holznagel/*Beckmann*/*Müller*, HdB-MultimediaR, Teil 10 Rn. 75.

[47] BGH, Beschl. v. 14.08.2008 – KVR 54/07 = BeckRS 2008, 20019 Rn. 27; Immenga/Mestmäcker/*Zimmer*, WettbR, AEUV Art. 101 Rn. 82.

[48] Zum unionalen Rechtstaatsprinzip und dem Verhältnismäßigkeitsgrundsatz: EuGH, Urt. v. 07.09.2006 – C-310/04 = BeckRS 2006, 70657 Rn. 97; *Payandeh*, JuS 2021, 481 (484); dieses findet auch auf juristische Personen Anwendung, vgl. hierzu: Dauses/Ludwigs/*Pieper*, HdB-EuWiRe, B. I. Rn. 151;

welche, ebenso wie die Tatbestandsvariante des „abgestimmten Verhaltens“ (s. unten), die Bildung eines nach außen tretendenden Gesamtwillens voraussetzen. Gerade an einer solchen Bildung fehlt es bei unverbindlichen, nicht umgesetzten Empfehlungen, sodass eine Gleichstellung mit Vereinbarungen, Beschlüssen i. e. S. und abgestimmten Verhaltensweisen unverhältnismäßig erscheint, sodass zumindest eine rechtliche Verbindlichkeit der Empfehlung zu fordern ist.

3.3.4 Abgestimmtes Verhalten

Die Tatbestandsvariante der „abgestimmten Verhaltensweisen“ fungiert als Auffangtatbestand und umfasst alle Koordinierungen zwischen Normadressaten, die keine Willenseinigung darstellen, jedoch bewusst eine praktische Zusammenarbeit an die Stelle unternehmerischer bzw. wettbewerbsmäßiger Risiken treten lassen, sofern diese keine voneinander unabhängigen Marktverhalten (z. B. auf bestimmte Marktsignale) darstellen.[49]

3.3.5 Zwischenstaatlichkeit und Spürbarkeit

Hinsichtlich des Erfordernisses der Zwischenstaatlichkeit und Spürbarkeit der Maßnahmen, kann auf die obigen Ausführungen verwiesen werden.

3.3.6 Bezwecken und Bewirken von Wettbewerbseinschränkungen

Die tatbestandlichen Verhaltensweisen müssen ferner eine Verhinderung, Einschränkung oder Verfälschung des Wettbewerbs innerhalb des Binnenmarkts bezwecken oder bewirken.

Wenn auch der Begriff des „Wettbewerbs“ durch die Rechtsprechung noch nicht abschließend definiert wurde, ist aufgrund der Hoffmann La-Roche-Entscheidung des EuGH[50], in welcher dieser den Schutz des Wettbewerb basierend auf Leistungen der Marktbürger durch die Kartellvorschriften hervorhebt, davon auszugehen, dass hierunter ein Leistungswettbewerb zu verstehen ist, in welchem mehrere Anbieter Leistungen auf einem Markt anbieten, deren Nachfrage durch Preis oder Qualität determiniert wird.[51] Geschützt ist zudem ausschließlich der rechtmäßige Wettbewerb, welcher zum fraglichen Zeitpunkt

---

[49] EuGH, Urt. v. 16.12.1975 – C-40–48/73, 50/73, 54–56/73, 111/73, 113 und 114/73 = NJW 1976, 470 (471); Urt. v. 23.11.2006 – C-238/05 = EuZW 2006, 753 (755 Rn. 32); Urt. v. 04.06.2009 – C-8/08 = EuZW 2009, 505 (506 Rn. 26); Geiger/Khan/Kotzur/Kirchmair/*Khan*, EUV/AEUV, AEUV Art. 101 Rn. 17 f.; *Herdegen*, Europarecht, § 22 Rn. 9.

[50] EuGH, Urt. v. 13.02.1979 – C-85/76 = BeckRS 2004, 73814 Rn. 91.

[51] Streinz/*Eilmansberger*/*Kruis*, EUV/AEUV, AEUV Art. 101 Rn. 40; s. auch: EuGH, Urt. v. 15.03.2007 – C-95/04 P = JuS 2007, 975 (976 Rn. 66); Calliess/Ruffert/*Weiß*,

aber noch nicht aktuell gegeben sein muss (potentieller Wettbewerb).[52] Die Rechtsprechung wendet den Wettbewerbsbegriff jedoch relativ flexibel an, etwa um einer Verengung der Tatbestandsvoraussetzungen entgegenzuwirken.[53]

Eine Einschränkung des Wettbewerbs ist nach der Rechtsprechung des EuGH dann gegeben, wenn durch das Verhalten objektiv Wettbewerbsbedingungen entstehen, die im Hinblick auf die konkreten Umstände (Art der Leistung, Bedeutung und Zahl der Beteiligten und Umfang des entsprechenden Marktes) nicht den normalen – d. h. im Vergleich zu einer gedachten Wettbewerbssituation ohne die entsprechende Maßnahme – Marktbedingungen entsprechen.[54]

Durch die Miteinbeziehung der Tatbestandsvariante des „Bezweckens" werden auch solche Maßnahmen erfasst, hinsichtlich derer eine Wettbewerbseinschränkung, welche zudem nicht ausschließlicher Zweck der Verhaltensweise sein muss, das Versuchsstadium nicht überschreitet, ohne dass es notwendig wäre, dass die Beschränkung auch tatsächlich eintreten müsste (dann: Bezwecken) oder die Akteure eine *eingetretene* Beschränkung tatsächlich beabsichtigten (dann: Bewirken).[55]

3.3.7 Fallgruppen

Solche verbotenen Verhaltensweisen können gemäß Art. 101 Abs. 1 HS 2 AEUV vor allem in der unmittelbaren oder mittelbaren Festsetzung der An- oder Verkaufspreise oder sonstiger Geschäftsbedingungen (Art. 101 Abs. 1 Bst. a AEUV), der Einschränkung oder Kontrolle der Erzeugung, des Absatzes, der technischen Entwicklung oder der Investitionen (Art. 101 Abs. 1 Bst. b AEUV), der Aufteilung der Märkte oder Versorgungsquellen (Art. 101 Abs. 1 Bst. c AEUV), der Anwendung unterschiedlicher Bedingungen bei gleichwertigen Leistungen gegenüber Handelspartnern, wodurch diese im Wettbewerb benachteiligt

---

AEUV Art. 101 Rn. 78; Grabitz/Hilf/Nettesheim/*Stockenhuber*, EU-Recht, AEUV Art. 101 Rn. 116.

[52] Dauses/Ludwigs/*Pieper*, HdB-EuWiRe, H. I. § 2 Rn. 46; Geiger/Khan/Kotzur/Kirchmair/*Khan*, EUV/AEUV, AEUV Art. 101 Rn. 21.

[53] Hierzu: Immenga/Mestmäcker/*Zimmer*, WettbR, AEUV Art. 101 Rn. 107.

[54] EuGH, Urt. v. 28.05.1998 – C-7/95 = BeckRS 2004, 77716 Rn. 87; Urt. v. 21.01.1999 – C-215/96 = BeckRS 2004, 75147 Rn. 33; Urt. v. 04.06.2009 – C-8/08 = EuZW 2009, 505 (506 f. Rn. 33); Dauses/Ludwigs/*Pieper*, HdB-EuWiRe, H. I. § 2 Rn. 46.

[55] EuGH, Urt. v. 30.06.1966 – C-56/65 = BeckRS 2004, 73331; Urt. v. 20.11.2008 – C-209/07 = BeckRS 2008, 71211 Rn. 21; Urt. v. 04.06.2009 – C-8/08 = EuZW 2009, 505 (506 f. Rn. 28, 30); Urt. v. 14.03.2013 – C-32/11 = EuZW 2013, 716 (718 Rn. 34, 37 ff.); Geiger/Khan/Kotzur/Kirchmair/*Khan*, EUV/AEUV, AEUV Art. 101 Rn. 26 f.; Immenga/Mestmäcker/*Zimmer*, WettbR, AEUV Art. 101 Rn. 129.

werden (Art. 101 Abs. 1 Bst. d AEUV) sowie in an den Abschluss von Verträgen geknüpften Bedingungen, dass die Vertragspartner zusätzliche Leistungen annehmen, die weder sachlich noch nach Handelsbrauch in Beziehung zum Vertragsgegenstand stehen (Art. 101 Abs. 1 Bst. e AEUV) bestehen.

Auf weitere mögliche, nicht als Regelbeispiel aufgeführte Fallgruppen wird unter IV. eingegangen, sofern diese in Hinblick auf die Fragestellung als einschlägig in Betracht kommen.

3.3.8 Ausnahmen nach Abs. 3

Von diesen Beschränkungen können nach Art. 101 Abs. 3 HS 1 AEUV Vereinbarungen oder Gruppen von Vereinbarungen zwischen Unternehmen, Beschlüsse oder Gruppen von Beschlüssen von Unternehmensvereinigungen sowie aufeinander abgestimmte Verhaltensweisen oder Gruppen von solchen ausgenommen werden, sofern die unter angemessener Beteiligung der Verbraucher an dem entstehenden Gewinn zur Verbesserung der Warenerzeugung oder -verteilung oder zur Förderung des technischen oder wirtschaftlichen Fortschritts beitragen. In diesem Fall sollen den betroffenen Unternehmen einerseits keine Beschränkungen auferlegt werden, die für die Verwirklichung dieser Ziele nicht unerlässlich sind, andererseits soll ihnen auch nicht die Möglichkeit eröffnet werden, für einen wesentlichen Teil der betreffenden Waren den Wettbewerb auszuschalten (Art. 101 Abs. 3 HS 2 AEUV).

3.3.9 Rechtsfolge

Als Rechtsfolge sieht Art. 101 Abs. 2 AEUV die Nichtigkeit von den nach Art. 101 AEUV verbotenen Vereinbarungen und Beschlüssen vor.

4 Ökosysteme im Kontext des EU-Kartellrechts

4.1 Anwendbarkeit des Art. 101 AEUV auf etablierte Ökosysteme

Ökosysteme können aufgrund ihres Agierens als Zusammenschluss einzelner Unternehmen (im kartellrechtlichen Sinne) mit gemeinsamen Interessen und einem gewissen Maße an Mitentscheidungsbefugnis hinsichtlich des Auftretens am Markt jedenfalls unter den Begriff der Unternehmensvereinigung subsumiert werden.

Hinsichtlich etablierter Ökosysteme im hier verstandenen Sinne erscheint eine Anwendbarkeit des Kartellverbotes jedoch zunächst dahingehend fraglich, als dass zu klären ist, ob der tatbestandliche „Wettbewerb“ das Anbieten *gleichartiger* Leistungen erfordert und ob die Teilnahme der einzelnen Anbieter am Markt nicht gegenseitig bedingt sein darf.

Im Rahmen der Ökosystematisierung geht es schließlich gerade um derartige Zusammenschlüsse, die bestehende wirtschaftlichen Leistungen (Waren oder Dienstleistungen) der Akteure entweder zu einem Produktbündel verbinden, wobei jedoch die individuellen Leistungen der Akteure als solche bestehen bleiben (vgl. Prämisse 3a), oder die im Zuge eines Synergieeffekts eine bisher nichtexistierende wirtschaftliche Leistung schaffen (vgl. Prämisse 3b). Die Mitglieder des Ökosystems werden sich damit gerade deshalb zusammenschließen, weil sie nicht oder nicht vollständig allein in der Lage sind, gewisse Kundenbedürfnisse abzudecken.

Im Vergleich zu der Art des Zusammenschlusses im Rahmen von „klassischen“ Kartellen, im Zuge dessen sich Akteure zusammenschließen, welche auf einem bestehenden Markt gleichartige Leistungen anbieten, verhält es sich bei Ökosystemen aber gerade so, dass die Akteure im *ökonomischen* Sinne *miteinander* mitunter geradeeben nicht zwangsläufig zueinander in einem Wettbewerbsverhältnis stehen. Dieser Gedanke lässt sich anhand eines vereinfachten Beispiels darstellen: Ein Ökosystem, verfolgt durch seinen „Orchestrator“ zentralisiert das Ziel, am Markt ein vollumfänglich nutzbares Kraftfahrzeug anzubieten. Es schließen sich daher Akteure zusammen, welche über fundierte Kenntnisse über die Entwicklung und Produktion bestimmter technischer Komponenten (Fahrwerk, Bremsen, Getriebe etc.) sowie Vertrieb, Buchhaltung u.ä. verfügen. Die Mitglieder stehen daher miteinander nicht in direktem Wettbewerb zueinander, da beispielsweise Mitglied A für die Entwicklung der Fahrwerkskomponenten, Mitglied B jedoch für die Entwicklung der Bremsen zuständig ist. Mitglied A und Mitglied B agieren daher grundsätzlich, für sich allein betrachtet, auf unterschiedlichen Märkten. Erst durch die Bildung des Ökosystems treten sie gemeinsam in den gleichen Markt ein, ihre Teilnahme am Markt wird jedoch, da A keine Bremsen und B keine Fahrwerkskomponenten liefern kann, durch die Zusammenarbeit miteinander sowie mit den restlichen Mitgliedern bedingt. Es stellt sich somit die Frage, ob überhaupt ein „Wettbewerb“ i. S. d. Norm vorliegt.

Wie oben dargelegt, basiert nach der Rechtsprechung der tatbestandliche Wettbewerb auf Leistungen der Marktbürger, d. h. er ist als ein Leistungswettbewerb zu verstehen, in welchem mehrere Anbieter Leistungen auf einem Markt anbieten, deren Nachfrage durch Preis oder Qualität determiniert wird.[56] Die Gleichartigkeit der Leistungen wird damit zumindest nicht explizit gefordert, sodass dies einer entsprechenden Auslegung nicht entgegenstünde.

Der offene Wortlaut der Tatbestandsvariante der Einschränkung oder Kontrolle der Erzeugung, des Absatzes, der technischen Entwicklung oder der Investitionen nach Art. 101 Abs. 1 Bst. b AEUV könnte unter Berücksichtigung der bisherigen Auslegung für eine Subsumtion der Ökosystemteilnehmer unter den Wettbewerbsbegriff sprechen. Die Variante erfasst gerade auch Arbeitsgemeinschaften von Unternehmen, d. h. die gemeinschaftliche Durchführung und Koordinierung unternehmerischer Vorhaben, jedenfalls dann, wenn diese entweder *allein* in der Lage wären sich am entsprechenden Markt zu betätigen, was im Hinblick auf technische Kapazitäten, finanzielle Ressourcen und den Umfang des unternehmerischen Risikos zu bewerten ist, und die Akteure miteinander auf dem entsprechenden Markt im Wettbewerb stehen.[57] Literatur[58] und Kommission[59] gehen zwar davon aus, dass im Falle bloß „komplementärer" Arbeitsgemeinschaften, d. h. dem oben skizzierten Zusammenschluss einzelner Akteure, Art. 101 Abs. 1 AEUV schon alleine deshalb nicht tangiert sei, da die Vorhaben von den einzelnen Mitgliedern der Arbeitsgemeinschaft nicht allein umgesetzt werden könnten.

Diese Ansicht kann nicht überzeugen, da sie den eigens ausgeführten Gedankengang nicht vollständig zu Ende führt. Gerade mit fortschreitendem Bestehen des Ökosystems wird zwischen den einzelnen Mitgliedern die Abhängigkeit voneinander derart stark werden, dass der Druck, im übergeordneten Wettbewerb (zu den anderen bestehenden Ökosystemen) weiterhin bestehen zu können, zu Zwangssituationen führen kann – verlässt ein Mitglied das Ökosystem, bricht es zusammen. Ebendiese Co-Abhängigkeit kann dazu führen, dass einzelne Mitglieder mit dem Beenden der Zusammenarbeit drohen können, um eigene Vorteile durchzusetzen. So wäre es einem Mitglied, welches de facto ein Monopol auf seinem eigenen Markt (z. B. Bremsenherstellung) bekleidet möglich, das gesamte Ökosystem (welches sich auf die Herstellung von Fahrzeugen in ihrer Gesamtheit spezialisiert hat) dazu zu bringen, beispielsweise die

---

[56] Streinz/*Eilmansberger*/*Kruis*, EUV/AEUV, AEUV Art. 101 Rn. 40; s. auch: EuGH, Urt. v. 15.03.2007 – C-95/04 P = JuS 2007, 975 (976 Rn. 66); Calliess/Ruffert/*Weiß*, AEUV Art. 101 Rn. 78; Grabitz/Hilf/Nettesheim/*Stockenhuber*, EU-Recht, AEUV Art. 101 Rn. 116.

[57] Dauses/Ludwigs/*Pieper*, HdB-EuWiRe, H. I. § 2 Rn. 97.

[58] So z. B. Grabitz/Hilf/Nettesheim/*Schroeder*, EU-Recht, AEUV Art. 101 Rn. 531 ff.

[59] Entscheidung der Kommission v. 08.08.1999 – 1999/574/EG, ABl. 1999 L 218, S. 24 Rn. 32; Bekanntmachung der Kommission v. 29.7.1968 – ABl. 1968 C 93, 14 Abschn. II.5.

Menge der insgesamt produzierten Fahrzeuge zu reduzieren, wenn der Monopolist die Menge der produzierten Einheiten auf seinem eigenen Markt zu reduzieren gedenkt. Es käme damit zu einer marktübergreifenden Wettbewerbsverzerrung. Derartige abgestimmte Verhaltensweisen aus dem Anwendungsbereich des Art. 101 AEUV auszunehmen, erscheint daher nicht vollziehbar.

Hierfür spricht gerade auch das Telos der Kartellvorschriften, wie es seitens des EuGH verstanden wird, namentlich der Schutz des Wettbewerbs basierend auf *Leistungen* der Marktbürger[60], spricht gerade für die Einbeziehung solcher komplementären Arbeitsgemeinschaften. Dies entspräche ebenfalls dem oben dargestellten Gedanken, den Tatbestand nicht zu stark zu verengen, um auf zukünftige ökonomische Entwicklung reagieren zu können.

4.2 Anwendbarkeit des Art. 101 AEUV auf neuerrichtete Ökosysteme

Einer Anwendbarkeit des Art. 101 AEUV auf neuerrichtete Ökosysteme steht nicht entgegen, dass die zuvor skizzierte, sich mit der Zeit entwickelnde Einflussmöglichkeit der Ökosystemmitglieder aufeinander im Zeitpunkt der Errichtung des Ökosystems noch nicht gegeben ist, da gerade auch ein potentieller Wettbewerb erfasst wird.[61]

5 Fazit

Im Ergebnis würde ein Wandel hin zu einer Ökosystematisierung mithin dazu führen, dass sich das europäische Kartellrecht der Herausforderung zu stellen hätte, dass es durch den Zusammenschluss mehrerer Akteure aus unterschiedlichen Stufen der Wertschöpfungskette zu einer „dauerhaften komplementären Arbeitsgemeinschaft“ zu marktübergreifenden Wettbewerbsverzerrungen käme.

Um auf eine derartige Entwicklung reagieren zu können, erscheint es daher geboten, auch komplementäre Arbeitsgemeinschaften in den Anwendungsbereich des Art. 101 AEUV miteinzubeziehen.

---

[60] EuGH, Urt. v. 13.02.1979 – C-85/76 = BeckRS 2004, 73814 Rn. 91.

[61] S. oben.

Literatur

BeckOK-InfoMedR/Bearbeiter: BeckOK Informations- und Medienrecht, 41. Edition, München 2023
Bechtold/Bosch/Brinker: EU-Kartellrecht, 4. Auflage, München 2023
Bieber/Epiney/Haag/Kotzur: Europarecht, 7. Auflage, Baden-Baden 2023
Calliess/Ruffert/Bearbeiter: EUV/AEUV – Das Verfassungsrecht der Europäischen Union Kommentar, 6. Auflage, München 2023
Dauses/Ludwigs/Bearbeiter: Handbuch des EU-Wirtschaftsrechts, 59. Ergänzungslieferung, München 2023
Dürig/Herzog/Scholz/Bearbeiter: Grundgesetz-Kommentar, 101. Ergänzungslieferung, München 2023
Geiger/Khan/Kotzur/Kirchmair/Bearbeiter: EUV/AEUV, 7. Auflage, München 2023
Grabitz/Hilf/Nettesheim/Bearbeiter: Das Recht der Europäischen Union, 80. Ergänzungslieferung, München 2023
Hasselblatt/Bearbeiter: Münchener Anwaltshandbuch Gewerblicher Rechtsschutz, 6. Auflage, München 2022
Herdegen: Europarecht, 24. Auflage, München 2023
Hoeren/Sieber/Holznagel/Bearbeiter: Handbuch Multimedia-Recht, 59. Ergänzungslieferung, München 2023
Jüchser/Klein: Datenschutz im Asset-Deal – Ein unternehmensbezogener Ansatz, NZI 2023, 370–374.
Immenga/Mestmäcker/Bearbeiter: Wettbewerbsrecht Band I, 6. Auflage, München 2019
Langen/Bunte/Bearbeiter: Kartellrecht Band II, 10. Auflage, München 2006
Loewenheim/Meessen/Riesenkampff/Kersting/Meyer-Lindemann/Bearbeiter: Kartellrecht, 4. Auflage, München 2020
MüKo-EUWettbR/Bearbeiter: Münchener Kommentar zum Wettbewerbsrecht Band I Teilband 1, 4. Auflage, München 2023
MüKo-BGB/Bearbeiter: Münchener Kommentar zum Bürgerlichen Gesetzbuch Band XIII, 8. Auflage, München 2021
Münch-HdB-GesR/Bearbeiter: Münchener Handbuch des Gesellschaftsrechts Band II, 5. Auflage, München 2019
Payandeh: Das unionsverfassungsrechtliche Rechtsstaatsprinzip, JuS 2021, 481–489
Saenger/Aderhold/Lenkaitis/Speckmann/Bearbeiter: Handels- und Gesellschaftsrecht, 2. Auflage, Baden-Baden 2011
Streinz/Bearbeiter: EUV/AEUV, 3. Auflage, München 2018
von der Groeben/Schwarze/Hatje/Bearbeiter: Europäisches Unionsrecht, 7. Auflage, Baden-Baden 2015

Mastering Execution: Navigating De-Risking Stages and Building a Business

Sebastian David Hersberger

1 Introduction

When it comes to entrepreneurship, I firmly believe that it is the embodiment of creativity, unwavering perseverance, and a willingness to take calculated risks. Academic literature on how to become a successful entrepreneur is readily available. However, acknowledging the fact that being able to make strategic decisions and execute them are two factors that define the success of an entrepreneur. Mastering such execution requires one to be able to use their skill set to take an idea from its initial phase and turn it into a well-managed business, simply said.

As a deep tech venture progresses, it must strategically mitigate a myriad of initial risks, Fig. 1 provides an overview about the main risks you face as a company. The evolution of these risks is systematically illustrated in this chapter, which delineates how, with increasing maturity, a startup diminishes each risk. Initially, all risks—including for example technological, regulatory, and market risks—are prevalent. As the company matures, it prioritizes and addresses these risks sequentially. Technological risks are quelled through rigorous product development, while regulatory risks are allayed via stringent compliance.

In the early stages of a business, certain risks are particularly pronounced, but as these risks are progressively addressed and resolved, they diminish, leading to a significant transformation in the company's risk profile as perceived by investors. This shift not only impacts the company's valuation but is also indicative of its developmental progress. Managing this evolution presents considerable

S. D. Hersberger (✉)
sh Ventures GmbH, Allschwil, Switzerland
e-mail: sebastian@hersberger.com

C. H. Hoffmann (ed.), *Artificial Intelligence, Entrepreneurship and Risk*, Technikzukünfte, Wissenschaft und Gesellschaft / Futures of Technology, Science and Society, https://doi.org/10.1007/978-3-658-45544-6_22

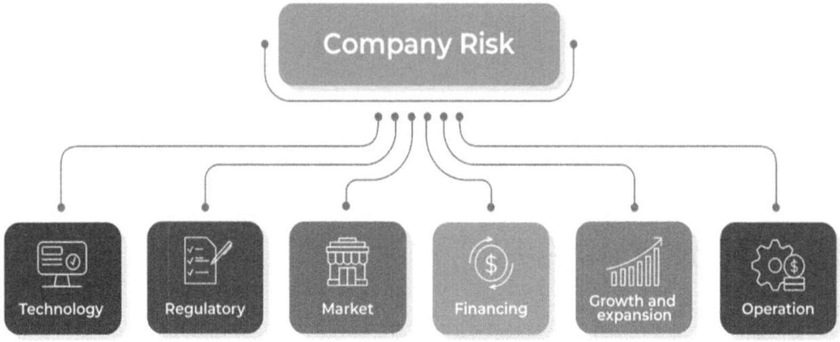

Fig. 1 Company risk

challenges and demands substantial adaptability and effort from both the founders and their team, as the requirements evolve across various domains of the business.

Before we proceed, it's important to know that successfully navigating this journey requires one to have sound decision-making capabilities. Additionally, one must be able to lead through examples. To take a business from a chaotic idea to a thriving venture, different skill sets pertaining to team development and alignment, business planning, managing creatives, and more are required. Mastering execution through all of these phases requires entrepreneurs to fully comprehend the early stages of the business and build a solid foundation.

2 Early Stages of a Business

Starting a business can be a hectic time, centered on transforming an idea into a successful enterprise. Many overlook this critical phase, rushing through without proper planning. Yet, this is the stage that forms the bedrock of your entire business. A solid foundation is crucial; without it, a business cannot stand firm.

That said, it's important for one to comprehend what the early stage of the business entails and the risks they might be exposed to. This phase generally consists of a vast array of activities pertaining to planning and development. This may include market research, product development, team building, and creating a business plan. This stage requires founders to focus on different factors like product development, regulation, product-market fit, business model validation, pricing, and financing.

A business may lack the necessary resources to properly assess or manage risks. Before exploring solutions to this issue, let's first examine the essential tasks that need attention in the early stages of a business. These operations include:

- *Market Research*
  Market research is the process of how one can assess the viability of a product or service the business offers. When conducting market research, you basically identify your target audience and aim to get their opinions on the product or service you have to offer. You can do this research by using surveys, focus groups, or other methods like interviews.
- *Business Plan*
  A business plan serves as a key document, detailing strategic and financial objectives and the methods to attain them. This comprehensive document encompasses extensive details on marketing, pricing, and distribution strategies. Additionally, it features financial projections. Crucial in a business's early stages, it assists in securing investor funding and fostering collaborations with potential business partners.
- *Team Building*
  Building a team means bringing a group together to pursue a shared objective. It's essential that all team members understand and agree on what is expected of them. This alignment helps guarantee everyone is contributing the right level of effort to reach the goals. Moreover, it's important to cultivate an environment where team members trust, support, and respect one another, which is key to encouraging teamwork.
- *Product Development*
  Product development is the entire process of taking the product all the way through the concept to the delivery phase. This includes ideation, planning, manufacturing, and more. However, during this phase, it's essential to incorporate the feedback you receive from your early customers. This feedback can also be used to create minimum viable product (MVP) standards.

Early-stage startups often lack the resources necessary for operational and marketing needs. As far as the risks and challenges are concerned, you need to remember that the risks and challenges prevalent in the early stages may vary among startups. Despite this, these risks will generally occur with regard to:

- Maintaining the quality of the product.
- Entering the market.

- Finding business partners.
- Finding skilled and talented workers.
- Building a team of core employees.
- Competing with larger players in the industry.
- Managing finances to ensure sustainable operations.
- Adhering to legal requirements.

Focus is the key to success. As the business grows, it'll transition into different phases, and the challenges within each phase will differ. Later in this chapter, we will get into risk management strategies and navigate the stages of de-risking. But first, let's focus on comprehending the significance of continuous strategy and planning.

3 Role of Continuous Strategy and Planning

Many early-stage business owners mistakenly think that strategic planning is a one-time task to be done before action and rigidly followed thereafter. However, it's crucial to understand that a growing business evolves through different phases, each with distinct risks. This reality renders a static strategy ineffective.

Businesses must, therefore, engage in ongoing strategizing and planning, adapting to real-time changes while being mindful of limited resources. A significant challenge for startups is that while they can plan for certain objectives, they might not always have the means to achieve them. This often necessitates compromises and strategic decision-making. However, without careful consideration, such decisions can impede progress.

Conducting a quadrant analysis, which assesses the importance and urgency of tasks, can be a valuable tool in decision-making, helping to ensure that choices lead to the desired outcome, see Fig. 2.

Business owners should focus on working 'on' the business, not just 'in' it. This means prioritizing strategic thinking and planning, which is often neglected in the hustle of daily operations.

When it comes to de-risking, it's important to understand that an idea may appear risk-free or low-risk on paper, but the result may be different when executed. This can cause severe problems for businesses in their infant stage. Due to such realities, constant strategizing and planning are necessary. When planning for your business and developing strategies in its initial phases, you need to:

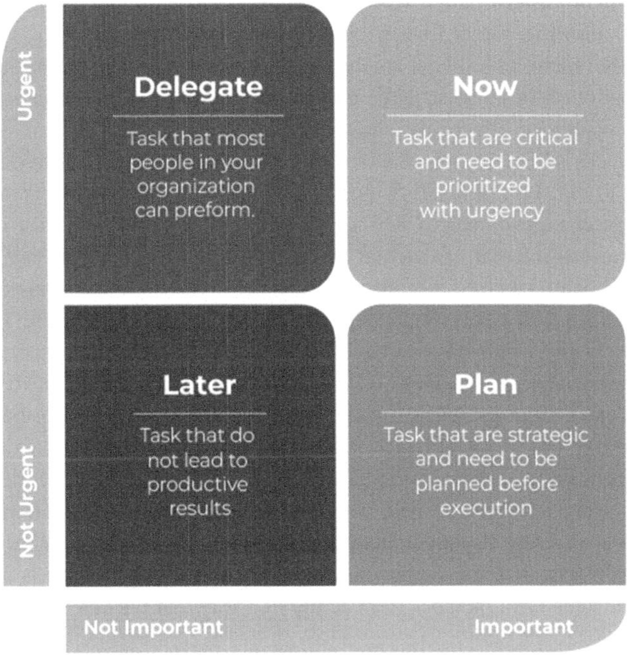

Fig. 2 Prioritization quadrant

1. **Identify risk**—de-risking requires one to have a proactive approach, meaning that potential risks should be identified prior to the execution. Such risks may prevail due to monetary constraints, unexpected consumer demands, tech-related limitations, or other similar factors. Identifying such risks before execution can help businesses minimize their impact and eliminate them altogether.
2. **Establish realistic goals**—when it comes to risk management, you need to know that developing realistic goals is critical. Having goals that you can attain and measure allows you to stay on track and also helps you avoid unnecessary risks. However, when you're establishing your goals, it's important to consider different factors, such as time and resource constraints and potential obstacles.
3. **Create a detailed plan**—developing a comprehensive plan for growth operations can also help reduce risk as it considers the milestones, deliverables, resources, and deadlines associated with each risk. The comprehensiveness

of the plan ensures that it acts as a roadmap for your strategy and planning efforts, allowing you to foresee obstacles and minimize risk.

4. **Allocate resources**—this is another major element that can aid in de-risking. As mentioned, businesses, while in their infancy, have limited resources. Allocating resources effectively allows you to minimize the risk of delays or bottlenecks. However, when allocating resources, it's important to do so with regard to the skill, availability, and workload of the team members.
5. **Manage change**—changes during execution are a given, and they can arise from various internal or external factors. Without a plan for managing these changes, your business becomes vulnerable to risks. It's crucial to have a well-defined and easily implementable change management plan. This plan should lay out clear steps for requesting and implementing necessary changes. However, when putting these changes into action, it's important to evaluate their impact on your overall goals, budget, and resources to ensure smooth integration.
6. **Communicate clearly**—oftentimes, the most overlooked aspect of de-risking. When trying to eliminate risk, you need to incorporate clear communication protocols as a part of your strategy and planning. This not only helps you gain regular insights but also helps you ensure team alignment, which is necessary as it can keep each member from acting in a way that puts the entire business at risk.
7. **Leverage agile methods**—one of the best things you can do to de-risk effectively is to use agile planning and management methodologies. Agile methodologies promote collaboration and transparency among team members as they can be used to break down large projects into smaller, more manageable chunks. It helps eliminate risk because the process creates a necessity for constant review and adjustments.
8. **Develop mitigation strategies**—proactive risk planning involves preparing for potential risks before they actually happen. Businesses develop mitigation strategies to have an action plan in place, ready to be activated as soon as a risk materializes. Such strategies typically include contingency plans and monitoring systems. It's crucial to evaluate these strategies in advance, considering their impact on the business's overall goals, budget, and resources, to ensure they are effective and ready for implementation.

When planning strategically for your business, you'll soon acknowledge that many of the risks and challenges you're facing stem from the scarcity of resources you have. To address the risks and challenges based on this insight, you'll need to access capital. One approach here is to enter the market, increase revenue, and build capital over time. However, this approach can limit your ability to scale as it increases your exposure to risk since it can't be addressed until you have secured the needed capital.

Seeking the required funding from investors or potential business partners, on the other hand, can be a more feasible option. This not only gives you the monetary resources needed to scale but also allows you to address risks your business is exposed to. However, to do this, you need to have a solid business case that can be used as part of your investment pitch. You see, the business case basically provides justification for why you've started the venture.

Before moving forward, it's essential to grasp that the core principle of a business case is to convincingly show that the venture's benefits surpass its costs. This underscores the importance of dedicating sufficient time to craft a compelling case. As you develop your business case, keep in mind that it should encompass five key elements:

- **Strategic context**—information pertaining to why the change is necessary.
- **Economic analysis**—information about the return on investment (ROI) based on the investment options that were appraised.
- **Commercial insights**—details pertaining to the sourcing and procurement strategy of the business.
- **Financial case**—an analysis of affordability with regard to the time frame.
- **Management**—details about the roles, governance structure, life cycle, and decision-making processes.

Writing a business case can be challenging for those who don't have adequate knowledge. However, given the importance of the document, it's essential that you learn how to create one. Provided below is a simple four-step method I use to create a compelling business case.

1. **Identify the problem or opportunity**
   The first and most important thing you need to do when creating a business case is to identify a problem or opportunity. When you're identifying the problem or opportunity, it's essential that you provide all the necessary details, the root cause, and the time needed to fix it.
2. **Determine solutions**
   Once you've identified the problem, you need to come up with alternative solutions that address the problem. However, it's important to include the costs and benefits associated with each solution. In addition, you also need

to determine any risks that you may face and the overall feasibility of each option.

3. **Choose an option**
   When you identify potential options, you need to identify the most feasible one. However, this is something that can be challenging as what's feasible to you as a business might not be of value to the customer. What this means is that your decision needs to be based on both feasibility and perceived customer value. Using the decision matrix, see Fig. 3, can help you choose options that meet both these requirements.
4. **Outline implementation**
   After choosing your preferred option, the next step is to detail your approach. This involves providing comprehensive information on how you intend to implement the chosen solution and develop it into a successful business venture.

Now that you're well aware of why constant strategizing and planning are necessary when a business is in its initial phases, let us move on to the next challenge: building a team.

Fig. 3 Decision matrix

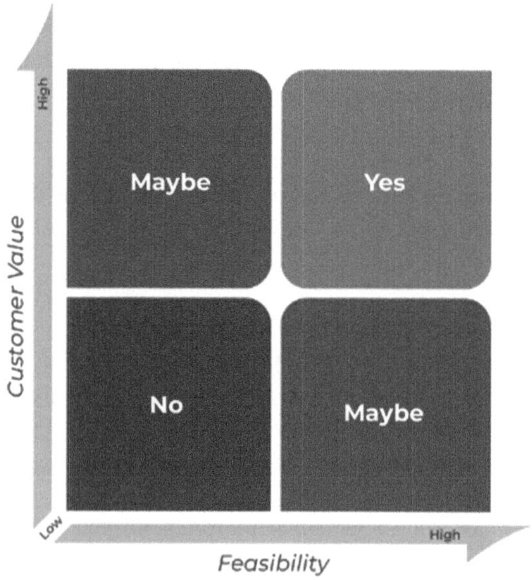

4 Getting Started with a Founding Team

Most people hear the word *"team"* and automatically think that they need to create different departments within the company and hire lots of employees. However, it's important to ask if all that is actually necessary in the early stage of a startup. Truth be told, I've always found the first team of a startup is built with just a few members who are core to the business.

However, it's important to understand that when you're building the founder team for a startup, individuals within that team need to be skilled in key areas. These areas include product development, business management, and commercialization—a multidisciplinary team is key for successful execution.

Having a member who's skilled in product development will allow you to come up with innovative solutions to customer pain points in a viable manner. Having team members skilled in commercialization will aid in scaling your venture, generating new revenue streams, and boosting overall revenue. Meanwhile, those with business management expertise will assist in organizing your company and managing your initial employees. Once these foundational members are in place, you can then move on to expanding your team further. If some skills are not represented in the founder team, make sure you hire them as first employees.

There is a vast array of academic literature on psychological traits that make founders successful or govern their decision-making. Research studies have measured the personality traits and risk attitudes of entrepreneurs against the goals and aspirations they bring to the table.

5 Building a Team While the Business is in Infancy

Successful businesses in any industry owe much of their success to skilled professionals who drive the company towards its goals. A business, while in its early stage, experiences quite a lot of changes and uncertainties. It's important for you to know that both these factors can either directly or indirectly impact the end goal. Any significant changes to the end goal render the efforts of certain members invaluable without them being at fault.

An imbalance and sense of unfairness among team members can hinder effective collaboration. Additionally, the inexperience of team members poses another risk to team building. It's crucial to understand that collaboration is a key factor in forming an effective team.

However, members who don't have adequate experience often find it hard to actively listen to and communicate with others. This makes it difficult for them to work together and negatively impacts the overall productivity of the team. In addition, another significant risk that can negatively impact team building is the members' lack of adaptability and flexibility.

As mentioned, businesses in their early stage face rapid changes. When you couple this with limited resources, it means that team members may have to wear more than one hat at a time as they are expected to fulfill varying duties. This can be challenging with inexperienced teams as they may not possess the skills needed to adapt to changing circumstances.

When you look at these challenges, one of the main things that will likely strike out is the inexperience or lack of skills members are likely to have. In addition, some team members may not have the required motivation or will to work in a team or contribute towards the startup's growth. Simply put, they'll just be there to do their jobs.

Understanding that some team members might lack the necessary will or skills is important, but as a business founder, you likely have both. Often, those who initiate the right business for themselves possess the essential will and core skills required. This implies your role includes providing motivation and guidance to each team member.

It's crucial to recognize that a one-size-fits-all training session may not be the most effective approach. Instead, identifying and catering to the specific needs of each team member is key. For this purpose, I often use the skill-will matrix to analyze team members effectively, see Fig. 4.

This matrix basically allows you to categorize team members based on their willingness and skill level. Once you've done that, you'll know what type of support you need to provide to members of each category. In addition, when it comes to building a team, you need to:

1. **Define the purpose and goals**
   Defining the purpose is the first and most crucial step. This involves laying out the overall mission and objectives to be achieved. Providing this clear direction helps team members understand their roles and tailor their efforts accordingly.
2. **Find the right people**
   One of the largest challenges for businesses in their early stages is the difficulty of finding team members with the right skills and experience. Nonetheless, it's essential to remember that not every team member needs to

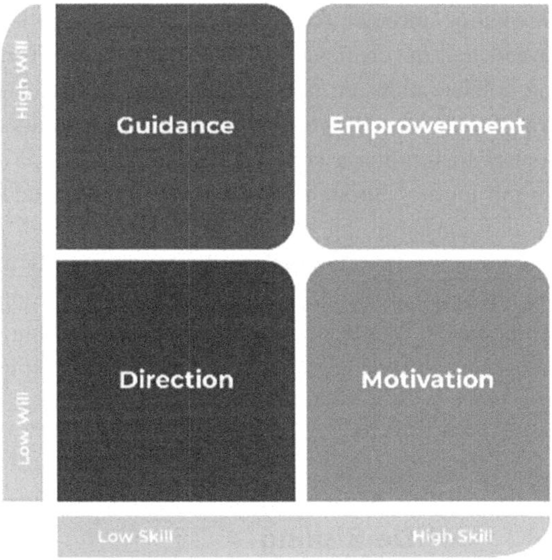

Fig. 4 Skill-will matrix

be an expert. In some cases, having individuals with average or above-average capabilities across various domains can be more beneficial.

3. **Nurture collaboration**
   To have an environment where team members collaborate seamlessly with others, you need to ensure that they are comfortable sharing their opinions and ideas. You can do this by encouraging communication, providing support, and promoting creativity.
4. **Provide training and support**
   Providing your team members with the right training and support is the next crucial step. To effectively do this, first identify what each member needs. As mentioned earlier, the skill-will matrix can be a useful tool for this assessment. Supporting your team's growth as professionals not only benefits their development but also enhances the value of their contributions to your business.
5. **Measure progress**
   Lastly, one of the most important things you need to do is to monitor their progress and growth. This will not only help identify members who are growing and becoming more valuable but will also help you determine how effective your support and guidance really are.

Navigating through the intricacies of team building in early-stage businesses reveals that the challenges of managing change, fostering adaptability, and ensuring collaboration are not merely obstacles but essential elements of the team's growth and the startup's development. This evolution from handling initial uncertainties to forming a cohesive, dynamic team is a microcosm of the company's overall journey through various stages of risk and investment. For founders, the focus is on nurturing this transformation through precise motivation, guidance, and training, each tailored to the unique needs of team members. Emphasizing strength, resilience, and adaptability, this comprehensive approach to team development is crucial in navigating the shifting risk profiles and investment landscapes, setting a robust foundation for the company's sustained growth and success.

6 Risk Management and Different Phases of De-Risking

The transition of a business from its early stage through to the late stage, growth equity phase, and ultimately the growth buyout phase presents a complex and dynamic evolution, as depicted in Fig. 5. The following section navigates the shifting landscape of risks, from technological to regulatory, market, and operational, highlighting how they transform as the business matures. Key to this journey is understanding de-risking—the strategic process of minimizing risks to enhance the startup's appeal to investors. We will explore how these risks and investor profiles vary across different funding stages, from seed to later stage funding rounds, and how industry-specific challenges influence them. This analysis aims to equip entrepreneurs with the knowledge to navigate these stages adeptly, leveraging risk management for strategic growth and development.

In the early stages, technology risks loom large. For instance, a new artificial intelligence startup might face significant challenges in algorithm development or data acquisition. These technological hurdles are prominent but gradually mitigate as the startup refines its core offerings. This phase is also riddled with regulatory and market risks. A clean energy company, early in its life cycle, might struggle with securing patents and navigating the complexity of environmental regulations while also trying to carve out a market niche against well-established energy conglomerates.

Progressing into the late-stage venture, the focus shifts as operational, growth, and financing risks intensify. Consider a mobile gaming company that has

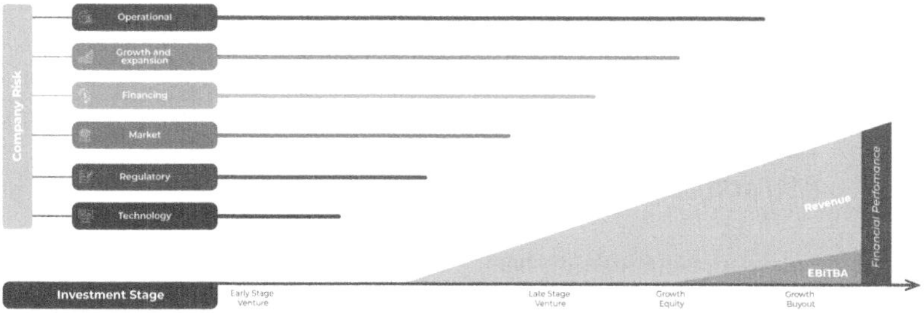

Fig. 5 Phases of de-risking

successfully launched several games. As it enters this phase, the company's concerns may pivot towards scaling its user base, diversifying game offerings, and managing the substantial financial injections needed to fund these initiatives.

During the growth equity phase, Fig. 5 illustrates a dilution in technology risks—signaling that companies typically have established technological frameworks. Nonetheless, the battle with market risks continues unabated. A digital health company, for example, might have honed its telemedicine platform but now faces the challenge of adapting to changing healthcare policies and patient adoption rates.

As the business approaches the growth buyout stage, the emphasis magnifies on financial metrics such as revenue and EBITDA, as evidenced by the expanding orange gradient in Fig. 5. This underscores the shifting priorities toward achieving and demonstrating solid financial health and profitability. A mature cybersecurity firm at this stage may focus on optimizing its sales cycle, customer retention rates, and operational efficiencies to showcase its financial prowess to potential acquirers or public market investors.

Effective de-risking, as a business matures, involves not just mitigating known risks but also anticipating new ones that may emerge. A company that started as a software developer, for example, might find itself needing to transition into a service-oriented model as it scales, which introduces a whole new set of operational risks related to service delivery and client management.

De-risking is an ongoing, strategic process that evolves with the company's lifecycle. Early on, it's about securing technological capability and market entry. In the middle stages, the focus is on building robust operational frameworks and securing growth capital. And in the final stages, the spotlight turns to fine-tuning financial performance and preparing for exit strategies. Each phase demands a detailed understanding of the prevailing risks, strategic foresight, and an ability to secure the right kind of investment to shepherd the company through to the next phase of its evolution.

7 The Impact of De-Risking on Investment Attractiveness

De-risking plays a pivotal role in shaping the investment attractiveness of a company through each stage of its evolution, influencing the type and engagement level of potential investors. In the initial seed and early stages, where the risks are most pronounced, angel investors and early stage focussed venture capitalists are often the primary contributors. They are drawn to the high-risk environment by the potential for substantial returns and the opportunity to guide innovative ideas to market fruition. Their investment decisions are driven by the business's ability to mitigate technological uncertainties and capture market opportunities, which, in turn, can significantly reduce the perceived risk and increase the startup's valuation.

As the company matures and moves into later funding rounds, the de-risking focus shifts towards operational efficiency, market penetration, and financial stability, attracting a different class of investors. Growth-stage investors, including later-stage venture capitalists and private equity firms, typically look for businesses with established market fit, clear operational strategies, and the ability to scale rapidly. The effectiveness of a company's de-risking measures in earlier phases can greatly enhance its appeal to these investors, who bring not only larger capital injections but also strategic acumen and a network of resources to support further scaling.

For companies entering the Series C, D, or pre-IPO stages, de-risking becomes synonymous with preparing for market leadership or a successful exit. The investor pool at this stage may include not only late-stage venture capitalists and private equity firms but also hedge funds and investment banks. These investors are interested in companies with de-risked profiles, showcasing stable revenue streams, strong governance, and solid market positioning. The degree to which a company has successfully navigated and mitigated risks in its earlier stages directly impacts these investors' assessments of growth potential and exit strategies, making de-risking a critical component of a company's growth narrative and investment proposition.

8 Having an Evolving Skill Set

De-risking a startup crucially hinges on a founder's ability to adapt and evolve their skillset. Initially, a founder might excel as a creative thinker who handles daily operations, but as the company grows, they must transition into a professional capable of managing a structured organization.

Often overlooked, this aspect of de-risking is critical. Building a company from the ground up doesn't automatically equip one with the skills to sustain it at its peak. Scaling the heights is one challenge; maintaining that altitude is quite another. Throughout the various stages of a startup's life cycle, the necessary skills for effective management evolve. Early on, managing a small team might require just effective communication and networking. But as the business expands and moves into later stages, these skills alone will no longer suffice to handle a growing team.

Under such circumstances, you'll need to have the skills needed to create an organizational structure. However, to develop an evolving skill set, you need to understand how it differs from a static one. You see, the static skill set is something that's often seen as taking the specialist route. Where such an approach allows you to become an expert at something, it keeps you from having the flexibility that's needed to turn a start-up into a thriving business venture. Working towards an evolving skill set, on the other hand, allows you to be more adaptable and ensures that you're better equipped to leverage opportunities.

That said, some of the most sought-after skills that you need to develop to de-risk a startup may include financial management, problem-solving, and organizational management. When it comes to financial management, most people think that it's all about expenses and profit. Although this might be the essence of the skill, it is not all there is to it. Having the ability to manage business finances is something that should evolve as your business progresses through different stages. In the initial stage, it might be about profit and loss. However, once those initial stages pass, you'll be required to streamline and optimize cash flow. As far as problem-solving is concerned, you've understood by now that each stage of the business will come with a set of challenges that are unique to it. However, having problem-solving skills will allow you to address these challenges despite the unfamiliarity you have with them. In addition, you also need to have organizational management skills, as the structure of your business will change when you transition through different phases. Given that, you must be able to design and implement a management framework and monitoring and reporting protocols in line with your business requirements.

While having such skills is necessary, the secret to de-risking a startup does not lie in having them. It lies within your ability to execute these skills as you lead others and hire the right talent.

9 Executing with Efficiency

One of the most prevalent risks in a start-up environment is the lack of efficient execution. The fact of the matter is that most founders are capable of developing strategies, but they're unable to execute them as needed to achieve the desired results. Having the ability to successfully execute your strategy allows you to ensure team alignment and streamline operations.

Executing with efficiency might seem pretty simple at first, but it does require you to have a planned approach. To do this, you must gather information, identify resources, develop SMART goals, assign responsibilities, and monitor progress. Let's look at each of these in detail to get a better understanding.

1. **Gathering information**
   Gathering information marks the first crucial step in the execution process, which usually intertwines with the planning phase. It's essential to focus on identifying:
   - What do you want to achieve?
   - What is your budget?
   - What is your expected return?
   - How long will the completion take?
   - How many milestones are involved?
   - Are there any expected setbacks?
2. **Identifying resources**
   Determining the resources at your disposal is an essential next step. Understand that these resources can differ from one plan to another but generally encompass employees, material resources, software, and capital. Alongside identifying these resources, considering backup options for each is crucial to enhance the sustainability of your plans.
3. **Creating goals**
   When you're developing goals, you need to use the SMART framework. This framework basically states that your goals can not be generic. They need to be specific. In addition, they also need to be measurable, meaning that you need to identify performance metrics.

Furthermore, they need to be attainable with the resources you have and need to be relevant to your purpose. Lastly, you also need to set a timeframe for these goals by mentioning when certain milestones should be achieved.

4. **Developing a framework**
   Once you've identified what your goals are, the next thing you need to do is create a framework that'll help you achieve them. When you're developing such a framework, you must ensure that it clearly identifies each step of the initiative. In addition, you also need to make sure the framework includes guidelines and fundamental procedures your team can follow and is communicated in an easy-to-understand manner.
5. **Assign responsibilities**
   Assigning responsibilities comes next, which involves defining the specific role each team member will play. Doing so will create team alignment and allow you to improve operational efficiency. However, before assigning responsibilities, you need to ensure that you're well aware of the strengths and weaknesses of each member and are assigning roles based on these factors. Having people do what they're good at will allow you to improve the overall quality of your operations and output.
6. **Monitoring progress**
   Once everything is up and running, you need to regularly monitor the progress to ensure that each initiative is completed as required. When measuring the progress of an initiative, there are various factors that you need to consider. These factors include:
   - **Timeliness**—you need to have specific deadlines and milestones for each initiative to ensure that it's completed in a timely manner.
   - **Budget**—you also need to project and track expenses for different initiatives, as it can help ensure that the project stays within the allocated budget.
   - **Quality**—how you measure quality can differ from one initiative to another. However, it's important to remember that it should always be measured with the end user in mind.

By following the steps mentioned above, you'll be able to execute your plans with greater efficiency. However, it's important for you to remember that measuring how effective your executions really are is essential. To do this, you can look at the number of milestones you've achieved and whether or not your project meets quality and monetary requirements. Success in project execution is strongly dependent on the choice of leadership style.

10 Leadership and Culture

To succeed in a startup environment, the right leadership and culture are critical. We'll delve into each of these aspects, starting with startup culture. Characterized by creativity, risk-taking, and innovation, it demands that founders craft solutions tailored to specific market needs.

What you need to understand is that a start-up culture essentially promotes thinking outside the box and taking risks. Given this, developing de-risking strategies becomes even more essential. Leadership is one of the most important aspects of startup cultures. Leaders basically set the tone for the startup environment and foster creativity and innovation.

In addition, they're also required to create a positive work environment where their employees feel valued and respected. Having the ability to develop such a work environment helps employees be more engaged and motivated at work, which then transitions into more productivity and better results. It can also help you lower turnover and build a team.

Choosing the right leadership style for your organization can be challenging. However, what you need to understand is that how you want to lead is something that depends on clarity and empowerment. You must determine the amount of clarity an employee has and how much you want to empower them to choose the right leadership style for you (Fig. 6).

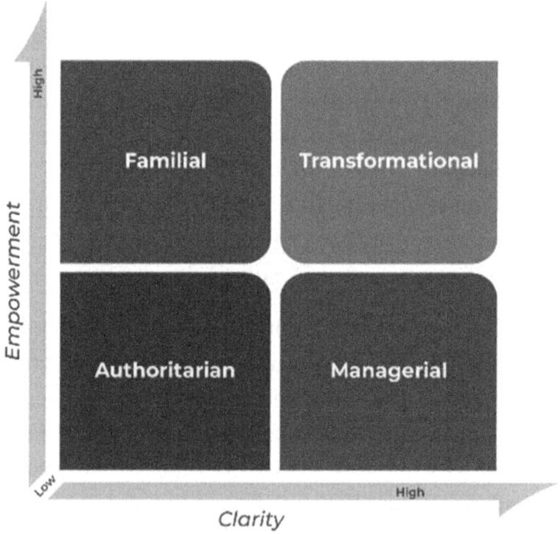

Fig. 6 Leadership style matrix

Using the leadership style matrix is an effective way to determine the kind of leader you need to be. This tool allows you to select a leadership style based on the degree of clarity and empowerment required for one or more employees. It offers four distinct styles to choose from, which include:

1. **Authoritarian leadership**—this type of leadership style basically emerges in a startup environment where both clarity and empowerment are low. It's important to understand that this leadership style is suitable for crisis situations. However, excessive authoritarian leadership can limit creativity and innovation, both of which are essential for success in a startup environment.
2. **Managerial leadership**—this type of leadership style is used when the need for clarity is high, however, the same amount of empowerment is not needed. A managerial leadership style can foster trust since the required clarity is provided. However, excessive use of this leadership style can lead to micromanagement, which decreases both autonomy and innovation.
3. **Familial leadership**—this style is suitable in environments where clarity is not paramount, and employees are being empowered with regard to their duties. Using a familial leadership style can help you build relationships with employees and create a sense of loyalty. You need to know that excessive use of this leadership style can lead to consequences similar to that of an authoritarian leadership style.
4. **Transformational leadership**—this is one of the most sought-after leadership styles in modern-day businesses, as it can be used for both clarity and empowerment. These leaders are always willing to provide employees with the necessary motivation and assistance to those who need it, and it can help foster trust and creativity. However, it may create the need for constant guidance or dependency on the leader.

Choosing a leadership style requires assessing your startup's environment and its current phase, as this can significantly aid in de-risking. For instance, an authoritative style may be suitable for the initial stages where not much clarity is prevalent. However, shifting over to transformational leadership is a better option in the later stages so you can empower employees and make business operations more sustainable.

11 The Success Factor

By now, we've learned a lot about de-risking and the different phases of the startup lifecycle. Before implementing what we've learned, you need to know that with de-risking, success does not lie within one factor. In fact, it lies within using different factors we've learned in combination. Some of the things you can do differently to de-risk your startup include:

- **Gaining Clarity on Your Market Offering:**
  Achieving a clear understanding of your product or service is the cornerstone of de-risking your startup. Dive deep into understanding the unique features of what you're offering, the specific problems it addresses, and the unique selling points that set it apart in the market. Develop a comprehensive go-to-market strategy that includes detailed customer profiles, competitive analysis, and a clear value proposition. This clarity will guide all aspects of your business, from product development to marketing and sales.
- **Managing Growth with Precision:**
  Growth management is a delicate balance that can significantly impact the risk profile of your startup. Avoid the pitfalls of uncontrolled rapid growth or stagnation by setting well-defined, realistic growth targets. These should be based on a thorough analysis of your early-stage performance, market potential, and resource availability. Develop a phased growth plan that allows for gradual scaling, ensuring that your infrastructure, team, and operations can adapt and expand sustainably.
- **Preventing Burnout through a Healthy Work Culture:**
  Burnout is a critical risk in the startup environment, both for founders and employees. To mitigate this, cultivate a culture that emphasizes work-life balance and mental well-being. Implement policies that encourage regular breaks, flexible working hours, and a supportive atmosphere. Investing in the mental health and overall well-being of your team not only prevents burnout but also fosters a more engaged, productive, and creative workforce.
- **Securing Adequate Capital and Resources Strategically:**
  Access to capital and resources is a vital element of startup success and risk mitigation. In the early stages, focus on securing funding from sources that align with your business values and growth objectives. This could include venture capital, angel investors, grants, or crowdfunding. Alongside financial capital, prioritize the acquisition of key resources such as technology, talent, and strategic partnerships. Efficient resource management and smart capital allocation are essential for navigating risks and driving sustainable growth.

12 Key Takeaways

As you start to de-risk your startup, it's important to understand that the methods and strategies for de-risking will change as your business progresses through different stages. This requires you to continuously develop and adapt your skill set to effectively manage the varying aspects of de-risking throughout your startup's growth.

To sum it all up, gaining commercial and economic insights and developing strategic context can help you effectively de-risk your business. As we close off this chapter, let's look at some of the key things we've learned.

- **Establishing a Solid Foundation**: A strong foundation is the bedrock of a successful startup. This involves setting up a clear business model, understanding your target market, and aligning your goals with market demands. It's about creating a resilient structure that can withstand early-stage challenges and uncertainties.
- **Focus in Early Stages**: The initial phase should concentrate on comprehensive market research to understand customer needs, building a skilled and cohesive team, and developing a product that fulfills market gaps. This stage sets the tone for future growth and scalability.
- **Building a Competent Team**: It's essential that your founding team possesses a blend of skills in product development, technological prowess, strategic business management, and market commercialization. This diversity ensures a well-rounded approach to tackling early-stage challenges.
- **Product Development and MVP Standards**: Engaging with early customers and integrating their feedback is key in shaping your product. This customer-driven approach helps in defining the standards for a Minimum Viable Product (MVP), ensuring it resonates with the market.
- **Continuous Strategizing and Planning**: Ongoing strategic planning and risk identification are crucial. This means not just setting goals but also anticipating potential challenges and preparing to tackle them. It involves being agile and adaptable in your approach.
- **Evolving Risk Management**: Recognize that the strategies for managing risks will evolve as the startup progresses through its lifecycle. Each phase brings different challenges, and your risk management tactics should adapt accordingly.
- **Understanding Investor Profiles**: Different stages of the startup attract different types of investors. Early stages may appeal to risk-tolerant investors,

while later stages may attract those looking for more stability. Understanding this dynamic can guide your funding strategies.

- **Developing an Evolving Skill Set**: As the startup grows, so should your skills. This includes enhancing your financial management skills for better resource allocation, honing problem-solving abilities to tackle unexpected challenges, and improving organizational management to lead a growing team effectively.
- **Adapting Leadership Styles**: Leadership styles must evolve with the startup's growth. Early stages may require hands-on leadership, while later stages might need a more delegatory approach to manage larger teams and operations.
- **Mitigating Risks Effectively**: Effective risk mitigation involves a clear understanding of your business environment, managing growth to avoid over-expansion, preventing team burnout through sustainable practices, and securing adequate capital and resources for continuous operation and growth.

Further reading

Bratterud, H., Burgess, M., Fasy, B. T., Millman, D. L., Oster, T., & Sung, E. (Christine). (2020). The Sung Diagram: Revitalizing the Eisenhower Matrix. Diagrammatic Representation and Inference, 12169, 498–502. https://doi.org/10.1007/978-3-030-54249-8_43.

Drakeman, D., & Oraiopoulos, N. (2020). The Risk of De-Risking Innovation: Optimal R&D Strategies in Ambiguous Environments. California Management Review, 62(3), 42–63. https://doi.org/10.1177/0008125620915289.

De-Risking Business. (n.d.). Chemical & Engineering News. Retrieved November 20, 2023, from https://cendev.acs.org/articles/89/i28/De-Risking-Business.html.

Fluid Leadership: Inviting Diverse Inputs to Address Complex Problems. (n.d.). Apps.dtic.mil. Retrieved November 20, 2023, from https://apps.dtic.mil/sti/citations/tr/AD1027504.

Fulmer, R. M. (1997). The evolving paradigm of leadership development. Organizational Dynamics, 25(4), 59–72. https://doi.org/10.1016/s0090-2616(97)90037-1.

Kane, K. (2011). Revisiting the behavioral matrix for leadership and team development. Organization Management Journal, 8(1), 23–30. https://doi.org/10.1057/omj.2011.6.

Kennedy, D. R., & Porter, A. L. (2021). The Illusion of Urgency. American Journal of Pharmaceutical Education, 86(7), 8914. https://doi.org/10.5688/ajpe8914.

Kerr, S. P., Kerr, W. R., & Xu, T. (2018). Personality Traits of Entrepreneurs: a Review of Recent Literature. Foundations and Trends® in Entrepreneurship, 14(3), 279–356. https://doi.org/10.1561/0300000080.

Leadership Style Influence in Execution in Matrix Organizations: A Case Study of Naval Base Kitsap, Fleet and Family Readiness—ProQuest. (n.d.). www.proquest.com.

Retrieved November 20, 2023, from https://www.proquest.com/openview/4b5c90dcf28f7f252503658cdd924427/1?pq-origsite=gscholar&cbl=18750&diss=y.
Manafe, M., Ohara, M., Gadzali, S., Harahap, M., & Ausat, A. (2023). Exploring the Relationship Between Entrepreneurial Mindsets and Business Success: Implications for Entrepreneurship Education. Journal on Education, 5(4), 12540–12547. https://doi.org/10.31004/joe.v5i4.2238.
Metcalf, L., & Benn, S. (2012). Leadership for Sustainability: An Evolution of Leadership Ability. Journal of Business Ethics, 112(3), 369–384. https://doi.org/10.1007/s10551-012-1278-6.
Obolensky, N. (2010). Complex Adaptive Leadership: Embracing Paradox and Uncertainty (1st ed.). Routledge. https://doi.org/10.4324/9781315264929.
També Bearpark, N. and Demetis, D. (2022), "Re-thinking de-risking: a systems theoretical approach", Journal of Money Laundering Control, Vol. 25 No. 1, pp. 27–49. https://doi.org/10.1108/JMLC-04-2021-0030.
Triantaphyllou, E. (2000). Multi-Criteria Decision Making Methods. Applied Optimization, 5–21. https://doi.org/10.1007/978-1-4757-3157-6_2.
Tripathi, N., Oivo, M., Liukkunen, K., & Markkula, J. (2019). Startup ecosystem effect on minimum viable product development in software startups. Information and Software Technology, 114, 77–91. https://doi.org/10.1016/j.infsof.2019.06.008.
Winkel, D., Wilcox, J., & Teckchandani, A. (2020). The 60-Minute MVP. Entrepreneurship Education and Pedagogy, 3(4), 371–386. https://doi.org/10.1177/2515127419870293.

Dr. Sebastian David Hersberger serves as the Managing Director & Founder of sh ventures GmbH, an investment and investment advisory boutique with a specialized focus on Venture Capital. He holds a PhD in Economics from the University of Basel in Switzerland and works since 2021 as an external lecturer at the Faculty of Business and Economics at the University of Basel, teaching a course on Venture Capital and Start-ups. Sebastian brings over 12 years of experience in management consulting and banking to the table. His previous roles include serving as the Head of Strategy and Innovation at a Swiss regional bank. In 2019, he transitioned his career trajectory towards the dynamic realm of start-ups, holding various C-level positions in BioTech, Tech, and Climatech ventures over the last few years. He is currently the CEO and Co-Founder of Yuon Control AG, and, in addition to this, he works with several start-ups as an advisor.

Managing Entrepreneurial AI Ethics Risks

Erich Prem

1 Introduction

1.1 A New Artificial Intelligence Infrastructure for Innovation

Recent advances in artificial intelligence (AI) made it an enabler of new business opportunities and improved business processes (e.g. Giuggioli & Pellegrini 2022, Prem 2022, Shepherd & Majchrzak 2022). This ranges from new products and services to innovative business models and improved decision-making to better performance through increased automation. The recent improvements in the performance of generative AI (GenAI) resulted in a new dynamic of applying AI in a wide range of areas for which automation was difficult or impossible until recently. Businesses and laymen are turning to AI for the creation, analysis and improvement of texts or use it as a natural language interface to program code, create images and presentations, and as a personal assistant for just about anything from fitness to language skills. Unsurprisingly, worldwide investments

E. Prem (✉)
Institute of Philosophy, Philosophy of Media and Technology, University of Vienna, Vienna, Austria
e-mail: prem@eutema.com

eutema GmbH, Vienna, Austria

C. H. Hoffmann (ed.), *Artificial Intelligence, Entrepreneurship and Risk*, Technikzukünfte, Wissenschaft und Gesellschaft / Futures of Technology, Science and Society, https://doi.org/10.1007/978-3-658-45544-6_23

in AI are estimated to break the US$ 500 billion mark[1] in 2023 and numerous start-ups were founded to harvest novel AI opportunities.[2]

At the same time, there are enormous concerns about GenAI and AI in general and its wide-spread application. These concerns are debated in mass media as much as in specialised scholarly publications. They range from questions about safety and security of AI systems to the ethics of their application, their social impact, unbalanced power relations, threats to democracy, and even geopolitical concerns (cf. Werthner et al. 2024). It is to be expected that these debates also affect business leaders. Indeed, a recent Chief Risk Officer Outlook by the World Economic Forum (WEF 2023) describes AI as *"a potentially disruptive inflexion point for societies [that] creates a complex and uncertain environment in which organizations must operate."* While many risk managers and business leaders exhibit confidence in understanding the risks emerging from AI-based applications, they are concerned that managing those risks provides significant challenges for the organisation. This is exacerbated by the fast development of AI technology which leads to a situation in which business managers are increasingly unsure about which technology to try out first and on which tools or services to focus. Despite a certain popularity of the "Silicon Valley" narrative that regulation hinders innovation (Zuboff 2018), many risk managers would in fact welcome clearer rules and regulations (WEF 2023, p. 15 ff).

1.2 Aims and Methodology

This Chapter aims to link the ethics risks that have been identified for AI in the literature with business risks and to propose strategies for addressing them. The philosophical underpinning and mission are that AI-driven businesses should align with ethics and social welfare considerations while at the same time accepting that they operate under economic boundary conditions including considerations of shareholder value. Floridi (2023, p. 89) describes this as a "normative cascade" from ethics to law and business finally impacting customers. This normative cascade should be used constructively and productively rather than come into force solely when customers or regulators protest against unethical business practices.

---

[1] IDC, Worldwide semiannual artificial intelligence infrastructure tracker. https://www.idc.com/getdoc.jsp?containerId=IDC_P37251 (Last accessed December 2023).

[2] AI adoption skyrocketed over the last 18 months. In: Harvard Business Review, September 27, 2021.

This research is based on previous work in the philosophy and ethics of AI, in business management, and in AI research. It roots in previous work on AI principlism (Prem 2023) and links approaches to AI risk management with experiences in other fields of engineering and medicine that have successfully managed significant risks including risks of life. This suggests using risk management approaches and established operational practices from these fields and applying them more broadly to AI risks. In this Chapter, the focus is on managing residual risks potentially remaining even after making AI models as safe and compliant as possible.

This Chapter is organized as follows. After a review of related work in the following subsection, we discuss ethical issues of AI relevant to the entrepreneur and businesses in general in Sect. 2. This section presents principlism as the currently prevailing approach to address AI ethics issues. Section 3 lists business risks emerging from these ethical risks and links them with risk mitigation strategies assuming that many AI risks cannot be completely avoided. Section 4 investigates industries that have established practices to deal with risks and ethical issues, in particular medicine, pharmaceutics, and aeronautics with the aim to derive useful practices for the field of AI. The Chapter concludes with relevant learnings and recommendations for the future of AI businesses.

1.3 Related Work

A well detailed list of AI risks and their technical sources is provided in Steimers and Schneider (2022). It is concerned with a taxonomy and sources of AI risks as well as safety risks—mostly from an engineering perspective. It addresses risk management based on ISO risk management standards and adopts them for applications in AI but does not primarily address those risks from a business perspective. Clarke (2019) argues that business ethics on its own provides only little guidance for an organisation to manage AI responsibly and suggests a focus on risk management instead, very much in line with the approach presented here. It includes a risk management process model to include perspectives from multiple stakeholders. It also organises various existing sets of AI ethics principles into 50 useful principles along 10 relevant themes (cf. Sect. 2). The prototype approach presented here can be used to complement and concretize these rather general principles.

A process model addressing review and approval stages to manage AI risks is presented in Baquero et al. (2020). It includes an example for managing bias risks in four different stages of the AI model development process (i.e. ideation, data sourcing, model development, and industrialization—monitoring—maintenance). With a strong orientation to regulatory proposals for AI, Giudici et al. (2024) propose a statistically oriented risk management framework and key indicators. It addresses metrics for *sustainability* (i.e. the resilience of the AI model outputs under anomalous events or cyberattacks), *accuracy* of the model outputs, *fairness* (i.e. the absence of bias), and *explainability* (i.e. understandability of the model outputs by people).

A broader approach to managing risks emerging from AI systems can be found in Schuett (2023). It proposes "three lines of defence" as a risk management framework based on IIA (2013) and later updates. Its main contribution is a responsibility framework defining first line roles (risk management in the provision of services and products), second lines roles (expertise and monitoring), and third line roles (audits). In the application to AI, Schuett identifies the first line with AI research and product development. The second line assesses the first line and provides complementary expertise, e.g. legal or risk-specific expertise. Schuett also suggests that the second line supports the first line with policies and procedures and takes responsibility for monitoring. The third line assesses the other two to provide independent review. The approach can be an interesting governance set-up for realising the AI risk management proposal in this Chapter.

The idea to take inspiration from medicine and aeronautics for AI risk management is not new. Ozlati and Yampolskiy (2017) address safety standards and artificial general intelligence risks to humanity focusing on a modified Delphi method for risk management. Similarly for the case of automated vehicle safety, Goodall (2016) points out that successful strategies for managing difficult ethical decisions were developed in situations such as organ donation, radiation exposure, and industrial safety standards and that developers should seek to learn from these application fields.

The article by Mannes (2020) discusses approaches to mitigating societal AI risks. Apart from discussing AI regulation, it is noteworthy for describing the task of risk communication to engage a broader public and build trust. Public relation risks are also addressed here in Sect. 3. Furthermore, Cheatham et al. (2019) is an early effort to describe how individuals, organizations, and society are affected by advanced analytics. It proposes technical and management controls for some key risks, e.g. monitoring and risk tiering for potentially unethical use cases, privacy protections and data quality metrics for dealing with inaccurate or protected data, transparency requirements for addressing bias etc.

2 Entrepreneurial Ethics in the Age of AI

A significant portion of the business literature that describes and evaluates the entrepreneurial promises of AI remains only shallow about AI's potential pitfalls and the significant amount of criticism and concerns that scholars have raised. This is surprising for at least three reasons. *Firstly,* there is a massive wave of scholarly literature, for example in the field of ethics of AI. The concerns range from a lack of human autonomy to state sovereignty, from threats against human rights and fundamental values (e.g. Freedom of Expression) to unfair treatment of customers and trends towards monopolistic business practices. However, the ethical issues are rarely addressed from a business perspective and often only result in a call to adhere to ethical 'principles' (cf. below). *Secondly,* there are already examples where ethics has turned into a showstopper for AI, i.e. cases where AI-related applications failed or had to change significantly due to ethical concerns and public debates. The case of algorithmic selection of clients depending on their re-employment probability for the Austrian Federal employment services was massively debated in public and led to its abolishment (Allhutter et al. 2020). Another drastic example is the resignation of the Dutch government following a massive public fairness debate after an erroneous algorithm assessed child benefits (Rao 2022). *Thirdly,* concerns about potential negative impacts of AI applications have given rise to regulation such as the EU AI Act that will mean new organizational requirements, changing procedures, and therefore costs for businesses. In summary, questions debated in the ethics of AI are hugely relevant for any entrepreneur wishing to exploit the potential of AI for their activities.

Business ethics

Ethics is the philosophical discipline that addresses questions of morale. The traditional focus of ethics concerns questions of how to act vis-à-vis other humans. In the words of Bernard Gert, "[m]*orality is an informal public system applying to all rational persons, governing behaviour that affects others, and includes what are commonly known as the moral rules, ideals and virtues and has the lessening of evil and harm as its goal.*" (Gert 2005). Therefore, for the largest part, ethical issues of AI systems arise where those systems directly or indirectly affect humans. Contemporary conceptions of ethics also include actions that may affect animals, the environment, or machines (e.g. robots), however, in the following we focus on the traditional view and on actions that affect people.

Ethics has long been the subject of debate in business and management science. Historically, much of business ethics focused on profit distribution and the role of the entrepreneur for social welfare, e.g. fair financial compensation for workers (Harris et al. 2009). Business ethics has been criticised for remaining vague and difficult to put into practice (Clarke 2019). Recently, frameworks for Corporate Social Responsibility (CSR) have broadened the spectrum of ethical debates and environmental considerations also entered the field of business ethics. However, there is still a need to better understand what type of problem ethical challenges of AI precisely are for companies as the ethics of AI is still sometimes regarded a mostly technical issue that only few experts relate to business ethics. The situation is further complicated as there is no clear line that distinguishes ethics from politics. What is acceptable behaviour in a society changes over time and has often been at the heart of political debates. The politics of AI has become the centre of debates around the world given its potential societal influence (Crawford 2021, Cockelbergh 2022). For example, the Digital Humanism movement focuses on topics such as bias, power distribution, colonialism, and language—all of which are relevant to AI and in many ways related to AI-driven business practices (Werthner et al. 2021, 2024).

AI affecting people

AI systems become the subject of ethical concern and political debate where their actions affect humans. AI systems may *directly* affect people when they take decisions about people (famously called *algorithmic decision making* in the EU legal system).[3] Typical examples include decisions about granting bank loans, personnel selection, insurance risks. Where AI systems control machines and take decisions in automated driving, people may be exposed to safety risks. Similarly, AI algorithms may influence human behaviour in cases such as buying goods and services following algorithmic recommendations and people may be nudged into certain behaviours such as exercise routines or dietary choices. In these cases, AI algorithms could support unfair practices that are disadvantageous to customers or other businesses. Similarly, AI may have an influence on humans based on selecting news, online articles, and comments in online social media. Such influence can be considered unethical, for example, if it effectively limits freedom of expression or aims to clandestinely exert political influence. In many of these cases, AI-based decision-making is based on computer models of humans. These models can then be used for classification, prediction, or control of human behaviour (Fig. 1).

---

[3] Art.22 of the EU's General Data Protection Regulation is entitled "Automated individual decision-making including profiling".

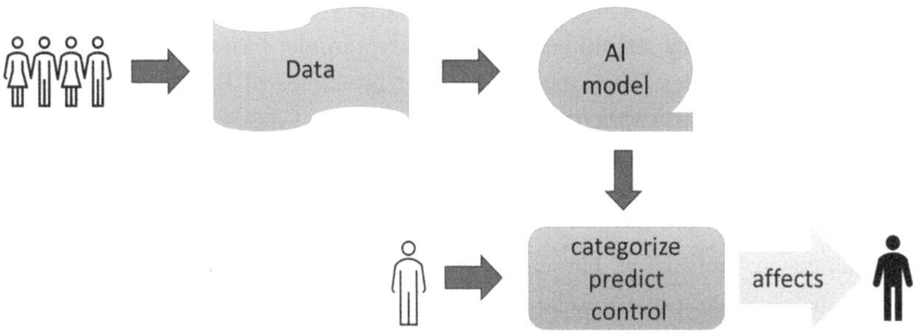

Fig. 1 AI models built from user data or other information create ethical risks where they affect people, for example by categorization, prediction, or control

AI systems may also affect people more *indirectly*, for example when systems tend to perpetuate trends that existed in the past. AI systems may impact at the individual, at group, or societal level. Some AI systems could be systematically disadvantageous to people less represented in the training data. At the societal level, AI can lead to new business models, for example support the individualisation of insurance risks previously only manageable for large populations. Some of these issues are well explored, e.g. bias related to certain parts of the population or dealing with protected properties, e.g. people with diseases or minorities. Others, however, are still rarely discussed such as the consequences of new AI technology for the insurance sector and the degree to which the individualisation of risks is ethical or positive from a social welfare perspective.

Principlism

For the most part, the scholarly literature about AI ethics is concerned with *ethical principles* (Prem 2023). The central idea is to support practitioners in choosing ethical actions with a set of principles to which they should adhere. In philosophy, this has been termed principlism. It dates to the work of Tom Beauchamp and James Childress who proposed four principles for guiding biomedical researchers in 1974.[4] Principlism was motivated by medical experiments that would be characterized unethical today, e.g. the experiments on inmates of Nazi concentration camps and other infamous misguided medical experiments such as the Tuskegee syphilis study (e.g. Reverby 2009).

The four principles of Beauchamp and Childress are *autonomy, non-maleficence, beneficence, and justice*. For the case of AI, many ethical principles

[4] See Beauchamp and Childress (2019) for a recent account.

were proposed. There is some argument about how many and which principles there are and how they should be grouped. For example, Floridi (2023, pp. 57ff) suggests five principles. They include the four principles of Beauchamp & Childress to which he adds *explicability*. The latter is very AI-specific and especially relevant for modern AI architectures based on machine learning. These high-level principles include other concepts or principles (cf. Floridi 2023):

- Beneficence: well-being, human dignity, sustainability
- Non-maleficence: privacy, security
- Autonomy: human decision making
- Justice: prosperity, solidarity, fairness
- Explicability: intelligibility, accountability

It should be emphasized again (Prem 2023) that principles lack information about *how* to realize an AI system. They can only inform about abstract properties that require a great deal of generalization, assessment, and informed decision-making between technical alternatives. The path from *what* to *how*, i.e. from principles to practice can be hugely challenging as evidenced, for example, by the case of explainable AI or *XAI*. This has developed into a whole subfield of machine learning, hence clearly indicating that principles can be very difficult to apply. Similarly, the fields of privacy-preservation or AI security are research areas in their own right and it is simply not easy to make AI systems conform to "privacy" or "security". This already implies a remaining or *residual* risk which businesses need to address.

3 AI Business Risks

Despite its many promises, business leaders often lack trust in AI and have reservations including low confidence in the technology.[5] This may be due to the necessary technology and management changes being perceived as too challenging, its relative novelty and lack of experience, a lack of experienced staff that understands both the technological and ethical risks, or reservations about the degree of acceptance by customers. AI pushes managers to their limits and many managers agree that the development of AI outpaces the ability to manage its ethical and societal risks (WEF 2023). This also explains why some have argued

[5] https://www.consultancy.uk/news/36083/half-of-business-leaders-have-reservations-about-ai.

that the development of new AI technologies should be paused or slowed until its potential impact was better understood—following a controversial debate about pausing AI development in academia (cf. Clarke 2023, Ienca 2023).

In the following, I focus on AI risks from a business perspective while the scholarly literature often assesses AI risks at the societal or individual level. Societal risks include, for example, the risk of a developing *super intelligence* that poses existential risks to humankind. Perhaps more realistically, societal risks include a geopolitical power shift to certain regions and governments in the world, an AI arms race, or an increase in economic inequality. At the individual level, there are risks such as deskilling due to delegation of skilful tasks to machines, an increased dependence on AI (also a societal risk) and a loss of human connection due to AI-driven communication (Marr 2023). Many of the risks for individuals such as consumers are also risks for companies, e.g. lack of transparency, unfair discrimination, or security and safety risks. Important business risks emerging from AI technology include the following (cf. Table 1). These risks are likely to remain even after careful AI systems engineering as *residual risks* that a company should consider.

Table 1 Selected business risks for AI-driven businesses

Type of risk	Examples
Innovation risks	Non-acceptance of new technologies, missed opportunities through technology lock-in
Technology risks	Lack of robust and correct functioning, threats from new and better technologies, quality issues, residual risks due to lack of proven correctness
Security risks	Cybersecurity attacks, data leaks, unpredicted and dangerous behaviour
Public relations risks	Lack of credibility towards consumers, trustworthiness of products and services, public perception of unfair treatment and bias
Regulatory risks	Changes in regulatory environment that impact on value proposition, production costs, work environments etc.
Legal risks	Challenges of addressing regulatory requirements, lawsuits emerging from legal conflicts
Human resources risks	Deskilling of staff, lack of staff at required skilled levels
Market risk	Unfulfilled user expectation due to hyped AI technology perception

These risks are naturally linked with each other and do not form a disjunct set. For example, a quality issue in an AI model's prediction may lead to an unfulfilled customer expectation and could also reduce the credibility of the service more generally. This may lead to a problem in the public perception or could even lead to lawsuits or proposals for new regulation.

The above risks can be more directly linked to ethics issues and ethics principles. Table 2 provides an overview of ethics issues, possible reactions from business partners and consumers, and the resulting business risks. It also includes some possible risk mediation or avoidance strategies which are also further motivated in Sect. 4.

The various risks emerging from potential AI ethics issues are diverse but many result in similar business risks such as negative public perception, potentially costly complaint management or legal procedures. An important risk mitigation strategy is the *early identification* of a materialised risk. In many cases, costs of managing AI ethics issues will continue to grow if they remain undetected or unmanaged. A simple reason is that more customers may be affected or that more customers will learn about an ethics issue. This can be achieved through *monitoring* of AI systems as well as early detection systems such as effective consumer complaint management that provides feedback for marketing and engineering. Ethics issue identification will benefit from a solid understanding of the risks as unidentified risks may take much longer to be clearly identified (e.g. quality issues or biases). Similarly, AI-focused companies should be clear about their risk management objectives and the range of technical, legal, social, financial, and management strategies to react to AI ethics issues. The next section takes inspiration from other risky businesses that have established risk management procedures and could serve as a source of inspiration for addressing and managing AI ethics risks.

Table 2 Examples of business risks emerging from AI ethics issues and possible strategies for their mitigation

Ethical issue (principle)	Reaction from public, partners & consumers	Business risk	Risk avoidance and mitigation strategies
Lack of transparency (transparency)	Distrust, resistance	Sales loss, missed opportunities, legal and regulatory issues	Open data and process, model card (information), processes for providing explanations, human intervention and oversight
Bias, discrimination (fairness)	Distrust, complaints	Negative public perception, complaint management	Debiasing, diversity measures, testing, user information, industry standards (e.g. fairness)
Privacy infringement (privacy)	Consumer complaints, resistance, distrust	Complaint management, lawsuits (e.g. GDPR), sales loss	Safe data handling practices, improved privacy technologies, minimization of data needs, preparation for data losses
Security risks (non-maleficence)	Compensation requests, distrust	Legal procedures and lawsuits, negative public perception, complaint management	Quality assurance, testing, monitoring, early detection, maintenance
Regulatory non-compliance	Distrust, complaints, public inquiry	Negative public perception, legal costs	Compliance processes, monitoring, audits, early detection, maintenance

(continued)

Table 2 (continued)

Ethical issue (principle)	Reaction from public, partners & consumers	Business risk	Risk avoidance and mitigation strategies
Misinformation, manipulation, system abuse	Public complaints & calls for action, political attention	Negative public perception, PR costs, change of technology or business model	Monitoring, early detection, legal procedure, public statements, contract management
Concentration of power (own)	Distrust, monopoly action	Legal procedures, limitation in choice of partners, premium prices, service restrictions	Establish relationship and communication with regulator

4 Managing AI Risks

In the following, AI risks are considered a case of general risk management for several frameworks and methods have been proposed including ones that are relevant in standardization (cf. Leighton 2020). Some of these frameworks can become very complex, consist of many steps, and include structures and governance issues. In the following we present a simplified approach consisting of five steps which are:

1. AI risk **policy**: determination of an overall approach to AI risks.
2. AI risk **analysis:** listing and assessing AI risks.
3. AI risk **detection**: understanding the AI risks of the application.
4. AI risk **avoidance**: taking measures to decrease the likelihood of AI risks.
5. AI risk **mitigation**: taking measures to reduce the impact of AI risks that materialize.

4.1 AI Risk Strategy

Risk policy

Enterprise risk management requires an overall risk policy. This includes determining acceptable and unacceptable AI-related risks and the degree to which AI risks need to be avoided or mitigated. (Also see Steimers and Schneider (2022) regarding tolerable risks.) The company AI risk policy will often be a top-level management decision as it needs to be aligned with the company's overall strategy, especially its mission and values. An AI risk policy needs to consider costs on the side of materializing risks as well as the costs of risk management (risk avoidance and risk mitigation), but a policy is usually not solely cost-driven. Decisions regarding unacceptable risks, for example, will often be a consequence of a more general corporate social responsibility framework.

As many companies are experimenting with AI-based services, they may not yet have an explicitly formulated AI risk strategy. Hence, such a strategy may only be implicit and a consequence of go/no go management decisions for new AI-based products and services. Also, the design of AI risk policies can be challenging given the lack of information about AI risks, their prevalence or likelihood and the possible costs.

Risk assessment and analysis

Businesses planning to deploy AI systems should perform a rigorous risk analysis. In some cases, this may be required by regulation (e.g. the EU AI Act). In other cases, it may simply be wise to do so. As a first step, a risk assessment should clarify the possible risks of an AI system. It concerns questions such as: What could go wrong? What would be the impact on the system, the costumer, the company, and on other stakeholders? The assessment should include a range of scenarios, from likely scenarios to the worst-case scenario. A proper risk assessment should include an assessment of the likelihood of the scenarios and where possible costs associated with the consequences or for fixing the resulting problems. Typically, neither a risk likelihood nor a fully detailed cost analysis may be feasible in many cases. This is partially due to the nature of AI systems, especially for foundational models or other machine learning systems for which a certain margin of error may remain even after excessive testing and validation.

For the case of ethics risks, an assessment may not always be straightforward (cf. Steimers & Schneider 2022) and may require a good degree of knowledge about typical problems in AI ethics. Usually this needs to be paired with a solid understanding of the technical aspects of AI systems—a combination that is not widely available today and will require the formation of a new generation of AI ethics experts that has only just started. For many AI-based applications, there may be few or no ethical risks. For some, such as in cases of AI-based machine automation, there may be straightforward health and safety risks. However, it can also be a challenge to detect ethical issues arising from AI-based products and services. This requires an understanding of the technology and of its potential impacts on people. A range of guidelines have been developed to identify risks, especially for the main ethical principles such as fairness, autonomy etc. (Steimers & Schneider 2022).

AI risk and incident detection
It is inevitable given the current immaturity of the field that new issues will arise with novel applications and that some issues may only be detected based on feedback from the stakeholders including the users. This applies both to new or undetected risks and to the materialisation of risks (i.e. damage) which also poses the risk of repeated damages if no appropriate measures are taken. Early incident detection has not yet been widely addressed in the literature, although it can be vital given the early stages of AI technology and its continued development. It is also a reason for documentation duties included in proposed AI regulations (e.g. the EU Act). Section 4 returns to early detection with the example of the pharmaceutics industry.

AI risk avoidance and mitigation
Risk avoidance through technical means is an important strategy, however it has its limits in the availability of current AI technology. As mentioned before, some ethics risks and how to avoid them are the subject of challenging ongoing research (e.g. Prem 2023). In many cases it is either impossible or too costly to guarantee that AI systems are fully aligned at all times with all ethics requirements. It is therefore indispensable that companies put in place risk mitigation strategies, i.e. strategies to minimize the impact of risks where they materialize.

There is a broad range of possible AI ethics risks and incidence mitigation strategies:

- *Early detection:* Often the early detection of incidents can help address problems and save costs as well as further consequences or more incidents. Relevant tools include monitoring, feedback channels for users, documentation, and reporting.
- *Minimising financial consequences:* This may include preparations for rapid discontinuation of services, insurance for business and customers, switching to different or previous systems.
- *Ensuring consumer trust:* Provision of updates and repairs, procedures for maintenance, information of consumers such as handbooks and self-help guidelines.
- *Minimising negative public perception:* Companies should be prepared to address incidents publicly. This can include an explanation what happened, communication of how the situation was addressed and how it will be avoided

in the future. It can include an investigation and public statements from top management executives.

- *Policy links:* AI businesses should establish functioning links with policy makers and industry associations, in particular in areas of identified risks. These links can be useful to steer regulation, to establish commonly accepted industry standards and to address incidents at a more general level beyond that of just a single company.

AI risks may be unavoidable, but AI is not the only field that needs to deal with residual risks. A range of industries has successfully established practices for managing risky systems. The next section investigates some of these practices to suggest that they can be fruitfully adopted for AI.

4.2 Learning from Risky Businesses

There is huge experience with significant risks for humans in areas such as aeronautics, medicine, and pharmaceutics. These industries provide products and services that require utmost safety and are therefore heavily regulated. Over time these industries have learned from many mistakes and ethical meanders, some of which had catastrophic character. Today, these industries follow well-defined procedures to ensure consumer safety due to legal and regulatory requirements or based on internal procedures.

Prototype situations in medical ethics

Ethical considerations may have their longest history in the field of medicine where ethical principles have already been described in antiquity by Hippocrates. In addition, the medical profession has developed processes and structures that support ethical decision making, for example with ethics boards and reviews. Also, a significant part of medical work is subject to legal regulation, contracts, in-house rules, and professional standards. Modern medicine poses significant ethical challenges, partially due to technical innovation and progress in areas from reproductive medicine to organ transplantation. While bioethical principles such as those stipulated by Beauchamp and Childress are still important in many cases, especially in research and for rarer or special cases, medical ethics has benefited from the development of fairly standardized practices in prototypical situations. The ethical challenge of which patient on a waiting list for a donor organ or an important surgery should be served first is often answered following a set of clear rules and based on often standardized decisions of special review boards. In many countries these decisions follow regulation and often there are centralised bodies. In many cases, prototypical cases of challenging ethical decision making are used in teaching bioethics (e.g. Freeman and McDonnel 2001). The prototypes, rules, and bodies may not be perfect and clearly cannot have solutions laid down for every possible case, but they have evolved into broadly accepted practices that save time, provide clarity, and are usually justifiable from an ethics perspective. It should be added that application context plays an important role in medical ethics (Thomasama 1984). This did not stop medical ethics from developing accepted answers for many practical situations and from using these cases as guidelines in comparable situations.

It may be argued that the list of situations for which medicine has developed clear ethical practices is relatively small compared to the vast range of possible situations that AI-supported business may encounter. This may be true in general given the wide range of possible situations for which a business may deploy an AI model. However, there will still be typical situations where AI solutions will be deployed and where prototypical circumstances may help in facilitating ethically sound approaches. This could include, for example, standardized insurance cases (e.g. life, car, or household insurance), dos and don'ts of credit worthiness assessments, and even principles for shopping recommendations. However, all of these are not yet available and will have to be developed in the future.

Monitoring, processes, and early warning in pharmaceutics

The research, production, and distribution of drugs is one of the most strictly regulated business sectors. It is also severely monitored, examined, and controlled given the potentially catastrophic impact that drugs or quality issues in their production can have on individuals or even on large parts of the population. There are several interesting aspects of quality controls that can provide useful pointers for businesses (and regulators) in the AI sector including monitoring, follow-up, processes, bodies, and auditing.

A precise description of the quality control system in pharmaceutics is beyond the scope of this paper. In short, there are global reporting and monitoring systems that involve both the private and the public sector. These systems are designed to ensure a constant monitoring of the quality of pharmaceutical products and an early detection of potential quality issues or new and unwanted side-effects. Beyond just monitoring the results, i.e. drugs, pharmaceutics is one of the industries in which processes are designed to avoid errors as much as possible. In addition, there is auditing at the process level to ensure that production and quality processes are established that minimize errors at all levels of production. On top of that, there are clear rules for monitoring and reporting quality issues once drugs reached the consumer, for documentation, and follow-up to mention just some key aspects.

On the downside, the quality and auditing standards of the pharmaceutical industry are extremely costly processes. They may neither be feasible nor commercially viable for large parts of AI-based businesses. However, there are still important lessons to be learned.

- *Process-level controls:* Drugs are not simply tested as an end product. There are clear and formal rules with strict controls for the production processes that aim to ensure that drugs are produced to the highest quality standards and that deviations are detected and addressed. This is not the case in standard AI application development, where much of model training is trial and error and although there may be some procedural controls in place, it is far from common standard practice or audited procedures today.
- *Auditing:* The level of internal and external auditing—both from consultants and regulators—is very high throughout the industry. Audits are frequent and concern many different processes. Internal quality control is performed by teams that receive special training. There are significant threats to businesses not following up on significant observations from auditors including the shutdown of production lines or sites.
- *Quality standards:* There are very clear rules about what are acceptable quality deviations and documentation of quality issues as well as follow-up procedures. There are detailed standards at the international level and nationally as well as formalised industry practices. There are procedures to ensure that incidents are identified as early as possible and that there is effective follow-up. A whole network of actors from medical experts to pharmacies and laymen is used for this type of early detection. This is accompanied with reporting and follow-up duties. A lighter version of such an approach would clearly be useful also for AI businesses to ensure the detection of issues and adequate reaction from industry.
- *Specialised staff:* Managing quality is performed by staff that is specifically trained for this job. This usually requires knowledge both in pharmaceutics, in production and process engineering, and of the regulatory environment on top of an understanding of business issues. Again, this is very similar to the

situation in AI where more staff knowledgeable in the technology and in ethics issues and policies will be required.[6]

In theory, the pharmaceutical industry could also provide an example for the AI industry in establishing a culture that is open to making and addressing errors. Admittedly, this may in practice still often be an issue also for the pharmaceutical industry as there are huge challenges for international companies to overcome cultural barriers towards a completely open identification and debate of quality problems in production.

Whistleblowing and double-checking in aeronautics
Aeronautics is another case of an industry that has embraced safety standards like not many industries. This has resulted in extremely low and historically declining rates of accidents and catastrophic incidents. Obviously, aeronautics is central to a whole branch of engineering that deals with safety–critical systems. It has developed and established systems and tools, rules, regulations and certification schemes and bodies that make airplanes some of the safest systems available, but also make their development and production costly and slow. This means that in many everyday (i.e. non-safety–critical) situations, the strict standards from safety–critical systems theory may not be affordable. However, there are still important lessons to be learned.

- *Monitoring and maintenance:* The airline industry has developed clear and rather concise but useful routines for checking and maintaining the safety of their operations. From short checks before every flight to routine complete overhauls of planes, there are well-established practices to ensure that planes are safe.
- *Error handling:* There is a system of reporting incidents and potential safety issues and a continuous improvement of safety that has evolved over decades of aviation. There are systematic investigations into detected issues and after incidents. They lead to information, warnings, recommendations, or even the grounding of aircraft. This has significantly contributed to making air travel safer over time.
- *Redundancy:* There is hardly any industry where the principle of redundancy is as established as in aeronautics. This includes to not fully rely on the actions

[6] It is therefore alarming when responsible AI teams are the first to be let go in situations of downturns as their competences are needed for managing AI risks, cf. https://www3.weforum.org/docs/WEF_Chief_Risk_Officers_Outlook_2023.pdf

of others and to perform systematic double-checking as well as safe fall-back procedures in case of system failures. Nearly everywhere in aeronautic processes there is a tacit assumption that someone else could have made a mistake when performing a safety critical action. This ranges from the cross-checking of airplane doors before and after a flight to the non-flying pilot's hand resting on the other pilot's hand throttling the engine for take-off.
- *Safety culture:* Apart from many standards and examinations, technical checks and maintenance duties, aeronautics has established its own safety and error culture.[7] It ranges from a principle of "safety first" to reporting and whistle-blowing, for example in programs that facilitate the confidential reporting of errors, shortcomings, and malfunctions, their potential causes and how to address them.[8]

Apart from these examples, certainly more sectors such as the automotive industry, the legal system, or even insurance can provide relevant practices for managing challenging AI ethics risks. This requires further research and adoption of good practice to the AI sector.

5 Discussion

5.1 Beyond Principles

Much of AI ethics is concerned with principles or rules from which the ethicality of actions is to be derived. This includes a large part of the literature that is focused on ethical dilemmas and how to derive solutions for them from principles and rules. It is, however, questionable if this general approach will ultimately be useful for the design of AI systems and for establishing practices in companies that build them. It is very well possible that the enormous breadth of situations for which people use IT systems makes it impossible to provide clear guidelines following just principles and rules (or even virtues). Instead, it may be necessary to decide on ethical aspects of IT systems in a much more bottom-up and individualised fashion, and therefore, politically. It is possible that there is a need to

[7] Helmreich (2000) lists lessons from aviation recommended for medical practice.

[8] An example is the Confidential Aviation Incident Reporting (CAIR) instituted by the Australian Transport Safety Bureau (ATSB). https://skybrary.aero/articles/confidential-aviation-incident-reporting-cair (Last accessed 12/2023).

debate the ethicality of AI systems considering their precise aim and application context rather than deriving it from just a few principles or rules.

This may not be necessary for every single system as a good way to approach the problem could consist in defining *classes* of problems for which we understand what the issues are. This is in practice how we have been addressing challenging ethical questions in the past. Many consumer rights have emerged from debates about similar situations about broken products, hidden faults, belated delivery etc. In a similar fashion, its is conceivable to address important types of AI applications in a way that is similar to established practices in other fields. This does not mean that principles become irrelevant. They can still provide important input to debates and act as guidelines for new cases. Over time, practices of chatbot ethics, ethics of recommender systems, ethics of autonomous cars, etc. can emerge that are not in contradiction with principles and are the results of public debate or political agreement. There may be standard understandings for certain types of principles, e.g. for fairness, in certain types of situations, e.g. for assessing creditworthiness or selecting candidates for universities. Fairness is a good example as it is not just a mathematical function and requires social agreement and acceptance (cf. Steimers & Schneider 2022, pp. 10–11, Floridi 2023, p. 104).

The case of medicine demonstrates that a solid practice of dealing with challenging ethical situations is possible. It roots in principles, but also in clear governance structures and procedures. Teaching medical ethics is routine for students of medicine. All these practices are important pointers for addressing ethical business risks of AI. Staff that has a background in both AI or data science and ethics will increasingly become available as more and more curricula include ethical topics in the training of engineers. This is hugely relevant for being able to address AI ethics business risks. The fields of pharmaceutics and aeronautics have important lessons to teach, especially regarding early identification of materialised risks, how to address them as well as auditing and documentation practices. Regulators in Europe have chosen to follow an approach to regulating AI that is similar to some of these procedures, although the EU regulation makes it obligatory mainly for high-risk AI systems. In line with the recommendations of Buehler et al. (2021) lessons to be learned from high-risk industries for AI include *solid documentation, standardized practices*, and *independent review,* even if procedures can be lighter in many cases than for aeronautics or pharmaceutics.

5.2 The Way Forward

To advance beyond the current focus on ethics principles, a debate about ethics standards in prototypical situations is required. In many cases this will not just be ethics debates but elaborate political discussions about how to live a good digitally supported life and how to preserve and advance individual, social, environmental, and democratic standards. For example, this includes debates about what should be considered fair in a particual AI application scenario. These practices could over time also be rather different in different regions, cultures, or even communities within a country. It is perfectly possible that practices of AI-based risk assessment in the insurance sector will be different in Europe and the USA, for example. Similarly, more research is needed to learn from other disciplines and apply them to AI businesses. A main research challenge here is to achieve comparatively high standards at lower costs, simply because AI products and services are expected to be everywhere.

AI-active businesses can design many of the recommendations presented here, for example early warning systems, auditing, and process-level controls that ensure safe and ethical practices. Others, however, will benefit from joint industry or even political debate. Businesses should motivate industry associations and lobbying bodies to jointly develop and debate principles for prototypical situations and subsectors, clarify acceptable and non-acceptable behaviours in business standards, contracts, codes of practices, and clarify existing or shape new regulations and technical standards. They should work to design, support, and collaborate with training programmes that support engineers with knowledge about the intricacies of ethical issues in AI (and its politics). Finally, whole sectors need to embrace a positive and productive error culture that facilitates early identification and whistleblowing, double checking and independent auditing for learning from errors of the past to make future AI systems safe, trustworthy for consumers and less risky for businesses.

References

Allhutter D., Mager A., Cech F., Fischer F., Grill G. (2020) Der AMS-Algorithmus. (The AMS algorithm.) ITA project report 2020-02. https://epub.oeaw.ac.at/ita/ita-projektberichte/2020-02.pdf.

Baquero, J. A., Burkhardt, R., Govindarajan, A., & Wallace, T. (2020). Derisking AI by design: How to build risk management into AI development. McKinsey & Company.

https://africaplc.com/wp-content/uploads/2020/04/derisking-ai-by-design-build-risk-management-into-ai-dev.pdf.
Beauchamp T., Childress J. (2019). Principles of biomedical ethics: marking its fortieth anniversary. The American Journal of Bioethics, 19(11), 9–12.
Buehler K., Dooley R., Grennan L., Singla A. (2021) Getting to know – and manage – your biggest AI risks. McKinsey & Company. https://www.mckinsey.com/capabilities/quantumblack/our-insights/getting-to-know-and-manage-your-biggest-ai-risks (Last accessed December 2023).
Cheatham, B., Javanmardian, K., & Samandari, H. (2019). Confronting the risks of artificial intelligence. McKinsey Quarterly, 2(38), 1–9. https://www.sipotra.it/wp-content/uploads/2019/05/Confronting-the-risks-of-artificial-intelligence.pdf.
Clarke, L. (2023). Call for AI pause highlights potential dangers. Science, 380(6641), 120–121. https://www.science.org/doi/pdf/https://doi.org/10.1126/science.adi2240.
Clarke R. (2019) Principles and business processes for responsible AI. Computer Law & Security Review, Vol. 35 (4), pp. 410-422, https://doi.org/10.1016/j.clsr.2019.04.007.
Cockelbergh M. (2022) The political philosophy of AI. Polity Press, Cambridge, UK.
Crawford K. (2021) The atlas of AI. Yale University Press.
Floridi L. (2023) The ethics of artificial intelligence. Oxford University Press.
Freeman J.M., McDonnel K. (2001) Tough decisions: Cases in medical ethics. Oxford University Press.
Gert B. (2005) *Morality: Its Nature and Justification.* New York: Oxford University Press.
Giudici P., Centurelli M., Turchetta S. (2024) Artificial Intelligence risk measurement. In: Expert Systems with Applications, Vol. 235 (121220), https://doi.org/10.1016/j.eswa.2023.121220.
Giuggioli G., Pellegrini M. (2022) Artificial intelligence as an enabler for entrepreneurs: a systematic literature review and an agenda for future research. International Journal of Entrepreneurial Behavior & Research, Vol. 29 (4), pp. 816–837. https://doi.org/10.1108/IJEBR-05-2021-0426.
Goodall N.J. (2016) Away from Trolley Problems and Toward Risk Management. Applied Artificial Intelligence, Vol. 30 (8), pp. 810–821, https://doi.org/10.1080/08839514.2016.1229922
Harris J.D., Sapienza H.J., Bowie N.E. (2009) Ethics and entrepreneurship. Journal of Business Venturing 24, pp. 407–418.
Helmreich R. L. (2000) On error management: lessons from aviation. BMJ (Clinical research ed.), 320(7237), pp. 781–785. https://doi.org/10.1136/bmj.320.7237.781.
Ienca, M. (2023). Don't pause giant AI for the wrong reasons. Nature Machine Intelligence, 1–2. https://www.nature.com/articles/s42256-023-00649-x.
IIA (2013) IIA position paper: the three lines of defense in effective risk management and control. https://perma.cc/NQM2-DD7V.
Leighton J. (2020) Chapter 5 - Risk Management Framework. In Leighton J. (Ed.) Security Controls Evaluation, Testing, and Assessment Handbook, 2nd Edition, Academic Press, pp. 31–41, https://doi.org/10.1016/B978-0-12-818427-1.00005-7.
Mannes, A. (2020) Governance, Risk, and Artificial Intelligence. AI Magazine, Vol. 41 (1), pp. 61–69. https://doi.org/10.1609/aimag.v41i1.5200.

Marr, B. (2023) The 15 biggest risks of artificial intelligence, Forbes, June 2, 2023 https://www.forbes.com/sites/bernardmarr/2023/06/02/the-15-biggest-risks-of-artificial-intelligence/.

Ozlati, S., & Yampolskiy, R. (2017, March). The formalization of AI risk management and safety standards. In Workshops at the Thirty-First AAAI Conference on Artificial Intelligence. https://cdn.aaai.org/ocs/ws/ws0371/15175-68337-1-PB.pdf.

Prem E. (2023) From Ethical AI Frameworks to Tools: A review of approaches. In: AI and Ethics. https://link.springer.com/content/pdf/10.1007/s43681-023-00258-9.pdf.

Prem E. (2022) A knowledge-based perspective of strategic AI innovation management. In: Tanev S., Blackbright H. (Eds.) Artificial Intelligence and Innovation, World Scientific, February 2022. https://doi.org/10.1142/9781800611337_0003.

Rao R. (2022) The Dutch tax authority was felled by AI – What comes next? In: IEEE Spectrum, 9 May 2022, https://spectrum.ieee.org/artificial-intelligence-in-government.

Reverby S. M. (2009). Examining Tuskegee: The infamous syphilis study and its legacy. University of North Carolina Press.

Shepherd, D.A. and Majchrzak, A. (2022). Machines augmenting entrepreneurs: Opportunities (and threats) at the nexus of artificial intelligence and entrepreneurship. Journal of Business Venturing, Elsevier, Vol. 37 (4).

Schuett J. (2023) Three lines of defense against risks from AI. In: AI & Society. https://doi.org/10.1007/s00146-023-01811-0.

Steimers A., Schneider M. (2022) Sources of Risk of AI Systems. In: International Journal of Environmental Research and Public Health, Vol. 19 (3641). https://doi.org/10.3390/ijerph19063641.

Thomasama D.C. (1084) The context as a moral rule in medical ethics. Journal of Bioethics, Vol. 5, pp. 53–79. https://link.springer.com/article/https://doi.org/10.1007/BF01103648.

Wach, K., Duong, C.D., Ejdys, J., Kazlauskaitė, R., Korzynski, P., Mazurek, G., Paliszkiewicz, J., & Ziemba, E. (2023). The dark side of generative artificial intelligence: A critical analysis of controversies and risks of ChatGPT. Entrepreneurial Business and Economics Review, 11(2), 7–30. https://doi.org/10.15678/EBER.2023.110201.

WEF (2023) Chief risk officers outlook. Centre for the New Economy and Society, July 2023. https://www3.weforum.org/docs/WEF_Chief_Risk_Officers_Outlook_2023.pdf.

Werthner H., Prem E. Lee E., Ghezzi C. (Eds.) Perspectives on digital humanism. Springer, 2021.

Werthner H., Ghezzi C., Kramer J., Nida-Rümein J., Nuseibeh B., Prem E., Stanger A. (Eds.) Introduction to digital humanism: A textbook. Springer Nature, 2024.

Zuboff, S. (2018) The age of surveillance capitalism: the fight for a human future at the new frontier of power. New York: Public Affairs.

On the Role of Ethics and Sustainability in Business Innovation

Maria Fay and Frederik F. Flöther

1 Background

Analysis of the Fortune 500 list over time shows that new companies and business models have been rapidly emerging in recent years.[1] Those companies on the list that have already existed for several decades have generally undergone multiple fundamental shifts in business area and strategy, consider for instance IBM and Toyota. Hence, it is imperative that organizations continually introduce new ideas, methods, and even ways of thinking in order to stay competitive and relevant. This applies to established organizations as well as startups and new institutions.

Establishing and maintaining such an influx of novelty raises many questions that demand careful analysis. Innovation adoption frameworks have been developed, such as the TOE framework,[2] that cover technological, organizational, environmental, and in some cases even social factors. Topics that have been addressed in these categories respectively include, for example:

---

[1] Stangler, Dane, and Samuel Arbesman. "What Does Fortune 500 Turnover Mean?." *Available at SSRN 2078162* (2012).

[2] Baker, Jeff. "The technology–organization–environment framework." *Information Systems Theory: Explaining and Predicting Our Digital Society, Vol. 1* (2012): 231–245.

M. Fay
University of Münster, Münster, Germany
e-mail: maria.fay@uni-muenster.de

F. F. Flöther (✉)
QuantumBasel, Arlesheim, Switzerland
e-mail: frederik.floether@quantumbasel.com

C. H. Hoffmann (ed.), *Artificial Intelligence, Entrepreneurship and Risk*, Technikzukünfte, Wissenschaft und Gesellschaft / Futures of Technology, Science and Society, https://doi.org/10.1007/978-3-658-45544-6_24

- Security
- Enablement
- Regulation
- Trust.

There has also been some work around exploring the connections between ethics and innovation. For example, in considering the relationship between ethics and innovation it was argued that they are much more entangled and have a much greater degree of overlap, in a positive rather than a conflicting sense, than might seem to be the case from a first glance.[3] In addition, a theoretical model was developed for the governance of innovation processes, and it was argued that companies will intrinsically consider ethical aspects as they build long-term strategies.[4] While the mechanics and optimal processes to engender business innovation have been considered,[5] inclusivity has been explored in this context,[6] and innovation ethics have been studied in conjunction with international development,[7] a broader exploration of the ethics associated with innovation policies is required.

Furthermore, in recent years there has been a surge of global awareness of and interest in sustainability. Here, business sustainability is defined to be doing business without negatively impacting the environment, community, or society as a whole.[8] Sustainability and ethics have close ties, the former being grounded in ethical commitment to the wellbeing of both current and future generations.[9]

However, extensions of frameworks around technology adoption in organizations with ethical as well as sustainability dimensions have not yet been deeply

---

[3] Fontrodona, Joan. "The relation between ethics and innovation." In *Social Innovation: Solutions for a Sustainable Future*, pp. 23–33. Berlin, Heidelberg: Springer Berlin Heidelberg, 2013.

[4] Schumacher, E. Günter, and David M. Wasieleski. "Institutionalizing ethical innovation in organizations: An integrated causal model of moral innovation decision processes." *Journal of Business Ethics* 113, no. 1 (2013): 15–37.

[5] Pinchot, Gifford, and Ron Pellman. *Intrapreneuring in action: A handbook for business innovation*. Berrett-Koehler Publishers, 1999.

[6] Herrera, Maria Elena Baltazar. "Innovation for impact: Business innovation for inclusive growth." *Journal of Business Research* 69, no. 5 (2016): 1725–1730.

[7] "The Ethics of Innovation," Stanford Social Innovation Review, last modified August 5, 2014, https://ssir.org/articles/entry/the_ethics_of_innovation.

[8] "WHAT DOES "SUSTAINABILITY" MEAN IN BUSINESS?," Harvard Business School Online, last modified October 10, 2018, https://online.hbs.edu/blog/post/what-is-sustainability-in-business.

[9] Kibert, Charles J. "The ethics of sustainability." (2010).

explored. This is the focus of the present chapter. The remainder of the text is structured as follows. First, pertinent ethical considerations for the development of novelty in organizations are discussed. Second, light is cast upon the variegated role of sustainability in organizational innovation processes. Third, recommendations are formulated with regard to the inclusion of ethics (E) and sustainability (S) dimensions in technology adoption and business innovation beyond the traditional aspects of technology (T), organization (O), and the environment (E), thus resulting in a "TOEES" framework. Finally, conclusions are drawn.

2 Ethical Considerations

Introducing, or even just fostering, innovation in an organization comes with priority setting, opportunity costs, and (often) changes. Therefore, values and ethical dimensions have to be elucidated.[10] These do not have to be thought of as "brakes" to innovation; on the contrary, they may be conducive to novel ideas[11] and represent great opportunities with regard to product, service, and organizational innovation.[12] Given their varied nature, it is not possible to consider every possible dimension and situation. In this section, a number of quite different scenarios are discussed in order to illustrate the range and complexity of the questions.

2.1 Space and Time for Innovation

Consider a company that wants to nurture entrepreneurship and innovation culture. Often good ideas come through impromptu face-to-face interactions among people, at the (proverbial) water cooler. A straightforward way to increase the frequency and impact of such interactions is creating an environment that promotes exchange and creativity, provides comfort, and takes many of the daily "problems" off employees' shoulders (free meals, gym access, evening events, ...).

[10] "Ethics Framework for Innovation," Médecins Sans Frontières, accessed December 22, 2023, https://msf.org.uk/sites/default/files/ethics_framework_for_innovation.pdf.

[11] Apsalone, Madara, and Ricardo Martín Flores. "Ethical behavior and organizational innovation: Analysis of small and medium-sized enterprises in Latvia." *Economics and Business* 32, no. 1 (2018): 74–88.

[12] Brusoni, S. and Vaccaro, A., 2017. Ethics, technology and organizational innovation. *Journal of Business Ethics*, *143*, pp. 223–226.

This is how many companies have proceeded, Silicon Valley campuses being prominent examples.

Further consider a company's headquarters where every single need is met, from supermarkets to childcare facilities to cinemas to accommodation. Taking the idea this far, not unlike the campus in the (in)famous "The Circle" novel/movie, the center of employees' lives are effectively relocated to the workplace. Such a shift is even quantifiable, the number of hours per week spent by employees on the campus being a simple measure.

Is this ethical? It very much depends on a given employee. Some may welcome such an environment with open arms. Others would perhaps prefer more time "at home", particularly given the recent shift towards hybrid work models.[13] It is also a matter of a given company's expectations towards its employees, as peer pressure and culture in general make a great difference.

2.2 Work Product—A Product of...Work?

What is "work", when employees are so dedicated to what they do that they do not treat it as work and willingly spend free time on those initiatives? What if one's hobbies result in a work product? Most employment contracts regulate the results of work in the field where a person is employed as well as the reporting of side activities—which all makes perfect sense to avoid conflicts of interest and intellectual property (IP) rights infringement. Any organization would like to hire professionals who are experts in their respective fields, are intrinsically motivated, and are constantly developing and learning. On the way to leveraging and developing their expertise, those employees might want to teach a course at a university, become a mentor, present keynotes, or write a book. Given that role and company changes are becoming more frequent, it is becoming more difficult to determine if a given piece of thought leadership is the product wholly or partially due to an individual's time in a certain role. How will IP laws change in the future with AI coming into play?

[13] Sokolic, Danijela. "Remote work and hybrid work organizations." *Economic and social development: Book of proceedings* (2022): 202–213.

2.3 Human Response to (Technological) Changes and Evolution of Jobs

Naturally, changes in the environment typically lead to changes in the feelings and actions of humans. Changes in technology are no exception,[14] and people may welcome or dread them, perhaps even both at the same time to a degree; medical innovation is a case in point.[15] On the one hand, new technology may lead to novel tools that one can play with and which perhaps even make work easier. This goes hand in hand with learning opportunities and chances for stimulating interactions with others who already know about or perhaps are also learning about the new tool.

On the other hand, technological changes generally engender drastic changes in the job market, reducing demand for certain roles while increasing demand for other occupations. Thus, one must keep the question of how introduction of new technology affects the workforce always top of mind. There is early evidence that digitalization leads to increased employment of high-skilled workers and reduced employment of low-skilled workers.[16] There are entirely new job types which are required, prompt engineers possibly being one example.[17] Given the pace of technological development (as well as the abstractness of, for instance, quantum technology[18]), all workers should be open, and ideally intrinsically motivated, as concerns life-long learning. As such, introduction of new ideas, methods, and technology as well as general business innovation should go hand in hand with fostering a culture of enablement and learning.

---

[14] Hefner, Philip. *Technology and human becoming*. Fortress Press, 2003.

[15] McKneally, Martin F., and Abdallah S. Daar. "Introducing new technologies: protecting subjects of surgical innovation and research." *World Journal of Surgery* 27 (2003): 930–934.

[16] Balsmeier, Benjamin, and Martin Woerter. "Is this time different? How digitalization influences job creation and destruction." *Research Policy* 48, no. 8 (2019): 103765.

[17] Zamfirescu-Pereira, J. D., Richmond Y. Wong, Bjoern Hartmann, and Qian Yang. "Why Johnny can't prompt: how non-AI experts try (and fail) to design LLM prompts." In *Proceedings of the 2023 CHI Conference on Human Factors in Computing Systems*, pp. 1–21. 2023.

[18] Vermaas, Pieter E. "The societal impact of the emerging quantum technologies: a renewed urgency to make quantum theory understandable." *Ethics and Information Technology* 19 (2017): 241–246.

2.4 Example of Artificial General Intelligence (AGI)

Consider the rise of artificial intelligence (AI). After decades of research, AI has now achieved mass awareness as a result of various breakthroughs that have been enabled by factors such as greater data availability, enhanced computational power, and improved algorithms. While movie-like humanoid systems and robots are likely still quite some way off, AGI, that is general-purpose systems approaching or exceeding human intelligence,[19] now seems no longer quite such an illusory prospect. Sparks of AGI are arguably already seen in today's advanced generative AI systems.[20] This has induced public debate, and the EU AI Act represents one of the first regulatory frameworks around AI.[21]

The ethical questions surrounding AI, in general and in the context of business innovation, are manifold. For instance, consider creative work. Previously, it was (relatively) clear that novel output was the result of one or more humans. Today, an engaging document, image, audio file, or video could be the result of machine or human-machine processes. In both cases, systems were likely trained with data sets from diverse sources. As such, many people contribute, via their data which is used to train new AI systems, incrementally to many novel outputs. Clearly, this raises difficult ethical questions as to how such outputs may be used and attributed and who should be rewarded for successes. What contribution percentage belongs to the people sharing the data, the people designing the systems, the people providing generative AI prompts, the people leveraging the raw system outputs… and the system itself? There is an evident risk of "responsibility gaps".[22] To complicate (ethical) matters further, some of these outputs may be of great interest to the public, a case in point being medical innovation.

Another area where AI impacts business innovation ethics is in democratized access. AI, and other emerging technologies such as quantum computing, are believed to have revolutionary potential; as such, a global race has started between

---

[19] Wang, Pei. "On defining artificial intelligence." *Journal of Artificial General Intelligence* 10, no. 2 (2019): 1–37.

[20] Bubeck, Sébastien, Varun Chandrasekaran, Ronen Eldan, Johannes Gehrke, Eric Horvitz, Ece Kamar, Peter Lee et al. "Sparks of artificial general intelligence: Early experiments with gpt-4." *arXiv preprint arXiv:2303.12712* (2023).

[21] "EU agrees 'historic' deal with world's first laws to regulate AI," The Guardian, last modified December 9, 2023, https://www.theguardian.com/world/2023/dec/08/eu-agrees-historic-deal-with-worlds-first-laws-to-regulate-ai.

[22] Hoffmann, Christian Hugo, and Frederik F. Flöther. "Why Business Adoption of Quantum and AI Technology Must Be Ethical." https://doi.org/10.1017/qut.2024 (2023).

organizations and countries in order to be the first to fully develop and take advantage of these technologies.[23] Should early-mover organizations and countries be permitted to reap all these benefits as a result of their likely significant investments? Or is some form of democratized access, leading to a more level playing field, a better ethical approach?

3 Sustainability Considerations

Sustainability and innovation are closely interlinked, and more and more companies acknowledge that. This includes not only bringing in sustainability aspects into new products (by making them more circular or reducing emissions) but also creating new roles. Holcim—a Swiss-based building materials giant—has introduced the role of Chief Sustainability and Innovation Officer (CSIO) in 2021.[24] In one of her keynotes in 2022, Magalie Anderson (Holcim's CSIO at the time) explained that this allows the company to truly introduce sustainability into every product and process, from sustainable innovation in design and R&D to production to recycling. The results speak for themselves: in 2022, Holcim became the first company in its market segment to achieve a "double A" rating from the Climate Disclosure Project in both climate and water and has been driving social efforts with technology such as "building" the world's first 3D-printed school in Malawi.[25] Therefore, it seems that linking sustainability and innovation also on the organizational level drives clear benefits.

Furthermore, it is of utmost importance to consider sustainability holistically. There is a reason the United Nations (UN) has chosen 17 sustainable development goals (SDGs)[26] as they are addressing various aspects of how development can be sustainable. They range from improving education to ensuring access to healthcare and clean water to preserving biodiversity. A new product—such as

[23] Cave, Stephen, and Seán S. ÓhÉigeartaigh. "An AI race for strategic advantage: rhetoric and risks." In *Proceedings of the 2018 AAAI/ACM Conference on AI, Ethics, and Society*, pp. 36–40. 2018.

[24] "LAFARGEHOLCIM EXPANDS MAGALI ANDERSON'S ROLE TO CHIEF SUSTAINABILITY AND INNOVATION OFFICER," Holcim, last modified March 18, 2021, https://www.holcim.com/media/media-releases/magali-anderson-role-chief-sustainability-innovation-officer-3d-printing-xtreee-cpc.

[25] "Holcim Sustainability Performance Report 2022," Holcim, accessed December 16, 2023, https://www.holcim.com/sites/holcim/files/2023-02/24022023-sustainability-holcim-fy-2022-report-en-3114693266.pdf.

[26] "THE 17 GOALS," United Nations, accessed December 1, 2023, https://sdgs.un.org/goals.

1-dollar glasses[27]—might help address poverty and improve healthcare; still, its production and recycling will have a certain carbon footprint, consuming water and energy. Taking a holistic perspective, would this product be sustainable?

Based on such considerations and given that sustainability is very much multifaceted, it is necessary to illuminate the intersection of sustainability and technology adoption from different angles. Below are several sustainability dimensions that play an important role in an innovation context.

3.1 Sustainability Impact of Innovation

The phrasing of this subtitle deliberately highlights the "sustainability" impact without referencing a specific environmental or product carbon footprint due to the reasons listed above. When considering innovation and new ideas, it is important, for instance, to estimate to what degree a novel product would be more sustainable than previous ones, and, in addition, what impact a changed process could create. There are multiple methods allowing calculations of this type to be carried out, such as a lifecycle assessment (LCA) or product carbon footprint measurement. Still, hardly one of them takes a holistic approach and goes beyond climate or environmental effects, for instance considering social impact or inequality reduction aspects, biodiversity changes, and responsible consumption. While there are alternative approaches, such as the social and environmental life cycle assessments (S-LCA, E-LCA)[28] for assessing the contributions of products to the 61 out of 169 targets set by the UN SDGs, the feasibility of scaling implementation on the product level at every company can be rightfully questioned.

In the end, a lot comes down to what the companies that manufacture those products and create innovation are driven by. They are all part of a global market economy, and they have to comply with the market mechanisms that exist as well as regulations. But what if every company and every decision-maker would go just one step further? Perhaps they could come up with a recycling process for the products which they produce or try to reduce their biodiversity footprint (even if it is not yet required by law). Reflecting on what has been said earlier

[27] Carvalho, Luísa Cagica, and Adriana Backx Noronha Viana. "Social Innovation as a Promoter of the Welfare: The Case of One Dollar Glasses in Brazil." In *Global Campaigning Initiatives for Socio-Economic Development*, pp. 1–10. IGI Global, 2019.

[28] Eberle, Ulrike, Julius Wenzig, and Nico Mumm. "Assessing the contribution of products to the United Nations' Sustainable Development Goals: a methodological proposal." *The International Journal of Life Cycle Assessment* 27, no. 7 (2022): 959–977.

in this chapter, it is a question of ethics and of personal and company values. Is it ethical "just" to do what you were hired to do—or is it more ethical to go the extra mile? Is it ethical to expect extra miles from others?

3.2 Innovation for Sustainability

Creative products and ideas meanwhile contribute significantly to sustainability progress. Consider the following perspectives and examples.

"Technology innovation lens"

It is no coincidence that "climate tech" and "green tech" are two terms that have been trending in the press the last few years. Every industry has thousands of brilliant innovation examples that contribute to sustainability, especially through research and technology improvements.

A lot is being done on the forefront of materials research, from biodegradable recycled plastics for consumer products and textiles to mushroom growth for fabric production and as an alternative to carbon-intensive concrete, plastic, and foam in building insulation.

Another example is food technology that can make an impactful contribution by addressing malnutrition and hunger. It is crucial to make advances and scale alternative proteins, develop precision fermentation along with other technologies, and enable more communities to have reliable access to food.

Finally, energy technology and renewables represent a rapidly growing area. Studying, for instance, the list of global climate technology unicorns[29] shows that many energy and mobility companies are represented in the top 20, with Northvolt, a battery technology company, being the frontrunner. Clearly, whether it is about designing new batteries or meals, technological breakthroughs are key enablers on the road towards renewable materials and sustainable processes.

"Circular economy lens"

Innovation not always manifests in new products. The circular economy is transitioning from traditional linear "take-make-waste" thinking to closing the loop by extending product lifecycles through reuse, refurbishment, and remanufacturing. Hence, in the case of remanufacturing, when a product comes to its end of life (for instance, a mobile phone that no longer works because its battery is

[29] "Global Climate Tech Unicorns", HolonIQ, accessed December 21, 2023, https://www.holoniq.com/climatetech-unicorns.

dead), one returns it to the original manufacturer or recycler. They then dismantle the product and take out valuable components that could still be reused for repair or remanufacturing ("the process of bringing a used product to like-new condition through replacing and rebuilding component parts"[30]). Is this innovative? Well, it requires a new process (such as product takeback and its quality verification) with new skills and tools related to it, new design guidelines (and sometimes changing product specifications), and further updates. Is this sustainable? It allows one to reduce (even avoid) raw material (as well as water and energy) consumption and waste. As a result, this constitutes yet another example of innovation for sustainability.

Furthermore, the circular economy is also about refusing (so not producing or buying in the first place), rethinking, and reducing consumption.[31] As an example, consider a refillable water bottle that one carries to work or fitness or while traveling. In a way, this is a small piece of innovation introduced into one's daily routine / drinking process, which allows one to refuse using plastic water glasses, rethink consumption, and perhaps keep the bottle for a long time (for instance, if personal meaning is attached to it).

"Frugal innovation lens"

Frugal innovation is a perfect example of how new ideas and products can address sustainability on multiple levels. One of the classical definitions and key premises for frugal innovation is "achieving more with less".[32] This pertains to products that are simple and affordable, address real needs of people, and are easy to produce and use. Consider the problem of early breast cancer screening in countries like India. Traditional mammography equipment (large, expensive) is usually available at bigger hospitals. Thus, living in a rural area, often one cannot afford to travel to a hospital just for a checkup and needs to wait for a long time to get an appointment. This leads to the fact that most women in such areas do not undergo prophylactic examinations until it is too late. What if there were an easy-to-use-and-produce handheld device that could be straightforwardly brought to rural areas and employed for regular screenings of women living there to catch early signs of disease?

---

[30] Haynsworth, H. C., and R. Tim Lyons. "Remanufacturing by design, the missing link." *Production and Inventory Management* 28, no. 2 (1987): 24–29.

[31] Potting, José, Marko P. Hekkert, Ernst Worrell, and Aldert Hanemaaijer. "Circular economy: measuring innovation in the product chain." *Planbureau voor de Leefomgeving* 2544 (2017).

[32] Honegger, Angela. "The Role of Non-Profit Organizations in Accelerating Frugal Innovation." PhD diss., Universität St. Gallen, 2023.

This was the idea by UE LifeSciences, a company offering iBreastExam.[33] Company representatives travel to distant areas to train more people on using this simple hand-held system. Although the device cannot achieve the precision of professional mammography equipment, it can detect most knots (even small ones). In case a malicious knot is suspected, a woman can then travel to a larger hospital for further testing. Therefore, just one such novel idea can address a tremendous healthcare problem for large populations, especially in areas with great poverty—and therefore make a significant contribution to sustainable development. Such simple and scalable solutions could become major enablers across the globe as the world attempts to race towards maintaining a suitable climate.

"Ecosystem innovation lens"

Sustainability demands cooperation from various parties. One cannot calculate product footprints alone: for this, one needs data from one's suppliers and partners for Scope 3 emissions (which represent emissions that an organization only indirectly affects in its value chain).

How is collaboration becoming relevant in the innovation context? First, it is about jointly defining problems and developing solutions. For instance, this is what the Catena-X[34] automotive network is doing on the industry level. Such initiatives can help ensure acceptance, adoption, and ease of use by relying on common standards and platforms. Second, crowdsourcing and crowdfunding are powerful. There are many great ideas and businesses whose first steps happened on Kickstarter or similar platforms—and early feedback together with other advantages such platforms offer is a crucial element of creating a sustainable product. Third, even without being directly connected to others one can be a part of sustainability ecosystems and exchange ideas. Consider projects such as Local Happinez. This is a platform that allows users to find and support a local sustainable initiative of their choice next time they are on vacation.[35] Such a solution is simple but effective, and there are both direct and indirect benefits to it. A direct benefit is what the goal of the project is and what the community gets out of it: a cultivated coral reef or renovated water pipelines or harvested crop. One should also not underestimate the indirect benefits, which include what participants in those projects learn from one another, making them more conscious

[33] "iBreastExam," UE LifeSciences, accessed December 16, 2023, https://www.uelifesciences.com/ibreastexam.

[34] "Standards and methods for saving carbon dioxide", Catena-X Automotive Network, accessed December 22, 2023, https://catena-x.net/en/benefits-pros/sustainability.

[35] Local Happinez, accessed December 21, 2023, https://localhappinez.com.

of the world's ecosystem and the impact of decisions they make every day. Therefore, such projects have a strong educational effect, contributing to sustainability awareness, and thus also ethical dimensions, altogether.

3.3 Sustainability of Innovation Lifecycles

A first question arises: *when is it time for innovation?* When is it time for existing methods and ideas to be replaced by more novel ones? This taps into psychology and philosophy and is closely interlinked with management practices; it applies not only to product novelty but also to practices and processes. Consider company reorganizations. How ready are employees for such changes? How many are wondering why it is time again for a change if new structures are not yet in place from previous changes? Still, often companies fail to change in time, for example by not opening a new division to design an original product, and instead elect to focus on established practices instead. The concept of organizational unlearning[36] could be helpful here to understand how and when organizations need to "forget" old practices. By creating situational awareness, providing temporal and spatial freedom, encouraging an error-forgiving culture, and reducing the influence of old knowledge over time, companies can create room for new knowledge and learning.[37] In this way, the spirit of innovation can be sustained and flourish.

Now the next question comes: *how can one ensure that innovation persists?* New technology and methods should not just be implemented but also accepted, scaled, and further improved. While multiple frameworks around innovation management exist, one that has earned its reputation in the field of technology innovation is highlighted here (future business innovation will likely often be at least partially driven by technology innovation). This is the technology-organization-environment (TOE) framework, as originally proposed in 1990[38]

---

[36] Klammer, Adrian. "Embracing organisational unlearning as a facilitator of business model innovation." *International Journal of Innovation Management* 25, no. 06 (2021): 2150061.

[37] Klammer, Adrian, Thomas Grisold, and Stefan Gueldenberg. "Introducing a 'stop-doing' culture: How to free your organization from rigidity." *Business Horizons* 62, no. 4 (2019): 451–458.

[38] R. DePietro, E. Wiarda, M. Fleischer, The context for change: Organization, technology and environment, in L.G. Tornatzky, M. Fleischer (Eds.), *The Processes of Technological Innovation*, Lexington Books, Lexington, MA., 1990, pp. 151–175.

and as employed in later applications and adaptations.[39] It defined three key dimensions (technology, organization, environment) that have impact on technology adoption and scalability. For instance, within a company, its size and structure as well as communication processes and slack built over time (c.f. the aforementioned organizational unlearning) influence the speed and success of technology adoption. In terms of the external environment (the focus here lies on the market and policies rather than climate), market structure or government regulations could also be of major importance. A prominent recent example is the historic EU AI Act,[40] which regulates development and usage of AI and thus impacts a great many startups and companies across industries. Linking this with the original question around sustaining innovation in an organization, one needs to navigate between all those dimensions and stay on top of developments in a given company and market, which is becoming more and more difficult over time. Ironically, AI itself may also help overcome this barrier.[41]

4 Recommendations

Ethics and sustainability need to become integral parts of frameworks used in innovation and technology management. Taking the TOE example further, while TOE includes the "environment" dimension, this refers to contextual information about the external task environment and not (explicitly) to sustainability angles. There is therefore a need for more holistic approaches which, building on the earlier considerations from this chapter, include the dimensions of ethics and sustainability. Fig. 1 illustrates an adaptation of the TOE framework which has been extended with ethics and sustainability ideas, thus resulting in the "TOEES" framework.

---

[39] Oliveira, Tiago, and Maria Fraga Martins. "Literature review of information technology adoption models at firm level." *Electronic Journal of Information Systems Evaluation* 14, no. 1 (2011): pp. 110–121.

[40] "Artificial intelligence act: Council and Parliament strike a deal on the first rules for AI in the world," European Council, accessed December 17, 2023, https://www.consilium.europa.eu/en/press/press-releases/2023/12/09/artificial-intelligence-act-council-and-parliament-strike-a-deal-on-the-first-worldwide-rules-for-ai.

[41] Theodorou, Traianos-Ioannis, Alexandros Zamichos, Michalis Skoumperdis, Anna Kougioumtzidou, Kalliopi Tsolaki, Dimitris Papadopoulos, Thanasis Patsios et al. "An AI-enabled stock prediction platform combining news and social sensing with financial statements." *Future Internet* 13, no. 6 (2021): 138.

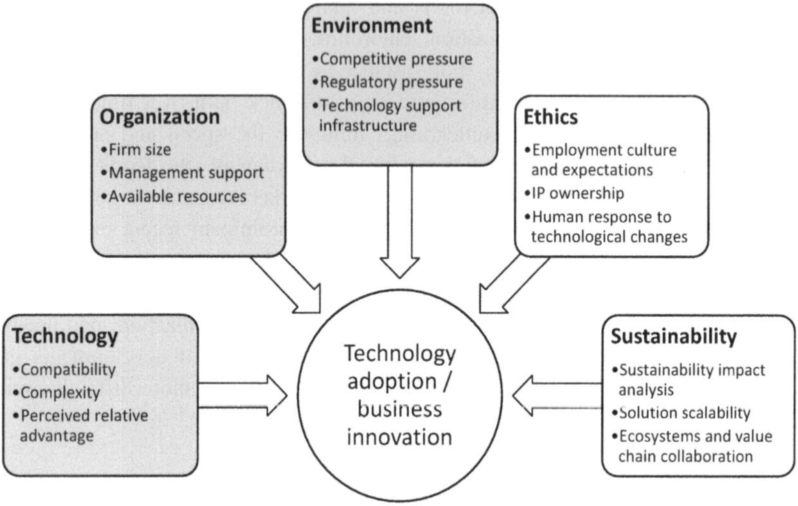

Fig. 1 Adapted TOE framework[42] that has been extended to explicitly include the dimensions of ethics and sustainability, thus becoming the "TOEES" framework

Furthermore, systems thinking is crucial to ensure a holistic understanding of innovation and its impact; responsible innovation requires exploration of individual, organizational, and system levels.[43] System dynamics as a discipline has been around for decades[44]; given the increasingly interconnected and complex world, it is further gaining in relevance. In 1972 the "Limits to growth" report[45] utilized a computer model to simulate the consequences of interactions between the Earth and human systems. Moreover, systems thinking has been described as "a framework for seeing interrelationships rather than things, for seeing patterns

[42] "Looking for a theory to explain the impact of security on technology adoption decisions," Spartakan, last modified February 26, 2016, https://spartakan.wordpress.com/2016/02/26/information-security-technology-adoption-decisions-toe.

[43] Long, Thomas B., Edurne Iñigo, and Vincent Blok. "Responsible management of innovation in business." (2020).

[44] Checkland, Peter. "Systems thinking." *Rethinking Management Information Systems* (1999): 45–56.

[45] Meadows, Donella H., Dennis L. Meadows, Jørgen Randers, and William W. Behrens. "The limits to growth: a Report for the Club of Rome's Project on the Predicament of Mankind," *Universe Books,* 1972.

rather than static snapshots".[46] It is extremely difficult to account for all the possible consequences of decisions that one makes and their ethical implications or the innovation one introduces and how this influences sustainability in a broader sense. Modeling those interactions in real life might not always be realistic or even possible; however, reflecting on the points outlined in this chapter can help explore at least some of the pertinent considerations.

Consider the case of Amazon Echo in terms of an anatomical map that requires human labor, data, and planetary resources for the unit to function.[47] In turn, such a unit, powered by AI, becomes part of a great breadth of decisions and processes; as with any technology, it may help with "good" or "bad" acts.[48] Clearly, the ethics and sustainability of a new product such as an Amazon Echo requires a broad analysis and, hence, systems thinking is very apt.

Finally, leveraging communities and ecosystems is key when it comes to ethical and sustainable business innovation. Sustainability is impossible to achieve alone: one must have suppliers on board who follow the principles of ethical sourcing and reduce the environmental footprint of their products. One also has to have consumers on board who "vote" for one's sustainable product with their money. Thus, one needs partners, including academia, startups, and the crowd, to make even more impactful innovation happen.

5 Conclusion

Ultimately, technology adoption and business innovation are highly complex, depending on many aspects that may also change over time. The rise and fall of companies, including the degree to which they are able to remain at the forefront of innovation, are testament to this. Particularly in recent years, many ethical and sustainability dimensions have gained importance in this regard. Today, they must be considered as businesses find their footing in an age with hybrid work models, accelerating AI, and mandates necessitating sustainable business models embedded in circular economies. These changes take time to internalize. Awareness and discourse thus represent a sensible first step; the "TOEES" framework outlined in this chapter is meant as an enabler in this regard. Clearly, this must

[46] Senge, Peter M. "The fifth discipline." *Measuring Business Excellence* 1, no. 3 (1997): 46–51.

[47] "Anatomy of an AI System," accessed December 19, 2023, https://anatomyof.ai.

[48] Shulevitz, Judith. "Alexa, should we trust you." *The Atlantic* (2018): 555–576.

then be complemented with concrete actions and plans, which may leverage systems thinking. Only organizations that apply the latest ideas to remain flexible, nimble, and proactive in how they treat business innovation… will (continue to) be successful in achieving business innovation.

Dr. Maria Fay is AI Advisor at SAP Switzerland AG and Guest Researcher at the University of Münster, focusing on AI ethics and sustainability. She holds a PhD in Business Economics and has over a decade of experience in business and technology consulting. Maria is a host of the "Apropos… Nachhaltigkeit" podcast where current sustainability topics are discussed. Moreover, Maria has been a guest lecturer for digital business models and engages in mentoring and volunteering work. Maria has worked and lived in Switzerland, Liechtenstein, Germany, and Russia.

Dr. Frederik F. Flöther is Chief Quantum Officer at QuantumBasel, the first commercial quantum hub of Switzerland, and a member of the University of Basel Center for Quantum Computing and Quantum Coherence (QC2). Prior to that, he was over 7 years with IBM, most recently responsible for healthcare and life sciences at IBM Quantum. Frederik researches the technological applications as well as philosophical implications of quantum computing and AI. He was elected to the IBM Academy of Technology and appointed Master Inventor and Qiskit Advocate. Frederik holds a PhD in physics, with a focus on photonic quantum computing, and a MA, MSci, and BA from the University of Cambridge as well as a professional certificate in quantum computing applications from MIT. In total, he has authored over 40 filed patents, peer-reviewed publications, white papers, and book chapters. Frederik has lived/worked in Switzerland, USA, United Kingdom, Belgium, and Germany.

The Power of Social Connections and Collective Intelligence in the Age of AI

Serge Pelap

1 Introduction

Welcome to a world where your morning coffee is made by a "smart" machine that "knows" exactly how you like it. This machine doesn't just brew coffee, it crafts a unique experience tailored to your individual preferences. It remembers your bean choices, whether you prefer a floral Arabica or a spicy Robusta, and grinds them to the ideal grain size for your palate. Customization goes further with precise temperature and intensity settings, capturing every nuance of flavor, whether you like a lightly tart or intensely roasted coffee. The machine can even adjust the amount of milk foam for the perfect balance according to your taste. Every cup it brews isn't just coffee, it's your coffee, brewed with details that suit your unique desires, reflecting your personality in every sip.

In this world, your glasses are much more than a simple vision accessory, they have become windows on the emotions of those close to you. These "smart" glasses are able to detect variations in your best friend's mood. They analyze facial expressions, tone of voices, and other subtle cues to let you know about emotional changes. If your friend feels happy, sad, angry or lies to you, your glasses discreetly alert you.

An essay and personal short story by Dr. Serge Pelap, a beautiful mind.

S. Pelap (✉)
Baoba, Zurich, Switzerland
e-mail: serge.pelap@baoba-group.com

C. H. Hoffmann (ed.), *Artificial Intelligence, Entrepreneurship and Risk*, Technikzukünfte, Wissenschaft und Gesellschaft / Futures of Technology, Science and Society, https://doi.org/10.1007/978-3-658-45544-6_25

We are in an age where digital technology and particularly AI are no longer just part of our lives, they are practically orchestrating and shaping them. And this also affects how we interact with others.

Gone are the days when socializing meant sitting around a fire and sharing stories of the day's hunt, or of my childhood in Cameroon where, during the long school holidays, in the evenings we would spend time around the fire, my grandmother preparing our dinner and our elders and particularly my uncle Gaston told us tales which always ended with a moral of the story, a lesson, a wisdom to remember. Now our gatherings have turned into group chats, our stories into posts and tweets, relayed by algorithms that choose what we see and hear, influencing our perception of the world and those around us.

So, this evolution raises crucial questions: How do these technologies, in particular AI, redefine our social connections? How do they influence our ability to work together, to create collective intelligence? And more importantly, how can we, as individuals and as a society, use this shift to forge deeper, meaningful connections, not to drive ourselves further away from each other? It is important to emphasize that in our world, more connected than a spider's web on a humid morning, a paradox emerges: we are more and more disconnected and we feel alone, as if we were lost in a crowd.

This essay, through a multidimensional exploration, seeks to answer these questions. We will explore the transformative power of social connections, the essence of collective intelligence, and the profound impact of AI on these fundamental aspects of our humanity. By looking at real-world cases, like the coffee machine that knows your preferences or glasses that can read emotions, we will begin to understand how AI can both enrich and complicate our social experience.

Through this journey, we hope to offer enlightening perspectives and reflections on how we can navigate this digital age, not by passively undergoing the changes it brings, but by actively embracing them to create a future where technology enriches, rather than diminishes, our deep human connections.

2 The Power of Social Connections

October 25, 2017, will remain engraved as an exceptional day in our family memory. While the night still enveloped the world in its cloak of darkness, at least in Switzerland, the first contractions announced the imminent arrival of our third daughter. The house, usually quiet at this hour, came alive with a new urgency, mixed with excitement and a gentle apprehension. The hospital bags, carefully prepared and placed near the door for weeks, were finally going to be useful.

We left the house in a feverish hurry, my wife breathing deeply with each contraction, while I carried the bags into the car. The streets were deserted, the city asleep, unaware of the magic that was taking place. As I drove to the hospital, my mind was torn between focusing on the road and worrying about my wife and our unborn child. In my haste, the speedometer slightly exceeded the 50 km/h limit, an excess of only 5 km/h, but enough to be flashed by a speed camera. At that moment, the only thing that mattered was my wife's safety and the imminent arrival of our baby.

The hospital was quiet when we arrived, a calm that contrasted with the storm of emotions within us, despite this being our third childbirth experience. It was just as intense as the first time. The hours that followed were a mixture of pain, waiting, and intense anticipation. Then, suddenly, in an explosion of life, Julie-Rose was born. Her first cry pierced the air, a sound so powerful and so fragile at the same time. There she was, so small and so perfect, her eyes instinctively searching for her mother's face.

In Julie-Rose's eyes, there was an immensity of possibilities, an invitation to a journey of mutual discovery. Every little gesture, every breath, was the start of a story, our story. To hold her in my arms, to feel the warmth of her skin, was to suddenly understand the full extent of our vulnerability and our strength as human beings. It was the beginning of an indescribable connection, a bond that went far beyond words.

A few hours later, the time had come for Myriam and Anne to meet their new little sister. I returned home to pick them up, their palpable excitement filling the car on the drive back to the hospital. When they first saw Julie-Rose, their eyes lit up with pure wonder. Their sweetness and affection were evident, every gesture and gentle word towards their sister showing unconditional love.

The issue of speeding was of little concern to me in this moment of pure joy. But later, when I explained the circumstances to the police, they showed unexpected understanding and leniency. The sanction was rescinded, a tacit recognition that certain situations, certain moments in life, transcend the usual rules.

Julie-Rose's birth was a powerful reminder of the importance of social connections from the first breath of life. These bonds, which begin with a look, a touch, a voice, are the foundations of our humanity. They are what makes us unique, what allows us to survive and thrive. This story of her coming into the world is a celebration of these bonds, a recognition of their immense power and their immeasurable beauty.

Within the animal kingdom, human beings are not distinguished by their brute strength, nor by their speed, nor even by their resistance. Our particularity lies in our remarkable ability to establish complex relationships and build social bonds. This unique ability is the cornerstone of our existence and the key to our survival as a species. These social ties, whether family, friends, professional or otherwise, are much more than a simple survival strategy. They give meaning to our lives, shape our identities and influence our behaviors and attitudes.

Unlike many mammals who are born ready to face the challenges of their environment, humans come into the world in a state of complete vulnerability, unable to care for themselves and therefore survival depends entirely on parental care. A human baby is born with a largely immature brain, requiring years of growth and learning within its community. This prolonged dependence contrasts sharply with, say, a fawn, which can stand and walk shortly after birth. This difference illustrates the need for a rich and stimulating social environment for human development.

Our social needs, which are also vital, go well beyond simple physiological needs, as the tragic experiment carried out by Emperor Frederick II, also known as the Forbidden Experiment, illustrates so well. In the thirteenth century, fascinated by the origins of human language, Emperor Frederick II conducted an experiment where infants were raised without social interactions in an attempt to discover what language they would naturally speak. It was permitted to feed the babies, bathe them, clean them, but under no circumstances cuddle or talk to them. The outcome was fatal, the children all died. This sad experience highlights the imperative need for social bonds for the development and survival of human beings.

This story highlights that social connections are much more than just a survival strategy. They are the foundation of our emotional, intellectual and cultural development. In every interaction, every gesture of tenderness, every word exchanged, we strengthen the social fabric that is essential not only to our individual well-being, but also to the prosperity of our species.

In summary, this section demonstrates that the strength of humanity lies in our ability to form social bonds. This ability, born from our vulnerability and dependence from the earliest moments of our existence and many other reasons, is what has allowed us to thrive in diverse, sometimes unfriendly, environments and face challenges that threaten our existence as a species. Human relationships are at the heart of our existence, allowing us not only to survive, but also to shape the world around us. And this is just the prelude to an even greater phenomenon that shapes our lives: our collective intelligence. It is through our relationships that we learn, grow and, together, create something greater than the sum of our individualities.

3 Collective Intelligence: The Human Advantage

While Sect. 2 focused on the importance of social bonds in both individual and collective development and the very survival of our species, Sect. 3 explores another crucial aspect of our interactions: the collective intelligence. This intelligence, born from our ability to collaborate, share knowledge and work together towards common goals, is what has propelled humanity to extraordinary achievements. Unlike most animals who rely primarily on their physical abilities or instincts to survive, humans have developed an approach based on collaboration and innovation to overcome their challenges.

3.1 The First Manifestations of Collaboration

Our ancestors, tens of thousands of years ago, living in an unforgiving world, quickly understood that their survival depended on their ability to work together. Imagine a hunting scene in the prehistoric era, a small group of hunter-gatherers gathers together. Their task: to track down game, a gigantic mammoth, a prey that will ensure their survival. They have neither the strength of a lion nor the speed of a horse, but a plan developed together, a strategy born from collective experience and the sharing of knowledge. Every movement is calculated, every role essential. In this ancestral ballet, survival depends on the union of their minds and their efforts.

These first manifestations of collective intelligence were a direct response to the existential problems of the time. In addition to hunting, gathering food, building shelter, protecting against predators and natural elements required group cooperation and creativity. The solutions to these challenges lay not in individual physical strength, but in the group's ability to work together and innovate.

3.2 The Dawn of Agriculture and Early Societies

The First Agricultural Revolution, which occurred around 10,000 years ago, marks a fundamental turning point in the history of humanity. This period of profound change saw our ancestors move from a nomadic lifestyle, based on hunting and gathering, to a sedentary lifestyle, centered on agriculture and livestock. This transition took place in several regions of the world almost simultaneously, suggesting a remarkable ability to observe, innovate and adapt to new environments.

In the Fertile Crescent, for example, communities began growing grains like wheat and barley. They discovered how to irrigate their fields, thereby increasing the yield of their crops. These early farmers also domesticated animals such as goats and sheep, not only for their meat, but also for their milk, wool, and to help with farm work.

These innovations were not isolated achievements. They were the result of a continuous exchange of techniques and ideas within and between these communities. Knowledge was passed down from generation to generation, with each new discovery adding to a common pool of agricultural know-how.

This sedentarization had profound effects on the social and economic structure. With the ability to produce surplus food, societies were able to support a growing population. Specialized trades emerged, such as potters, blacksmiths and weavers, creating the foundations of a complex and more collaborative economy. Food surpluses enabled the development of trade, both within communities and with neighboring peoples.

The first forms of governance and legal systems emerged to manage these increasingly complex societies. Likewise, new forms of cultural and artistic expression emerged, reflecting a society that had more time and resources to devote to more than just survival.

Ultimately, the First Agricultural Revolution not only changed the way humans interacted with nature, but also the way they interacted with each other. It laid the foundations of the first civilizations, proving that collective intelligence, applied to agriculture and resource management, was a powerful engine of change and progress.

3.3 The Era of Great Constructions

The transition to an agricultural lifestyle, with its stable, sedentary societies, paved the way for some of the greatest architectural feats in human history. Among these achievements, the pyramids of Egypt stand as an eternal symbol of collective intelligence and human ingenuity. Built around 2560 BCE, these imposing structures illustrate man's extraordinary capacity to design, organize and carry out projects of previously unimaginable scale.

The Great Pyramid of Giza, in particular, remains one of the wonders of the ancient world. Its construction required the collaboration of tens of thousands of workers, from the quarrymen extracting the stone blocks to the workers and engineers supervising the assembly of the structure. Each block, weighing on average more than two tons, had to not only be quarried and transported long distances, but also carefully cut and adjusted to fit precisely its place in the pyramid.

The construction of this Great Pyramid, which took approximately 20 years, demonstrates careful planning and organization. This required an advanced understanding of geometry, astronomy and physics, as well as exceptional coordination between hundreds of groups of workers. The collective effort extended far beyond the construction site, involving entire communities that provided food, housing and other essential resources to support such a massive workforce.

The pyramids were not only feats of engineering, they were also deeply rooted in the culture and religious beliefs of ancient Egypt. Their design reflects a complex understanding of the universe, a desire for immortality and a devotion to the pharaohs, considered gods on Earth. The ability of Egyptians to unite their people around a common project testifies to the influence of collective intelligence not only on the physical level, but also in the sphere of cultural and spiritual identity.

The pyramids of Egypt remain a lasting testament to what collective intelligence can accomplish. They symbolize the capacity of human societies to transform grandiose visions into concrete realities, through collaboration, innovation and perseverance. More than a simple expression of power or wealth, they embody the spirit of cooperation and determination that has always been at the heart of our greatest collective exploits.

3.4 From the Renaissance to the Industrial Revolution

The Renaissance, spanning approximately from the fourteenth to the seventeenth century, represents an era when collective intelligence took on a new lease of life. In Europe, cities like Florence, Paris and London are becoming crucibles of innovation. In these urban centers, artists, scientists and inventors interact in a stimulating environment, inspiring each other and propelling a wave of cultural and intellectual progress.

Imagine the bustling workshops, where artists not only paint or sculpt, but also engage in discussions about anatomy, physics and engineering. These transdisciplinary interactions led to discoveries and works of art that defined this era. The Renaissance thus became a symbol of human potential, where the union of knowledge and skills brought about an era of unprecedented progress and fulfillment.

The end of the eighteenth century marked the beginning of the Industrial Revolution, a time when collective intelligence was applied on an industrial and global scale. This period saw the emergence of revolutionary inventions such as James Watt's steam engine and the electric telegraph. These innovations were not the fruit of isolated genius, but rather the result of an accumulation of millennia-old knowledge and collaboration between various experts and workers.

Factories, with their complex production lines, have become symbols of this new era, demonstrating how organization and collaboration can dramatically increase productivity and efficiency. The impact of these changes has been monumental, leading to rapid urbanization, unprecedented economic growth and the emergence of new social classes as well as growing inequalities. We will not develop this aspect here.

The Industrial Revolution also played a key role in the globalization of the world. The development of railways and steam shipping brought the continents closer together, while the telegraph allowed almost instantaneous communication across long distances. This interconnectivity has strengthened collective intelligence by enabling an exchange of ideas and resources on a scale previously unimaginable.

From the Renaissance to the Industrial Revolution, human collective intelligence has been a driving force for change and innovation. These periods laid the foundations for our modern world, demonstrating how collaboration, the exchange of ideas and the desire to push the boundaries of what is possible can radically transform societies. The spirit of cooperation and innovation born during these eras continues to inspire us and shape our future.

3.5 The Modern Era: Globalization of Collective Intelligence

The advent of the twentieth century and the rapid acceleration of technological innovation marked an era of transformation in the way collective intelligence operates on a global scale. Today, in a hyper-connected world, ideas and knowledge no longer know borders, circulating freely and quickly across continents and cultures.

The emergence of the Internet and communications technologies has revolutionized the way we share and access information. Researchers in Asia can collaborate in real time with colleagues in America, Africa or Europe. Online platforms, discussion forums, and social media have created spaces where knowledge and experiences are freely exchanged, allowing ideas to spread and evolve at unprecedented speed. However, we will see in the next Sect. 4 that there is the other side of the coin, a dark reality looming in the background.

4 Artificial Intelligence and Its Impact on Social Interactions

As we dive into the realm of artificial intelligence (AI), it is essential to clearly define what we mean by intelligence. We must recognize that intelligence is not about what we know or what we can do. Intelligence is characterized by understanding, the ability to learn new things, adapt to new contexts and improvise in the face of the unexpected. These aspects of adaptation, improvisation and deviation form the very heart of true intelligence.

4.1 The Journey from AI to the Pursuit of Human Intelligence

AI, in its attempt to simulate human intelligence, has come a long way since the pioneering work of Alan Turing. From its conceptual foundations to current developments in deep "learning" and neural networks, AI has evolved to appear to mimic the human capacity for learning and innovation. However, the question remains: can it really mimic human intelligence?

Can AI Really Understand? Can we speak of understanding without self-awareness? Take OpenAI's ChatGPT for example: while it's a powerful tool for generating text responses, its method relies on statistical prediction of words rather than true understanding.[1] When it generates sentences, it simply calculates which string of words is statistically likely, but that doesn't mean anything that it understands the text it generates itself.

ChatGPT's effectiveness in forming coherent sentences may give the illusion of understanding, but this is similar to someone who does not speak Swahili, an African language, but uses a probability guide to form sentences. Despite the

[1] Comment by the editor: As a philosopher by training, I cannot simply disregard the question what "understanding" amounts to. The philosophical or, simply, curious readers among you, I invite to look for and find answers in Hoffmann, C.H. (2022). "The Quest for a Universal Theory of Intelligence: The Mind, the Machine, and Singularity Hypotheses". Berlin: De Gruyter.

appearance of logic in certain sentences, the lack of real understanding is revealed both by the formation of perfectly coherent sentences which amaze and by the formation of sentences of disconcerting incoherence. This analogy highlights the current limitations of AI in its quest to mimic human intelligence: although it may appear to mimic certain aspects of human intelligence, there remains a significant gap between the two that is important to highlight.

And that's without mentioning the aspect of efficiency. Show a two-year-old child a first, then a second cat and this child will be able to recognize any cat that comes his way, even a cat in the dark. This is far from being the case for AI which must go through "learning" of hundreds of thousands of images of cats to be able to "recognize" one, and even less all cats.

This analysis leads us to rethink our perception of AI. Although it offers incredible possibilities for the improvement of various fields, it is essential to recognize what AI truly is. This understanding is crucial to use it responsibly and to prevent potential risks. Should we not also reformulate the terms used to designate what we mean by AI today?

4.2 From Isolation to Digital Connection

At the heart of a constantly evolving world, technology and artificial intelligence (AI) in particular are proving to be a dualistic force, capable of weaving close bonds between individuals while having the potential to distend them. Like a knife used in the kitchen to create culinary masterpieces or, in a darker context, like a bladed weapon, AI embodies this ambivalence.

Several years after leaving Anguissa high school, an establishment located in a dynamic district of Yaoundé, capital of Cameroon, I noted with nostalgia how much I had lost contact with many of my classmates. At the time we left high school, more than fifteen years after the start of the commercialization of cell phones, these devices remained a rarity in our country. Even access to landlines was a luxury reserved for a privileged minority, and the lack of a practical addressing system made the desire to maintain or reunite with friends a near impossible task. Even the few named streets, such as Avenue Germaine or Boulevard Ahmadou Ahidjo, disappeared into a labyrinth of anonymous homes, the postal addresses of businesses being the only milestones in this blur of location. Most of us then dispersed across the four corners of Cameroon, leaving behind the bonds formed during our seven years of complicity in high school.

However, like a sunrise dispelling the shadows of the night, the advent of the Internet has changed this picture. Social networks and messaging applications, such as Facebook, WhatsApp and Telegram, have become the beacons illuminating my path to unexpected reunions. Thanks to these tools, I was able to reconnect with friends and comrades spread across the world. Together, we created a digital haven for former students of Anguissa high school, where daily exchanges rekindled the flames of friendship.

4.3 The Dark Side of Technology

However, a discordant note gradually appeared in this symphony of reunion. Technologies, once celebrated as vectors of rapprochement, now seem to distance us from each other. This transformation is not the result of chance, but the result of deliberate design. In pursuit of ever greater profits, disregarding the well-being of users reduced to the status of merchandise, the creators of these platforms have made an insidious change.

With the integration of artificial intelligence into social networks, a subtle but profound transformation has taken place in our interaction with these platforms. It's no longer just us actively browsing content; now, every click, every view, and even our attention span are meticulously analyzed. AI, working in the shadows, learns from our behaviors, our preferences and our interactions to draw up user profiles with surprising precision. This information is not only used to personalize our experience on these platforms; they are also used to target with formidable efficiency the advertisements presented to us in order to maximize the profitability of advertisers who are their real customers and us simple products. Even more worrying, they exploit our psychological weaknesses to subtly influence us, manipulating our perceptions and beliefs without us being aware. It creates parallel universes where everyone lives according to their own perception of reality, contributing to an increased fragmentation of society. This compartmentalization into personalized realities thus aggravates social polarization, making dialogue and mutual understanding between different points of view increasingly difficult.

These algorithms, hungry for clicks and screen time, shape our interactions. Deep conversations and authentic connections become diluted in an ocean of ephemeral content, creating an illusion of closeness while exacerbating our feelings of isolation. So here we are, trapped in a digital mirage, where the promise of connection turns into shared solitude. This solitude within this hyper-connected world presents a disturbing paradox. In this world increasingly consumed by digital interactions, we find ourselves navigating an endless stream of bytes and pixels. The irony is inevitable: the more "connected" we become online, the more disconnected we feel in real life. Loneliness has become the hidden epidemic of our time, and yet we continue to isolate ourselves behind screens, trading the depth of human contact for the breadth of digital reach.

The statistics are concerning: A study by Cigna in 2021 shows that 63% of young adults between 18 and 24 years old in the United States frequently experience loneliness. In Europe, daily loneliness is a reality for approximately 30 million people, a condition that is mirrored worldwide, impacting close to two billion people as recent research by Meta-Gallup indicates. Generation Z ranks as the most isolated group, with 73% regularly feeling lonely, according to Cigna.

The tragic story of a person who died in his apartment in Nantes, France, whose body was not discovered until 11 years later, is a striking example of the isolation in our modern societies. This story and many others highlight the dark reality of this deep-seated evil in our communities, even in densely populated urban areas.

Social isolation and increasing disconnection in our digitalized society also have alarming repercussions on our health. The implications are serious: isolation increases the risk of heart disease by 29%, that of stroke by 32% and increases the chance of developing dementia by 50%. Even more worryingly, there is a direct link between loneliness and a 64% increased risk of Alzheimer's disease, highlighting the severe consequences of isolation on mental and cognitive health.

As we can see, the introduction of artificial intelligence into our lives, while offering fascinating prospects for the future, raises crucial questions. Let us analyze a few more cases.

4.4 Technological Mediation

Let's return to the example of "smart" glasses mentioned previously, which can detect and interpret the emotions of the people around us. At first glance, this technology appears to be a promising advancement, providing valuable assistance in understanding the emotional reactions of others. However, as we dig deeper, we discover an ethical and psychological dilemma.

These "smart" glasses, armed with AI algorithms capable of reading facial expressions and providing instant feedback on the emotional state of the interlocutors, could theoretically improve our ability to react with empathy. However, this technological mediation in the recognition of emotions poses the risk of making us less able to develop and exercise our own emotional judgment. If we begin to rely on devices to "read" emotions for us, our intrinsic ability to perceive and understand the feelings of others without technological assistance may atrophy.

Another case of technological mediation. Within the digital theater where artificial intelligence (AI) orchestrates our lives, a conversation with a former classmate unfolds, revealing the subtle nuances and paradoxes of this era of assisted communication. The exchange, seemingly fluid, carries echoes of unexpected complexity, mixing human authenticity with algorithmic accuracy.

Me: When I receive his message, the AI promptly offers, "I'm doing well, how about you?" The suggestion seems perfect, polite, without the shadow of a typo or verbal clumsiness that would have betrayed human spontaneity. I agree, seduced by the ease, but part of me wonders what we are losing in this digital perfection.

Friend: "Everything is fine, thank you. And how are your children?" His message arrives, impeccable, a courtesy that I believe to be the fruit of AI. An inner voice whispers, sowing doubt: is this a real concern on his part or simply the echo of an algorithmic suggestion? The absence of any imperfection, of any distinctive sign of humanity, like those little mistakes we all make when typing, leaves room for ambiguity.

In our past exchanges, these little imperfections were windows into our emotions, subtle clues perhaps revealing haste, joy, or suppressed sadness. Today, AI filters and polishes our messages to such an extent that these clues are lost, replaced by smooth, but perhaps also soulless, communication.

Me: "They're growing too fast, thanks for asking!" I respond, following the AI's suggestion, while feeling a nostalgia for the messier but more human exchanges of the past. Those moments when a typo or awkward word could elicit a smile or a concerned question, unknowingly weaving the tighter threads of our connection.

As our conversation progresses, orchestrated by suggestions from the AI, I find myself analyzing each message, looking for signs of an authentic human presence behind the text. The conversation, although fluid, seems to navigate the smooth surface of a frozen lake, where the depth of the underlying waters, rich in emotions and nuances, remains invisible and unexplored.

In this world where AI becomes the scribe of our interactions and we mere validators, a question persists: are we sacrificing the very essence of our communication, these charming and revealing imperfections, for cold efficiency and impersonal perfection? Could technology, in its quest to assist us, distance us from the imperfect but deeply human art of truly connecting with one another?

This dilemma, illustrated by this simple conversation between friends, highlights a broader challenge of our digital age: balancing the undeniable benefits of AI with preserving authentic, imperfect but vibrant communication that celebrates our common humanity in all its richness and complexity.

4.5 AI and the Erosion of Empathy

The widespread use of such technologies could have particularly marked consequences on younger generations. Children and adolescents, as they become accustomed to interacting with and through AI devices, may find it more difficult to cultivate natural empathy without an intermediary. Learning the complex nuances of human interactions, which occurs through observation, experience and imitation, risks being compromised in a world where machines take over the interpretation of emotions.

The example of "smart" glasses or the mediation of AI in our daily exchanges, through automatic responses or content suggestions, serve as an illustration of the need to balance technological advances with development and preserving our social and emotional skills. It reminds us that while AI may augment our experience of the world in unimaginable ways, we must remain vigilant to ensure that this augmentation does not come at the expense of our humanity. Encouraging interactions that foster authentic empathy is essential to maintaining the rich and nuanced social fabric that characterizes our human experience.

4.6 AI and Atrophy of Intellectual Faculties

At the same time, our growing reliance on predetermined AI solutions for decision-making and problem-solving could weaken our critical thinking and creativity. By letting AI navigate the complexities of life for us, we risk seeing our ability to think independently and innovatively atrophy.

It is therefore important to recognize this risk and seek a balance where AI serves as a tool, without supplanting our active engagement in the thought process. By preserving our space for personal reflection and curiosity, we can use AI as a lever for our creativity, not as a substitute for our intelligence.

4.7 Collective Intelligence: A Treasure in Danger

We cannot fail to emphasize that our collective intelligence, this treasure accumulated through centuries of collaboration and sharing of knowledge, is also threatened. AI, by promising unparalleled efficiency and precision in data analysis and decision-making, could reduce our incentive to collaborate intellectually. Discussion forums, brainstorming, and passionate debates that sparked innovative ideas could give way to passive consultation of the "answers" provided by AI.

This shift toward less intellectual collaboration and authentic interactions with AI not only means lost opportunities for collective and individual intelligence to flourish, but also a decrease in the diversity of perspectives. Like an impoverished ecosystem, our collective intellectual landscape risks becoming less rich, less resilient in the face of the complex challenges of our time.

5 Conclusion

As we contemplate the impact of artificial intelligence (AI) on certain aspects of our lives, it is necessary to reflect on the role that our choices and motivations play in the direction this technology takes. AI, in its essence, is neutral, it is a blank canvas, it is the mirror of our desires, our ambitions, and sometimes, our shadows. This technological advance, carrying with it the potential to redefine the contours of our existence, confronts us with unprecedented ethical and moral dilemmas, revealing the dichotomy between AI's potential to unite and divide, to elevate and diminish, to enrich and impoverish our social fabric.

This essay aimed not only to raise awareness, but also to inspire responsibility for the future that we are shaping. The design and use of AI should not be driven solely by a quest for immediate gains or selfish interests, but fueled by a long-term vision, considering the repercussions on our social fabric, our common well-being and our intellectual heritage.

The history of AI and its influence on our lives is a story still being written, marked by our current choices. Faced with this powerful technology, we are called to reflect not only as creators and users, but also as guardians of humanity.[2] Before us lies the task of balancing innovation with integrity, of reconciling progress with the preservation of our fundamental values.

Ultimately, the question before us is not simply whether AI can enrich our humanity, but rather whether we, as a society, will choose to let our noblest nature prevail in our use of this technology. The answer to this question will shape our common future and determine the legacy we leave to future generations.

[2] Comment by the editor: I like this reference to "guardians of humanity" which reminds me of William MacAskill's fabulous book "What We Owe the Future" (2022).

The manufacturer's authorised representative in the EU is Springer Nature Customer Service Centre GmbH, Europaplatz 3, 69115 Heidelberg, Germany. If you have any concerns regarding our products, please contact ProductSafety@springernature.com

Printed and bound by CPI Group (UK) Ltd, Croydon, CR0 4YY
30/09/2026
02225560-0003